Applying Machine Learning Techniques to Bioinformatics:

Few-Shot and Zero-Shot Methods

Umesh Kumar Lilhore
Galgotia University, India

Abhishek Kumar
Chandigarh University, India

Sarita Simaiya
Galgotia University, India

Narayan Vyas
Chandigarh University, India

Vishal Dutt
Chandigarh University, India

IGI Global
PUBLISHER of TIMELY KNOWLEDGE

A volume in the Advances in Bioinformatics and Biomedical Engineering (ABBE) Book Series

Published in the United States of America by
IGI Global
Medical Information Science Reference (an imprint of IGI Global)
701 E. Chocolate Avenue
Hershey PA, USA 17033
Tel: 717-533-8845
Fax: 717-533-8661
E-mail: cust@igi-global.com
Web site: http://www.igi-global.com

Library of Congress Cataloging-in-Publication Data

Names: Lilhore, Umesh, 1982- editor. | Kumar, Abhishek, 1989- editor. | Simaiya, Sarita, 1986- editor. | Vyas, Narayan, 1998- editor. | Dutt, Vishal, Dr., editor.
Title: Applying machine learning techniques to bioinformatics : few-shot and zero-shot methods / edited by Umesh Lilhore, Abhishek Kumar, Sarita Simaiya, Narayan Vyas, Vishal Dutt.
Description: Hershey, PA : Medical Information Science Reference, [2024] | Includes bibliographical references and index. | Summary: "This book aims to provide a rigorous and in-depth examination of the intersection of bioinformatics, few-shot learning, and zero-shot learning"-- Provided by publisher.
Identifiers: LCCN 2023053174 (print) | LCCN 2023053175 (ebook) | ISBN 9798369318225 (hardcover) | ISBN 9798369318232 (ebook)
Subjects: MESH: Machine Learning | Agriculture | Computational Biology--methods | Data Visualization
Classification: LCC S494.5.D3 (print) | LCC S494.5.D3 (ebook) | NLM S 494.5.D3 | DDC 630.285--dc23/eng/20231229
LC record available at https://lccn.loc.gov/2023053174
LC ebook record available at https://lccn.loc.gov/2023053175

This book is published in the IGI Global book series Advances in Bioinformatics and Biomedical Engineering (ABBE) (ISSN: 2327-7033; eISSN: 2327-7041)

British Cataloguing in Publication Data
A Cataloguing in Publication record for this book is available from the British Library.

All work contributed to this book is new, previously-unpublished material. The views expressed in this book are those of the authors, but not necessarily of the publisher.

For electronic access to this publication, please contact: eresources@igi-global.com.

Advances in Bioinformatics and Biomedical Engineering (ABBE) Book Series

Ahmad Taher Azar
College of Computer & Information Sciences, Prince Sultan University, Riyadh, Saudi Arabia & Faculty of Computers and Artificial Intelligence, Benha University, Benha, Egypt

ISSN:2327-7033
EISSN:2327-7041

Mission

The fields of biology and medicine are constantly changing as research evolves and novel engineering applications and methods of data analysis are developed. Continued research in the areas of bioinformatics and biomedical engineering is essential to continuing to advance the available knowledge and tools available to medical and healthcare professionals.

The **Advances in Bioinformatics and Biomedical Engineering (ABBE) Book Series** publishes research on all areas of bioinformatics and bioengineering including the development and testing of new computational methods, the management and analysis of biological data, and the implementation of novel engineering applications in all areas of medicine and biology. Through showcasing the latest in bioinformatics and biomedical engineering research, ABBE aims to be an essential resource for healthcare and medical professionals.

Coverage

- Computational Biology
- Genetics
- Genomics
- Biomechanical Engineering
- Robotics and Medicine
- Molecular Engineering
- Prosthetic Limbs
- Biostatistics
- Biomedical Sensors
- Tissue Engineering

IGI Global is currently accepting manuscripts for publication within this series. To submit a proposal for a volume in this series, please contact our Acquisition Editors at Acquisitions@igi-global.com or visit: http://www.igi-global.com/publish/.

The Advances in Bioinformatics and Biomedical Engineering (ABBE) Book Series (ISSN 2327-7033) is published by IGI Global, 701 E. Chocolate Avenue, Hershey, PA 17033-1240, USA, www.igi-global.com. This series is composed of titles available for purchase individually; each title is edited to be contextually exclusive from any other title within the series. For pricing and ordering information please visit http://www.igi-global.com/book-series/advances-bioinformatics-biomedical-engineering/73671. Postmaster: Send all address changes to above address.

Titles in this Series

For a list of additional titles in this series, please visit: www.igi-global.com/book-series/advances-bioinformatics-biomedical-engineering/73671

Biomedical Research Developments for Improved Healthcare
Pranav Kumar Prabhakar (Lovely Professional University, India)
Medical Information Science Reference • copyright 2024 • 393pp • H/C (ISBN: 9798369319222) • US $360.00 (our price)

Quantum Innovations at the Nexus of Biomedical Intelligence
Vishal Dutt (AVN Innovations Pvt. Ltd., India) Abhishek Kumar (Department of CSE, UIE, Chandigarh University, Punjab, India) Sachin Ahuja (Chandigarh University, India) Anupam Baliyan (Geeta University, India) and Narayan Vyas (AVN Innovations Pvt. Ltd., India)
Medical Information Science Reference • copyright 2024 • 287pp • H/C (ISBN: 9798369314791) • US $330.00 (our price)

Physiology and Function of Glial Cells in Health and Disease
Bilal El-Mansoury (Faculty of Sciences, Chouaib Doukkali University, Morocco) Omar El Hiba (Chouaib Doukkali University, Morocco) and Arumugam Radhakrishnan Jayakumar (University of Miami, USA)
Engineering Science Reference • copyright 2024 • 589pp • H/C (ISBN: 9781668496756) • US $325.00 (our price)

Applications of Synthetic Biology in Health, Energy, and Environment
Muhammad Arshad (International Islamic University, Islamabad, Pakistan)
Medical Information Science Reference • copyright 2023 • 454pp • H/C (ISBN: 9781668465776) • US $325.00 (our price)

Isolation, Characterization, and Therapeutic Applications of Natural Bioactive Compounds
Ajeet Singh (ICAR-Indian Institute of Wheat and Barley Research, India)
Medical Information Science Reference • copyright 2022 • 401pp • H/C (ISBN: 9781668473375) • US $345.00 (our price)

AI-Enabled Smart Healthcare Using Biomedical Signals
Rahul Kumar Chaurasiya (Maulana Azad National Institute of Technology, Bhopal, India) Dheeraj Agrawal (Maulana Azad National Institute of Technology, Bhopal, India) and Ram Bilas Pachori (Indian Institute of Technology, Indore, India)
Medical Information Science Reference • copyright 2022 • 322pp • H/C (ISBN: 9781668439470) • US $435.00 (our price)

IGI Global
PUBLISHER of TIMELY KNOWLEDGE

701 East Chocolate Avenue, Hershey, PA 17033, USA
Tel: 717-533-8845 x100 • Fax: 717-533-8661
E-Mail: cust@igi-global.com • www.igi-global.com

Table of Contents

Detailed Table of Contents

This chapter presents a novel method for improving zero-shot learning in ML by using fully connected weighted bipartite graphs. Problems with generalizability and adaptability plague zero-shot learning, a method that lets models identify and categorize things or ideas without any explicit training. To overcome these obstacles and greatly enhance machine learning models' ability to absorb and comprehend unknown input, this chapter investigates how fully linked weighted bipartite graphs may be integrated. A thorough introduction to zero-shot learning is provided at the outset of the investigation. It describes the method's value in the machine learning field while drawing attention to the problems with and restrictions on current approaches. Anyone involved with machine learning, whether as a researcher, practitioner, or hobbyist, will find this chapter to be an invaluable resource. It lays out the theory and some practical considerations for improving zero-shot learning with fully connected weighted bipartite graphs.

The integration of bioinformatics with contemporary machine-learning algorithms is transforming sustainable practices and conservation activities in biology and agriculture. Plant disease identification is an area where few-shot learning (FSL) excels because of data scarcity. This study applies FSL to computational biology to tackle agricultural and environmental concerns. Bioinformatics has a significant influence on sustainable farming and research, according to the report. The chapter introduces few-shot learning, and shows how it may address the lack of labelled data in several disciplines. Case studies,

In the rapidly evolving field of bioinformatics, the surge in biological data necessitates innovative approaches for practical analysis and comprehension. This chapter presents an in-depth exploration of the potential applications of deep learning and transfer learning methods, showcasing their transformative impact on various facets of bioinformatics research. Deep learning methodologies significantly enhance the capacity to extract intricate patterns and features from diverse biological datasets, fostering a more profound understanding of genomic, proteomic, and metabolomic dynamics. Additionally, transfer learning assumes a pivotal role in optimizing model performance, particularly in scenarios where labeled data is scarce, by leveraging knowledge acquired from pre-trained models. The integration of deep learning and transfer learning techniques in bioinformatics yields valuable insights, serving as a guiding framework for researchers and practitioners aiming to harness the full potential of these methodologies in deciphering the complexities of biological systems.

Modernization and advancement in technology have developed a new field called Health Informatics. It is an interdisciplinary that accommodates bioinformatics, artificial intelligence (AI), and machine learning (ML) applications. Additionally, a large amount of data that travels around the global network of cloud computing has coined the term internet of things (IoT), which has fostered a new wave of innovation in the healthcare sector. This chapter provides insights into how recognition systems, self-monitoring systems, big data analysis and interpretation, etc. use AI and IoT-based models to accelerate the capabilities of providing advanced healthcare outcomes. It not only unlocks futuristic smart healthcare, but also platforms a revolutionary milestone in the medical world. Moreover, the chapter addresses the challenges of security, scalability, cost-effectiveness, standardization, and quality care encountered by the healthcare sector.

This chapter take a fascinating voyage across the complex bioinformatics landscapes. The authors begin by defining bioinformatics as the essential connection that connects unprocessed biological data to useful knowledge, laying the groundwork for an investigation into the core fields of proteomics, genomics, and data mining. The story flows naturally, showing how feature refinement techniques, machine learning's predictive power, and deep learning's revolutionary ability to understand intricate biological processes all work together in harmony. Applications in the real world emerge, demonstrating the palpable effects

through modeling. The chapter also addresses the challenges faced by researchers when integrating these diverse fields. Nevertheless, optimism prevails in the realm of bioinformatics research, driven by the ever-expanding wealth of biological data and the potential for the development of more sophisticated machine learning models.

This chapter delves deeply into few-shot learning, a rapidly developing area crucial in driving bioinformatics forward. First, the basics of few-shot learning are laid forth, emphasizing the field's applicability to bioinformatics. Case studies showcasing real-world applications in areas as varied as protein structure prediction, drug development, and genomic analysis provide a deep dive into several few-shot learning approaches like meta-learning and transfer learning. The chapter also provides an in-depth analysis of recent developments, highlights current difficulties, and suggests exciting new avenues for exploration. This chapter highlights the rising significance of few-shot learning in bioinformatics and provides insights into its potential to benefit biomedical research.

ML has revolutionised bioinformatics' difficult biological data analysis. Pattern recognition and biological process categorization help ML systems diagnose diseases, predict protein structures, and investigate gene expression. Few-shot learning and bioinformatics are effective at optimising results with limited resources, overcoming biological dataset access issues. ML and bioinformatics advance precision medicine and drug discovery while improving biological understanding. This chapter examines bioinformatics ML methods like supervised classification, clustering, and probabilistic graphical models to find new insights. Text mining, systems biology, evolution, proteomics, and genomics use deterministic and stochastic heuristics for optimisation. Understanding bioinformatics methods and modern ML technologies while understanding implementation challenges is the study goal. Few-shot learning is highlighted to show its importance. ML and bioinformatics together improve our knowledge and solve real problems, improving research methods in biological and medical sciences.

This chapter delves into the transformative synergy of few-shot learning and healthcare, elucidating its impact on medical procedures. Anchored in machine learning fundamentals, it establishes a core framework through a review of algorithms. Addressing challenges of small healthcare datasets, the chapter highlights the pivotal role of few-shot learning. Innovative methods like multimodal integration and federated learning enhance model robustness, offering insights into complex healthcare scenarios.

Preface

As editors of this comprehensive reference book, *Bioinformatics, Few-Shot Learning, and Zero-Shot Learning in Biological Sciences*, we are delighted to present a collection that delves into the fascinating intersection of cutting-edge data science methodologies and the intricate world of biological studies. Our esteemed team—Umesh Kumar Lilhore, Abhishek Kumar, Sarita Simaiya, Narayan Vyas, and Vishal Dutt—brings together expertise from diverse geographical and professional backgrounds, uniting to provide a holistic exploration of the evolving landscape of bioinformatics and machine learning in the biological sciences.

The motivation behind this book is rooted in the profound impact that data science techniques, particularly few-shot learning and zero-shot learning, have had on our understanding of biological systems, personalized medicine, and biomarker identification. Our goal is to offer readers, ranging from seasoned researchers to budding scientists, a comprehensive analysis of these emerging disciplines. We embark on a journey through applications, common challenges, and ethical considerations, emphasizing the integration of artificial intelligence in the processing of genomic data and its pivotal role in drug discovery.

Bioinformatics, Few-Shot Learning, and Zero-Shot Learning in Biological Sciences is crafted to be accessible to a broad audience, providing information and insights necessary to navigate the dynamic realms of bioinformatics and machine learning within the biological sciences. Through in-depth case studies, practical implementations, and thought-provoking discussions on ethical AI usage, we aim to empower our readers to grasp the nuances of these evolving fields.

In the context of biological sciences, this book offers a rigorous examination of the interplay between bioinformatics, few-shot learning, and zero-shot learning. We strive to cultivate an in-depth familiarity with applications, challenges, and ethical concerns in these areas, equipping researchers, scientists, and readers with the knowledge and techniques to harness the potential of data science for advancing biological research, personalized medicine, and biomarker discovery.

Our target audience spans academics and practitioners interested in bioinformatics, computational biology, and machine learning. Additionally, this material is designed for upper-level undergraduates and graduates seeking to expand their understanding of these topics. This book is a must-read for anyone intrigued by the convergence of data science and biology/healthcare, including healthcare professionals, biotechnologists, and academics.

The structure of the book is organized to guide the reader seamlessly through the intricate landscape of bioinformatics and machine learning, covering topics such as Few-Shot Learning and Zero-Shot Learning Paradigms, Biomarkers and Personalized Medicine, Bioinformatics in Agriculture and Ecology, AI-Driven Drug Discovery, Deep Learning & Transfer Learning Techniques, Ethical Considerations, and Future Directions.

ORGANIZATION OF THE BOOK

Chapter 1: This chapter introduces a novel approach to enhance zero-shot learning in machine learning by utilizing fully connected weighted bipartite graphs. It addresses issues of generalizability and adaptability in zero-shot learning, providing a comprehensive exploration of how fully linked weighted bipartite graphs can be integrated to significantly improve machine learning models' ability to comprehend unknown input. The chapter offers both theoretical foundations and practical considerations, making it an invaluable resource for researchers, practitioners, and enthusiasts in the field of machine learning.

Chapter 2: This chapter explores the integration of bioinformatics with contemporary machine-learning algorithms, focusing on the transformative impact of few-shot learning in agriculture and ecology. With a spotlight on plant disease identification and the scarcity of labeled data, the chapter demonstrates how few-shot learning excels in addressing data limitations. It includes case studies illustrating the widespread use of few-shot learning in ecological surveillance, environmental programs, and crop management, while also discussing ethical considerations in the application of machine learning in ecological systems and agriculture.

Chapter 3: The third chapter emphasizes the crucial role of big data in addressing complex real-world problems, particularly in healthcare and medicine. It delves into the significance of classification algorithms as essential tools in decoding challenging problems, such as detecting deadly diseases. The chapter provides insights into how different classification algorithms perform in the healthcare system, offering a practical understanding of their applications and benefits.

Chapter 4: This chapter navigates the ethical and legal challenges posed by artificial intelligence in healthcare. Addressing issues such as privacy invasion, discrimination, and human judgment, the chapter underscores the potential catastrophic repercussions for patients in the event of healthcare errors. It calls for comprehensive measures to ensure the protection of all parties involved, emphasizing openness, privacy in algorithms, and cybersecurity.

Chapter 5: Examining the health risks associated with 5G technology radiation, this chapter sheds light on the potential dangers posed by high-frequency waves. It explores the impact of radiation on human health, detailing its role in causing chronic diseases and suppressing the immune system. The chapter aims to raise awareness about the consequences of 5G technology radiation on human well-being, providing valuable insights for readers concerned about the health implications of modern technology.

Chapter 6: This interdisciplinary chapter explores the integration of advanced machine learning techniques with bioinformatics to revolutionize drug discovery. From detailing the historical journey of drug discovery to discussing current challenges, the chapter showcases how machine learning algorithms can analyze biological data efficiently. Case studies highlight successful applications of these techniques in identifying new therapeutic compounds and personalizing treatment, while also addressing regulations and guidelines for robust and interpretable models.

Chapter 7: Focusing on the theoretical foundations and challenges of few-shot and zero-shot learning, this chapter explores advancements in these techniques. While highlighting the efficiency and adaptability of these AI systems, the chapter delves into intrinsic challenges such as model bias, domain adaptation problems, and computational restrictions. By comparing and contrasting these methods, the chapter aims to enhance understanding of their potential use in diverse real-world contexts.

Chapter 8: This chapter provides an in-depth exploration of the potential applications of deep learning and transfer learning methods in bioinformatics. Showcasing their transformative impact on various aspects of bioinformatics research, the chapter emphasizes their capacity to extract intricate patterns from diverse biological datasets. It serves as a guiding framework for researchers and practitioners seeking to harness the full potential of deep learning and transfer learning in deciphering the complexities of biological systems.

Chapter 9: Focusing on the intersection of health informatics, bioinformatics, artificial intelligence, and the Internet of Things (IoT), this chapter provides insights into how AI and IoT-based models accelerate advancements in healthcare. It explores recognition systems, self-monitoring systems, big data analysis, and interpretation, offering a glimpse into futuristic smart healthcare. The chapter addresses challenges such as security, scalability, cost-effectiveness, standardization, and quality care, providing a comprehensive overview of the evolving healthcare landscape.

Chapter 10: This chapter takes readers on a fascinating journey across the complex bioinformatics landscapes, defining bioinformatics as the essential connection between unprocessed biological data and useful knowledge. It explores the core fields of proteomics, genomics, and data mining, showcasing the synergy of feature refinement techniques, predictive power of machine learning, and revolutionary capabilities of deep learning. While highlighting real-world applications, the chapter acknowledges challenges such as data integration nuances, interpretability of machine learning models, ethical dilemmas, and computing demands of big data.

Chapter 11: This thorough analysis delves into the ethical implications of patient data sharing in the healthcare industry. Examining the intricate relationships between security, consent, privacy, and societal welfare, the chapter addresses the potential advantages and challenges of data sharing. Practical advice is offered on striking a balance between the benefits of data sharing and concerns about prejudice, discrimination, and access, providing a comprehensive framework for navigating the complex ethical terrain of patient data sharing.

Chapter 12: This chapter explores the transformative synergy of deep learning and transfer learning techniques in bioinformatics. Detailing the journey from neural networks to deciphering intricate biological data, the chapter emphasizes the role of transfer learning in optimizing model performance. It showcases how cross-domain knowledge transfer expedites progress in bioinformatics applications, from predicting protein structures to identifying disease biomarkers.

Chapter 13: Diving into the intersection of deep learning and transfer learning in bioinformatics, this chapter provides a detailed assessment of deep learning principles and their application to challenges in bioinformatics datasets. It explores the concept of transfer learning, utilizing knowledge learned in one domain to enhance performance in another. The chapter highlights the effectiveness of transfer learning and its applications in various domains, contributing to a transformative worldview.

Chapter 14: This chapter focuses on the application of few-shot and zero-shot learning techniques in biomarker prediction tasks for personalized medicine. It offers a comprehensive exploration of biomarker types and applications, showcasing how few-shot and zero-shot techniques could enhance biomarker discovery and validation. The chapter presents a promising paradigm for more efficient biomarker modeling, potentially accelerating progress towards personalized medicine.

Chapter 15: This chapter examines recent developments in both bioinformatics and machine learning, exploring the potential advantages of employing machine learning to enhance critical bioinformatics tasks. Delving into the analysis of genomic sequences and protein structures, the chapter addresses challenges faced by researchers when integrating these diverse fields. It concludes with an optimistic outlook on the future of bioinformatics research, driven by the expanding wealth of biological data and the development of more sophisticated machine learning models.

Chapter 16: This chapter provides a deep exploration of few-shot learning, a rapidly developing area crucial in advancing bioinformatics. It introduces the basics of few-shot learning and its applicability to bioinformatics, showcasing real-world applications in protein structure prediction, drug development, and genomic analysis. The chapter highlights recent developments, current difficulties, and suggests exciting new avenues for exploration, emphasizing the rising significance of few-shot learning in bioinformatics.

Chapter 17: Examining the revolutionary impact of machine learning on the analysis of difficult biological data, this chapter explores the applications of supervised classification, clustering, and probabilistic graphical models in bioinformatics. It delves into various bioinformatics methods and modern machine learning technologies, addressing implementation challenges and highlighting the importance of few-shot learning in improving research methods in biological and medical sciences.

Chapter 18: The final chapter delves into the transformative synergy of few-shot learning in healthcare, elucidating its impact on medical procedures. Grounded in machine learning fundamentals, the chapter establishes a core framework through a review of algorithms. It addresses challenges related to small healthcare datasets and highlights the pivotal role of few-shot learning. The chapter introduces innovative methods like multimodal integration and federated learning, offering insights into complex healthcare scenarios and providing avenues for a deeper understanding and implementation in medical imaging.

We believe that this book will serve as a valuable resource, offering not only theoretical insights but also practical knowledge through real-world case studies. As editors, we extend our gratitude to the contributors and hope that this compilation stimulates curiosity, sparks discussions, and propels the reader towards innovative applications at the forefront of bioinformatics and machine learning in the biological sciences.

CONCLUSION

As editors of *Bioinformatics, Few-Shot Learning, and Zero-Shot Learning in Biological Sciences*, we take great pride in presenting this comprehensive reference book that delves into the cutting-edge intersections of bioinformatics and machine learning. The chapters within this collection encapsulate a wealth of knowledge, insights, and practical applications that reflect the collaborative efforts of esteemed contributors from diverse backgrounds.

From the innovative methods of improving zero-shot learning with fully connected weighted bipartite graphs to the transformative synergy of deep learning and transfer learning in bioinformatics, each chapter explores distinct facets of the dynamic relationship between data science and biological studies. The integration of machine learning algorithms in drug discovery, the ethical considerations in AI-driven healthcare, and the potential health risks associated with 5G technology radiation are just a few examples of the multifaceted exploration within these pages.

The diverse range of topics covered in this book aims to cater to a broad audience, including seasoned researchers, practitioners, educators, and curious minds seeking to deepen their understanding of the evolving landscape of bioinformatics and machine learning. The presented case studies, theoretical foundations, and practical applications provide a rich tapestry of knowledge that contributes to the ongoing dialogue within these fields.

We hope that this compilation serves not only as a valuable resource for academics and professionals but also as an inspiration for further exploration and innovation. As the fields of bioinformatics and machine learning continue to evolve, the insights shared in these chapters lay the groundwork for future advancements and collaborations.

Our sincere gratitude goes to the dedicated contributors who have shared their expertise, making this book a reflection of the collective wisdom of the scientific community. It is our belief that the information presented within these pages will spark discussions, fuel curiosity, and propel readers toward new horizons in the ever-evolving realms of bioinformatics and machine learning within the biological sciences.

May this book contribute to the ongoing pursuit of knowledge, ethical considerations, and transformative applications in the dynamic and intertwined fields of bioinformatics and machine learning.

Editors,

Umesh Kumar Lilhore
Galgotia University, India

Abhishek Kumar
Chandigarh University, India

Sarita Simaiya
Galgotia University, India

Narayan Vyas
Chandigarh University, India

Vishal Dutt
Chandigarh University, India

Chapter 1
Advancing Zero-Shot Learning With Fully Connected Weighted Bipartite Graphs in Machine Learning

V. Dankan Gowda
https://orcid.org/0000-0003-0724-0333
BMS Institute of Technology and Management, India

Rama Chaithanya Tanguturi
https://orcid.org/0000-0002-9923-7360
PACE Institute of Technology and Sciences, Ongole, India

Neha Patwari
Thakur College of Engineering and Technology, Mumbai, India

S. B. Sridhara
Amruta Institute of Engineering and Management Sciences, Bidadi, India

Sampada Abhijit Dhole
Bharati Vidyapeeth College of Engineering for Women, Pune, India

ABSTRACT

This chapter presents a novel method for improving zero-shot learning in ML by using fully connected weighted bipartite graphs. Problems with generalizability and adaptability plague zero-shot learning, a method that lets models identify and categorize things or ideas without any explicit training. To overcome these obstacles and greatly enhance machine learning models' ability to absorb and comprehend unknown input, this chapter investigates how fully linked weighted bipartite graphs may be integrated. A thorough introduction to zero-shot learning is provided at the outset of the investigation. It describes the method's value in the machine learning field while drawing attention to the problems with and restrictions on current approaches. Anyone involved with machine learning, whether as a researcher, practitioner, or hobbyist, will find this chapter to be an invaluable resource. It lays out the theory and some practical considerations for improving zero-shot learning with fully connected weighted bipartite graphs.

DOI: 10.4018/979-8-3693-1822-5.ch001

1. INTRODUCTION

Definition of Zero-Shot Learning (ZSL) in Machine Learning: Through a machine learning technique in which models are trained to recognize and classify objects that have not previously been introduced. ZSL is unlike other machine learning strategies, which teach a class by using labeled samples from each one. It uses transfer techniques to make inferences about the characteristics and attributes of unseen classes--by observing differences between seen classes.

Importance and Applications of Zero-Shot Learning: When it is inconvenient or impossible to do so, collecting a large volume of training data for each category becomes important. ZSL plays an especially critical role in such cases. One particularly important area of use is in environmental monitoring, and natural language processing using image recognition or voice recognition when new classes continually emerge. The field of biodi-versity protection is one example. ZSL can identify species that were not among the training data before (N Madapana and J Wachs, 2021). Computer scientists can thus create flexible algorithms that learn from examples and don't need human intervention to classify changing data.

Challenges in Zero-Shot Learning and Its Distinction from Traditional Machine Learning Methods: Zero-Shot Learning (ZSL) presents its own unique challenges that are unlike those of traditional machine learning methods. Y. Feng, X. Huang (2022) One of the most important problems is the semantic gap; here it means that data are too far apart from raw features and high-level conceptual understanding needed for classification information--category unseen categories. ZSL In ZSL, this gap is even bigger. The model must predict about classes never encountered during training time at all. Another major obstacle is the propensity to teach in seen classes; typical training methods typically cause the model's results not only influenced by data of seen samples, but often at unseen sample expense accurate classification. Since this reflects a bias, in ZSL it makes asking the classifier to generalize between seen and unseen classes without overfitting to previously-used data another central problem.

In contrast, the main problem faced by traditional machine learning methods is that the seen data (labeled) must be large in volume for each class studied. These conventional techniques are simply concerned with reducing error on this available labeled data. But this method leaves them brittle, unable to adopt flexibly when faced with new and unknown data. But ZSL requires models to reason from seen classes and expand that knowledge for an unseen one. That is a completely different strategy, with more intricate relationships between data (N. Bendre K Desai P Najafirad 2021). Thus, ZSL distinguishes itself by its inherent need for models to possess the capability to interpret and classify data beyond the constraints of the training dataset, pushing the boundaries of what is achievable in machine learning.

In the modern world, digital image processing significantly influences various technical fields. This processing encompasses various forms, such as enhancing images, extracting image attributes, or detecting objects(B. Jiang, Y. Lu, B. Zhang and G. Lu, 2023). Image enhancement, a subset of this process, focuses on reducing noise, enhancing contrast, and filtering. Referred to as low-level image processing, it aims to improve image quality by modifying intensity levels. While enhancing an image generally elevates its quality, it's important to note that this process can sometimes lead to the loss of critical information from the original image.

Image attributes, such as edges, contours, and objects within the image, are critical components in digital image processing. The process of image attribute extraction encompasses several key techniques, including segmentation, object detection and extraction, edge detection, and object classification(W. Cho and E. Kim, 2022). The outcome of this extraction is typically a symbolic representation detailing the object's position and shape. This extraction process facilitates a basic level of image analysis. The attributes identified through this process can undergo further analysis for a more comprehensive understanding of the original image.

Computer vision (CV) is an area that is closely related to high-level image processing. While computer vision includes image processing as a subset, computer vision goes beyond by trying to mimic how the human visual system works. The opsins—a network of millions of cone cells—make the human eye photosensitive(S. Venkatakiran and N. S. Reddy, 2023). Upon exposure to light photons, each opsin changes form and reacts to one of three main hues. This causes the brain to receive and interpret neural signals. The human eye's object identification skills are intricately designed.

The main objective of computer vision is to have computers see at this level, so they can understand and respond to 2D and 3D visual content much like humans. Computer vision utilizes a mix of machine learning, image processing, and other approaches to detect patterns and understand pictures, similar to how the human eye uses the cornea and other parts to understand light and identify things(U. K N and R. V M, 2022). The discipline is facing a huge barrier in its quest to give robots visual skills similar to humans.

Light is the primary means by which the human eye identifies things. After passing through the cornea, this light reaches the cones and rods of the eye. Through the optic nerve, the cones and rods transform light into electrical impulses that are sent to the brain. In figure 1, we can see how the visual cortex of the brain takes in these electrical impulses and uses them to build a visual representation of the picture(M. R. G. and H. Anandaram, 2022). This procedure exemplifies the multi-organ and intricately-designed human visual system.

Despite the remarkable capabilities of the human eye in perceiving images, it is not infallible and can be deceived. Certain images can create illusions, giving the impression of representing objects in

Figure 1. A concise representation of the human visual system

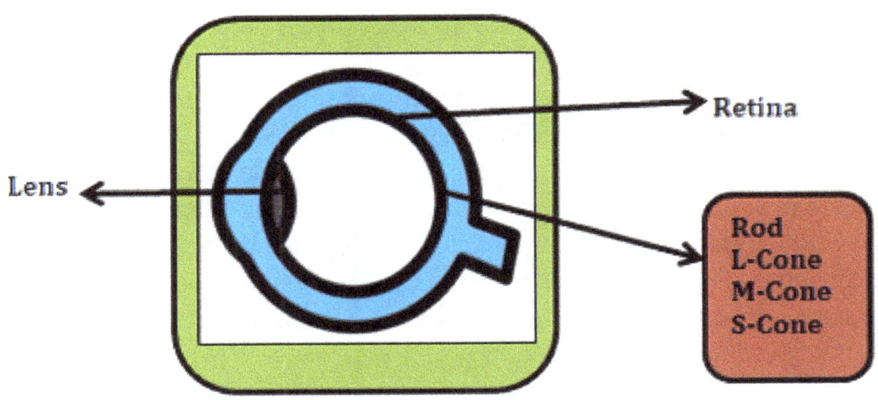

Figure 2. An illustration of highly detailed fish species

(a) an instance blue whale

(b) an instance of killer whale

ways that defy logical or physical representation. This aspect of visual perception highlights both the sophistication and the limitations of human vision.

Many artists skillfully employ techniques in their artwork to deceive the eye, creating images of objects that defy physical representation(Revanna C and Parismita Sarma, 2022). This illusionary art often lacks the dimensions of depth and motion, intrinsic to real-world perception, leading to various ambiguities. Human vision, being binocular and dynamic, is naturally adept at perceiving depth and motion(S. Reddy P, P. S. Patwal, 2022). However, pictorial images are inherently static and flat, lacking these dimensional qualities. This contrast between the dynamic, three-dimensional perception of the human eye and the static, two-dimensional nature of images results in perceptual paradoxes, challenging our visual understanding and interpretation.

Conventional machine learning techniques often rely on accurately labeled images, but the process of labeling can be complex. While general object classes are easily recognized and labeled by most people, precise labeling of specific or fine-grained classes, such as different fish species, requires the knowledge

Figure 3. A general classification of image-based zero-shot learning (ZSL) approaches

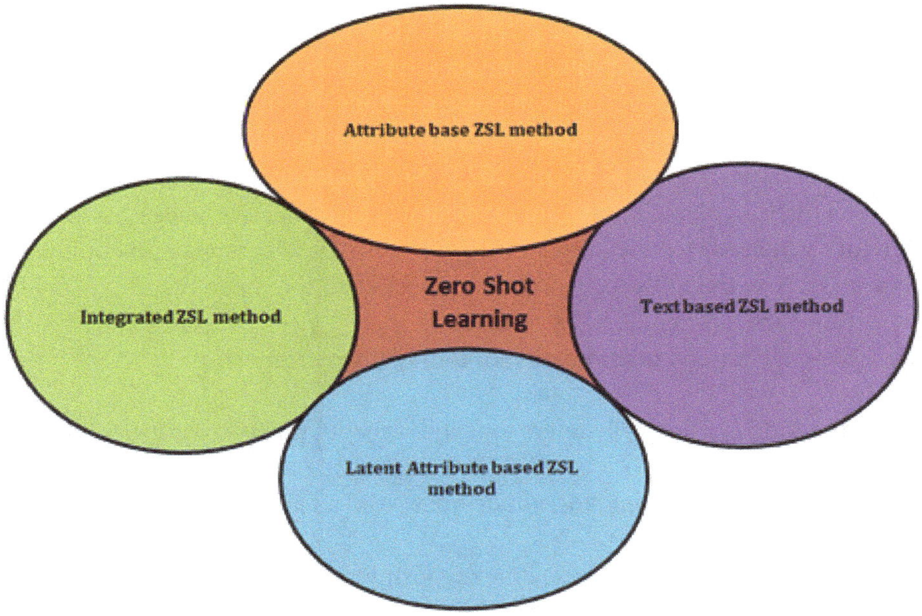

of a subject expert(S. R. Kawale, S. P. Diwan, 2022). For example, various fish species might be indistinct to an average person who may simply categorize them all as fish. In a more specific case, a blue whale and a killer whale, though distinct species as perhaps illustrated in a hypothetical 'figure.2', might be undifferentiated by someone without specialized marine biology knowledge, being labeled generically as whales. It's only experts in fish species who can accurately identify and differentiate between species like the blue whale and the killer whale from an image(B. Kameswara Rao and Naziya Hussain, 2022). This highlights the nuanced challenges in labeling detailed object classes for machine learning purposes.

The requirement of subject matter experts for labeling fine-grained object classes in machine learning is indeed a costly and time-intensive process. This concept explores the idea of creating a model where the dataset, training, and testing object classes are entirely distinct from one another(M. Swathi Pai and B. Naveen K, 2020). For instance, if a machine is trained exclusively on images of horses, could it then have the capability to recognize zebras, a related but different species it has never been trained on? This challenge is about pushing the boundaries of machine learning models towards greater adaptability and generalization.

Assuming some prior information about the unseen item, the basis for human identification of an unseen object class is often the connection between known (seen) and unknown (unseen) classes(K. B. Naveen and G. N. Pai, 2020). The common features of the visible and invisible things provide the groundwork for this association.

For instance, consider someone who has seen horses but has never encountered a zebra. However, they know that a zebra is quite similar to a horse, with the primary difference being the distinctive stripes on a zebra's body(P. Ramesh Naidu, N. Guruprasad, 2020). This process represents how humans, by relating new information to existing knowledge (the associative learning method) and arranging objects in relation bito one another according to common characteristics (by analogy), can even recognize and classify objects which they have never directly observed.

Snaven et al., 2015; Deng and Kulis, 2016; Nihashi Matsuoka, Takahashi & Iwata-Yamada. Prasad and G. Poornima, 2022). However, one significant gap in these classifications is the lack of discussion about how linguistic information plays a role in these methods. To mitigate this problem a special and deep-sense section has been adopted, based on the kind of prior knowledge or extra information used in ZSL methods (H. Anandaram, and B. K Verma 2023). Such balanced attachments between visible and invisible classes are largely dependent upon supplementary data, which is naturally of different kinds. It can be characteristics generated from external resources like WordNet or texts; it could also potentially be latent attributes (such as the POS-tagger) through backward inference based on some higher level structured information such as annotation by humans.

Accordingly, ZSL methods can be grouped into four distinct categories:

Attribute-Based ZSL (A ZSL): This type refers to methods beginning with properties that humans have annotated.

Text-Based ZSL (T ZSL): These techniques, which use text as an aid, are used to link visible and invisible classes.

Latent Attribute-Based ZSL (Lt ZSL): Those drawing visible and invisible classes together on the basis of latent characteristics are in this category.

Integrated ZSL (I ZSL): hatched out of the many sources they draw from (WordNet hierarchy, word-2vec, GloVe and so on), these methods establish connections between visible and invisible classes.

Each of these categories reflects a different approach in harnessing side information for zero-shot learning, as conceptualized in the classification scheme depicted in the hypothetical 'figure.3'. This categorization offers a more nuanced understanding of how ZSL methods leverage different types of linguistic and attribute-based knowledge to bridge the gap between known and unknown classes.

2. BASICS OF GRAPH THEORY

Introduction to Graph Theory Basics Relevant to Zero-Shot Learning: Combinatorial mathematics' foundational field, graph theory, studies mathematical structures called graphs that represent pairwise interactions between things(K. Prasad and B. Ashreetha, 2023). In the context of zero-shot learning, graph theory becomes instrumental in modeling complex relationships and patterns within data. Concepts like nodes (or vertices) representing data points, and edges symbolizing the relationships or connections between these points, are foundational in understanding data structures and algorithms in machine learning.

Bipartite Graphs and Their Properties: For any given set of vertices, there exists an edge that links a vertex in the other set to a vertex in the first set; this sort of graph is called a bipartite graph. Put simply, no edges connect vertices in the same collection(R. Kishore Kumar and M. S. C. Senthil Kamalesh, 2023). When there are two separate classes of things and the only connections are between objects in different classes, bipartite graphs shine.

Key properties of bipartite graphs include:

Two-Colorability: So there is a way to color the bipartite graph with two colors, such that no adjacent vertices have the same color.

Matching: One of the most important concepts in bipartite graphs is "matching"; it refers to a set of edges such that there are no duplicate vertices and can be used about any issues involving extradition resources.

Partitioning: Researchers find this property extremely useful for categorizing and segregating data in machine learning. This is because vertices can be divided into two classes--class 0, which has no other way to get there than from the origin; class1 vertexes have a second gateway: themselves or their neighbors.

Weighted Graphs and Their Significance in Machine Learning: Some graphs have weights on their edges.The weight quantifies the strength or capacity of a connection (Sadashiva V. Chakrasali, A. Azhagu Jaisudhan Pazhani, 2023). In the field of machine learning, these weights can indicate importance, similarity or probability of transition between nodes. They are therefore essential for algorithms in which calculations depend on relationships ranging from a specific point to another.

In the context of zero-shot learning, weighted graphs can be used to:

Model the Strength of Relationships: Weights can reflect how close the various classes or features are related to each other.

Facilitate Knowledge Transfer: So weighted graphs help to transfer knowledge from seen classes to unseen.

Optimize Learning Algorithms: Weighted edges can direct learning algorithms to emphasize crucial links, enhancing the speed and accuracy of predictions.

Understanding of bipartite and weighted graphs thus forms the basis for expanding zero-shot learning models (Hombalimath, 2023). With such a method in place we can now consider deeper relationships within data. The interaction between these graph theory concepts and machine learning techniques is the theme of this chapter, which will create conditions for new forms of ZSL.

3. CONNECTING GRAPH THEORY WITH MACHINE LEARNING

Application of Graph Theory in Machine Learning: This insight makes graph theory central to machine learning, for it enables the development of models and analysis which bring together complex linkages and patterns. The field of machine learning makes use of graphs for:

Model Interconnections: Represent the complex relationships between various data points, attributes or classes.

Cluster and Classify Data: Apply graph-based algorithms to clustering and classi fication tasks.

Feature Extraction: Which is important in many applications of machine learning (such as social network analysis, recommendation systems and bioinformatics) where the input to model building consists mainly of graphs represented by data. Additionally you must extract and identify key features from this kind of data.

Neural Network Architectures: Also affect the design of neural network architectures, such as Graph Neural Networks (GNN's) that process graph-structured data directly.

Role of Fully Connected Weighted Bipartite Graphs in Zero-Shot Learning (ZSL): Fully connected weighted bipartite graphs can help enrich ZSL. This type of graphtoward to social tennis has two distinct sets of vertices (representing seen and unseen classes) with weighted edges connecting every vertex in one set to every other vertex. Their role in ZSL includes:

Encoding Relationships: Their weights are the strength or similarity of these relationships.

Facilitating Knowledge Transfer: The weighted links here further promote the transfer of knowledge between seen and unseen classes, doing so by exploiting the similarities coded in graph structures.

Improving Prediction Accuracy: These graphs add the degree of similarity between classes to their predictions, increasing the accuracy with which unseen examples can be predicted.

Previous Research and Applications: A number of research studies and applications in machine learning have used similar methods.

Recommendation Systems: Bipartite graphs representing the user-item relationships have been used in models of information recommender systems, with users and items being two disjoint sets of vertices.

Natural Language Processing (NLP): For example, in NLP the graph-based model has been applied to tasks such as semantic analysis (where words or phrases are regarded as vertices and have edges formed of relations between them.

Computer Vision: Graph-based methods have been used in image classification and object recognition for modeling relations between different classes of objects or within images, respectively.

Specifically in ZSL: On the attribute-class relationship, researchers have used weighted bipartite graphs to study it. One set represents different classes and another sets represent attributes (Namitha A R, Veera Sivakumar Chinamuttevi 202) Each record contains a positive integer representing how strongly an attribute is associated with its class.

This marriage of graph theory, especially that using fully connected weighted bipartite graphs in ZSL may be able to provide just such a technically advanced means for coping with the natural complexities that accompany these two domains. The framework of this chapter explores these intersections, revealing that graph theory can help improve our comprehension and application of ZSL models.

4. REVOLUTIONIZING ZERO-SHOT LEARNING WITH FULLY CONNECTED WEIGHTED BIPARTITE GRAPHS

The main emphasis of this chapter is on how fully connected weighted bipartite graphs can overcome the limitations that ZSL faces by providing a richer and more orderly way to represent, describe, and analyse connections between visible classes. Ramesh Naidu and M. Guruprasad (2021). In these graphs, visible classes and invisible ones are separated into two sets of vertices respectively; edges have weights representing the strength of links (G. A., Bashiardes anicee Naik, 2014). With labelled data for unseen classes naturally hard to obtain, this method allows a more effective knowledge transfer from visible to the unknown.

The theoretical underpinning of this approach lies in the principles of graph theory and knowledge transfer in machine learning:

Graph Theory Principles: The bipartite nature of these graphs means that relationships are explicitly modeled between different sets (seen and unseen classes) separated by a horizontal line, while

Figure 4. The procedure for identifying an image belonging to a class that a human has never encountered before

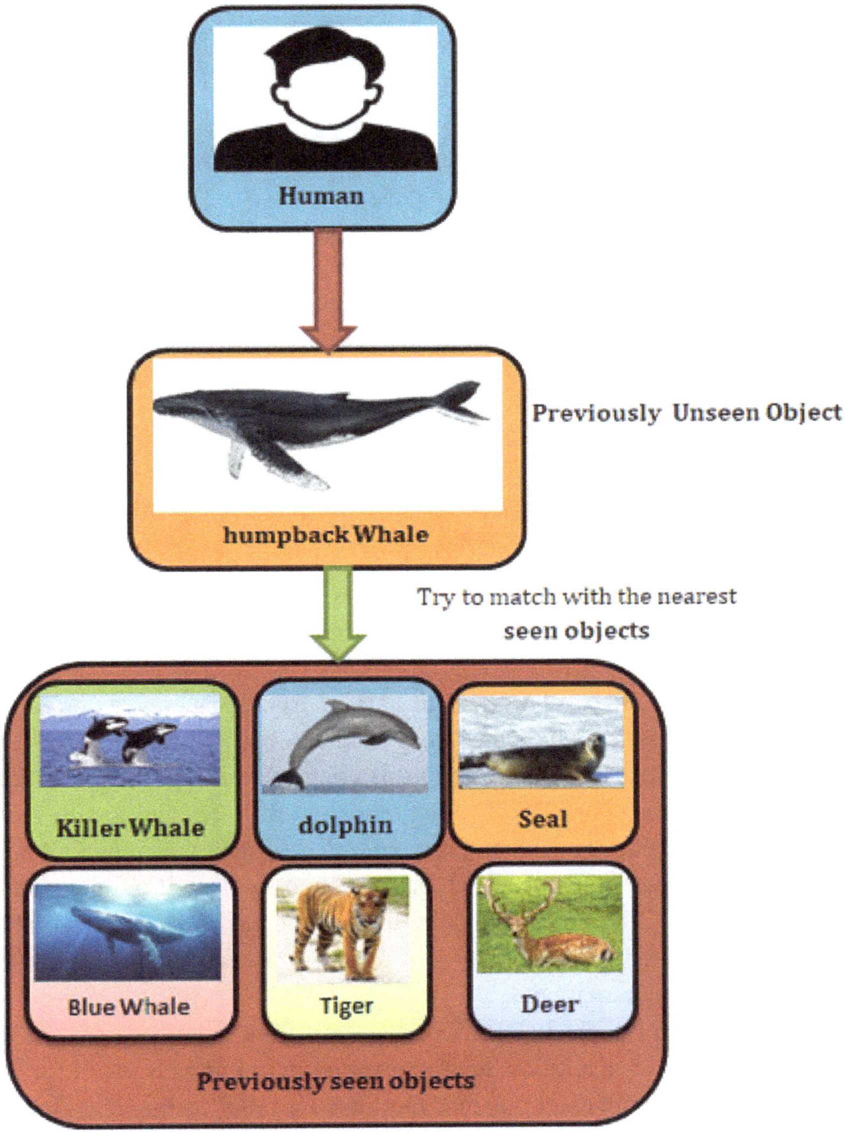

weights on the edges mean that varying degrees of similarity or relevance between each pair can be expressed.

Knowledge Transfer in ZSL: Because attribute differences between seen and unseen classes can be systematized, they just need to provide a way of measuring these attributes (K. Jeevan and B. M

Sathisha 2020). This structure is the fully connected weighted bipartite graph, which encodes complex relationships in a quantitative manner.

Addressing ZSL Challenges: This approach does indeed solve many of the problems in ZSL as outlined at the beginning.

Bridging the Semantic Gap: Weighted edges in the graph represent how much features relate to classes. We map from feature space to semantic space more clearly this way.

Mitigating Bias Towards Seen Classes: By explicitly representing unseen classes in the graph and quantifying their relationships with seen classes, the model can better generalize to unseen classes, reducing the inherent bias towards seen classes.

Enhancing Generalization: The structured representation in the form of a graph allows for more effective generalization from seen to unseen classes(B. Ram Vishal and K. A. M, 2023). The weights on the edges can be adjusted or learned in a way that optimizes this transfer of knowledge, making the model more robust to the variations inherent in unseen classes.

It is indeed the brain, not the eyes, that interprets images. Humans have the remarkable ability to recognize familiar objects from any viewpoint, even from angles they have never encountered before, except in cases of highly unusual projections(D. Palanikkumar, A. Y. Begum, 2023). However, the recognition of unseen objects by humans fundamentally relies on their relationship to previously seen objects. When encountering an unseen object, its various characteristics—such as shape, color, texture, and movement—are mentally compared with known objects. This comparison process, which might be represented in a hypothetical 'figure.4.', involves correlating these characteristics with those of familiar objects from our memory.

The process of matching these characteristics allows us to form conclusions about the nature of these unseen objects. The information about an object's characteristics, often referred to as semantic information or semantic space, plays a crucial role in this cognitive process(A. Nareshkumar and B. Gururaj, 2022). Semantic information encompasses the inherent attributes and features of objects, enabling us to establish correlations between seen and unseen objects.If an unseen picture is included in the training classes, a classifier may use supervised classification to place it in one of the observed classes(R. Kavitha, and V. Hariram, 2023). Nevertheless, in the absence of a reciprocal link between the unseen class and the training classes, the classifier will normally assign the picture to the closest seen class if it comes across an unseen class image.

Using this feature of supervised classifiers, the Zero-Shot Learning (ZSL) issue may be addressed. Each testing class in ZSL has a relationship to one or more training classes since the two sets of data are considered homologous. As an example, the "humpback whale" testing class in the widely-used AWA2 dataset for ZSL research is strongly connected to three training classes: "killer whale," "blue whale," and "dolphin." The size, shape, color, and texture of the creatures that make up these groups all contribute to establishing this connection.

One way to make this procedure even better is to use the mapping that exists between the training and testing classes(S. Archana Shreee, G. Jeevitha Sai, 2023). As a result, the classifier is able to recognize the testing picture as a "humpback whale," even though this class has never been applied before.

In Zero-Shot Learning (ZSL), the semantic space of the classes may be used to build the relationship between the training and testing classes(M. Kranthi, and R. C. Tanguturi, 2023). Here, semantic information—usually supplied on a class-by-class basis instead of for specific pictures—is vital. Attribute

Figure 5. Shared characteristics between humpback whales and killer whales within the AWA2 dataset

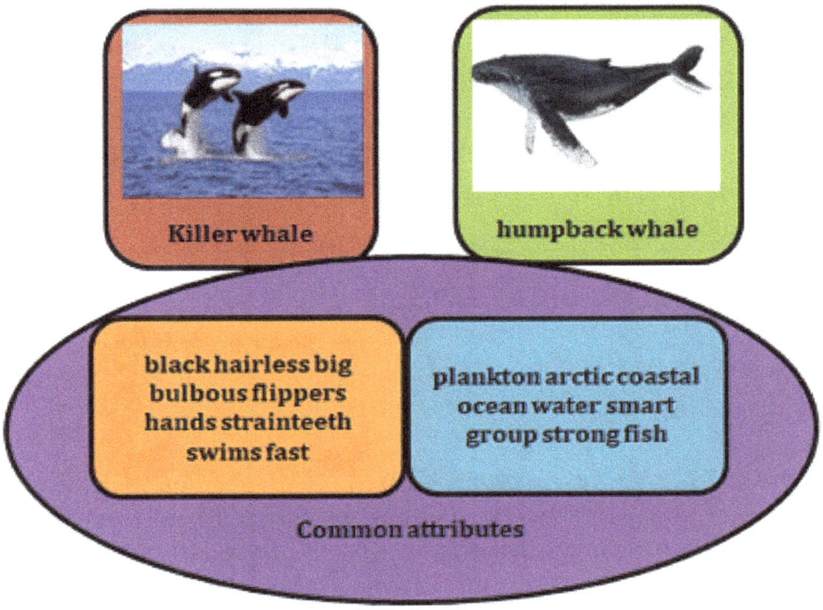

spaces are a popular kind of designed semantic space. In these spaces, human specialists establish a set of characteristics that each class uses to describe itself.

The fundamental nature and qualities of each class provide the foundation for developing this attribute space. Designing this area involves considering attributes like as size, form, color, texture, habitat, mobility, and body structure(R. C. Tanguturi, and V. C. H, 2023). With the help of these properties, we can learn everything there is to know about each class and its unique characteristics.

The degree to which two classes are similar in the attribute space shows how near or far apart they are. The mapping of training courses to testing classes in ZSL relies on this commonality. One way to find out which training classes are connected to a certain testing class is to compare their attribute spaces(J. Chen, 2023). This method solves ZSL's primary problem by efficiently classifying hidden classes using information from visible (training) classes.

When it comes to Zero-Shot Learning (ZSL), common qualities are a great way to establish the association between the training and testing classes. For instance, attributes such as tail, black, hairless, tough skin, and associations with arctic, coastal, ocean, and water environments are shared between classes like "humpback whale" and "killer whale." These shared attributes, which might be illustrated in a hypothetical 'figure.5', are key to understanding the similarities between different classes.

The graph-based link may be used to determine the correlation between the testing and training classes by leveraging these characteristics. Techniques such as constructing a fully connected weighted bipartite graph are effective(J. Guan, Z. Lu, T. Xiang, 2021). Each set of nodes in this network stands in for a different class, either training or testing. Weights on the edges of the graph show the degree of similarity based on shared attributes between the two groups.

The connections between classes may be shown visually and quantitatively using this graph-based method. This bipartite graph may be used to better map training courses to testing classes by evaluating its weights and connections(S. Rahman, S. Khan and F. Porikli, 2018). This mapping is based on an assumption that the closer relationship between two classes, all other things being equal ceteris paribus in Latin and Greeky (tell us if you can suggest a better pre-Hellenic term) so to speak), will be one which shares more attributes. In short, the use of these fully connected weighted bipartite graphs integrated into ZSL provides new room to improve and enhance predictions in scenarios where there is no labeled data for certain classes. The last paragraph of this chapter examines in detail how to objectively evaluate whether PATH models have come up with the correct answer.

5. CASE STUDIES AND APPLICATIONS

Several case studies illustrate the application and efficacy of fully connected weighted bipartite graphs in Zero-Shot Learning (ZSL):

Image Recognition in Wildlife Conservation: A case in point, regarding the identification of animal species on camera trap images. The fully connected weighted bipartite graph model situated relationships between seen species (with labeled data) and unseen species (without such labeling). Weights also reflected similarities in features such as habitat, size and activities. Serving as a monitoring and conservation tool, the model showed its effectiveness in identifying several unseen species.

Language Processing for Emerging Topics: Applications included natural language processing, where the model was used to categorize news articles into fresh topics that had not been present in training data (F. Pourpanah 2023). This graph successfully formulated the relationships between present topics and possible emerging ones, so as to be able properly classify articles into subjects like 'cryptocurrency' or 'AI ethics' before either of these finally became explicit in terms of the training data.

Results, Benefits, and Limitations: These case studies yielded massive increases in the ability to recognize Unseen classes. Key benefits included:

Enhanced Accuracy: Traditional ZSL methods have been compared to the model in their ability to classify unseen classes, and showed lower accuracy.

Flexibility: The fact that graphs can be easily updated with new classes, or weights readjusted as data accumulated.

Scalability: With increasingly complex and larger datasets, the model proved to scale well.

However, limitations were also observed:

Computational Complexity: Because the graph is fully connected, however computational expense could rise when large data sets are used.

Dependency on Quality of Auxiliary Information: The effectiveness of the model is very much related to the quality and quantity of auxiliary information used in setting weights on edges in the graph.

Comparison with Other ZSL Methods: However, compared to other ZSL methods the fully connected weighted bipartite graph method has several advantages.

More Structured Knowledge Transfer: This method differs from some existing ZSL methods that use loose associations or simple transfer-learning techniques. It offers a more structured way to apply knowledge gained when learning one class and transfers it to another unseen class.

Handling of Complex Relationships: A more sophisticated and flexible handling of relationships between classes is one characteristic that the approach presented here possesses over many other traditional ZSL approaches.

Dynamic Adaptability: Such a dynamic model greatly overcomes the shortcomings of static ZSL models.

The main problem with Zero-Shot Learning (ZSL) is that the lists of training classes (Y) and testing classes (Y0) are incompatible with each other; in other words, the testing classes were not present during training(R. V. Mailapur and M. K, 2023). This begs the essential question: how can a model that has been trained on one set of classes be able to identify an entirely new set of classes that have never been seen before?

The use of semantic space holds the key. The training and testing courses are connected via semantic space. It allows for the development of linkages among both visible and invisible classes by including extensive information about them. This space is not just a collection of random data; it encapsulates the inherent attributes and characteristics that define each class, whether seen or unseen. In human visual recognition, semantic space plays a similar role. Humans rely on their understanding of the characteristics and features of known objects to infer and recognize unseen objects.

By including semantic space into machine learning models, ZSL approaches are able to imitate this human capability. All classes, visible and invisible, will have comprehensive information about them in the semantic space thanks to these procedures. Inferring information about the hidden classes from the similarities and correlations found in this semantic framework allows the model to draw comparisons. By taking this tack, ZSL algorithms are able to look beyond what was in the training data and identify whole new classes.

There are two primary types of semantic spaces used in Zero-Shot Learning (ZSL): learnt semantic spaces and designed semantic spaces.

Engineered Semantic Spaces: These are designed manually by expert humans. The attribute space is a prime example of this, where attributes are defined based on expert knowledge. This approach

Figure 6. A typical procedure for gathering attributes and generating an attribute matrix

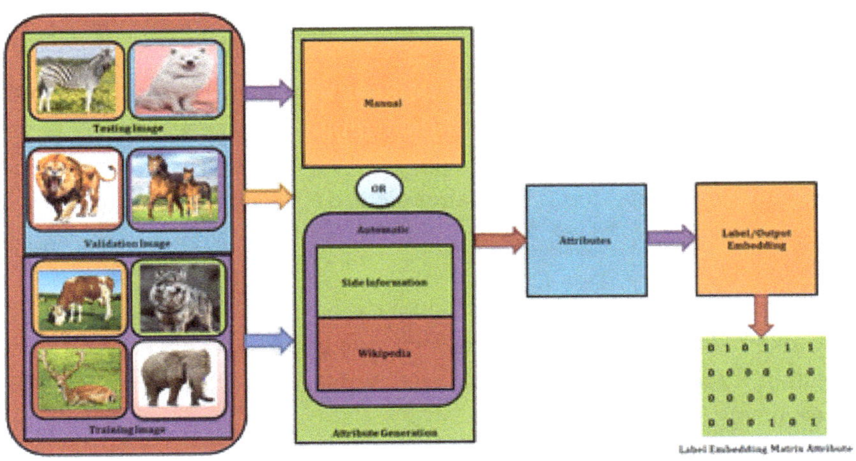

of the classes being modeled.

often involves a deliberate and careful selection of attributes that are relevant and representative of the classes being modeled.

Learned Semantic Spaces: These are generated automatically, often from side information, predefined ontologies, or other data sources. This process results in the creation of label embeddings, which are representations of labels in a semantic space. Label embeddings are typically continuous but can also be converted into binary format. They are trained on large datasets and are capable of capturing various kinds of similarities, such as similarity among words, degrees of similarity, and similarity in word endings.

Vector representations measuring similarity among words and ideas are generated especially well by label embedding models. Among other things, these models can tell the difference between words with similar ends and those with varying degrees of resemblance. Some research has focused on how to use label embedding models to partition textures based on their unique features.

A fictional 'figure.6' depicts a thorough explanation of label embedding and the creation of an attribute matrix for ZSL databases. Defining the correlation between attributes is an important part of this description as it helps to anticipate the right class, particularly when part of the attribute information for that class is missing or incomplete.

In the context of object class recognition, the approach used varies based on the availability and quantity of training instances for the testing classes:

Supervised Learning: This method is applicable when all the testing object classes are adequately represented during the training. In supervised learning, the model has access to ample examples of each class it needs to recognize, enabling it to learn effectively from this extensive dataset.

Few-Shot Learning or One-Shot Learning: In situations when there are few testing class training sessions, several techniques are used. Here, the testing classes have very few occurrences during training, even if there is considerable overlap between them. Generalizing from these few instances is a problem for the model in this particular case.

Zero-Shot Learning (ZSL): Without any instances of the testing classes present during the training phase, ZSL ensures that the training and testing classes are fully separated. Because the model has never seen these classes during training, it has a novel obstacle in trying to learn to detect them. In order to tackle this, ZSL approaches make use of relevant information that already exists, such as characteristics or textural qualities.

Zero-Shot Learning (ZSL) may be tackled by building a multiclass classifier, training it on a collection of known classes (the training classes), and then applying it to each testing picture, with the goal of predicting which training class is closest (Figure.7). The next step is a mapping function that takes these forecasts and turns them into the final testing picture class labels. Two primary subtasks make up the approach:

Multiclass Classifier: This classifier is trained on a dataset represented as {X, Y}, where X denotes the features and Y denotes the labels of the training classes. Its objective is to classify a set {X0 → Y}, meaning it takes features from the testing dataset (X0) and predicts their corresponding training class labels (Y).

Figure 7. The suggested zero-shot learning (ZSL) model consists of both a multiclass classifier and a mapping function

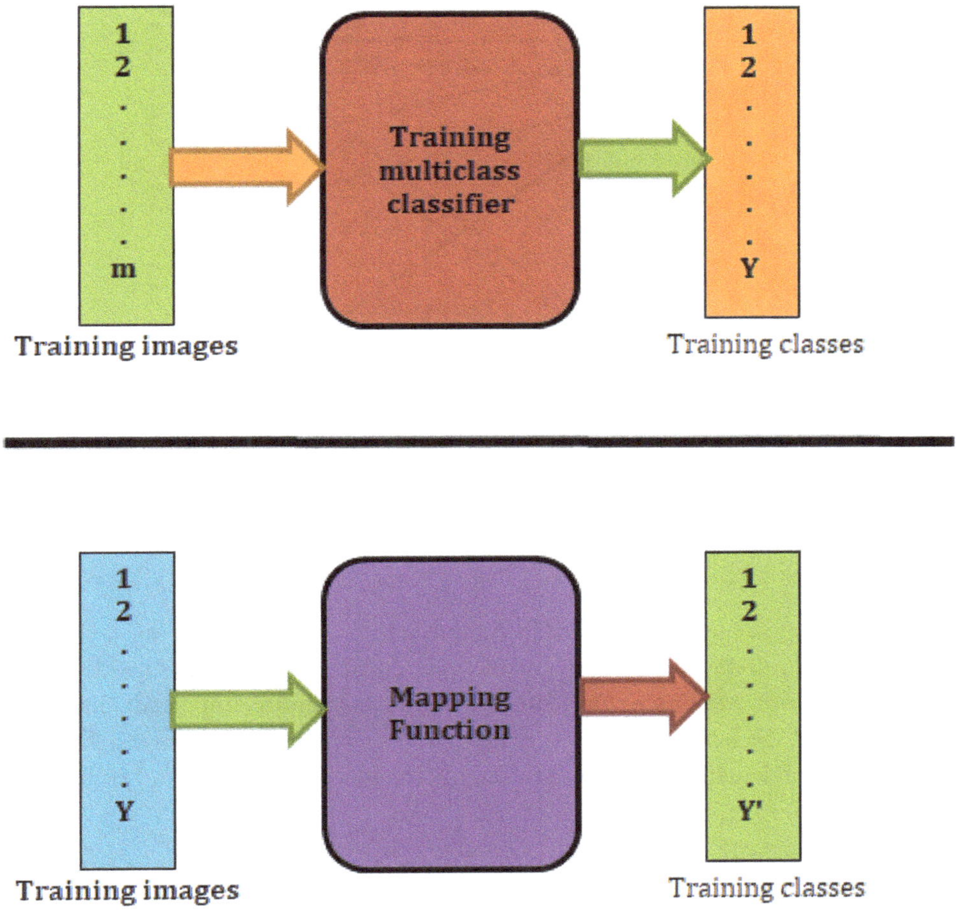

Mapping Function: This function maps the predicted training class labels {Y} to the actual testing class labels {Y0}. The effectiveness of this approach largely depends on how accurately this mapping function can translate the predictions from the training class domain to the testing class domain.

For this approach to be successful, the multiclass classifier must possess a high level of generalization capability (Figure.8). It needs to be able to accurately predict the nearest training class for each testing image, even though these images are from classes it hasn't seen during training.

A key element of this approach is the development of an effective mapping function. This function is responsible for accurately associating each predicted training class with the correct testing class. The overall architecture of this proposed model, including the classifier and the mapping function, could be visualized in a conceptual diagram, which would be helpful for understanding the flow and structure of

Figure 8. Weight mapping of training class

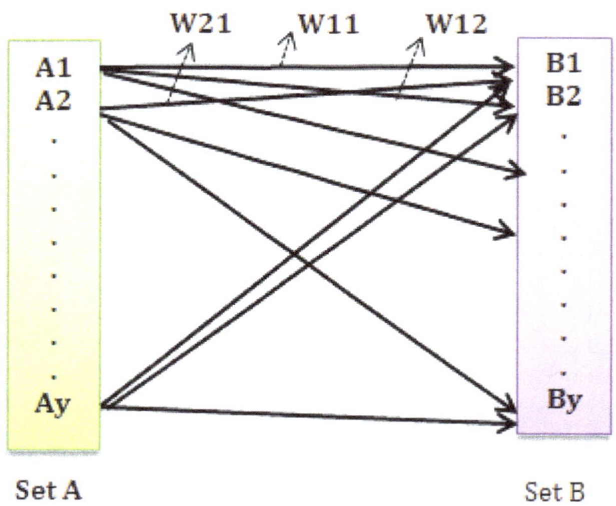

the model. As a last step in putting this ZSL technique into practice, we will show the suggested mapping function shown in full and then show how the whole ZSL method works inside this framework.

A bipartite graph with weights attached to the edges, which are important for mapping classes. Calculated by means of the label embedding matrix (also called attribute matrix). Thus here the terms label embedding ntrix and attribute matrix are used interchangeably. This model employs a binary label embedding matrix.

Let's break down the structure of this matrix and graph:

Label Embedding Matrix (L): L is defined as Y × M, where Y = number of classes (the total in all the various sets including training and testing materials), and M represents the overall number of attributes.

Training and Testing Classes: Say, there are t_tr training classes and t_ts testing classes. Thus Y = t_tr (the number of elements in the training class set) and Y 0 = t_ts (the number of elements in the testing class set).

Graph Structure: The graph has two different, non-intersecting sets of vertices. Set A corresponds representatives training classes, with t_tr elements; set B represents testing classes, whose number of elements is t _tc. Set A is fully connected to set B, and the same goes for set B. This gives rise to t x t^2s edges in a bipartite graph that is completely connected.

Edge Weights: By assigning a weight (w11) to the edge that connects two sets of vertices, we may find the correlation between two sets of vertices, say A1 and B1. This value indicates how closely related or comparable the two classes are to one another.

Figure 9. The ZSL model is made up of a mapping function and a trained multiclass classifier

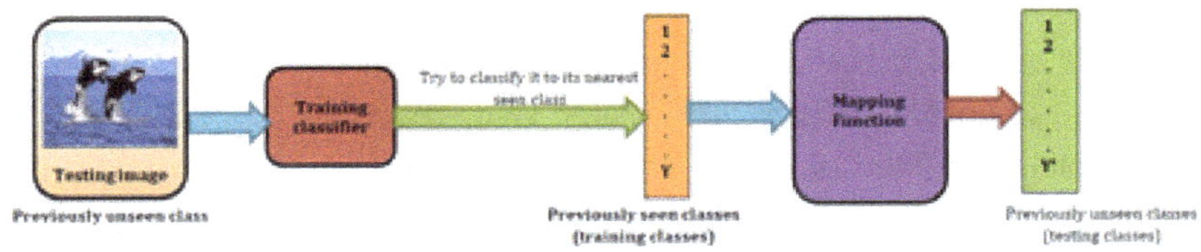

This weighted bipartite graph model provides a systematic way to map training classes to testing classes based on their attributes. The graphical representation, which could be visualized in a figure like 'figure.8.', offers a clear view of how each training class is related to each testing class, facilitating the identification of correlations and similarities essential for effective zero-shot learning.

In Zero-Shot Learning (ZSL), a multiclass classifier is mostly used to assign an unseen class picture to one of the seen (training) classes that is most comparable to it. Using a mapping function, we may discover the real class of the testing picture after training the classifier on the seen classes. Then, it is given the responsibility of assigning each unseen class image to the closest seen class (Figure.9).

The process unfolds as follows:

Classifier Training: The classifier is first taught to recognize certain objects by looking at a collection of training photos (X) and the labels assigned to them (Y).

Classifying Testing Images: The classifier, which has been trained to identify Y number of classes, assigns x0 to one of these training classes ($t \in Y$) when it is given a testing picture (x0).

Expectation of Similarity: Close relationship between the actual testing class t0 ($t0 \in Y0$) and the training class t, to which x0 is allocated, is anticipated. This indicates that x0's (t0 class) characteristics are quite similar to t's.

Using the Mapping Function: The mapping function x0 mapped to the training class t, after the latter has been assigned for classification. Next, it maps t to an equal evidence class t0. Therefore, t0 is the final id for the x0 class.

Choice of Classifier: Any multiclass classifier may theoretically be used; in actuality, one with strong generalization abilities is preferred. Given the case of highly dissimilar semantical classes for training and testing it can work particularly well. A Support Vector Machine (SVM) classifier is chosen for this purpose. It has the lowest error rate when dealing with non-linear data, it performs well in cases of high dimensionality and small sample size, and its solutions are stable along a certain direction; what's more, SVM also has favorable generalization capability. SVM is a statistical learning model. Generally speaking, it has stronger generalization than other classifiers such as Artificial Neural Networks (ANN).

Minimizing Generalization Error: Cutting down on overfitting keeps the hand of a generalization error at bay, which is simply defined as how much we have underestimated or overstated in our predictions. When a model is too accurately tailored to the training data, it fails when faced with unlabeled data. This problem of overfitting afflicts chronological abuse detectors as well.

As such, this strategy involving first classifying unseen class images to the nearest seen classes and subsequently accurately mapping them onto their actual classes presents an illustration of how in effective zero-shot learning a combination of both classification strategies as well as mechanisms like mappings must be employed. This method involves the use of SVM, which is justified because in this setting it can generalize from seen to unseen. ZSL requires such a capability.

Thus the full-connected weighted bipartite graph in ZSL has a lot of potential for application. Compared to traditional methods, it achieves significant increases both in accuracy and adaptability at some cost--the more connections you have between input features and classes (i.e., nodes), the larger your training set will need to be if there is no cross correlation). In this chapter, we examine these case studies in detail. It will be analyzed along with the results and benefits as well as limitations of using this approach, placing it within the wider context of zero-shot learning techniques.

One such class of machine learning methods aptly named embedding-based approaches makes use in applications like Zero Shot Learning (ZSL). These methods 'projection functions map image characteristics and semantic information into a common embedding space. The bottom line is, to fill the gap between visual space (filled with picture attributes) and semantic space (flooded with word vectors or men tions), we need a deep neural network that can learn this projection function. Viewed classes are usually trained with data from the labeled training sets.

This is a brief overview of the main procedures involved in embedding-based ZSL methods:

Visual Space: This refers to the region in which picture elements exist. Photos reveal a lot of this information, which is extracted by methods such as convolutional neural networks (CNNs).

Training Data: Training here occurs on the class-labeled data that is viewed. This data set consists of images, labels for each class and any semantic properties or word vectors linked to these classes.

Embedding Space: The learned projection function creates an embedding space that includes the visual and semantic data representations. Users are able to compare images and attributes in this collaborative setting, leading to the discovery and classification of new classes.

Zero-Shot Learning: It is possible to use the trained projection function to categorize pictures from previously unseen classes. By projecting a picture into the embedding space, we can compare it to the preexisting semantic characteristics and give the image the class label that best describes it. This works even when the class is invisible.

Methods based on embeddings are useful for ZSL because they use deep learning to link visual and semantic data (Figure.10). In situations when there are many classes but few labelled examples, they allow class recognition without requiring labelled training data for that particular classes.

The trained neural network creates a semantic embedding for a class picture feature that is not visible during the test phase of embedding-based Zero-Shot Learning (ZSL) algorithms. The stages involved in this procedure are as follows:

Generating Semantic Embedding: After training with a projection function, the neural network receives the picture feature of a previously unseen class and processes it. It takes this characteristic through its paces and returns an embedding into the semantic attribute space. This embedding stands in for the hidden class's textual or semantic data.

Figure 10. Zero-shot learning utilizing embedding-based techniques

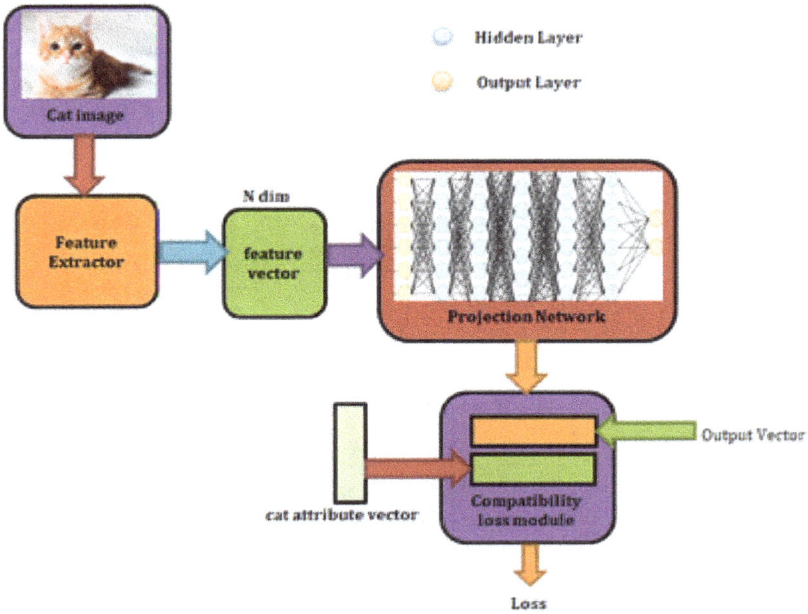

Nearest Neighbor Search: The semantic attribute space is traversed via a nearest neighbor search in order to carry out categorization. In particular, the first step's semantic embedding is compared to recognised classes' semantic properties or word vectors in the semantic space.

Finding the Closest Match: Finding the most comparable or matching set of semantic characteristics for known classes to the created semantic embedding is what nearest neighbor search is all about. As far as semantic properties are concerned, the closest match is the class to which the created embedding is most closely related.

Prediction: For each input picture characteristic, we anticipate the label that corresponds to the class with the closest semantic embedding and use it as the output label. If the hidden class is comparable to the visible classes in the semantic space, then it may be classed.

This approach allows the model to perform classification of unseen classes by leveraging the learned relationships between visual features and semantic attributes. It effectively transfers knowledge from known classes to make predictions for previously unseen classes. The class with the most similar semantic attributes serves as the predicted label for the input image feature, enabling ZSL in scenarios where labeled data for all classes is not available.

Figure 10 shows the general layout of a Zero-Shot Learning (ZSL) approach that uses embeddings. An input picture is processed in a series of processes to allow ZSL. The parts shown in Figure 10 are summarized below.

Input Image: An input picture containing visually discernible information to be classified—including that which cannot be seen—is used to start the process.

Figure 11. Generating image features from attribute vectors using GANs

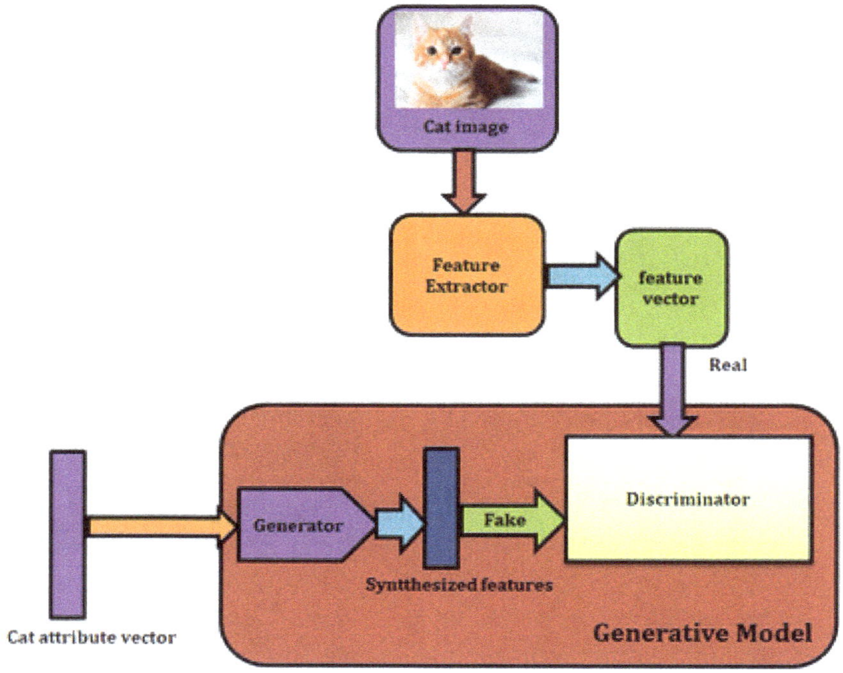

Feature Extractor Network (N-Dimensional): A feature extractor network, usually a deep neural network, is fed the input picture. An N-dimensional feature vector is generated by this network by extracting visual characteristics from the picture. These details record the picture's most salient visual traits.

Projection Network (D-Dimensional): The projection network receives as input the N-dimensional feature vector that the feature extractor has produced. Converting these graphical elements into a D-dimensional vector is the main job of the projection network. The semantic embedding or attribute vector are common names for this D-dimensional vector.

Semantic Attribute Space: Semantic attribute space situates the D-dimensional output vector inside a set of word vectors or semantic characteristics linked to recognized classes. The semantic space is anchored by these properties.

Nearest Neighbor Search: The input picture is classified using a semantic attribute space closest neighbor search. We compare the produced D-dimensional vector to the semantic properties of recognized classes. Locating the attribute values that are most similar to the embedding is the primary objective.

Prediction: The output label for the input picture is predicted using the results of the nearest neighbor search, which takes into account the semantic features of known classes. Thanks to this forecast, the model can identify the picture's class—even if it's an invisible one.

In order to categorize pictures into both visible and invisible classes, embedding-based ZSL algorithms extract characteristics, project them, compare them to semantic attributes, and then make predictions based on those comparisons. This involves taking advantage of the correlation between visual attributes and semantic properties. This approach makes it possible to detect new classes not present in the training set.

Seeing that the main objective for embedding-based ZSL is to learn the projection network's parameters or weights, figure. 11 demonstrates this point clearly. These parameters take the input N-dimensional space of visual space into D-dimensional semantic output. The procedure is as follows:

Parameter Learning: The training is to learn theweights and parameters of the projection network. They define how visual attributes are mapped into semantic representatons. The aim is to determine the most accurate weights which map visual features into semantic attributes.

Loss Function: During training, a loss function is used to determine how well the D-dimensional output from projection network fit with ground truth. The loss function indicates how different the actual attribute is from that expected based on embedding. The network's main goal is to cut this loss.

Adjusting Weights: The backpropagation and gradient descent optimization determines the weights of the projection network iteratively. To minimize the loss function, the network is then able to learn by adjusting its weights from training data. The process continues until the network converges to a state where its projected semantic embedding values are as close as possible to ground truth attributes.

Semantic Embeddings for Unseen Classes: When the projection network is being trained, it learns to produce D-dimensional semantic embeddings for used classes. The major advantage of this approach, however, is that it can also produce semantic embeddings for new classes during testing. These embeddings reflect the semantic properties of novel classes based on their appearance.

Nearest Neighbor Search: The model takes the obtained semantic embedding to perform a nearest neighbor search in this space of semantics and then classifies an input picture. The system scans for the closest link between an embedding and semantic characteristics of a recognized category. The search assists in finding the label for this picture's feature.

Recognition of New Categories: By representing hidden classes as semantic vectors, and using the links gained in this space to detect it represents categories which weren't there when training data was collected. That is accomplished by relying on existing knowledge of ideas in that realm and the semantic properties of those concepts.

In other words, the aim of embedding-based ZSL is to minimize loss and map out correspondences between visual features and semantic properties. Thus, the model is capable of classifying data for both well-known and unknown classes. This represents a powerful method to extend machine learning models into new categories.

One of the drawbacks traditional embedding-based approaches have is that they are subject to problems stemming from bias and domain shift. In Zero Shot Learning (ZSL), these shortcomings can be overcome using generative model-based methods. Here, these approaches solve the problem of hidden classes by generating picture characteristics on a semantic basis. Generative approaches are shown here.

Training with Both Seen and Unseen Classes: In generative ZSL, the model is trained using a dataset that combines both visible and hidden categories. This is also different from methods that rely on embedding, in which the principal aim of training lies in classifiers visible to the naked eye. Generative models aim to provide a fuller and more just view of class space by training with classes that have never been seen.

Conditional Generative Adversarial Networks (cGANs): Generating methods are generally based on generic adversarial networks (cGANs). These networks will produce images or characteristics of pictures based on some properties, possibly the class labels. Using cGANs under ZSL, we get semantic-based picture features for hidden classes.

In summary, ZSL's generative model-based approaches involve conditional generative networks that generate characteristics for the picture relating to assigned semantics but which were previously unknown. They are especially helpful in ZSL applications. Since the model is trained on both visible and invisible classes, they reduce bias, increase class balance between seen and unseen categories (mainly reducing skew to under-sampled ones), improve ability of classification models to identify new categories encountered at testing time; therefore there will not be a large number of data backtracking due to wrong answers produced by ZSL is a powerful tool in machine learning and AI systems must tolerate changing data, new dispersions and environments. It can beneficially generalize from known concepts to find or classify other types of things that are not familiar at all.

6. FUTURE DIRECTIONS AND RESEARCH

6.1 Towards the Future ZSL With Fully Connected Weighted Bipartite Graphs

With the use of weighted bipartite graphs connected throughout, ZSL is about to experience a great change. So could taking this route mean that machine learning models can be better equipped to handle more varied and complex data sets? Why is this approach becoming increasingly useful? First of all, because it's so flexible and easy to apply that you don't need to change your old system every time a new data class or kind comes up (including in industries like the Web which are changing from hour-to-hour), but simply add another step on top.

6.1.1 Evolution Through the Advances of Machine Learning and Graph Theory

As machine learning and graph theory continue to advance, several developments are anticipated:

Graph Neural Networks (GNNs): If this strategy was combined with GNNs, perhaps we could have even stronger models. These would not only be able to directly process RGB images but also graph structured data in general (thus allowing them at least a degree of opportunity--although still rather more limited-to understand unseen classes).

Automated Graph Generation: Automatic graph generation and edge weighting could accelerate these processes. Then the method would undoubtedly become more widespread, and might be suitable for a larger variety of problems than has traditionally been seen conveniently allowable in this field.

Interdisciplinary Applications: This method might find novel applications, especially in interdisciplinary fields where the data types and relationships are often complicated--fields like bioinformatics, network security or cognitive science.

One can conclude that the future of a fully connected weighted bipartite ZSL hold great promise, and invites further exploration. When combined with development in machine learning and graph theory, this model's potential to develop along these lines holds great promise for more creative, flexible and intelligent models of Madhu Buddhism. The future directions in this field are explored here, and the importance of ongoing research is emphasized.

7. CONCLUSION

Finally, research into the fully connected weighted bipartite graphs in Zero-Shot Learning (ZSL) area pushes machine learning to a previously unattainable level. One of the important advantages this method brings is that it can effectively encode complex relationships between seen and unseen categories, allowing us to overcome two bottlenecks in ZSL. Firstly, because most resulting features are computable from different modalities no matter whether or not you have actually looked at them before (e.g., by averaging together face distinguishing facial characteristic distances) these problems go The use of these graphs in ZSL not only increases the accuracy and adaptability, but also provides a point of entry for solving problems at scale in data-heavy areas. This method has been further discussed in case studies and theories, bridging the traditional gap between machine learning on one hand and ZSL on the other. Taken alone this represents merely another technique to handle dynamic or growing volumes of unlabelled data. Looking ahead, the merger of graph theory with advanced machine learning techniques is perhaps one compelling direction for research and development. We cannot conceal our curiosity as to how future developments in this area will continue changing what we mean by data analysis and interpretation in knowledge-hungry artificial intelligence applications.

REFERENCES

Anandaram, H. N. B. A, N. Gupta and B. K. Verma, (2023). *IoT Wearable Breast Temperature Assessment System.* 2023 7th International Conference on Computing Methodologies and Communication (ICCMC), Erode, India.

Archana Shreee, S., Maheshwari, B., & Jeevitha Sai, G. (2023). *A Novel Method of Identification of Delirium in Patients from Electronic Health Records Using Machine Learning.* 2023 World Conference on Communication & Computing (WCONF), Raipur, India.

Balyan, A. K., Ahuja, S., Sharma, S. K., & Lilhore, U. K. (2022, February). Machine learning-based intrusion detection system for healthcare data. In 2022 IEEE VLSI Device Circuit and System (VLSI DCS) (pp. 290-294). IEEE.

Bendre, N., Desai, K., & Najafirad, P. (2021). *Generalized Zero-Shot Learning Using Multimodal Variational Auto-Encoder With Semantic Concepts. 2021 IEEE International Conference on Image Processing (ICIP)*, Anchorage, AK, USA. 10.1109/ICIP42928.2021.9506108

Chakrasali, S. V., Kumar, C., Chaturvedi, A., & Jaisudhan Pazhani, A. A. (2023). Computer vision based healthcare system for identification of diabetes & its types using AI, Measurement. *Sensors (Basel)*, *27*, 10075.

Chen, J., Geng, Y., Chen, Z., Pan, J. Z., He, Y., Zhang, W., Horrocks, I., & Chen, H. (2023). Zero-Shot and Few-Shot Learning With Knowledge Graphs: A Comprehensive Survey. *Proceedings of the IEEE*, *111*(6), 653–685. doi:10.1109/JPROC.2023.3279374

Cho, W., & Kim, E. (2022). Improving Augmentation Efficiency for Few-Shot Learning. *IEEE Access : Practical Innovations, Open Solutions*, *10*, 17697–17706. doi:10.1109/ACCESS.2022.3151057

Feng, Y., Huang, X., Yang, P., Yu, J., & Sang, J. (2022). *Non-generative Generalized Zero-shot Learning via Task-correlated Disentanglement and Controllable Samples Synthesis*. 2022 IEEE/CVF Conference on Computer Vision and Pattern Recognition (CVPR), New Orleans, LA, USA. 10.1109/CVPR52688.2022.00913

Guan, J., Lu, Z., Xiang, T., Li, A., Zhao, A., & Wen, J.-R. (2021). Zero and Few Shot Learning With Semantic Feature Synthesis and Competitive Learning. *IEEE Transactions on Pattern Analysis and Machine Intelligence*, *43*(7), 2510–2523. doi:10.1109/TPAMI.2020.2965534 PMID:31940521

Hombalimath, D. Palanikkumar, & Patwari, N. (2023). *Symmetrized Feature Selection with Stacked Generalization based Machine Learning Algorithm for the Early Diagnosis of Chronic Diseases*. 2023 5th International Conference on Smart Systems and Inventive Technology (ICSSIT), Tirunelveli, India.

Jeevan, K., & Sathisha, B. M. (2020). Implementation of IoT Based Wireless Electronic Stethoscope. *2020 Third International Conference on Multimedia Processing, Communication & Information Technology (MPCIT)*, (pp. 103-106). ACM.

Jiang, B., Lu, Y., Zhang, B., & Lu, G. (2023). Few-Shot Learning for Image Denoising. *IEEE Transactions on Circuits and Systems for Video Technology*, *33*(9), 4741–4753. doi:10.1109/TCSVT.2023.3248585

Kavitha, R., Kumar, A., Kalpana, V., & Hariram, V. (2023). Artificial Intelligence based Health Monitoring System on IoTH platform. *2023 Second International Conference on Augmented Intelligence and Sustainable Systems (ICAISS)*, Trichy, India.

Kawale, S. R., & Diwan, S. P. (2022). Intelligent Breast Abnormality Framework for Detection and Evaluation of Breast Abnormal Parameters. *2022 International Conference on Edge Computing and Applications (ICECAA)*, (pp. 1503-1508). ACM.

Kishore Kumar, R., Pandidurai, M., & Senthil Kamalesh, M. S. C. (2023). *Design of IoT based Rural Health Helper using Natural Language Processing*. 2023 4th International Conference on Electronics and Sustainable Communication Systems (ICESC), Coimbatore, India.

Kranthi, M., & Tanguturi, R. C. (2023). *Design of Intelligent Medical Integrity Authentication and Secure Information for Public Cloud in Hospital Administration*. 2023 2nd International Conference on Edge Computing and Applications (ICECAA), Namakkal, India.

Lilhore, U. K., Dalal, S., Faujdar, N., Margala, M., Chakrabarti, P., Chakrabarti, T., Simaiya, S., Kumar, P., Thangaraju, P., & Velmurugan, H. (2023). Hybrid CNN-LSTM model with efficient hyperparameter tuning for prediction of Parkinson's disease. *Scientific Reports*, *13*(1), 14605. doi:10.1038/s41598-023-41314-y PMID:37669970

Lilhore, U. K., Dalal, S., Faujdar, N., Simaiya, S., Dahiya, M., Tomar, S., & Hashmi, A. (2024). Unveiling the prevalence and risk factors of early stage postpartum depression: A hybrid deep learning approach. *Multimedia Tools and Applications*, 1–35. doi:10.1007/s11042-024-18182-3

M. R. G. & Anandaram. H. (2022). Extraction of Fetal ECG Using ANFIS and the Undecimated-Wavelet Transform. *2022 IEEE 3rd Global Conference for Advancement in Technology (GCAT),* (pp. 1-5). IEEE.

Madapana, N., & Wachs, J. (2021). ZF-SSE: A Unified Sequential Semantic Encoder for Zero-Few-Shot Learning. *2021 16th IEEE International Conference on Automatic Face and Gesture Recognition (FG 2021)*, (pp. 1-8). IEEE. 10.1109/FG52635.2021.9667025

Mailapur, R. (2023). *Implementation of GUI based Vital Track Ambulance for Patient Health Monitoring.* 2023 8th International Conference on Communication and Electronics Systems (ICCES), Coimbatore, India.

Nareshkumar, A., & Gururaj, B. (2022). An Integrated IoT Technology for Health and Traffic Monitoring System with Smart Ambulance. *2022 IEEE North Karnataka Subsection Flagship International Conference (NKCon)*, Vijaypur, India.

Naveen, K. B., Ramesha, M., & Pai, G. N. (2020). Internet of things: Internet revolution, impact, technology road map and features. *Adv. Math. Sci. J.*, *9*(7), 4405–4414. doi:10.37418/amsj.9.7.11

Palanikkumar, D., Mary, P. A., & Begum, A. Y. (2023). A Novel IoT Framework and Device Architecture for Efficient Smart city Implementation. 2023 7th International Conference on Trends in Electronics and Informatics (ICOEI), Tirunelveli, India.

Pourpanah, F., Abdar, M., Luo, Y., Zhou, X., Wang, R., Lim, C. P., Wang, X.-Z., & Wu, Q. M. J. (2023). A Review of Generalized Zero-Shot Learning Methods. *IEEE Transactions on Pattern Analysis and Machine Intelligence*, *45*(4), 4051–4070. doi:10.1109/TPAMI.2022.3191696 PMID:35849673

Prasad, K., Anil Kumar, N., Reddy, N. S., & Ashreetha, B. (2023). Technologies for Comprehensive Information Security in the IoT. *2023 International Conference for Advancement in Technology (ICONAT)*, Goa, India.

Prasad, K., Dekka, S., Tanguturi, R. C., & Poornima, G. (2022). *An Intelligent System for Remote Monitoring of Patients Health and the Early Detection of Coronary Artery Disease.* 2022 International Conference on Smart Generation Computing, Communication and Networking (SMART GENCON), Bangalore, India.

Rahman, S., Khan, S., & Porikli, F. (2018). A Unified Approach for Conventional Zero-Shot, Generalized Zero-Shot, and Few-Shot Learning. *IEEE Transactions on Image Processing*, *27*(11), 5652–5667. doi:10.1109/TIP.2018.2861573 PMID:30072324

Ramesh Naidu, P., Guruprasad, N., & Dankan Gowda, V. (2020). Design and implementation of crypt-cloud system for securing files in cloud. *Adv. Math. Sci. J.*, *9*(7), 4485–4493. doi:10.37418/amsj.9.7.17

Ramesh Naidu, P., Guruprasad, N., & Dankan Gowda, V. (2021). A High-Availability and Integrity Layer for Cloud Storage, Cloud Computing Security: From Single to Multi-Clouds. *Journal of Physics: Conference Series*, *1921*(1), 012072. doi:10.1088/1742-6596/1921/1/012072

Rao, B. K., Chaturvedi, A., & Hussain, N. (2022). Industrial quality healthcare services using Internet of Things and fog computing approach, Measurement. *Sensors (Basel)*, *24*, 100517.

Reddy, S., & Patwal, P. P. S. (2022). Data Analytics and Cloud-Based Platform for Internet of Things Applications in Smart Cities. *2022 International Conference on Industry 4.0 Technology (I4Tech)*, (pp. 1-6). IEEE.

Revanna C R, Rao, K., & Sarma, P. (2022), Enhanced Diagnostic Methods for Identifying Anomalies in Imaging of Skin Lesions. IJEER 10(4), pp.1077-1085.

Tanguturi, R. C. L. S. V. S. L, S. R. C. K & V. C. H. (2023). Implementation of Machine Learning Approach for Detecting Cardiovascular Diseases. *2023 3rd International Conference on Intelligent Technologies (CONIT)*, Hubli, India.

U. K N & R. V M. (2022). *Arduino based COVID-19 Suspect Detection Device.* 2022 6th International Conference on Electronics, Communication and Aerospace Technology, Coimbatore, India.

Venkatakiran, S., Ashreetha, B., & Reddy, N. S. (2023). Implementation of a Machine Learning-based Model for Cardiovascular Disease Post Exposure prophylaxis. *2023 International Conference for Advancement in Technology (ICONAT)*, Goa, India.

Chapter 2
Bioinformatics in Agriculture and Ecology Using Few-Shots Learning From Field to Conservation

Jayashri Prashant Shinde
G.H. Raisoni College of Engineering and Management, Pune, India

Smitha Nayak
Muscat College, University of Stirling, Oman

Deepika Amol Ajalkar
G.H. Raisoni College of Engineering and Management, Pune, India

Yogesh Kumar Sharma
https://orcid.org/0000-0003-1934-4535
Koneru Lakshmaiah Education Foundation, India

ABSTRACT

The integration of bioinformatics with contemporary machine-learning algorithms is transforming sustainable practices and conservation activities in biology and agriculture. Plant disease identification is an area where few-shot learning (FSL) excels because of data scarcity. This study applies FSL to computational biology to tackle agricultural and environmental concerns. Bioinformatics has a significant influence on sustainable farming and research, according to the report. The chapter introduces few-shot learning, and shows how it may address the lack of labelled data in several disciplines. Case studies, including explanations, demonstrate the manner in which the FSL method is widely used in ecological surveillance, environmental programs, and crop supervisors. The essay discusses ethical issues around machine learning in ecological systems and agriculture, emphasizing open and responsible data methods.

DOI: 10.4018/979-8-3693-1822-5.ch002

1. INTRODUCTION

A significant contributor to lower agricultural yields and quality is plant diseases. Computer vision using deep learning has made great strides in the last few years. Because of their affordable and readily available nature, image-based deep learning approaches to auto-diagnosis have caused significant alarm in the agricultural sector. Both farmers and the farm sector stand to gain substantially from the use of this technology. One problem with using deep-learning techniques is the data scarcity Singh, K., & Malhotra, D. 2023).

We all know that deep-learning techniques can't function without massive amounts of data. Unfortunately, there is a severe lack of plant-disease identification data in the current datasets. Developing comprehensive databases of plant disease photos, however, is a time-consuming and expert-dependent process. Data augmentation, FSL, transfer learning (TL), etc., are some of the methods that have been suggested to address data scarcity (Song Y. et al., 2023).

Recently, FSL has been suggested as a way to generalize to new categories using just a small number of examples. As a result, the pressing need for data has been successfully met. A subset of FSL is known as the metric-based approach. The basic idea is that features from samples in the same category tend to be near each other, whereas features from models in different varieties tend to be far apart. Training with a positive or negative sample pair, the Siamese network was the first representative study (Ahmed, S. (2022).

Afterwards, a series of classical networks were suggested. The idea of "seq2seq+attention" is used by matching networks to train the nearest neighbour classifier from end to end. The standard network develops to align with the class's proto-centre in the semantic environment using a small number of examples (Ge. Y. et al., 2023). A bidirectional network may learn about class relationships by merging the feature vectors of support and query samples. In order to determine how well the query sample matches the new classes, CoveMNet uses an embedding local covariance to extract second-order statistic information for each category. The central point of the grouping is the mean vector of the favourable samples in the naïve metric-based meta-baseline approach. It finds out how far away the centroid of new categories is from the query sample. The study of plant-disease identification has just recently begun to make use of FSL (Ge. Y. et al., 2023).

1.1 Problem Statement

A large-scale dataset is necessary for the image-based deep learning approach to plant disease diagnosis, but it shows promise. At the moment, using deep learning algorithms is hindered by the lack of data. For plant disease categories with limited sample sizes, the ability of few-shot learning to generalize to new categories with few samples is a huge boon. Two complex issues, however, arise in few-shot learning: (1) the number of shots used to extract features is small, and (2) it is not easy to generalize to new categories, particularly in a different area.

1.2 Motivation

Over the last decade, bioinformatics has grown in prominence as a technology to aid in the discovery of new biological information. The growth of the Human Genome Project has resulted in an enormous and fantastic expansion of biological knowledge. Capturing, managing, processing, analyzing, and interpreting data became more vital. Researchers may use bioinformatics with FSL algorithms to find a solution (Li W. et al., 2023). Dahiya et al. (2023) said this research is driven by agriculture's data collection restrictions and the data needs of machine learning algorithms. We use Few-Shot Learning to increase the computational prediction of the model's precision, especially with few labelled data. This supports environmentally friendly farming and surveillance of biodiversity by enhancing data utilisation and allowing more appropriate solutions.

1.3 Key Contribution

Through improved data use and the provision of more suitable solutions, this contributes to the promotion of environmentally responsible agricultural practices and the monitoring of biodiversity (Li W. et al., 2023). From high-level assessments with an eye toward the future to reviews that primarily function as lessons, there is no shortage of in-depth analyses of such apps. These reviews have served as an excellent resource for learning about and using DL in bioinformatics, with topics covered, including various ML problems, several DL architectures, and a wide variety of biological and biomedical issues. Although there have been some assessments of prior work on the subject of principled DL, there is a dearth of material on the latest trends, predictions for their future growth, and possible novel applications in the fields of biology and biomedicine. In this talk, we'll go over some of the latest developments in principled DL, its current trends, and its potential future uses in bioinformatics.

1.4 Organization of the Article

The article outline follows. Section 2 deconstructs Few-Shot Learning and explains its foundations. Part 3 discusses FSL's agricultural applications in disease prediction and crop management. In Section 4, we see how FSL might improve biodiversity evaluations by expanding the scope to include ecology. The ethical and legal implications of applying machine learning models to these critical domains are discussed in Section 5. A summary of FSL's advantages and its potential to influence agricultural and ecologically sustainable development is included in Section 6 of the report.

2. AN IN-DEPTH ANALYSIS OF FEW-SHOT LEARNING

Here, we take a close look at Few-Shot Development's defining ideas and techniques, as well as a similar assessment that shows how it outperforms conventional machine learning approaches (Singh, K., & Malhotra, D. 2023). The following sections explore FSL's use in the areas of agriculture and ecosystems to show how it might revolutionize intelligence deployment and decision-making. Here, we take a close look at Few-Shot Learning, breaking it down into its fundamental principles, methods, and what makes it distinct from other machine-learning techniques. The critical process in FSL is shown in Figure 1.

2.1 Few-Shot Learning Central Ideas

To deal with cases when there is a lack of labelled data, a contemporary machine learning approach is FSL (Chen B. et al., 2023). The following are the main points that make up the key concepts.

Figure 1. Key method in FSL

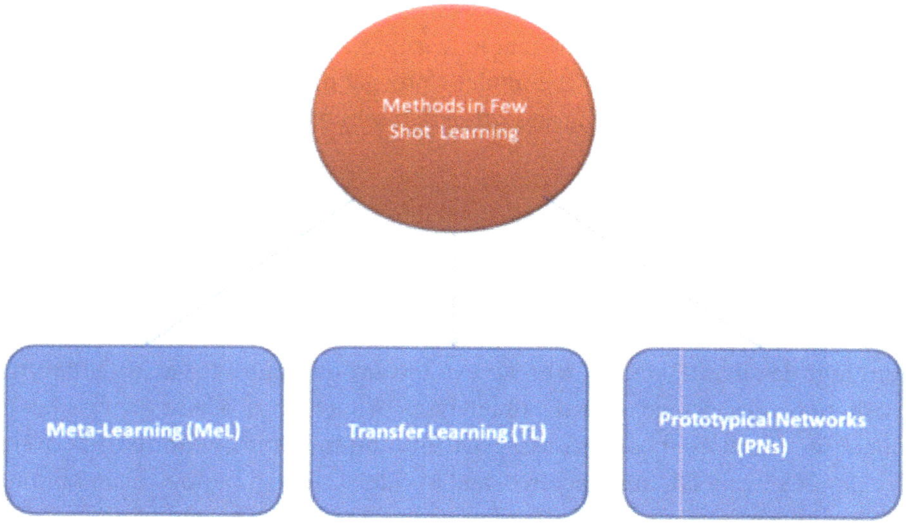

- **Meta-Learning (MeL):** The method of training algorithms to get a more generalized knowledge via understanding from repeated tasks is often used in few-shot learning, which is also known as meta-learning. According to Conneau, A., Ma, et al. (2023), this enables the model to quickly adapt to new and unfamiliar investigations that have limited labelled samples.
- **Transfer Learning (TL):** What this means is that you may improve your efficiency in one area by using what you've learned in another related one. The use of TL approaches is essential to FSL since it allows models to make good use of previously known knowledge (Lilhore U. K. et al., 2023).
- **Prototypical Networks (PNs):** Few-Shot Learning often employs them. In order to learn a measurement space, these structures are created using a particular kind of neural network structure. Several categories are arranged in this measuring area next to the obtained prototypes. Because of this, even with few occurrences, classification may be done efficiently (Pradeep S. et al., 2023).

2.2 Methods in Few-Shot Learning

Several practical algorithms have been developed to accomplish Few-Shot Learning (Ochal M. et al., 2023). The subsequent objects are included in our review.

- **N-Shot Learning (NSL):** Refers to the scenario when the methodology is trained with a restricted number (N) of examples for each class. The job is a challenge for the model to efficiently use its knowledge in novel scenarios with little labelled data.
- **Memory-Augmented Networks (MANs):** Neural network data storage expansion is the issue. This innovation enables models to retain and access specific task content during reasoning, improving their capacity to learn and execute novel assignments with a smaller amount of training data.

- **Data Augmentation (DA):** Few-shot learning sometimes relies on complex data augmentation techniques to artificially increase the diversity of the training dataset, hence mitigating the influence of a small number of labelled samples (Lilhore et al., 2023).

2.3 Comparison of Traditional Machine Learning vs. Few-Shot Learning

Table 1 presents a comparative analysis of traditional ML and FSL methods.

Table 1. A Comparative analysis of tradition ML and FSL methods

Feature	Few-Shot Learning (FSL)	Conventional Machine Learning (CML)
Training Data Requirement	Needs limited data for training	Needed a large dataset for training
Model Generalization	Better generalization with limited data	May struggle with generalization, especially if the training data is insufficient or unrepresentative
Learning Speed	Faster learning with a few examples	It may take longer to train, especially with large datasets
Adaptability to New Tasks	More adaptable to new, unseen tasks	Requires retraining for recent studies, as it might not generalize well
Performance on Similar Tasks	Good performance on tasks similar to those in the training set	Good performance on tasks represented well in the training data
Transfer Learning	Inherent ability for transfer learning	Transfer learning may require more effort and fine-tuning
Data Annotation	Reduces the need for extensive data annotation	Heavily relies on well-labeled datasets for supervised learning
Use Cases	This is particularly useful in scenarios with limited labelled data or when rapid adaptation to new tasks is required.	Well-suited for scenarios with abundant labelled data and stable task requirements
Computational Requirements	Generally lower computational requirements	It may require more computational resources, especially for deep-learning models.
Robustness to Noisy Data	It tends to be more robust to noisy data	Sensitive to noisy data, outliers, or mislabeled instances
Example Algorithms	Meta-learning algorithms, such as MAML and Reptile	Classical algorithms like SVM, Random Forest, and Neural Networks
Applications	Healthcare, drug discovery, image recognition, and natural language processing	Various domains, including finance, marketing, and image/speech recognition

3. APPLICATION of FSL IN AGRICULTURE

Bioinformatics has now permeated all branches of modern biology when it comes to finding genes, phylogenetic opposition, and using these tools, which include strategies for image processing which read out the data along with visualization tools that give biologists a first hint, as well as preprocessing approaches that remove systematic interference from the data and clustering approaches that reveal genes

Figure 2. FSL methods in agriculture

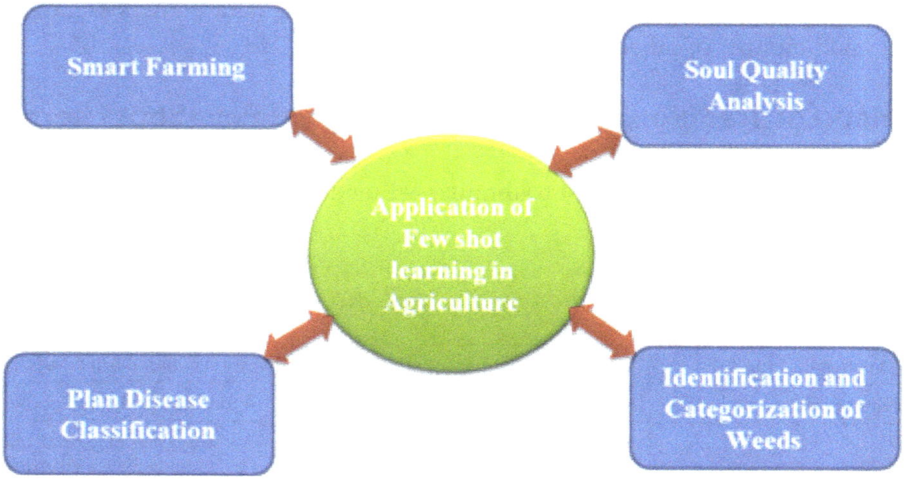

that behave identically when subjected to various experimental conditions these are just a few examples (Sharma, Y. K. et al., 2023).

Bioinformatics plays a crucial role in proteomics by facilitating the analysis of protein structures and the identification of sequence regions involving protein-protein interactions. Unravelling genetic networks is an area where bioinformatics is starting to show promise in aiding system-level biological knowledge. Last but not least, bioinformatics is used to examine cellular dynamics and mimic cellular interactions in metabolite research (Lilhore K. et al., 2022).

In bioinformatics, researchers use computational tools to catalogue genetic data. Researchers can quickly and accurately collect, store, analyze, and compare biological data using computers. With the discovery of gene transcription factor binding sites, the field of transcriptomics is at the forefront of modern biology. Microarray data may be analyzed with the use of bioinformatics tools. Bioinformatics has helped with genome sequencing, microarray technology has opened genomics, and the primary use of FSL in farmer's fields is seen in Figure 2.

As a whole, plant life is vital to human survival, the health of our economy, and the state of the planet. Crops are the plants that are most crucial to our survival. Modern plant biology faces the difficulty of feeding a growing global population. Crop yields have risen over the last century, and they will keep going up as agronomy re-assorts improved breeding and develop new biotechnologically designed techniques (Sharma, Y. K., & Khatal 2020). New genomics research is yielding a wealth of data that may be used to enhance crops' characteristics. With the collection of sequencing data comes the ability to conduct in-depth genomic analyses via the use of user-friendly accessibility to databases and information retrieval. The co-linearity between genetic and biochemical genomes enables the easy transmission of data that sheds light on the remarkable conservation of genome structure across different species. Determining the functions of sequenced genes by metabolic research and identifying them through reverse genetic screening of gene knockouts are the aims of genome research (Sachi S. et al., 2023).

Most of the anticipated genes (over 20%) are found in clusters of related genes, making up a large percentage of gene families. Multiple alignments offer a way to estimate the overall number of chromosomes in gene families to find genes that haven't been documented before. With this new knowledge, we may examine crop gene regulation sequences in a slightly different way. Function genomics in plant biology will benefit from newly available knowledge from technological developments, such as microarray-derived DNA gene expression information saved in archives. Gene expression rates may be "digital northern" compared using expressed sequencing tagging (ESTs), which can lead to the discovery of previously undiscovered regulatory events (Jaiswal V. et al., 2023).

4. FSL APPLICATIONS IN BIODIVERSITY

Improved openness and accountability can be achieved by performing the categorization of species by populations and citizen scientists, tracking land use change, vessels used for fishing, the effects of different biodiversity policies, and optimizing biodiversity-positive strategies for the primary industries (Iquebal M. A. et al., 2015).

The perspective of FSL applications for India's ecological sectors has been strengthened by the substantial breakthroughs in ML and the transformation brought about by electronic and satellite communication. India is home to 7% of the world's forests and ranks as the eighteenth largest biologically diverse region (Sharma, Y. K., & Rokade, M. D. 2019).

However, the country's abundant biodiversity is still in danger from the rapid expansion of agricultural lands, cities, and construction projects. So, if we want to monitor, manage, and conserve animal resources and biodiversity, we should consider implementing cutting-edge technology like AI into our forests. The United Nations projected that there will be over eight million species, including plants and animals, on the planet in 2020. There seems to have been an acceleration in the pace of global extinction of species in the past decade of a million years, and the present rate is twenty to hundreds of times greater than usual. Approximately one million of these species are in danger of becoming extinct in the next couple of centuries.

According to the data, we are on the verge of experiencing the sixth major extinction on Earth. Furthermore, every fourteen minutes, elephants are butchered in Africa, and every eleven hours, rhinos are slain. In addition, approximately fourteen thousand species are considered endangered in the United States. After millions of decades, it seems that the majority of the species managed to stay alive. As a result, poaching, deforestation, and hunting have a substantial effect on biological diversity.

Creative methods to preserve and safeguard the diversity of our species may be uncovered by advanced technology like AI. By using AI in biodiversity, a multitude of signals may be identified, allowing for the detection of potential dangers and the provision of real-time information. For this reason, AI is also considered a friend to efforts to protect natural resources and biodiversity (Pradeep, S., & Sharma, Y. K. 2019).

5. FSL IN AGRICULTURE

Recently, deep learning has paved the way for a new direction in machine learning. Achieving a state where computers can learn and analyze at a human level, allowing them to comprehend and react to images, sounds, and language, is the ultimate goal. One of these, the convolutional neural network, has become quite popular. The two primary steps in deep learning are training and testing. In order to train a CNN for a specific job, the training procedure entails doing hundreds of iterations using the labelled

Algorithm 1. Utilising few-shot learning in bioinformatics for agricultural and ecological applications

Step 1: Data Collection
1.1 Gather diverse agricultural and environmental datasets.
1.2 Label plant, species, illness, and ecological datasets.
Step 2: Data prep
2.1 Standardize and prepare data for uniformity.
2.2 Increase dataset variety and correct class imbalances.
2.3 Sample partitioning into training, verification, and testing.
Step 3: Few-Shot Learning Model Initialization
3.1 Choose a Few-Shot Learning framework like Traditional Networks or Meta-Learning.
3.2 Initialize the model using random or pre-existing weights.
Step 4: Model Training
4.1 Train the Few-Shot Learning percentage using the training sample.
4.2 Apply meta-learning with transfer learning to train the model.
4.3 Validate the model
Step 5: Performance evaluation
5.1 Apply crop testing sample dataset
5.2 Measure accuracy, precision, recall, and F1-score

training set, which is also called training data, to fine-tune the network's properties. A model that has been trained is generated at the end. The approach to testing uses the test set, often called test data, to find out the extent to which the model worked.

One of the most important aspects of agriculture and plant attributes that impact production is plant disease. Reducing economic costs when illnesses emerge requires quick diagnosis of disease types and related treatment. It used to be that specialists in plant disease identification had to rely on their eyes alone. Efficiency, precision, and identification speed may all be significantly improved by employing few-shot learning. Currently, few-shot learning has become the gold standard for identifying plant pathogens in the agricultural products and plant properties sectors and spotted cotton leaves using a few-shot learning approach grounded on metric learning.

The goal of FSL is to find and name the item in issue using a small number of labelled samples from each category of agriculture plants and crops. A variety of methods have been investigated for the few-shot identification of named entities. Shortening the time it takes to get through training to fine-tuning is the goal of the quick learning-based approach.

5.1 Algorithm for Bioinformatics in Agriculture Using Few-Shot Learning

Using Few-Shot Learning, Algorithm 1 shows bioinformatics basics in agriculture and ecology (Wang Y. et al., 2020).

6. CONCLUSION AND FUTURE DIRECTIONS

Within ecology, FSL (Functional Species Loss) is recognised as a crucial partner in efforts to promote and protect biodiversity. Identifying and tracking species with limited labelled data effectively tackles the difficulties of biodiversity monitoring. FSL's capacity to identify uncommon species and adjust to ever-changing environmental circumstances enhances conservation endeavours, facilitating effective and adaptable management approaches.

In conclusion, it is crucial to delineate future paths that can improve the incorporation of sophisticated learning methods in the fields of bioinformatics, food production, and ecology:

- **Interdisciplinary Collaboration**: Interdisciplinary Interaction refers to the cooperation and integration of knowledge and expertise from different academic disciplines or fields of study. Promote enhanced collaboration among data scientists, researchers in biology, ecologists, and agricultural scientists. Associations between other academic disciplines can result in a comprehensive comprehension of biological phenomena and the creation of customized machine-learning technologies.
- **Integration of Explainable AI (XAI):** Investigate the incorporation of Transparent AI methodologies in bioinformatics, the agricultural sector, and ecology. Improving the comprehensibility of machine learning algorithms is essential for establishing credibility among researchers, professionals, and the wider audience.
- **Principles of Ethical and Sustainable AI: Consider** ethical concerns when creating and implementing machine learning algorithms. Develop protocols for ethical data acquisition, model disclosure, and equitable portrayal, particularly in vulnerable fields such as ecology and preservation.
- **Iterative Model Enhancement:** Allocate resources towards researching to optimise the ongoing learning capacities of machine learning algorithms. Dynamic models that adjust to changing conditions and assimilate fresh inputs in real time will yield more precise and dependable forecasts.
- **Public involvement and instruction:** Facilitate active participation of the community in observing biodiversity and farming. Instruct agriculturalists, citizens, and scientists on the advantages and moral implications of machine learning implementations. Promote a collective commitment to sustainable practices.
- **Examine integrating machine learning:** With emerging technologies such as Blockchain and edge computing. This may enhance bioinformatics, agricultural and ecological data preservation, traceability, and decentralized computing.

We realize that bioinformatics, the farming industry, and ecology are still evolving. Technological advances and a more substantial commitment to ethics, sustainability, and environmental protection are possible in the future. Using cutting-edge technology may help us adapt to this changing world.

REFERENCES

Ahmed, S. (2022). *Classification of Plant Disease from Leaf Images Using Few-Shot Learning* [Doctoral dissertation, Department of Computer Science and Engineering (CSE), Islamic University of Technology (IUT), Board Bazar, Gazipur, Bangladesh]

Baik, S., Choi, M., Choi, J., Kim, H., & Lee, K. M. (2023). Learning to learn task-adaptive hyperparameters for few-shot learning. *IEEE Transactions on Pattern Analysis and Machine Intelligence*. PMID:37030677

Balyan, A. K., Ahuja, S., Sharma, S. K., & Lilhore, U. K. (2022, February). Machine learning-based intrusion detection system for healthcare data. In *2022 IEEE VLSI Device Circuit and System (VLSI DCS)* (pp. 290-294). IEEE.

Chen, B., Chen, T., Zeng, X., Zhang, W., Lu, Q., Hou, Z., & Helal, S. (2023). DFML: Dynamic federated meta-learning for rare disease prediction. *IEEE/ACM Transactions on Computational Biology and Bioinformatics*. PMID:37022085

Conneau, A., Ma, M., Khanuja, S., Zhang, Y., Axelrod, V., Dalmia, S, & Bapna, A. (2023, January). Fleurs: Few-shot learning evaluation of universal representations of speech. In *2022 IEEE Spoken Language Technology Workshop (SLT)* (pp. 798-805). IEEE. 10.1109/SLT54892.2023.10023141

Dahiya, N., Sharma, Y. K., Rani, U., Hussain, S., Nabilal, K. V., Mohan, A., & Nuristani, N. (2023). Hyper-parameter tuned deep learning approach for effective human monkeypox disease detection. *Scientific Reports*, *13*(1), 15930. doi:10.1038/s41598-023-43236-1 PMID:37741892

Gaur, L., Solanki, A., Wamba, S. F., & Jhanjhi, N. Z. (Eds.). (2021). *Advanced AI techniques and applications in bioinformatics*. CRC Press. doi:10.1201/9781003126164

Ge, Y., Guo, Y., Das, S., Al-Garadi, M. A., & Sarker, A. (2023). Few-shot learning for medical text: A review of advances, trends, and opportunities. *Journal of Biomedical Informatics*, *144*, 104458. doi:10.1016/j.jbi.2023.104458 PMID:37488023

Iquebal, M. A., Jaiswal, S., Mukhopadhyay, C. S., Sarkar, C., Rai, A., & Kumar, D. (2015). Applications of bioinformatics in plant and agriculture. *PlantOmics: the omics of plant science*, 755-789.

Jadhav, D., Sharma, Y. K., & Arora, D. P. (2020, April). Profound Learning Approach for Shot Boundary Location. In *2nd International Conference on Communication & Information Processing (ICCIP)*. SSRN. 10.2139/ssrn.3645409

Jadhav, M., Kumar Sharma, Y., & Bhandari, G. M. (2019, December). Currency identification and forged banknote detection using deep learning. In *2019 International Conference on Innovative Trends and Advances in Engineering and Technology (ICITAET)* (pp. 178-183). IEEE. 10.1109/ICITAET47105.2019.9170225

Jaiswal, V., Saurabh, P., Lilhore, U. K., Pathak, M., Simaiya, S., & Dalal, S. (2023). A breast cancer risk prediction and classification model with ensemble learning and big data fusion. *Decision Analytics Journal*, *8*, 100298. doi:10.1016/j.dajour.2023.100298

Khalid, M. N., Abdullah, A., Ijaz, Z., Naheed, N., Hamad, A., Sheir, M. A., & Khan, M. D. (2021). Application and Potential Use of Advanced Bioinformatics Techniques in Agriculture and Animal Sciences. *Ind. J. Pure App. Biosci*, *9*(3), 237–246. doi:10.18782/2582-2845.8702

Kumar, N., Verma, H., & Sharma, Y. K. (2023). Adversarial Attacks on Graph Neural Network: Techniques and Countermeasures. In Concepts and Techniques of Graph Neural Networks (pp. 58-73). IGI Global.

Li, P., Liu, F., Jiao, L., Li, S., Li, L., Liu, X., & Huang, X. (2023). Knowledge transduction for cross-domain few-shot learning. *Pattern Recognition*, *141*, 109652. doi:10.1016/j.patcog.2023.109652

Li, W., Wang, Z., Yang, X., Dong, C., Tian, P., Qin, T., Huo, J., Shi, Y., Wang, L., Gao, Y., & Luo, J. (2023). Libfewshot: A comprehensive library for few-shot learning. *IEEE Transactions on Pattern Analysis and Machine Intelligence*, *45*(12), 14938–14955. doi:10.1109/TPAMI.2023.3312125 PMID:37669193

Lilhore, U. K., Imoize, A. L., Lee, C. C., Simaiya, S., Pani, S. K., Goyal, N., Kumar, A., & Li, C. T. (2022). Enhanced convolutional neural network model for cassava leaf disease identification and classification. *Mathematics*, *10*(4), 580. doi:10.3390/math10040580

Lilhore, U. K., Manoharan, P., Sandhu, J. K., Simaiya, S., Dalal, S., Baqasah, A. M., Alsafyani, M., Alroobaea, R., Keshta, I., & Raahemifar, K. (2023). Hybrid model for precise hepatitis-C classification using improved random forest and SVM method. *Scientific Reports*, *13*(1), 12473. doi:10.1038/s41598-023-36605-3 PMID:37528148

Lilhore, U. K., Simaiya, S., Dalal, S., & Damaševičius, R. (2023). An intelligent waste classification model using hybrid CNN-LSTM with transfer learning for a sustainable environment. *Multimedia Tools and Applications*, 1–25.

Moein, M. M., Saradar, A., Rahmati, K., Mousavinejad, S. H. G., Bristow, J., Aramali, V., & Karakouzian, M. (2023). Predictive models for concrete properties using machine learning and deep learning approaches: A review. *Journal of Building Engineering*, *63*, 105444. doi:10.1016/j.jobe.2022.105444

Moein, M. M., Saradar, A., Rahmati, K., Mousavinejad, S. H. G., Bristow, J., Aramali, V., & Karakouzian, M. (2023). Predictive models for concrete properties using machine learning and deep learning approaches: A review. *Journal of Building Engineering*, *63*, 105444. doi:10.1016/j.jobe.2022.105444

Nayak, S., & Sharma, Y. K. (2023). A modified Bayesian boosting algorithm with weight-guided optimal feature selection for sentiment analysis. *Decision Analytics Journal*, *8*, 100289. doi:10.1016/j.dajour.2023.100289

Nayak, S., & Sharma, Y. K. (2023). A modified Bayesian boosting algorithm with weight-guided optimal feature selection for sentiment analysis. *Decision Analytics Journal*, *8*, 100289. doi:10.1016/j.dajour.2023.100289

Nayem, J., Hasan, S. S., Amina, N., Das, B., Ali, M. S., Ahsan, M. M., & Raman, S. (2023). Few Shot Learning for Medical Imaging: A Comparative Analysis of Methodologies and Formal Mathematical Framework. arXiv preprint arXiv:2305.04401. doi:10.1007/978-3-031-47772-0_4

NyamekeF. E.BinS.Ahiaklo-KuzR. K.PeprahR. O. Few-Shot Learning: A Step for Cash Crops Disease Classification. Available at SSRN 4080599. doi:10.2139/ssrn.4080599

Ochal, M., Patacchiola, M., Vazquez, J., Storkey, A., & Wang, S. (2023). Few-shot learning with class imbalance. *IEEE Transactions on Artificial Intelligence*, *4*(5), 1348–1358. doi:10.1109/TAI.2023.3298303

Pradeep, S., & Sharma, Y. K. (2019, February). A pragmatic evaluation of stress and performance testing technologies for web-based applications. In *2019 Amity International Conference on Artificial Intelligence (AICAI)* (pp. 399-403). IEEE. 10.1109/AICAI.2019.8701327

Pradeep, S., Sharma, Y. K., Lilhore, U. K., Simaiya, S., Kumar, A., Ahuja, S., Margala, M., Chakrabarti, P., & Chakrabarti, T. (2023). Developing an SDN security model (EnsureS) based on lightweight service path validation with batch hashing and tag verification. *Scientific Reports*, *13*(1), 17381. doi:10.1038/s41598-023-44701-7 PMID:37833379

Rokade, M. D., & Sharma, Y. K. (2021, March). MLIDS: A machine learning approach for intrusion detection for real-time network dataset. In *2021 International conference on emerging smart computing and informatics (ESCI)* (pp. 533-536). IEEE. 10.1109/ESCI50559.2021.9396829

Sachi, S., Singh, A. K., Jain, A., Devi, S., Sharma, Y. K., & Athithan, S. (2023, July). Hate Speech Detection Using the GPT-2 and Natural Language Processing. In *2023 Intelligent Methods, Systems, and Applications (IMSA)* (pp. 276-280). IEEE.

Sharma, Y. K., Athithan, S., Sachi, S., Singh, A. K., Jain, A., & Devi, S. (2023, July). Copy and Move Forged Image Detection by Deep Learning. In *2023 World Conference on Communication & Computing (WCONF)* (pp. 1-6). IEEE. 10.1109/WCONF58270.2023.10235115

Sharma, Y. K., & Khatal Sunil, S. (2019). Health Care Patient Monitoring using IoT and Machine Learning. In *IOSR Journal of Engineering (IOSR JEN).*

Shoaib, M., Shah, B., Sayed, N., Ali, F., Ullah, R., & Hussain, I. (2023). Deep learning for plant bioinformatics: An explainable gradient-based approach for disease detection. *Frontiers in Plant Science, 14*, 14. doi:10.3389/fpls.2023.1283235 PMID:37900739

Singh, K., & Kumar, A., Sharma, Y. K., & Rai, A. K. (2023). AIoT-based e-commerce. *AIoT Technologies and Applications for Smart Environments*, 215.

Singh, K., & Malhotra, D. (2023). Meta-Health: Learning-to-Learn (Meta-learning) as a Next Generation of Deep Learning Exploring Healthcare Challenges and Solutions for Rare Disorders: A Systematic Analysis. *Archives of Computational Methods in Engineering*, *30*(7), 1–32. doi:10.1007/s11831-023-09927-8

Singh, P., Singh, N., & Deka, G. C. (2021). Prospects of machine learning with Blockchain in healthcare and agriculture. In *Multidisciplinary functions of Blockchain technology in AI and IoT applications* (pp. 178–208). IGI Global. doi:10.4018/978-1-7998-5876-8.ch009

Song, Y., Wang, T., Cai, P., Mondal, S. K., & Sahoo, J. P. (2023). A comprehensive survey of few-shot learning: Evolution, applications, challenges, and opportunities. *ACM Computing Surveys*, *55*(13s), 1–40. doi:10.1145/3582688

Swapna, M., Sharma, Y. K., & Prasad, B. M. G. (2020). A survey on face recognition using convolutional neural network. In *Data Engineering and Communication Technology: Proceedings of 3rd ICDECT-2K19* (pp. 649–661). Springer Singapore. doi:10.1007/978-981-15-1097-7_54

Chapter 3
An Overview and Analysis of Machine Learning Classification Algorithms in Healthcare

Soumitra Saha
https://orcid.org/0000-0002-0012-9109
Chandigarh University, India

ABSTRACT

To decode a wide range of complex and challenging problems around us, we must utilize the data that already exists in our surroundings as effectively as possible. This data will be functional in diverse fields in everyday life, and the data plays the most crucial role in healthcare and medicine, finance and banking, and information technology. The first, foremost, and prime reason for forming or generating this big data is the increasing complexity of real-world problems, which takes considerable work to implement. For example, in healthcare and medicine, extensive data are needed to detect deadly diseases like cancer, cardiovascular diseases, and HIV/AIDS effectively. Classification algorithms are the essential and substantial machine learning algorithms used to decode numerous challenging real-life problems in the healthcare industry. Implementing the algorithm encounters fewer problems regarding time and space and comparatively better interpretability and scalability. Through this study, the authors have demonstrated how different classification algorithms perform in the healthcare system.

1. INTRODUCTION

These increasing technological developments of the 21st century have made our lives much more comfortable and stunning. Diverse data types have played the most critical role in this technological development. This advancement in technology in distinct fields uses different types of data depending on the task and judging the desired solution. Due to this promotion and spread of technology, the number of data in the world has increased, the use of data has increased, and a large amount of data is constantly being generated to solve various day-to-day practical problems. For example, in medical science, a considerable amount of data is being generated continuously as the types of diseases and variants of

DOI: 10.4018/979-8-3693-1822-5.ch003

these types increase alarmingly. A wide range of software platforms and apps are available, and some are explicitly designed to provide information to academics (Lobe et al., 2020). Collecting data from a suitable medium in the healthcare industry with the help of some revolutionary inventions like X-ray Machines, MRI (Magnetic Resonance Imaging) Machine, and CT (Computed Tomography) Scanner, making the data useful enough, and selecting the appropriate data to solve the specific health issues makes it possible to solve any complex problem efficiently. Nowadays, everything around us is wired up to a data source since we live in the data era (Sarker, 2021). Data collected from various sources are used to solve any tricky and complex problems around us. The widely used machine learning algorithms make a significant, influential, and indispensable contribution to solving these problems.

One of the main reasons for using machine learning algorithms to solve a variety of challenging and complex biological problems is that these algorithms can work straightforwardly, using relatively less time and less space, and can work towards achieving the desired results by solving a variety of existing internal problems and in most cases the best results can be achieved. For instance, using machine learning algorithms, predictions of anxiety, sadness, and stress were generated relatively simply in medical research (Priya et al., 2020). It can also be used to simultaneously select the all-important and irrelevant instances for a classification problem (Saha et al., 2022). Machine learning algorithms are divided into categories depending on the type and nature of the problem, the specific course of action to solve the problem, the nature of the data collected to solve the problem, and the target variables. The most important types of machine learning algorithms are supervised machine learning algorithms, unsupervised machine learning algorithms, semi-supervised machine learning algorithms, and reinforcement machine learning algorithms. Although there are various reasons, the type of machine learning algorithm to be used mainly depends on the type of problem being solved and the type of result to be achieved after solving it. Supervised machine learning algorithms are relatively more used than other algorithms when considering existing problems. One of the best and most used fields of the supervised machine learning algorithm is medical science, where for any specific disease prediction, multiple supervised algorithms are used to be comparable and generate better results on every occasion (Uddin et al., 2019).

Supervised machine learning algorithms work primarily on level data. Simply put, the class column is present in the data sets on which the supervised machine learning algorithm works. That is why it is worthwhile to use typical supervised learning techniques to make a prediction or classification about an outcome of interest (Jiang et al., 2020). This process is advantageous in the field of manufacturing industrial equipment as well as monitoring the performance in real-life situations (Ouadah et al., 2022), work in the field of sentimental analysis to detect the emotional attachment of a human being with the use of natural language processing (Shah et al., 2022) which is one of the most critical health issues however is usually not discussed much, and as well as other essential sectors worldwide. The presence of the class column makes the prediction task much more effortless. What kind of data or data from which fields exist in a data set, what is the nature of the data, and how consistent can the results be easily understood only because of the presence of the data set's level? Labels collect different types of information for the model used, and depending on that information, the model performs actions. Patterns are created from the labels to give a clear idea of what results the model will produce using the new data, as the data of different medical sectors are mostly interconnected. Later, it is considered whether the desired results have been achieved by comparing the levels with the newly generated results. Supervised Machine Learning Algorithm has various parts, and some of the essential algorithms are Classification Algorithm, Regression Algorithm, Neural Network, Ensemble Method, Etc.

Classification algorithms work with categorical data and generate results for a particular model based on the class level. A classification algorithm uses various features as input values and produces a specific class level as an output value. Classification algorithms are selected depending on the type of problem to be solved in a particular field, the type of data available to solve, and the expected result after solving. For instance, it is a critical problem for any educational institution to predict the student's performance, categorize each student according to their skills, and improve their performance in approaching examinations (Pallathadka et al., 2023). The classification approaches can also effectively estimate the education quality, which will be delivered through networks (Chen & Gadekallu, 2023). In the same way, for predicting the patient's physical conditions in any critical health issues, categorize each patient according to their health effect, collect the data from them, first compare the data with the existing one, properly utilize them, and finally try to improve the enhancement of the patient's condition. After the data processing, the classification algorithm generates patterns with the data collected from particular fields; a particular algorithm or model trains the data and then is used for prediction. This widely discussed algorithm is used in many essential fields, including fraud detection, image classification, medical diagnosis, natural language processing, spam filtering, Etc. The field's most widely used classification algorithms and problem considerations are K Nearest Neighbor, Naive Bayes Classifier, Random Forest, Decision Tree, Support Vector Machine, Etc. Each algorithm has advantages and disadvantages regarding the problem type and convenience.

It is imperative to understand how much noise exists in a particular data set, how many missing values there are, how much redundant data there is, how much less necessary data there is, and which data is being directly used to achieve the desired results. Doing these things is most important for the medical field as these issues are much more sensitive when considering patient health. Collecting data from different patients, the physical condition of the patient, different symptoms of different diseases, equipment used for data collection, and environmental conditions may vary. The classification algorithm is selected by considering various other parameters, including the nature of the data set, class values, type of desired result, and specific method to be adopted for the desired result. Different algorithms consider how well a data set will provide good results for solving a specific problem and how acceptable its results are. As a result, the best algorithm is chosen based on the comparison of different types of results. Since there are numerous problems to be solved and a large number of data sets and classification algorithms to solve those problems, finding out which particular algorithm can provide the best results for solving the problem is vital.

The following sections of this report are organized: Section II tries to show how the various classification algorithms used in different fields have performed. Section III describes the conclusion and discussion of what kind of work to do in the healthcare industry using these classification algorithms.

2. CLASSIFICATION TECHNIQUES FOR DISTINCTIVE ISSUES

This section will investigate various categorization approaches based on machine learning and widely used. In addition, we will analyze the learning processes of different algorithms associated with the algorithms that comprise the classification approaches. If we can accomplish this goal, it will be constructive, and we can use our skills to understand and explain the latter portion of the essay quite quickly and easily, which will be to our advantage. Figure 1 presents a taxonomy of the many classification methodologies used in ML systems in most real-world problems.

2.1 Decision Tree

A decision tree is a famous supervised machine learning algorithm where a tree-like model is used by which the probable outcomes of a problem are calculated. In a decision tree, the starting node is the root node, and the ending node is the leaf node. Interior nodes are indispensable and follow specific rules and regulations to reach the leaf node. The decision tree follows a recursive top-down approach where the root node is first selected to split the data, and the best-split point will diminish data impurity significantly. Further, the data is split into two groups, followed the same procedure, and continued until all the available data is classified.

In medical science and the healthcare system, Electrocardiograms (ECGs) for cardiovascular issues are innovative medical advances that provide early sickness analysis. Six distinctive heartbeat types are available, regular and arrhythmic, and for the best use, an optimal decision tree and an adaptive boosted optimum decision tree were proposed (Kumari & Sai, 2022). MIT- BIH arrhythmia database tested all the available models and showed better accuracy (98.77%) and performance than previous techniques. These models may detect heart arrhythmias by interpreting confusing and inadequate data. For studying ambulance offload delay (AOD) in Canada, a decision support tool is created using hospital and EMS provider data. They utilize a hybrid decision tree model using the CART algorithm and naive Bayes classifier to predict AOD severity within 1 to 5 hours (Li et al., 2022). The findings show that developed AOD categorization supporting decision-makers reduces delays, and the study becomes notable when a hospital setting uses administrative data creatively. For detecting collapsed building deaths, this author uses transfer learning-based deep learning. The author uses a customized data set with five class labels to fine-tune RESNet 50 to extract features. The random tree approach has 99.53% accuracy and 0.02 seconds of calculation time compared to earlier decision tree classification approaches (Seeja et al., 2023) because the AI-driven systems of mobile robots can quickly identify victims and provide medical aid in life-threatening circumstances, improving Search and rescue efforts. In order to discover and rectify Automatic Identification System (AIS) data errors, several Decision tree-based techniques are being tested, and this research will support Satellite AIS to monitor Marine traffic (Szarmach & Czarnowski, 2023). The correct detection of software development is highlighted by using computer experiment findings. The author replaced the domain overlap score function with decision trees in the sure independence screening and sparsifying operator (SISSO) materials discovery technique (Selvaratnam et al., 2023). With increased accuracy, the technique suggests solid-state chemistry such as perovskites, spinels, and rare Earth intermetallics. Training and test sets had accuracy values of 0.91 and 0.86 by the decision tree scoring technique. While decision tree evaluation is outsourced to a cloud server, the author discusses maintaining model privacy in this article (Chen et al., 2022). Using portable Cryptography tools based on secret-sharing protocols proposes a noble way to balance communication and computation expenses.

A novel rule representation approach was proposed here using referential intervals and decision tree categorization to build belief rule-based (BRB) systems (Fu et al., 2020). The suggested method covers huge BRB system sizes in complicated themes using decision tree analysis and rule extraction. The research also helps the BRB building technique, which optimizes rule parameters via decision trees and differential evolution. Machine learning is being used to pick secure relays in this work. The authors minimize computational complexity and feedback while maintaining an average secrecy rate by turning it into a multi-class classification issue and using a decision tree-based method (Wang & Liu, 2020). They suggest feature extraction for discrete inputs from quantized channel state data. Information loss causes uncertainty; hence, an approach uses a fuzzy decision tree for classification, and a fuzzification

Figure 1. Machine learning for classification

mechanism for transformation was proposed (Rabcan et al., 2019). The method is tested by identifying faulty gas turbine blades in an airplane engine, and the classification accuracy of 0.989 makes it superior to other methods for this work. In order to detect private illness in the mobile healthcare networks industry, a private decision tree classification with low resource consumption (PDTC-LRC)technique was being proposed (Alex et al., 2022). The author proposed an integer comparison protocol that is secure, and the protocol can be able to detect the illness quite safely and effectively in the context of resource-constrained. Also, the protocol can preserve better accuracy. A Secure and Privacy-Preserving Decision Tree Evaluation algorithm, namely SPP-DTE, was introduced by (Zhang et al., 2022) and is used in e-healthcare systems. The ultimate goal of the proposed method is that the patient data is stored in the cloud server, which needs to be more secure. Many methods, like matrix randomization and modified

Figure 2. Decision tree classifier

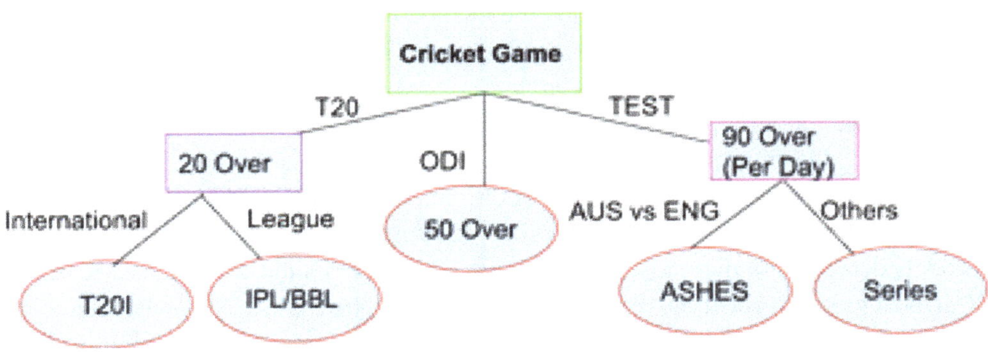

KNN calculation, are widely used to do the task, and the proposed method shows satisfactory results regarding these issues at the time of the diagnosis of the patient's illness.

2.2 GradientBoostingClassifier

The foundation of the gradient boosting classifier is the boosting algorithm. Multiple weak learners can work together with the help of boosting and creating more powerful learning techniques; in this case, decision trees are used most of the time. Each poor learner takes instruction in a predefined way, and this instruction depends on the predecessor's mistakes. The process goes on until an optimal solution occurs. The most beneficial thing about this classification is that it can minimize the loss function. As the process is very straightforward to understand, it can handle complicated data sets and is usable for real-world problems.

Histology-based Computer-aided diagnosis (CAD) for breast cancer detection was proposed (Abbasniya et al., 2022), and Categorical Boosting (CatBoost), Extreme Gradient Boosting (XGBoost), and Light Gradient boosting Machine (light GBM) are three boosting algorithms used for categorization. The (Inception-ResNet-v2)IRv2-CXL technique is suggested since it outperforms other cutting-edge methods on Breakhis. The author also signifies how successfully the CAD system categorizes breast cancer Histology pictures. New prediction method for underground excavation stability is introduced, known as PSO-GBDT, where the particle swarm optimization algorithm is used to optimize the gradient boosting decision tree's parameter (Zhou et al., 2023). PSO- GBDT outperformed eight classifiers and traditional methods with a classification rate of 0.93. Synthetic attention in bidirectional encoder representations from transformers (SA- BERT) model for aspect-based sentiment analysis (ABSA) was proposed by (Mewada & Dewang, 2023). The SA-BERT model and XG boost classifier surpass baseline approaches in accuracy and F1 scores in sentiment polarity classification. This method predicts emotion polarity, offers a dynamic aspect, and is more productive than the current model. Automated classification techniques can identify epilepsy (a brain disorder) using electroencephalogram (EEG) data, and in order to retrieve features, the method uses discrete Fourier transform and discrete wavelet transform (Sunaryono et al., 2022). Gradient boosting machines classify the signals following, and Genetic algorithms choose characteristics. The University of Bonn data set reveals that the method accurately detects epilepsy from EEG signals. However, additional research is needed to identify whether the method can be extended to other datasets. A classifier-based water quality prediction model was proposed by (Khan et al., 2022) using gradient boosting and principal component regression. Principal component analysis (PCA) is used to identify key water quality index (WQI) features and predict WQI. The weighted arithmetic index approach creates the WQI, and the method offers 95% prediction accuracy in principal component regression and 100% in gradient boosting classifier classification accuracy, where Gulshan Lake data outperforms leading models in research.

The dual ensemble model for anomaly-based intrusion detection systems (IDS) uses the Bagging and gradient boosting decision tree (GBDT) approaches by the author (Louk & Tama, 2023). The suggested approach yielded promising results, and additional modified GBDT algorithms were tested on other data sets. GBM and Bagging work better than other methods, and the research reveals that dual ensemble models detect more minutely, making them a good IDS solution. A classification technique using the XGBoost gradient boosting system was proposed (Uguz et al., 2020), where data is processed using electrocardiograms and wavelet changes to identify domain-specific and domain-neutral features. The author offers a unique beat-to-beat correlation method for arrhythmia identification with regular beats. A novel reject inference model for peer-to-peer (p2p) lending was introduced by (Xia, 2019), known as OD-LightGBM. The diagram uses outlier detection and the gradient-boosting decision tree technique to enhance prediction performance utilizing rejected data. The model handles challenging scenarios and outperforms Benchmarks with the help of real-world datasets. This method may help p2p lenders reduce information asymmetry and improve credit score accuracy. Breast cancer is now one of the most common diseases for females in recent times, and for that, a hybrid ensemble architecture was proposed by (Sanyal et al., 2021) to identify breast cancer. For that, the author used high-resolution histopathology photographs reliably. For extracting the features, the method combined many fine-tuned CNNs with XGBoost for classification techniques, which is very helpful in resolving the issues generated by the complex structure photos. To find out the combination and results of protein-protein interaction (PPI) prediction, a unique hybrid technique was proposed by (Mahapatra et al., 2021), where the author combines the extreme gradient boosting classifiers (XGB) alongside deep neural networks (DNN). With sequence-based features, the proposed DNN-XGB classifier can obtain high accuracy, and thus far, it is very worthwhile in the target prediction of medication and sickness etiology.

2.3 K-Nearest Neighbors

One of the prominent and widely used machine learning techniques in classification and regression is the K-Nearest Neighbours (KNN) classifier. It is a non-parametric learning technique that predicts similar data. The classifier from the training set finds the K number of similar type cases and then predicts the data labels with the k closest neighbors. The model will be more resilient to noise for a significant value of k; however, the result will be less accurate. In contrast, a smaller value of k brings the classifier more susceptible to noise. However, the result will be more accurate. The hyperparameter k controls the complexity of the model.

A distributed storage and computation k-nearest neighbor (D-kNN) algorithm is uncovered, where the author analyzes data for distributed computing and storage in cyber-physical systems (CPSS) (Zhang et al., 2020). Distributed computation speeds up k-nearest neighbor searches, storage nodes store massive data sets, and minimize time complexity. With massive data sets, deploying in cloud edge computing systems works well and is accessible in CPSS applications. FKNN, a fuzzy k-nearest neighbor framework with few adjustments, predicts whether college students will get a master's degree where FKNN, random forest, and a new chaos-enhanced sine cosine-inspired algorithm (CESCA) were used by (Lin et al., 2019). According to many trials, the CESCA-FKNN model outperforms all classifieds in accuracy. The article reveals that this adaptive FKNN Framework could anticipate student objectives and guide institutional talent development programs. A model describing a machine learning-based optical system for object detection distinguishes optical noise from the desired source using k-nearest neighbor was proposed by (Miranda-Vega et al., 2020). Despite poor illumination and weather, the optical scanning system-based structural health monitoring method has worked, and over 98% accuracy was achieved by classifying the reference source and optical noise using feature extraction. Regarding the pairwise distance metrics and belief function theory, a classifier named BPkNN introduced by the author (Jiao et al., 2019), a kNN-based classifier. Using pairwise distance measurements with belief function theory improves classification where BPkNN separates the distance metric into the pairwise distance to generate adaptive PkNN (pairwise kNN) sub-classifiers. In order to clear ambiguity, these sub-classifiers' findings are blended by using belief functions. The two-layer nearest neighbor (TLNN) rule was proposed in this work to enhance the k-nearest neighbor (kNN) classifier by (Wang et al., 2022). They proposed a k-two-layer nearest neighbor (kTLNN) classifier that tests classification skills using TLNN. In pragmatic experiments, KTLNN outsails kNN and seven other closest neighbor-based classifiers, and hence, the TLNN rule enhances classification accuracy by including two-level neighborhood information.

In order to parameter effects on classification, the article proposes a Fuzzy Parameterized Fuzzy Soft *k*NN (FPFS-*k*NN) algorithm (Memiş et al., 2022). In 35 UCI dataset trials, the approach outperforms state-of-the-earth kNN-based algorithms in accuracy, precision, recall, micro F score, and macro F-score. Statisticians demonstrate FPFS- kNN's advantages, and the study finds this needs further investigation using Pearson's correlation coefficient. Machine learning as a service reveals the risk of giving cloud service providers sensitive data. The authors use a homomorphic encryption scheme (TFHE) to perform the K-nearest neighbors algorithm on encrypted data while maintaining anonymity (Ameur et al., 2022). The proposed technique covers all k-NN algorithm stages, including class label assignment but server client contact. Pragmatic datasets show the approaches' usefulness; conversely, simulated data shows its scalability across many parameters.

Research gaps are introduced by user topic classification after surveying and classifying social media users using many categories. After training with ML-KNN, the classification model accurately classifies people into themes and situations (Huang et al., 2023). This research broadens the classification technique and helps with the multi-label categorization of social media. Parkinson's disease (PD) is one of the prominent diseases nowadays. Inconsistent data collection is a big issue in determining the harshness of PD among patients, and for that, a novel ensemble K-nearest neighbor (EnKNN) method was being introduced (Zhao et al., 2022). The proposed method can achieve an extraordinary average accuracy of 95.02%, which means the method may diagnose Parkinson's disease accurately while working with inconsistent data. Cutting-edge methods for diagnosing Alzheimer's disease in the early stage, namely the fuzzy k-nearest neighbor method based on the improved binary salp swarm algorithm (IBSSA-FKNN), were introduced (Lu et al., 2023). These methods parallelly used the fuzzy k-nearest neighbor and salp swarm approach for better feature selection. As a result of collecting the monumental features and using proper evaluation, the method achieves better accuracy regarding the other five approaches, and by using this, doctors can make better diagnoses for Alzheimer's patients.

2.4 Logistic Regression

Logistic Regression is one of the most common and widely used problems in classification and Regression. In the binary class classification task, the foremost task of Logistic Regression is to predict an instance's likelihood and that the instance belongs to which class. Logistic Regression can predict the probability of, and as a linear model in nature, it uses an activation function. The activation function accepts real numbers as input values, and at the same time as output values, the classifier generates values between 0 and 1. The model estimates the relationship between the independent and dependent variables, and this action occurs in the presence of new data.

A connected network regularized logistic regression (CNet- RLR) model to identify characteristics based on network connectivity was proposed by (Li & Liu, 2022). It generates feature selection as a convex optimization problem with inequality constraints to find solutions rapidly. It outperforms regularized logistic regression replacements and graph-embedded feature selection models in classification and interpretability. A logistic regression-based oppositional tunicate fuzzy C-mean cloud infiltration detection method known as LR- OTSFCM was proposed (Kanimozhi & Aruldoss Albert Victoire, 2022), where the approach has four steps: preprocessing, feature selection, data clustering, and cluster management.

Figure 3. K-Nearest neighbors classifier

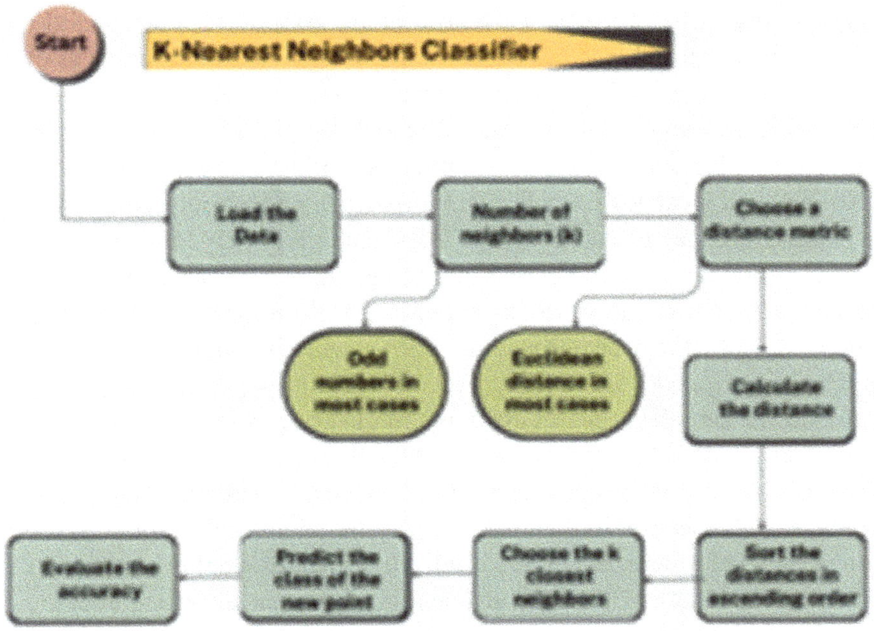

Figure 4. Logistic regression
(Bari Antor et al., 2021)

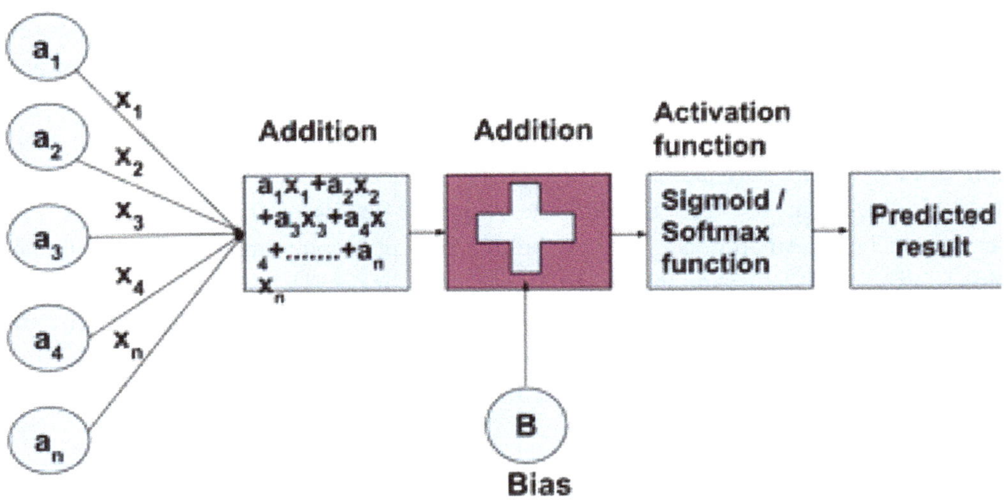

The recommendation has a more significant attack detection, and even though the research says Cloud Security has improved, the overview needs more information to analyze the approach. In the study of the public opinion on Rinca island's development in Indonesia, data was collected using Doc2Vec models and SVM or logistic regression classifiers (Hidayat et al., 2022). The study found that Twitter users most often opposed the proposed construction. The computers classified tweets as excellent, harmful, or neutral with over 75% accuracy. Despite social media data interpretation limitations, the research reveals how successful these methods are in determining public opinion. An ensemble transform approach for Bayesian inference in logistic regression was proposed (Pidstrigach & Reich, 2023), which extends the ensemble Kalman and feedback particle filters to include a cross-entropy loss function. Numerical examples demonstrate these tactics' efficacy, and Bayesian inference in Logistic regression is explained in the study, which applies to various fields. Binary Logistic regression models detect valvulopathies-related features and factors in this study (Sakly et al., 2023). Valvulopathies to cardiac abnormalities are linked to the Global Systolic Fraction of Ejection (GSFE), which strongly correlates with cardiopathy in the first mode. The second model found no statistically significant valvular abnormality risk variables. These results emphasize the relevance of GSFE as a marker of cardiac anomalies and the necessity for additional valvulopathy risk factor studies. Ordinal logistic regression (OLR) works well in classification and regression for depression prediction using behavioral markers. They provide a cutting-edge ordinal regression model selection tool called a greedy-based algorithm for partial proportional odds model selection (GREP) (Jayawardena et al., 2020). Root-mean-square deviation and Spearman's correlation show that OLR models with GREP efficiently may enhance speech processing and emotional computing depression prediction.

A logistic regression method with a cardinality limit was proposed by (Adeli et al., 2019) for choosing features and samples for vision-based applications. The technique addresses unstable data, worthless features, and uneven class distribution. The model provides a precise mathematical answer for enhancing logistic regression in visual identification. Low-rank Matrix approximation for efficient kernel Logistic regression (KLR) (Lei et al., 2019) was indicated in this research. The kernel matrix's computational complexity and logistic regression's limitations with large-scale data are examined. The system uses gradient descent, low-rank approximation, modified SMO, and Newton iteration. According to the findings of different experiments, KLR performs very well on state-of-the-art approaches. The proposed method speeds up KLR, improves classification accuracy and reduces kernel matrix processing and storage costs. To classify Alzheimer's disease, a novel multi-modal classification technique using the Logistic Regression classifier was proposed (Sheng et al., 2022). As a result of distinguishing between different stages of the disease and healthy patients, it has gained remarkable accuracy. The proposed method is very workable in the diagnostic field and provides valuable details on how Alzheimer's disease may occur in the human body. An approach with the help of patient blood parameters to determine whether the Hepatitis C Virus (HCV) is available in the human body or not (Li et al., 2022) was proposed. The Synthetic minority oversampling technique (SMOTE) is used here to select the imbalanced data, and the author uses the Artificial Bee Colony algorithm here to improve the two-stage technique. It converges the Random Forest and Logistic Regression classifier and builds an AI-based classifier model.

2.5 Naive Bayes

If anyone needs to work with classification problems, the Naive Bayes classifier is a fundamental but influential and worthwhile probabilistic machine learning technique. The naive Bayes classifier uses the simplest Bayesian network models; however, the assessment of the kernel density makes the classifier more realistic and accurate. However, the classifier assumes no feature interactions, and at the same time, the classifier can assume feature values independent of the other features. The real-world data are not readily usable all the time because of their diversity, and for that, they can violate the assumption. However, it is still a suitable classifier because it is straightforward and more efficient because of its simplicity.

Facebook pages influence people's opinions of firms and investors and their support. Thai-language online chats about startup agritech businesses are sentiment-analyzed using a naive Bayes classifier in the proposed method (Kewsuwun & Kajornkasirat, 2022). In order to ascertain Thai's opinions on agritech companies, the poll classifies responses as favorable or unfavorable. The model may be used to study consumer behavior, marketing, and information transmission for its adaptability. A privacy-preserving semi-fully distributed data paradigm was proposed using the Naive Bayes classification (Vu, 2022a) algorithm. Besides protecting data privacy and classification, the proposed safe multi-party computing solution allows machine learning models and shared knowledge discovery. The research shows it works effectively in actual world circumstances, which is a huge step towards tackling this data architecture's privacy preservation issues. Naive Bayes classifiers can perform regarding the climate as well. Covariance-based and Convergent cross-mapping-based weighted Naive Bayes classifiers for the air quality prediction in three Indian cities were recommended by (Sethi & Mittal, 2022). Both classifiers outperform Naive Bayes, SVMs, and neural networks in accuracy, precision, recall, error rate, and F1 score.

Naive Bayes classifications privacy protection is addressed by (Vu et al., 2022a) in this paper. This method uses secure multi-sum computation cryptography to classify accurate results while preserving sensitive data. It performs better than the privacy-preserving Naive Bayes classifier in experiments, proving its practicality. A parametric adjustable quantile function calculation is presented in this research. When assumptions are correct, the system adapts to varied distributions, including a novel Naive Bayes classifier technique (Redivo et al., 2023), and theoretical studies and computer simulations evaluate these strategies using real-world data. A Naive Bayesian classification-based signaling analysis technique is proposed to introduce the efficiency decrease caused by increasing data volume in 5G mobile networks (Wang et al., 2023a). Integrating signal analysis and machine learning improves signal classification in the proposed strategy. Important messages are user properties, and a supervised signaling feature classification model minutely categorizes them. The technique uses orthogonal Matrix transformation to strengthen attribute independence and address zero probability. Regularised Naive Bayes (RNB) was suggested by (Wang et al., 2020) to enhance the generalization of class-specific attributes weighted Naive Bayes (CAWNM). RNB improves the performance of the machine learning benchmark dataset by balancing generalizability and discriminative power with a regularization term. RNB outperforms Naive Bayes algorithms in producing the covariance matrix. A Naive Bayes under-sampling method was proposed to handle class imbalance in classification issues (Aridas et al., 2019). Primary tests and statistical testing suggest the strategy beats existing resampling strategies, and the mechanism reduces majority class bias and improves classification on imbalanced datasets. A novel heterogeneous network data categorization method was proposed by (Dong et al., 2019) where multinomial naive Bayes- classification integrates data about known and unknown neighbors. The analysis indicates that the proposed method outperforms traditional classifiers in Heterophilous networks. An android-based software was being developed to help Indonesian general practitioners diagnose heart disease issues (Madona et al., 2021). A portable ECG recording device is used to collect the data from the human body. With the device as a classification algorithm, the Naïve Bayes machine learning is used, and the algorithm gains

75% accuracy in the process of ECG signal evaluation. As the rate of heart and blood vessel disorders in that specific region is very high, further improvement of these techniques is badly needed, though very promising. The significance of skin in the human body is of great importance as it is the largest organ of the human body. Effective treatment is needed; machine learning algorithms can play a vital role in differentiating and classifying different skin types. A dynamic graph cut algorithm alongside the Naïve Bayes classifier can obtain better results regarding the traditional method and show the proposed method's effectiveness (Balaji et al., 2020).

2.6 Random Forest

The Random Forest algorithm is one of the most popular and powerful machine learning algorithms in classification and regression problems. Several decision trees work together to build an ensemble technique, which produces predictions in this method. If the majority of the trees from a class is chosen in a classification problem, then that is the output of the random forest algorithm. In the beginning, a random subset is produced using training data via the random forest method, and for building a decision tree classifier, these subsets are used many more times. Several decision trees are produced, and these combinations are used to get the final forecast.

Deep forests, random forests, and global pruning are used to predict vegetation indices from RGB photos in precision farming (Devyatkin, 2023), and the author's proposed research advises the concept. By exceeding them, it overcomes neural network-based image categorization and Normalized Difference Vegetation Index prediction weaknesses (NDVI). This method's competitive performance on small and medium-sized data sets makes vegetation index estimation more straightforward and accurate. A NearMiss-based under-sampling for unbalanced datasets and Random Forest classification (NM- RF) was proposed, a new protein interaction site prediction method (Li et al., 2021). The issue of categorizing residues with and without interfaces is introduced. The data set is balanced using NearMiss undersampling and Random forest binary classification, and NM- RF may anticipate protein interaction locations by correcting the class imbalance. A hybrid multi-label Random Forest (HML- RF) and multi-label classification approach was introduced (Jain et al., 2022). Multi-label decision trees underpin the HML- RF problem transformation and algorithm adaptation technique. Testing and statistical validation demonstrate its state-of-the-art level dependency methods on ten datasets. The suggested HML-RF and HML-RFws work well on existing approaches in accuracy hamming loss and zero-one loss on various datasets. Creating and deploying an integrated edge device for human sensing in an Indoor innovative environment was discussed by (Kianoush et al., 2023), where he employs the cumulative sum (CUSUM) method to integrate sensor data with the random forest algorithm to classify body movements, including ultrasonic (US) and infrared (IR).

The author addresses ensemble and multi-kernel learning for the automatic ECG heartbeat categorization method. Mixed-kernel-based extreme learning machines and random forest classifiers enhance accuracy and sensitivity over previous techniques proposed (Yang et al., 2021). An MIT-BIH-AR database study shows average positive predictive values of 93.9% and overall accuracy of 98.1%, exceeding past research by 4% and 6%. Random decision forest classifiers were used to distinguish polymer types and assess the abundance and size distribution (Hufnagl et al., 2019). According to the research, this model may solve the microplastic analysis problem, and label noise and the dimensionality curse are also highlighted. The radiomics classification problems focus on high-dimensional, small-sample-size, and multi-view issues, and the researchers suggest merging data from diverse perspectives using Random

Forest classifiers and dissimilarity (Cao et al., 2019). Their findings outperform multi-view learning and Radiomic's latest version in classification accuracy across many data sets. According to one study, Radiomics may help diagnose and cure sickness. Ensemble-margin-based random forests (EMRFs) improve classification performance in class imbalance and duplicate data situations (Feng et al., 2019). A unique sub-sampling approach using estimated ensemble margin values is employed in EMRFs. The method is compared to SVM, AdaBoost, and Random Forests, and EMRFs demonstrate their real-world value on balanced and unbalanced data sets. Class noise and imbalance are minimized using EMRFs to improve classification accuracy, which is done using ensemble margin theory. Guided regularized Random Forest (GRRF) feature selection was recommended by (Izquierdo-Verdiguier & Zurita-Milla, 2020) for processing massive Earth observation data. There is no need for a preset number of attributes or Guided Regularized Random Forest (GRRF) threshold. GRRF showed that Remote Sensing could be performed excellently of traditional models despite the enormous GEO database. In the working process of the heart disease field, one of the most critical and significant issues is to predict heart disease to tackle death as it is one of the major causes of high mortality rate. A recursion-enhanced random forest with an improved linear model (RFRF-ILM), a clinical data analysis technique, was proposed by (Guo et al., 2020) and is a deep learning-based model that generates remarkable accuracy. Breast cancer diagnosis is very familiar nowadays, and for breast cancer diagnosis improvement, a novel Hierarchical Clustering Random Forest (HCRF) technique was proposed (Huang & Chen, 2021). Representative decision trees with higher accuracy will be selected by the HCRF classifier, which is one of the most significant issues in traditional Random Forests. The accuracy of 97.05% and 97.76% is exceptional, evaluated using the Wisconsin Diagnosis Breast Cancer (WDBC) and Wisconsin Breast Cancer (WBC) databases, and the results show that HCRF outperforms traditional algorithms like the Decision Tree, Adaboost, and Random Forest.

2.7 Support Vector Machines

Support Vector machine is a supervised machine learning technique often used to solve complex problems because of its robustness. The ultimate goal of the support factor machine is to find the hyperplane that can separate two or multiple classes of data points. The hyperplane is nothing but a state line that can divide the feature space into two or multiple regions where all the points in one region entirely belong to one separate class, and all the other points in another region are completely from another class. Maximizing the margin between two or multiple classes is vital for the SVM algorithm because it is used to find the best hyperplane. The margin is the distance between the closest point of the class and the hyperplane.

Twin-bounded weighted relaxed support vector machines (WRSVMs) may solve classification challenges caused by imbalanced data and outliers proposed by (Alamdar et al., 2019). The approach combined twin-bounded SVMs' fast classification performance with WRSVM's outlier robustness. Different experiments on benchmark data sets show that the recommended approach is faster and more accurate than SVM-based techniques, as the skewed data and outliers in classification tasks are handled quickly and easily. Two new class imbalance learning (CIL) methodologies with fuzzy support vector machine FSVM algorithms (Yu et al., 2019) were integrated. The suggested techniques outperform well-known tactics on various datasets, especially those with complicated distributions. A Proximal Support Vector Machine (PSVM) approach to solving the Multiple Instance Learning issue was proposed by (Avolio & Fuduli, 2020). This strategy aims to construct a new instance-space model that combines SVM accuracy with PSVM computational efficiency. The model creates a hyperplane between two parallel onse to identify positive and negative bags. Two multiview SVMs for multiview learning, which can classify two-view and multiview problems, were introduced by (Xie & Sun, 2019). Experiments in different areas reveal that the recommended strategies outperform SVM-based methods in generalization.

The researcher recommends gene selection in Cancer classification utilizing microarray data using the BMO-SVM technique, which combines the Barnacles Mating Optimizer (BMO) and Support Vector Machines (Houssein et al., 2021). Since BMO-SVM has more excellent information on the superiority percentage that finds cancer-detecting genes better, this technique has been proven to discriminate normal from malignant tissues and different cancers. A framework for classifying spindle-shaped cell (SSC) components in synovial sarcoma (SS) malignancy images was proposed by (Arunachalam et al., 2022). With Discrete Wavelet Transform (DWT), the Framework can transform the Synovial Sarcoma (SS) cancer images into frequencies. Because of the adjustment of the hyperplane values in a systematic way, the Blockchain, IoT, and QDA plus SVM enhance SS cancer outcomes—reduced attributes increase categorization accuracy. A novel filtering algorithm that improves the SVM-based software defect prediction (SDP) model was introduced (Goyal, 2022). SDP is critical for high-quality software. The FILTER model tackles data sets with unequal counts of defective and non-faulty modules. The outcome proposed that FILTER improves SVM-based SDP models, resulting in more accurate error-prone module predictions. Data augmentation increased categorization accuracy, and the CNN models were improved using SVMs or support vector machines (Sonmez et al., 2022). SVM improved the model with the lowest performance from 98% to 99.66%, while CNN models achieved the highest accuracy of 99.66%. According to the database, support vector machines and deep learning can classify microalgae.

The researcher proposed the EMSVM paradigm, an enhanced multiclass support vector machine for digital forensics concentrating on file system data analysis and connection-solved cybercrime problems (Mohammad, 2022). EMSVM enhances classification performance by allowing multiclass classification and employing a unique perimeter selection mechanism. Unlike earlier machine learning methods, the study proposes that the model may be beneficial for assessing damning Digital information. An SVM-based photovoltaic PV array failure detection system was proposed by (Wang et al., 2023). The software may discover open circuits, short circuits, and irradiation issues by analyzing I-V characteristic curves. SVM input parameters include short circuit current, open circuit voltage, maximum power current, and maximum power voltage. The cloud-based medical computing system is not safe and secure all the time as most of the operation is now being operated through clouds and, for that, to protect the patient's private data, a multiclass-based support vector machine system for the medical diagnostic was proposed in this paper (Chen et al., 2022). To make the diagnosis accurate and, at the same time, to protect the privacy of the data, modern encryption techniques are being used here where effectiveness and security are the main highlighted parts of the proposed method in remote medical diagnostics. Early diagnosis of Parkinson's disease (PD) was focused on by (Haq et al., 2019), and to do the task, the author used a machine learning-based prediction technique. Support Vector Machine (SVM) and L1-norm feature selection techniques were used here. As a result, the author showed good sensitivity and satisfactory accuracy in identifying Parkinson's disease, and for accurate diagnosis, the author used voice recordings for selection and classification.

3. CONCLUSION

The article has shown the importance of data and how to apply it to challenging real-world problems in different fields. Massive data must be generated, and because of its increasing complexity, the data needs to be effectively translated using several machine-learning techniques. Like the others, this massive data's usability and importance are irreplaceable in medical science and healthcare. Classification algorithms have distinguished themselves as indispensable tools for solving many challenges within the wide range of available machine-learning algorithms. They provide a variety of advantages, such as the ability to achieve desired results with less complexity, durability, and far fewer issues with time and space requirements, which play the most critical role in diagnosing a particular disease and in the well-being of a patient. Comparing the outcomes of this widely used multipurpose algorithm made it possible to identify which algorithm performed best overall in pinpointing healthcare issues. At the same time, the findings of this study will also provide vital information on the suitability and effectiveness of different classification algorithms for dealing with real-world problems in the medical field. It is important to remember that how effectively machine learning algorithms work depends much on the kind and quality of the data utilized for a specific problem. Considerable attention should go into data collection, preprocessing, and feature engineering to attain the most significant algorithm performance. The importance of data in diverse fields, particularly in healthcare, and the role of classification algorithms in resolving challenging issues are highlighted in this paper.

ACKNOWLEDGMENT

I sincerely thank Dr. [UMESH KUMAR LILHORE], my adviser, for all his help and advice on this task. His knowledge and advice benefited me as I developed my study and wrote my article. I also want to thank Chandigarh University for letting me collaborate with Dr. UMESH. Finally, I appreciate the opportunity to communicate my study findings to a larger audience.

REFERENCES

Abbasniya, M. R., Sheikholeslamzadeh, S. A., Nasiri, H., & Emami, S. (2022). Classification of breast tumors based on histopathology images using deep features and ensemble of gradient boosting methods. *Computers & Electrical Engineering*, *103*, 108382. doi:10.1016/j.compeleceng.2022.108382

Adeli, E., Li, X., Kwon, D., Zhang, Y., & Pohl, K. M. (2019). Logistic regression confined by cardinality-constrained sample and feature selection. *IEEE Transactions on Pattern Analysis and Machine Intelligence*, *42*(7), 1713–1728. doi:10.1109/TPAMI.2019.2901688 PMID:30835210

Alamdar, F., Mohammadi, F. S., & Amiri, A. (2019). Twin bounded weighted relaxed support vector machines. *IEEE Access : Practical Innovations, Open Solutions*, *7*, 22260–22275. doi:10.1109/ACCESS.2019.2897891

Alex, S., Dhanaraj, K. J., & Deepthi, P. P. (2022). Private and energy-efficient decision tree-based disease detection for resource-constrained medical users in mobile healthcare network. *IEEE Access : Practical Innovations, Open Solutions*, *10*, 17098–17112. doi:10.1109/ACCESS.2022.3149771

Ameur, Y., Aziz, R., Audigier, V., & Bouzefrane, S. (2022, September). Secure and non-interactive k-NN classifier using symmetric fully homomorphic encryption. In *International Conference on Privacy in Statistical Databases* (pp. 142-154). Cham: Springer International Publishing. 10.1007/978-3-031-13945-1_11

Aridas, C. K., Karlos, S., Kanas, V. G., Fazakis, N., & Kotsiantis, S. B. (2019). Uncertainty based under-sampling for learning naive bayes classifiers under imbalanced data sets. *IEEE Access : Practical Innovations, Open Solutions*, *8*, 2122–2133. doi:10.1109/ACCESS.2019.2961784

Arunachalam, P., Janakiraman, N., Sivaraman, A. K., Balasundaram, A., Vincent, R., Rani, S., ... Rajesh, M. (2022). Synovial Sarcoma Classification Technique Using Support Vector Machine and Structure Features. *Intelligent Automation & Soft Computing*, *32*(2), 1241–1259. doi:10.32604/iasc.2022.022573

Avolio, M., & Fuduli, A. (2020). A semiproximal support vector machine approach for binary multiple instance learning. *IEEE Transactions on Neural Networks and Learning Systems*, *32*(8), 3566–3577. doi:10.1109/TNNLS.2020.3015442 PMID:32822307

Balaji, V. R., Suganthi, S. T., Rajadevi, R., Kumar, V. K., Balaji, B. S., & Pandiyan, S. (2020). Skin disease detection and segmentation using dynamic graph cut algorithm and classification through Naive Bayes classifier. *Measurement*, *163*, 107922. doi:10.1016/j.measurement.2020.107922

Bari Antor, M., Jamil, A. H. M., Mamtaz, M., Monirujjaman Khan, M., Aljahdali, S., Kaur, M., Singh, P., & Masud, M. (2021). A comparative analysis of machine learning algorithms to predict alzheimer's disease. *Journal of Healthcare Engineering*, *2021*, 2021. doi:10.1155/2021/9917919 PMID:34336171

Cao, H., Bernard, S., Sabourin, R., & Heutte, L. (2019). Random forest dissimilarity based multi-view learning for radiomics application. *Pattern Recognition*, *88*, 185–197. doi:10.1016/j.patcog.2018.11.011

Chen, F., & Gadekallu, T. R. (2023). Multi-dimensional data optimal classification algorithm for quality evaluation of distance teaching in universities. *Mobile Networks and Applications*, 1–11. doi:10.1007/s11036-023-02186-8

Chen, Y., Mao, Q., Wang, B., Duan, P., Zhang, B., & Hong, Z. (2022). Privacy-preserving multi-class support vector machine model on medical diagnosis. *IEEE Journal of Biomedical and Health Informatics*, *26*(7), 3342–3353. doi:10.1109/JBHI.2022.3157592 PMID:35259122

Chen, Y. C., Chang, C. C., Hung, C. C., Lin, J. F., & Hsu, S. Y. (2022). SecDT: Privacy-Preserving Outsourced Decision Tree Classification Without Polynomial Forms in Edge-Cloud Computing. *IEEE Transactions on Signal and Information Processing Over Networks*, *8*, 1037–1048. doi:10.1109/TSIPN.2022.3233193

Devyatkin, D. A. (2023). Estimation of vegetation indices with Random Kernel Forests. *IEEE Access : Practical Innovations, Open Solutions*, *11*, 29500–29509. doi:10.1109/ACCESS.2023.3261129

Dong, S., Liu, D., Ouyang, R., Zhu, Y., Li, L., Li, T., & Liu, J. (2019). Second-order markov assumption based bayes classifier for networked data with heterophily. *IEEE Access : Practical Innovations, Open Solutions*, *7*, 34153–34161. doi:10.1109/ACCESS.2019.2892757

Feng, W., Dauphin, G., Huang, W., Quan, Y., & Liao, W. (2019). New margin-based subsampling iterative technique in modified random forests for classification. *Knowledge-Based Systems*, *182*, 104845. doi:10.1016/j.knosys.2019.07.016

Fu, Y., Yin, Z., Su, M., Wu, Y., & Liu, G. (2020). Construction and reasoning approach of belief rule-base for classification base on decision tree. *IEEE Access : Practical Innovations, Open Solutions*, *8*, 138046–138057. doi:10.1109/ACCESS.2020.3012453

Goyal, S. (2022). Effective software defect prediction using support vector machines (SVMs). *International Journal of System Assurance Engineering and Management*, *13*(2), 681–696. doi:10.1007/s13198-021-01326-1

Guo, C., Zhang, J., Liu, Y., Xie, Y., Han, Z., & Yu, J. (2020). Recursion enhanced random forest with an improved linear model (RERF-ILM) for heart disease detection on the internet of medical things platform. *IEEE Access : Practical Innovations, Open Solutions*, *8*, 59247–59256. doi:10.1109/ACCESS.2020.2981159

Haq, A. U., Li, J. P., Memon, M. H., Malik, A., Ahmad, T., Ali, A., ... Shahid, M. (2019). Feature selection based on L1-norm support vector machine and effective recognition system for Parkinson's disease using voice recordings. *IEEE Access : Practical Innovations, Open Solutions*, *7*, 37718–37734. doi:10.1109/ACCESS.2019.2906350

Hidayat, T. H. J., Ruldeviyani, Y., Aditama, A. R., Madya, G. R., Nugraha, A. W., & Adisaputra, M. W. (2022). Sentiment analysis of twitter data related to Rinca Island development using Doc2Vec and SVM and logistic regression as classifier. *Procedia Computer Science*, *197*, 660–667. doi:10.1016/j.procs.2021.12.187

Houssein, E. H., Abdelminaam, D. S., Hassan, H. N., Al-Sayed, M. M., & Nabil, E. (2021). A hybrid barnacles mating optimizer algorithm with support vector machines for gene selection of microarray cancer classification. *IEEE Access : Practical Innovations, Open Solutions*, *9*, 64895–64905. doi:10.1109/ACCESS.2021.3075942

Huang, A., Xu, R., Chen, Y., & Guo, M. (2023). Research on multi-label user classification of social media based on ML-KNN algorithm. *Technological Forecasting and Social Change*, *188*, 122271. doi:10.1016/j.techfore.2022.122271

Huang, Z., & Chen, D. (2021). A breast cancer diagnosis method based on VIM feature selection and hierarchical clustering random forest algorithm. *IEEE Access : Practical Innovations, Open Solutions*, *10*, 3284–3293. doi:10.1109/ACCESS.2021.3139595

Hufnagl, B., Steiner, D., Renner, E., Löder, M. G., Laforsch, C., & Lohninger, H. (2019). A methodology for the fast identification and monitoring of microplastics in environmental samples using random decision forest classifiers. *Analytical Methods*, *11*(17), 2277–2285. doi:10.1039/C9AY00252A

Izquierdo-Verdiguier, E., & Zurita-Milla, R. (2020). An evaluation of Guided Regularized Random Forest for classification and regression tasks in remote sensing. *International Journal of Applied Earth Observation and Geoinformation*, *88*, 102051. doi:10.1016/j.jag.2020.102051

Jain, V., Phophalia, A., & Mitra, S. K. (2022). HML-RF: Hybrid multi-label random forest. *IEEE Access : Practical Innovations, Open Solutions*, *10*, 22902–22914. doi:10.1109/ACCESS.2022.3154420

Jayawardena, S., Epps, J., & Ambikairajah, E. (2020). Ordinal logistic regression with partial proportional odds for depression prediction. *IEEE Transactions on Affective Computing*.

Jiang, T., Gradus, J. L., & Rosellini, A. J. (2020). Supervised machine learning: A brief primer. *Behavior Therapy*, *51*(5), 675–687. doi:10.1016/j.beth.2020.05.002 PMID:32800297

Jiao, L., Geng, X., & Pan, Q. (2019). BP $ k $ NN: $ k $-nearest neighbor classifier with pairwise distance metrics and belief function theory. *IEEE Access : Practical Innovations, Open Solutions*, *7*, 48935–48947. doi:10.1109/ACCESS.2019.2909752

Kanimozhi, P., & Aruldoss Albert Victoire, T. (2022). Oppositional tunicate fuzzy C-means algorithm and logistic regression for intrusion detection on cloud. *Concurrency and Computation*, *34*(4), e6624. doi:10.1002/cpe.6624

Kewsuwun, N., & Kajornkasirat, S. (2022). A sentiment analysis model of agritech startup on Facebook comments using naive Bayes classifier. *International Journal of Electrical & Computer Engineering (2088-8708), 12*(3).

Khan, M. S. I., Islam, N., Uddin, J., Islam, S., & Nasir, M. K. (2022). Water quality prediction and classification based on principal component regression and gradient boosting classifier approach. *Journal of King Saud University. Computer and Information Sciences*, *34*(8), 4773–4781. doi:10.1016/j.jksuci.2021.06.003

Kianoush, S., Savazzi, S., Rampa, V., Costa, L., & Tolochenko, D. (2023). A Random Forest Approach to Body Motion Detection: Multisensory Fusion and Edge Processing. *IEEE Sensors Journal*, *23*(4), 3801–3814. doi:10.1109/JSEN.2022.3232085

Kumari, L. V., & Sai, Y. P. (2022). Classification of ECG beats using optimized decision tree and adaptive boosted optimized decision tree. *Signal, Image and Video Processing*, *16*(3), 695–703. doi:10.1007/s11760-021-02009-x

Lei, D., Tang, J., Li, Z., & Wu, Y. (2019). Using low-rank approximations to speed up kernel logistic regression algorithm. *IEEE Access : Practical Innovations, Open Solutions*, *7*, 84242–84252. doi:10.1109/ACCESS.2019.2924542

Li, L., & Liu, Z. P. (2022). A connected network-regularized logistic regression model for feature selection. *Applied Intelligence*, *52*(10), 11672–11702. doi:10.1007/s10489-021-02877-3

Li, M., Vanberkel, P., & Zhong, X. (2022). Predicting ambulance offload delay using a hybrid decision tree model. *Socio-Economic Planning Sciences*, *80*, 101146. doi:10.1016/j.seps.2021.101146

Li, M., Wu, Z., Wang, W., Lu, K., Zhang, J., Zhou, Y., Chen, Z., Li, D., Zheng, S., Chen, P., & Wang, B. (2021). Protein-protein interaction sites prediction based on an under-sampling strategy and random forest algorithm. *IEEE/ACM Transactions on Computational Biology and Bioinformatics*, *19*(6), 3646–3654. doi:10.1109/TCBB.2021.3123269 PMID:34705656

Li, T. H. S., Chiu, H. J., & Kuo, P. H. (2022). Hepatitis C Virus Detection Model by Using Random Forest, Logistic-Regression and ABC Algorithm. *IEEE Access : Practical Innovations, Open Solutions*, *10*, 91045–91058. doi:10.1109/ACCESS.2022.3202295

Lilhore, U. K., Dalal, S., Faujdar, N., Margala, M., Chakrabarti, P., Chakrabarti, T., Simaiya, S., Kumar, P., Thangaraju, P., & Velmurugan, H. (2023). Hybrid CNN-LSTM model with efficient hyperparameter tuning for prediction of Parkinson's disease. *Scientific Reports*, *13*(1), 14605. doi:10.1038/s41598-023-41314-y PMID:37669970

Lilhore, U. K., Dalal, S., Faujdar, N., Simaiya, S., Dahiya, M., Tomar, S., & Hashmi, A. (2024). Unveiling the prevalence and risk factors of early stage postpartum depression: A hybrid deep learning approach. *Multimedia Tools and Applications*, 1–35. doi:10.1007/s11042-024-18182-3

Lilhore, U. K., Manoharan, P., Sandhu, J. K., Simaiya, S., Dalal, S., Baqasah, A. M., Alsafyani, M., Alroobaea, R., Keshta, I., & Raahemifar, K. (2023). Hybrid model for precise hepatitis-C classification using improved random forest and SVM method. *Scientific Reports*, *13*(1), 12473. doi:10.1038/s41598-023-36605-3 PMID:37528148

Lilhore, U. K., Poongodi, M., Kaur, A., Simaiya, S., Algarni, A. D., Elmannai, H., Vijayakumar, V., Tunze, G. B., & Hamdi, M. (2022). Hybrid model for detection of cervical cancer using causal analysis and machine learning techniques. *Computational and Mathematical Methods in Medicine*, *2022*, 2022. doi:10.1155/2022/4688327 PMID:35572826

Lilhore, U. K., Simaiya, S., Pandey, H., Gautam, V., Garg, A., & Ghosh, P. (2022). Breast cancer detection in the IoT cloud-based healthcare environment using fuzzy cluster segmentation and SVM classifier. In Ambient Communications and Computer Systems [Singapore: Springer Nature Singapore.]. *Proceedings of RACCCS*, *2021*, 165–179.

Lilhore, U. K., Simaiya, S., Prasad, D., & Guleria, K. (2020). A hybrid tumor detection and classification based on machine learning. *Journal of Computational and Theoretical Nanoscience*, *17*(6), 2539–2544. doi:10.1166/jctn.2020.8927

Lin, A., Wu, Q., Heidari, A. A., Xu, Y., Chen, H., Geng, W., Li, C., & Li, C. (2019). Predicting intentions of students for master programs using a chaos-induced sine cosine-based fuzzy K-nearest neighbor classifier. *IEEE Access : Practical Innovations, Open Solutions*, *7*, 67235–67248. doi:10.1109/ACCESS.2019.2918026

Lobe, B., Morgan, D., & Hoffman, K. A. (2020). Qualitative data collection in an era of social distancing. *International Journal of Qualitative Methods*, *19*, 1609406920937875. doi:10.1177/1609406920937875

Louk, M. H. L., & Tama, B. A. (2023). Dual-IDS: A bagging-based gradient boosting decision tree model for network anomaly intrusion detection system. *Expert Systems with Applications*, *213*, 119030. doi:10.1016/j.eswa.2022.119030

Lu, D., Yue, Y., Hu, Z., Xu, M., Tong, Y., & Ma, H. (2023). Effective detection of Alzheimer's disease by optimizing fuzzy K-nearest neighbors based on salp swarm algorithm. *Computers in Biology and Medicine*, *159*, 106930. doi:10.1016/j.compbiomed.2023.106930 PMID:37087779

Madona, P., Zafitrah, Y., Sari, J. N., Zain, M. M., & Valzon, M. (2021, November). Classification of ECG Signals Using the Naïve Bayes Classification Method and Its Implementation in Android-Based Smart Health Care. In *2021 International Conference on Computer Science and Engineering (IC2SE)* (Vol. 1, pp. 1-7). IEEE. 10.1109/IC2SE52832.2021.9791475

Mahapatra, S., Gupta, V. R., Sahu, S. S., & Panda, G. (2021). Deep neural network and extreme gradient boosting based Hybrid classifier for improved prediction of Protein-Protein interaction. *IEEE/ACM Transactions on Computational Biology and Bioinformatics*, *19*(1), 155–165. doi:10.1109/TCBB.2021.3061300 PMID:33621179

Memiş, S., Enginoğlu, S., & Erkan, U. (2022). Fuzzy parameterized fuzzy soft k-nearest neighbor classifier. *Neurocomputing*, *500*, 351–378. doi:10.1016/j.neucom.2022.05.041

Mewada, A., & Dewang, R. K. (2023). SA-ASBA: A hybrid model for aspect-based sentiment analysis using synthetic attention in pre-trained language BERT model with extreme gradient boosting. *The Journal of Supercomputing*, *79*(5), 5516–5551. doi:10.1007/s11227-022-04881-x

Miranda-Vega, J. E., Rivas-López, M., & Fuentes, W. F. (2020). K-nearest neighbor classification for pattern recognition of a reference source light for machine vision system. *IEEE Sensors Journal*, *21*(10), 11514–11521. doi:10.1109/JSEN.2020.3024094

Mohammad, R. M. A. (2022). An enhanced multiclass support vector machine model and its application to classifying file systems affected by a digital crime. *Journal of King Saud University. Computer and Information Sciences*, *34*(2), 179–190. doi:10.1016/j.jksuci.2019.10.010

Ouadah, A., Zemmouchi-Ghomari, L., & Salhi, N. (2022). Selecting an appropriate supervised machine learning algorithm for predictive maintenance. *International Journal of Advanced Manufacturing Technology*, *119*(7-8), 4277–4301. doi:10.1007/s00170-021-08551-9

Pallathadka, H., Wenda, A., Ramirez-Asís, E., Asís-López, M., Flores-Albornoz, J., & Phasinam, K. (2023). Classification and prediction of student performance data using various machine learning algorithms. *Materials Today: Proceedings*, *80*, 3782–3785. doi:10.1016/j.matpr.2021.07.382

Pidstrigach, J., & Reich, S. (2023). Affine-invariant ensemble transform methods for logistic regression. *Foundations of Computational Mathematics*, *23*(2), 675–708. doi:10.1007/s10208-022-09550-2

Priya, A., Garg, S., & Tigga, N. P. (2020). Predicting anxiety, depression and stress in modern life using machine learning algorithms. *Procedia Computer Science*, *167*, 1258–1267. doi:10.1016/j.procs.2020.03.442

Rabcan, J., Levashenko, V., Zaitseva, E., Kvassay, M., & Subbotin, S. (2019). Application of fuzzy decision tree for signal classification. *IEEE Transactions on Industrial Informatics*, *15*(10), 5425–5434. doi:10.1109/TII.2019.2904845

Redivo, E., Viroli, C., & Farcomeni, A. (2023). Quantile-distribution functions and their use for classification, with application to naïve Bayes classifiers. *Statistics and Computing*, *33*(2), 55. doi:10.1007/s11222-023-10224-4

Saha, S., Sarker, P. S., Al Saud, A., Shatabda, S., & Newton, M. H. (2022). Cluster-oriented instance selection for classification problems. *Information Sciences*, *602*, 143–158. doi:10.1016/j.ins.2022.04.036

Sakly, H., Said, M., Jennane, R., & Tagina, M. (2023). Epidemiological Study of Cardiopathies and Valvulopathies using Binary Logistic Regression. *Journal of Data. Information & Management*, 1–13.

Sanyal, R., Kar, D., & Sarkar, R. (2021). Carcinoma type classification from high-resolution breast microscopy images using a hybrid ensemble of deep convolutional features and gradient boosting trees classifiers. *IEEE/ACM Transactions on Computational Biology and Bioinformatics*, *19*(4), 2124–2136. doi:10.1109/TCBB.2021.3071022 PMID:33819160

Sarker, I. H. (2021). Machine learning: Algorithms, real-world applications and research directions. *SN Computer Science*, *2*(3), 160. doi:10.1007/s42979-021-00592-x PMID:33778771

Seeja, G., Selvakumar, A. A., & Hency, V. B. (2023). A Novel Approach for Disaster Victim Detection Under Debris Environments Using Decision Tree Algorithms with Deep Learning Features. *IEEE Access : Practical Innovations, Open Solutions*, *11*, 54760–54772. doi:10.1109/ACCESS.2023.3281461

Selvaratnam, B., Oliynyk, A. O., & Mar, A. (2023). Interpretable Machine Learning in Solid-State Chemistry, with Applications to Perovskites, Spinels, and Rare-Earth Intermetallics: Finding Descriptors Using Decision Trees. *Inorganic Chemistry*, *62*(28), 10865–10875. doi:10.1021/acs.inorgchem.3c01153 PMID:37390482

Sethi, J. K., & Mittal, M. (2022). Efficient weighted naive bayes classifiers to predict air quality index. *Earth Science Informatics*, *15*(1), 541–552. doi:10.1007/s12145-021-00755-7

Shah, D., Singh, A., & Prasad, S. S. (2022, November). Sentimental Analysis Using Supervised Learning Algorithms. In *2022 3rd International Conference on Computation, Automation and Knowledge Management (ICCAKM)* (pp. 1-6). IEEE. 10.1109/ICCAKM54721.2022.9990320

Sheng, J., Wu, S., Zhang, Q., Li, Z., & Huang, H. (2022). A Binary Classification Study of Alzheimer's Disease Based on a Novel Subclass Weighted Logistic Regression Method. *IEEE Access : Practical Innovations, Open Solutions*, *10*, 68846–68856. doi:10.1109/ACCESS.2022.3186888

Sonmez, M. E., Eczacıoglu, N., Gumuş, N. E., Aslan, M. F., Sabanci, K., & Aşikkutlu, B. (2022). Convolutional neural network-Support vector machine based approach for classification of cyanobacteria and chlorophyta microalgae groups. *Algal Research*, *61*, 102568. doi:10.1016/j.algal.2021.102568

Sunaryono, D., Sarno, R., & Siswantoro, J. (2022). Gradient boosting machines fusion for automatic epilepsy detection from EEG signals based on wavelet features. *Journal of King Saud University. Computer and Information Sciences*, *34*(10), 9591–9607. doi:10.1016/j.jksuci.2021.11.015

Szarmach, M., & Czarnowski, I. (2023, June). Decision Tree-Based Algorithms for Detection of Damage in AIS Data. In *International Conference on Computational Science* (pp. 17-32). Cham: Springer Nature Switzerland. 10.1007/978-3-031-36024-4_2

Uddin, S., Khan, A., Hossain, M. E., & Moni, M. A. (2019). Comparing different supervised machine learning algorithms for disease prediction. *BMC Medical Informatics and Decision Making*, *19*(1), 1–16. doi:10.1186/s12911-019-1004-8 PMID:31864346

Uguz, D. U., Berief, F., Leonhardt, S., & Antink, C. H. (2020, September). Classification of 12-lead ECGs using gradient boosting on features acquired with domain-specific and domain-agnostic methods. In 2020 Computing in Cardiology (pp. 1-4). IEEE.

Vu, D. H. (2022). Privacy-preserving Naive Bayes classification in semi-fully distributed data model. *Computers & Security*, *115*, 102630. doi:10.1016/j.cose.2022.102630

Vu, D. H., Vu, T. S., & Luong, T. D. (2022). An efficient and practical approach for privacy-preserving Naive Bayes classification. *Journal of information Security and Applications, 68*, 103215.

Wang, J., Gao, D., Zhu, S., Wang, S., & Liu, H. (2023). Fault diagnosis method of photovoltaic array based on support vector machine. *Energy sources, part a: recovery, utilization, and environmental effects, 45*(2), 5380-5395.

Wang, S., Ren, J., & Bai, R. (2020). A regularized attribute weighting framework for naive Bayes. *IEEE Access : Practical Innovations, Open Solutions*, *8*, 225639–225649. doi:10.1109/ACCESS.2020.3044946

Wang, W., Duan, Y., Cao, L., & Jiang, Z. (2023). Application of improved Naive Bayes classification algorithm in 5G signaling analysis. *The Journal of Supercomputing*, *79*(6), 6941–6964. doi:10.1007/s11227-022-04946-x

Wang, X., & Liu, F. (2020). Data-driven relay selection for physical-layer security: A decision tree approach. *IEEE Access : Practical Innovations, Open Solutions*, *8*, 12105–12116. doi:10.1109/ACCESS.2020.2965963

Wang, Y., Pan, Z., & Dong, J. (2022). A new two-layer nearest neighbor selection method for kNN classifier. *Knowledge-Based Systems*, *235*, 107604. doi:10.1016/j.knosys.2021.107604

Xia, Y. (2019). A novel reject inference model using outlier detection and gradient boosting technique in peer-to-peer lending. *IEEE Access : Practical Innovations, Open Solutions*, *7*, 92893–92907. doi:10.1109/ACCESS.2019.2927602

Xie, X., & Sun, S. (2019). Multi-view support vector machines with the consensus and complementarity information. *IEEE Transactions on Knowledge and Data Engineering*, *32*(12), 2401–2413. doi:10.1109/TKDE.2019.2933511

Yang, P., Wang, D., Zhao, W. B., Fu, L. H., Du, J. L., & Su, H. (2021). Ensemble of kernel extreme learning machine based random forest classifiers for automatic heartbeat classification. *Biomedical Signal Processing and Control*, *63*, 102138. doi:10.1016/j.bspc.2020.102138

Yu, H., Sun, C., Yang, X., Zheng, S., & Zou, H. (2019). Fuzzy support vector machine with relative density information for classifying imbalanced data. *IEEE Transactions on Fuzzy Systems*, *27*(12), 2353–2367. doi:10.1109/TFUZZ.2019.2898371

Zhang, M., Chen, Y., & Susilo, W. (2022). Decision tree evaluation on sensitive datasets for secure E-healthcare systems. *IEEE Transactions on Dependable and Secure Computing*.

Zhang, W., Chen, X., Liu, Y., & Xi, Q. (2020). A distributed storage and computation k-nearest neighbor algorithm based cloud-edge computing for cyber-physical-social systems. *IEEE Access : Practical Innovations, Open Solutions*, *8*, 50118–50130. doi:10.1109/ACCESS.2020.2974764

Zhao, H., Wang, R., Lei, Y., Liao, W. H., Cao, H., & Cao, J. (2022). Severity level diagnosis of Parkinson's disease by ensemble K-nearest neighbor under imbalanced data. *Expert Systems with Applications*, *189*, 116113. doi:10.1016/j.eswa.2021.116113

Zhou, J., Huang, S., Tao, M., Khandelwal, M., Dai, Y., & Zhao, M. (2023). Stability prediction of underground entry-type excavations based on particle swarm optimization and gradient boosting decision tree. *Underground Space*, *9*, 234–249. doi:10.1016/j.undsp.2022.08.002

Chapter 4
Ethical and Legal Considerations in Machine Learning:
Promoting Responsible Data Use in Bioinformatics

Deepika Amol Ajalkar
G.H. Raisoni College of Engineering and Management, India

Yogesh Kumar Sharma
https://orcid.org/0000-0003-1934-4535
Koneru Lakshmaiah Education Foundation, India

Jayashri Prashant Shinde
G.H. Raisoni College of Engineering and Management, India

Smitha Nayak
Muscat College, Oman

ABSTRACT

Artificial intelligence (AI) poses a number of moral and legal challenges to modern civilization. These include invasions of privacy, discrimination, and the function of human judgment. The use of more recent digital technologies has sparked worries that they could introduce new forms of error and data breaches. For patients who fall prey to healthcare technique or protocol errors, the repercussions may be catastrophic. Keep this in mind at all times; patients often interact with doctors at times when they are feeling their weakest. The potential ethical and legal concerns raised by the widespread use of AI in healthcare settings are not yet adequately addressed by existing legislation. All parties participating in the process should be protected, and there should be openness and privacy in algorithms; also, there should be cybersecurity measures in place to address any vulnerability that may arise.

DOI: 10.4018/979-8-3693-1822-5.ch004

1. INTRODUCTION

The capacity of a computer or piece of software to mimic intelligent human behaviour, do calculations instantly, solve problems, and evaluate newly acquired information based on formerly considered data is known as machine learning (Patel et al., 2023). Artificial intelligence (AI) has far-reaching effects on numerous sectors and professions, such as medical services, fashion, sports statistics and operations, autonomous cars, the production process, and agribusiness. This advancement in technology is both a blessing and a curse since it will shape both the industry and humankind in the years to come (Wang et al., 2023).

Imaging and EMR, laboratory diagnosis, treatment, new medication discovery, prevention and precision medicine, storage and access to data for health organizations, extensive biological data analysis, speeding up processes, and artificial intelligence applications in healthcare have revolutionized the medical field. The scientific community, however, confronts a number of moral and legal quandaries (Padhi et al., 2023). Not all civilizations have access to AI, despite the outstanding achievements it has made in the area and its importance in improving therapy. Access to cutting-edge technology is still a problem in many developing and low-income nations. Issues with privacy and data security, social gaps, medical consultation, empathy, compassion, and ethical difficulties are just a few of the many challenges we confront while using AI (Uwinama et al., 2023).

1.1 Problem Statement

While other nations, including the United States and Canada, revised their privacy laws, the European Union (EU) was the first to pass the "General Data Protection Regulation" (GDPR). Data processors or controllers headquartered in the European Union are required by these laws to adequately safeguard natural persons' information by processing all personal data as well as the actions of foreign nations and enterprises (Cacciamani et al., 2023).

To prevent employers in the US from making discriminatory judgments based on people's genetic health information, there is a group called 'GINA'. The actual function of AI in healthcare is to speed up health research, evaluate customer health data and pictures from medical devices, and ultimately enhance diagnosis and results (Brauneck et al., 2023). Another essential use of AI in the realm of social media is the dissemination of health news and medical advice, particularly during pandemics. Despite these apparent benefits of AI, there is still a significant issue about the security of patients' data when robots are involved:

- Existing healthcare legislation is inadequate to guarantee the security of patients' personal health information.
- Robots employed in healthcare data collection represent a threat to patient privacy and security due to the possibility of unauthorized access and use of obtained information.
- Without users' knowledge or agreement, specific social networks collect and retain vast quantities of data, including information about their mental health, which may aid in the advertising, marketing, and sales efforts of these firms.

Pharmacy and biotech corporations buy consumer data from specific unlicensed and poorly regulated genetic testing and bioinformatics businesses (Detopoulou et al., 2023).

1.2 Contribution

These days, data-driven algorithms are all the rage in industries with a lot on the line, including healthcare, thanks to AI. But deciphering the logic and operation of these intricate and mysterious algorithms is proving to be an ever-increasing challenge for humans. In order to avoid the misuse of AI which might have disastrous effects on humans—and to ensure that AI can assist crucial choices in these areas, some ethical concerns must be resolved. Nevertheless, recommendations that comprehensively tackle ethical circumstances surrounding the use of AI technology in healthcare have received little study attention.

Our goal in conducting this literature study was to provide a synopsis of the present state of knowledge about ethical considerations and associated approaches to the use of AI in healthcare. It aims to help them make informed choices, promote consciousness of ethics, and contribute to establishing solid ethical frameworks that align with the constantly evolving field of machine learning in biological informatics. The primary objective is to foster a collective dedication within the community towards responsible data practices, guaranteeing that ethical principles are essential to the ongoing advancement of bioinformatics investigation.

1.3 Organization of the Article

The complete chapter is divided into several subsections. Section 2 covers Bioinformatics Ethical Principles, Section 3 Legal Strategies Regulating Data Use, Section 4 Difficulties Raised by Machine Learning Strategies, Section 5 covers best practices and Case studies, and Section 6 covers conclusion and future works.

2. BIOINFORMATICS ETHICAL PRINCIPLES

The integration of AI into every aspect of healthcare is daunting, if not impossible. Humans and medical robots may take some time to grow together because of the unique qualities of human emotions. It is not feasible for autonomous (robotic) systems to seek or provide advice from other professionals in the medical field, who should interact with one another. Conversely, "machine-human" medical connections, as opposed to "human-human," are less likely to be accepted by patients (Aminizadeh et al., 2023).

The recovery process of patients is significantly impacted by the level of empathy and compassion shown by doctors and nurses during therapy. The use of robotic doctors and nurses will make this impossible. Because they lack human qualities like compassion, patients will see a decline in empathy, friendliness, and proper conduct while interacting with robotic doctors and nurses (Kaushal et al., 2023). Figure 1 presents the fundamental principles of bioinformatics and ML.

- **Equity and Elimination of Prejudice:**
  Principle: Aim for equity in the development and execution of programmes to prevent biases that could unfairly affect specific demographics (Patel et al., 2023).
  Justification: Unregulated computational biases have the potential to exacerbate disparities. Bioinformatics professionals must proactively recognise and address biases to achieve fair results in investigations and programmes.
- **Advancing scientific knowledge through transparent and cooperative research:**

Figure 1. Bioinformatics and ML ethical principles

Principle: Encourage adopting open science practices by disseminating methodologies, information sets, and outcomes. Promote cooperation and unrestricted sharing of information (Demir, E 2022).

Justification: Open science improves the accessibility and replicability of investigation, speeds scientific advancement, and aids in the collaborative endeavour to tackle intricate biological problems.

- **Ethical Dissemination and Transmission:**

Principle: Perform accountability in disseminating and transmitting studies in bioinformatics, guaranteeing precise depiction and contextualization of discoveries.

Justification: Distorting or exaggerating information in periodicals can have significant and wide-ranging effects. Practitioners must give utmost importance to precision and lucidity when conveying research findings.

- **Lifelong Learning and Moral Consciousness:**

Principle: Continuously pursue education to remain updated on growing ethical issues and optimal practices in bioinformatics.

Justification: Due to the constantly shifting environment of bioinformatics and technology, professionals in this field must actively engage in ongoing education and maintain a robust ethical understanding to respond to emerging challenges effectively.

- **Community involvement and explicit agreement for the exchange of data:**

Principle: Actively involve the neighbourhoods, especially patients and the general society, and acquire explicit permission to disseminate biological facts.

Figure 2. Overview of essential strategies related to legal strategy regulation

3. LEGAL STRATEGIES REGULATING DATA USE

The implementation and deployment of a healthcare governance model was the focus of Reddy's group. Nevertheless, when it came to addressing ethical concerns, their attention was only directed towards the overarching model of governance for matters pertaining to the fundamental aspects of safety and the conscientious use of AI. To sum up, although academic studies and government policies have frequently discussed AI ethics in broad strokes, the healthcare industry has received far less focus (Lilhore et al., 2023).

Documenting, storing, updating, and communicating data is an essential part of ethical bioinformatics research. When it comes to fostering reliability in science, making peer review more straightforward, and confirming study results, transparent and repeatable research practices are crucial. Researchers must follow data management as well as research integrity standards to guarantee that their work is reliable and can be replicated. There are a few articles that go into detail about the ethical concerns with AI in healthcare, but none of them provide a comprehensive review of all the problems with AI in healthcare. (Wang et al., 2023). Table 1 presents a comprehensive review of ethical and legal considerations in ML and Bio-informatics research, and Figure 2 illustrates the essential strategies related to legal strategy regulation.

4. DIFFICULTIES RAISED BY MACHINE LEARNING STRATEGIES

Complex black-box models are the result of using advanced ML algorithms with DL and other methods; these models may lack transparency and explainability. Even the creators of black-box models have a hard time explaining the decision-making process of an AI system. Meanwhile, the users are presented

Table 1. Review of ethical and legal considerations in ML and bio-informatics research

Feature	Ethical Considerations in ML	Legal Considerations in Bioinformatics
Informed Consent	Emphasizes obtaining explicit, informed consent from individuals for data usage in a transparent manner	Adheres to legal frameworks requiring explicit consent for the collection, storage, and use of biological and genetic data
Privacy Preservation	Prioritizes protecting individual privacy, especially concerning sensitive health and genetic information	Complies with data protection laws and regulations, implementing measures to safeguard the confidentiality of genetic and health data
Transparency and Explainability	Advocates for transparent ML models, ensuring that decisions are explainable, understandable, and accountable	Aligns with legal requirements for transparency in bioinformatics processes, enabling individuals to understand how their data is used
Fairness and Bias Mitigation	Promotes fair and unbiased ML models, avoiding discrimination and ensuring equitable outcomes	Complies with legal mandates to prevent discrimination in the use of genetic and health data, promoting fairness and unbiased analysis
Data Security and Integrity	Stresses the importance of secure data storage, transmission, and processing to prevent unauthorized access or tampering	Aligns with legal standards for data security in bioinformatics, implementing measures to maintain the integrity and confidentiality of genetic and health data
Accountability and Responsibility	Encourages organizations to take responsibility for the consequences of ML applications and algorithms	Adheres to legal frameworks holding entities accountable for the ethical use of bioinformatics data and responsible handling of genetic information
Regulatory Compliance	Promotes compliance with ethical guidelines and codes of conduct relevant to ML applications	Aligns with legal requirements, such as HIPAA (Health Insurance Portability and Accountability Act) or GDPR (General Data Protection Regulation), in bioinformatics data handling
Data Ownership and Control	Acknowledges the importance of respecting individuals' ownership and control over their data	Complies with legal principles granting individuals control over their genetic and health data, including the right to access and manage their information
Stakeholder Engagement	Encourages engaging stakeholders, including the public, in discussions about the ethical implications of ML applications	Complies with legal requirements for involving stakeholders in decisions related to the use of genetic and health data, fostering public trust and collaboration
Continual Monitoring and Evaluation	Stresses the need for ongoing monitoring and evaluation of ML models to address evolving ethical concerns	Aligns with legal expectations for continual monitoring and evaluation of bioinformatics practices, adapting to emerging ethical and legal standards

with choices that aren't described (Dahiya et al., 2023). Legislation and the "black-box" character of ML often come into conflict in high-stakes domains, where stakeholders face severe repercussions for poor decision-making. Accuracy, explainability, transparency, and compatibility are only a few of the constraints that limit the actual adoption of AI in daily practice, especially in the healthcare sector, where lives are at stake (Onyema et al., 2023).

Therefore, it's critical to push for AI algorithms to be more explainable. Building trust and engagement with AI may be achieved via explainability, which is crucial for ethical AI. A number of social groups, including governments, media, and legal communities, have called for algorithms to be more transparent and explainable. In recent years, this idea has gained traction in the scientific community, leading to many attempts to develop AI systems that can be explained. However, apart from explainability, there are still many ethical problems with healthcare AI solutions, and they are slowly but surely becoming the most important variables affecting AI acceptance (Shivthare et al., 2023).

Concerned citizens and experts in the field have been seeking solutions to the ethical challenges posed by AI research and development. The US "Report on the Future of Artificial Intelligence", the European

Commission's "Ethics Guidelines for Trustworthy AI", and the Chinese government's "Beijing AI Principles" are all examples of rules and regulations. Of these regulatory organizations, the European Union (EU) is often considered to be at the forefront of developing AI-related ethical standards and laws. The primary tenet of the European Union's recommendations, in contrast to the other two sets, is to advance a "human-centred" strategy that adheres to European legislation and values (Tripathi, S., & Jha, M. 2022).

There is a worldwide relevance to the ethical issues that the EU framework seeks to solve. These rules are universally relevant and essential since they are founded on a fundamental-rights perspective. The majority of the UN Sustainable Development Goals (SDGs) are based on the power and duties that are inherent in these principles. Even beyond the European Union, this influences development policies in low- and middle-income nations. The ethical and legal considerations of artificial intelligence in healthcare are not addressed in any way by these general standards, which are applicable to all industries (Kunduru, A. R. 2023).

Numerous scholarly articles address more broad ethical concerns with AI, in addition to the rules and laws already stated. An examination of 22 AI ethical guidelines and suggestions for improving their relative ineffectiveness can be found in "The Ethics of AI Ethics: An Evaluation of Guidelines", "Artificial intelligence ethics guidelines for developers and users: clarifying their content and normative implications", and "The global landscape of AI ethics guidelines" (Swapna et al., 2020), which provides an overview of current AI ethical guidelines and strategies. None of these three papers deal with the ethics of AI in healthcare, despite the fact that they all provide helpful considerations of AI ethics in a broad context.

"The Ethics of AI in Health Care: A Mapping Review", "Ethical and Legal Challenges of Artificial Intelligence-driven Healthcare", and "A Governance Model for the Application of AI in Healthcare" are just a few examples of academic publications that address the ethical concerns surrounding AI in healthcare. From an epistemological, normative, and comprehensive vantage point, Morley's team mapped the ethical difficulties. Without providing a comprehensive analysis of how AI might impact ethical concerns in healthcare applications, Gerke's group just examined them from a legal standpoint (Kumar et al., 2023).

Data processing and interpretation in bioinformatics research often makes use of AI and ML methods. Questions of ethics arise when thinking about AI models' openness, objectivity, and comprehension, the possibility of bias in the information and algorithms, and the appropriate use of AI in decision-making. It is the responsibility of the research community to train AI models using impartial and relevant datasets and to interpret and evaluate the findings correctly (Swapna et al., 2020).

5. BEST PRACTICES AND CASE STUDIES

Following the FDA's clearance of an autonomous AI diagnostic system built on ML, the once-triumphant prospect of Machine Learning in healthcare activities is now a practical reality in clinical settings. Without human intervention or programming, these systems are able to learn from massive datasets and provide predictions.

5.1 Research Utilizing AI in the Field of Health

An essential area of AI-based health research is the use of data generated for electronic health records (EHR). Not preventing the spread of heterogeneous or low-quality data in the underlying information technology system and database might make such data challenging to utilize. Still, EHRs with AI may optimize clinical treatment, enhance quality of care, and conduct scientific research (Dahiya et al., 2023). When adequately designed and taught with sufficient data, AI may assist in extracting clinical best practices from electronic health records, bypassing the conventional methods of scientific publication, guideline formulation, and clinical support systems. AI may also assist new clinical practice models of healthcare delivery by examining patterns in clinical practice gathered from electronic health data (Nayak, S., & Sharma, Y. K., 2023).

5.2 The Role of AI in the Pharmaceutical Industry

We anticipate that AI will streamline and quicken the process of developing pharmaceuticals in the future. Through the use of robots and models of genetic targets, medications, organs, illnesses and their progression, pharmacokinetics, safety, and effectiveness, AI has the potential to transform drug development from a labour-intensive to a capita and data-intensive process (Pradeep, S. et al., 2023). The medication research and discovery process may be accelerated, made more cost-effective, and streamlined with the use of AI. Prior to this, AI was used to discover possible Ebola virus medications, but, as with any drug investigation, finding a lead molecule does not ensure the creation of a safe and effective treatment (Pradeep et al., 2023).

5.3 Ethical Concerns

The question of whether artificial intelligence "fits inside the current legal framework and the extent to which an entirely novel category using specific characteristics and consequences ought to be established" is often debated. There is great potential for AI to enhance healthcare when used in clinical practice, but there are also ethical concerns that need to be addressed. Four significant ethical concerns need to be resolved before artificial intelligence (AI) in healthcare can reach its full potential: Important considerations include (a) data usage permission, (b) data security and openness, (c) algorithmic biases and fairness, and (d) data privacy (Prasad Dhore et al., 2021).

The goal is to provide policymakers with the information they need to proactively address the morally challenging circumstances that arise when AI is enforced in healthcare settings. Most legal debates on AI have focused on the limits of algorithmic transparency. The need for responsible, fair, and open AI development and regulation has grown in response to the increasing use of AI in dangerous contexts. Most importantly, openness ensures that information is both easily accessible and understandable. It is common practice to obfuscate details about algorithm functioning (Singh, K. et al., 2023).

Claims have been made that robots with the potential to learn novel varieties of behaviour and function according to unfixed principles pose a danger to our ability to determine who is at fault. A growing chasm, which endangers "both the ethical framework of humanity as well as the underpinning of the responsibility idea in law," should make people nervous. The adoption of AI raises the possibility that we may be unable to hold anybody responsible for any harm that may occur. We can't tell how bad it is, and using machines will make it harder to determine who is at fault and who is responsible for making decisions (Ramesh et al., 2022).

The reasoning underlying an Artificial intelligence structure's (AIS) outcome may be concealed by modern computer technologies, rendering meaningful inspection impossible. Consequently, the method by which an AIS produces its results is "transparent." For a healthcare user without a background in technology, an AIS's technique may be impenetrable, yet for an information technology expert in that field, it may be as simple as pie. The goal of artificial intelligence systems (AISs) such as Watson, developed by IBM for cancer, is to aid clinical participants, who in turn impact clinical decision-making.

Based on its assessment, the AIS would provide treatment recommendations to the patient. There may be new stakeholder dynamics brought about by the potential use of AI to aid physicians in the future, which might alter clinical decision-making. A new paradigm in healthcare might be born out of the potential for AIS to assist doctors in the future, which has the potential to transform clinical decision-making. The secure implementation of new technology in healthcare settings is of utmost importance to clinicians, who include various health professionals such as physicians, nurses, and others (Kumar et al., 2023).

When it comes to their goals, potential construction methods, and potential applications, the new ML-HCAs cover much ground. From fully autonomous AI diabetes-related retinopathy prognoses in primary care environments to manually hidden and resource-allocated models, ML-HCAs span a broad spectrum of applications. Scientists conducting the study should explain how such results and forecasts may be included in the study. The scientific trial's budget and future research directions depend on this data (Kumar et al., 2023).

In order to keep patients safe, AI used in healthcare must adapt to a dynamic and unpredictable environment while upholding ethical standards. A simple but crucial part of determining the security of any medical software is the ability to test the program and identify any vulnerability. The method for software programs is analogous, for instance, to the additives and physiological processes of mechanical devices or pharmaceuticals. In contrast, ML-HCAs may hide their inner workings from assessors, physicians, and patients, creating a "black box" problem. Scientists conducting the study should explain how such results and forecasts may be included in the study. Both the scientific trial's budget and future studies may be better guided by this data (Nayak S. et al., 2023).

6. CONCLUSION AND FUTURE WORK

There has to be moral accountability for AI because of its growing role in healthcare. The use of suitable algorithms grounded on impartial real-time data is essential for preventing data bias. It is necessary to have inclusive development teams and to audit the methodology, particularly its system deployment, often. Although AI won't be able to replace human medical reasoning fully, it may certainly assist in improving decision-making. AI might conduct screening and assessment in the event that medical expertise is lacking in a setting with limited resources.

Because algorithms are involved, even the most rapid AI decisions are methodical in comparison with human decision-making. This means that actions will always result in accountability not from the machine itself but from those responsible for its creation and use, even if no legal consequences follow (due to the lack of adequate legal frameworks). While there are ethical dilemmas that may arise from AI's implementation, the technology is soon to usher in a new era of healthcare that relies on AI, and failing to do so may be both scientifically and ethically incorrect.

REFERENCES

Aminizadeh, S., Heidari, A., Toumaj, S., Darbandi, M., Navimipour, N. J., Rezaei, M., Talebi, S., Azad, P., & Unal, M. (2023). The applications of machine learning techniques in medical data processing based on distributed computing and the Internet of Things. *Computer Methods and Programs in Biomedicine*, *241*, 107745. doi:10.1016/j.cmpb.2023.107745 PMID:37579550

Ansori, A. N., Antonius, Y., Susilo, R. J., Hayaza, S., Kharisma, V. D., Parikesit, A. A., Zainul, R. J. K., Jakhmola, V., Saklani, T., Rebezov, M., Ullah, M. E., Maksimiuk, N., Derkho, M., & Burkov, P. (2023). Application of CRISPR-Cas9 genome editing technology in various fields: A review. *Narra J*, *3*(2), e184–e184. doi:10.52225/narra.v3i2.184

Brauneck, A., Schmalhorst, L., Kazemi Majdabadi, M. M., Bakhtiari, M., Völker, U., Baumbach, J., Baumbach, L., & Buchholtz, G. (2023). Federated machine learning, privacy-enhancing technologies, and data protection laws in medical research: Scoping review. *Journal of Medical Internet Research*, *25*, e41588. doi:10.2196/41588 PMID:36995759

Cacciamani, G. E., Chen, A., Gill, I. S., & Hung, A. J. (2023). Artificial intelligence and urology: Ethical considerations for urologists and patients. *Nature Reviews. Urology*, 1–10. PMID:37524914

Chavva, I. R., Crawford, A. L., Mazurek, M. H., Yuen, M. M., Prabhat, A. M., Payabvash, S., Sze, G., Falcone, G. J., Matouk, C. C., de Havenon, A., Kim, J. A., Sharma, R., Schiff, S. J., Rosen, M. S., Kalpathy-Cramer, J., Iglesias Gonzalez, J. E., Kimberly, W. T., & Sheth, K. N. (2022). Deep learning applications for acute stroke management. *Annals of Neurology*, *92*(4), 574–587. doi:10.1002/ana.26435 PMID:35689531

Dahiya, N., Sharma, Y. K., Rani, U., Hussain, S., Nabilal, K. V., Mohan, A., & Nuristani, N. (2023). Hyper-parameter tuned deep learning approach for effective human monkeypox disease detection. *Scientific Reports*, *13*(1), 15930. doi:10.1038/s41598-023-43236-1 PMID:37741892

Dahiya, N., Sharma, Y. K., Rani, U., Hussain, S., Nabilal, K. V., Mohan, A., & Nuristani, N. (2023). Hyper-parameter tuned deep learning approach for effective human monkeypox disease detection. *Scientific Reports*, *13*(1), 15930. doi:10.1038/s41598-023-43236-1 PMID:37741892

Demir, E. (2022, August). Big Biological Data: Need for a Reorientation of the Governance Framework. In *2022 IEEE Conference on Computational Intelligence in Bioinformatics and Computational Biology (CIBCB)* (pp. 1-7). IEEE. 10.1109/CIBCB55180.2022.9863047

Deshwal, D., Sangwan, P., Dahiya, N., Lilhore, U. K., Dalal, S., & Simaiya, S. (2023). [Detection using Hybrid CNN-RNN Architecture with Transfer Learning from X-Rays. *Current Medical Imaging*.]. *COVID*, 19.

Detopoulou, P., Voulgaridou, G., Moschos, P., Levidi, D., Anastasiou, T., Dedes, V., Diplari, E.-M., Fourfouri, N., Giaginis, C., Panoutsopoulos, G. I., & Papadopoulou, S. K. (2023). Artificial intelligence, nutrition, and ethical issues: A mini-review. *Clinical Nutrition Open Science*, *50*, 46–56. doi:10.1016/j.nutos.2023.07.001

Ettari, S. V., Roden, E., Ahuja, V., & Samadani, U. (2022). Oculogica: An Eye-Catching Innovation in Health Care and the Privacy Implications of Artificial Intelligence and Machine Learning in Diagnostics for the Human Brain. SMU Sci. & Tech. L. Rev., 25, 23.

Kaushal, A., Kaushal, D., & Pathak, Y. Practicing Medicine and Ethical Issues Applying Artificial Intelligence. In *Ethical Issues in AI for Bioinformatics and Chemoinformatics* (pp. 87–106). CRC Press.

Guleria, K., Sharma, A., Lilhore, U. K., & Prasad, D. (2020). Breast cancer prediction and classification using supervised learning techniques. *Journal of Computational and Theoretical Nanoscience*, *17*(6), 2519–2522. doi:10.1166/jctn.2020.8924

Herington, J., McCradden, M. D., Creel, K., Boellaard, R., Jones, E. C., Jha, A. K., Rahmim, A., Scott, P. J. H., Sunderland, J. J., Wahl, R. L., Zuehlsdorff, S., & Saboury, B. (2023). Ethical considerations for artificial intelligence in medical imaging: Deployment and governance. *Journal of Nuclear Medicine*, *64*(10), 1509–1515. doi:10.2967/jnumed.123.266110 PMID:37620051

Karabacak, M., & Margetis, K. (2023). Embracing Large Language Models for Medical Applications: Opportunities and Challenges. *Cureus*, *15*(5). doi:10.7759/cureus.39305 PMID:37378099

Kumar, N., Verma, H., & Sharma, Y. K. (2023). Adversarial Attacks on Graph Neural Network: Techniques and Countermeasures. In Concepts and Techniques of Graph Neural Networks (pp. 58-73). IGI Global.

Kumar, N., Verma, H., & Sharma, Y. K. (2023). Graph Convolutional Neural Networks for Link Prediction in Social Networks. In Concepts and Techniques of Graph Neural Networks (pp. 86-107). IGI Global. doi:10.4018/978-1-6684-6903-3.ch007

Kumar, R., Rani, S., & Khangura, S. S. (Eds.). (2023). *Machine Learning for Sustainable Manufacturing in Industry 4.0: Concept, Concerns and Applications*. CRC Press. doi:10.1201/9781003453567

Kunduru, A. R. (2023). Machine Learning in Drug Discovery: A Comprehensive Analysis of Applications, Challenges, and Future Directions. *International Journal on Orange Technologies*, *5*(8), 29–37.

Landau, A. Y., Ferrarello, S., Blanchard, A., Cato, K., Atkins, N., Salazar, S., Patton, D. U., & Topaz, M. (2022). Developing machine learning-based models to help identify child abuse and neglect: Key ethical challenges and recommended solutions. *Journal of the American Medical Informatics Association : JAMIA*, *29*(3), 576–580. doi:10.1093/jamia/ocab286 PMID:35024859

Lilhore, U. K., Manoharan, P., Sandhu, J. K., Simaiya, S., Dalal, S., Baqasah, A. M., Alsafyani, M., Alroobaea, R., Keshta, I., & Raahemifar, K. (2023). Hybrid model for precise hepatitis-C classification using improved random forest and SVM method. *Scientific Reports*, *13*(1), 12473. doi:10.1038/s41598-023-36605-3 PMID:37528148

Lilhore, U. K., Poongodi, M., Kaur, A., Simaiya, S., Algarni, A. D., Elmannai, H., Vijayakumar, V., Tunze, G. B., & Hamdi, M. (2022). Hybrid model for detection of cervical cancer using causal analysis and machine learning techniques. *Computational and Mathematical Methods in Medicine*, *2022*, 2022. doi:10.1155/2022/4688327 PMID:35572826

Mylrea, M., & Robinson, N. (2023). Artificial Intelligence (AI) Trust Framework and Maturity Model: Applying an Entropy Lens to Improve Security, Privacy, and Ethical AI. *Entropy (Basel, Switzerland)*, *25*(10), 1429. doi:10.3390/e25101429 PMID:37895550

Nayak, S., & Sharma, Y. K. (2023). A modified Bayesian boosting algorithm with weight-guided optimal feature selection for sentiment analysis. *Decision Analytics Journal*, *8*, 100289. doi:10.1016/j.dajour.2023.100289

Nayak, S., & Sharma, Y. K. (2023). A modified Bayesian boosting algorithm with weight-guided optimal feature selection for sentiment analysis. *Decision Analytics Journal*, *8*, 100289. doi:10.1016/j.dajour.2023.100289

Onyema, E. M., Lilhore, U. K., Saurabh, P., Dalal, S., Nwaeze, A. S., Chijindu, A. T., Ndufeiya-Kumasi, L. C., & Simaiya, S. (2023). Evaluation of IoT-Enabled hybrid Model for genome sequence analysis of patients in healthcare 4.0. *Measurement. Sensors*, *26*, 100679. doi:10.1016/j.measen.2023.100679

Padhi, A., Agarwal, A., Saxena, S. K., & Katoch, C. D. S. (2023). Transforming clinical virology with AI, machine learning and deep learning: A comprehensive review and outlook. *Virusdisease*, *34*(3), 1–11. doi:10.1007/s13337-023-00841-y PMID:37780897

Parashar, A., Parashar, A., Ding, W., Shabaz, M., & Rida, I. (2023). Data Pre-processing and Feature Selection Techniques in Gait Recognition: A Comparative Study of Machine Learning and Deep Learning Approaches. *Pattern Recognition Letters*, *172*, 65–73. doi:10.1016/j.patrec.2023.05.021

Patel, M. B., Patel, J., Patel, D., Prajapati, P., & Prajapati, J. Basic Ethical Issues in Bioinformatics and Chemoinformatics. In *Ethical Issues in AI for Bioinformatics and Chemoinformatics* (pp. 12–38). CRC Press. doi:10.1201/9781003353751-2

Pradeep, S., Sharma, Y. K., Lilhore, U. K., Simaiya, S., Kumar, A., Ahuja, S., Margala, M., Chakrabarti, P., & Chakrabarti, T. (2023). Developing an SDN security model (EnsureS) based on lightweight service path validation with batch hashing and tag verification. *Scientific Reports*, *13*(1), 17381. doi:10.1038/s41598-023-44701-7 PMID:37833379

Prasad Dhore, D. Y. K. S., & Kumbharkar, D. P. (2021). Resourceful Investigation of Malnutrition with Machine Learning Using CNN Procedure. International Journal of Advanced Research in Science. *Tongxin Jishu*, *3*(2), 12–19.

Ramesh, T. R., Lilhore, U. K., Poongodi, M., Simaiya, S., Kaur, A., & Hamdi, M. (2022). Predictive analysis of heart diseases with machine learning approaches. *Malaysian Journal of Computer Science*, •••, 132–148.

Sandhya, M. P., Raju, G. P. A., & Iyappan, M. (2023). Bioinformatics: Exploring the Past, Present and Future. *MULTIDISCIPLINARY*, *127*, 149.

Shivthare, S. K., Sharma, Y. K., & Patil, R. D. (2023). Multi-modal Biometric in Computer Vision. *Multi-modal Biometric and Machine Learning Technologies: Applications for Computer Vision*, 1-29.

Singh, K., Kumar, A., Sharma, Y. K., & Rai, A. K. (2023). AIoT-based e-commerce. *AIoT Technologies and Applications for Smart Environments*, 215.

Swapna, M., Sharma, Y. K., & Prasad, B. M. G. (2020). A survey on face recognition using convolutional neural network. In *Data Engineering and Communication Technology: Proceedings of 3rd ICDECT-2K19* (pp. 649–661). Springer Singapore. doi:10.1007/978-981-15-1097-7_54

Swapna, M., Sharma, Y. K., & Prasadh, B. M. G. (2020). CNN Architectures: Alex Net, Le Net, VGG, Google Net, Res Net. *Int. J. Recent Technol. Eng*, *8*(6), 953–960. doi:10.35940/ijrte.F9532.038620

Tang, A., Woldemariam, S., Roger, J., & Sirota, M. (2022). Translational Bioinformatics to Enable Precision Medicine for All: Elevating Equity across Molecular, Clinical, and Digital Realms. *Yearbook of Medical Informatics*, *31*(01), 106–115. doi:10.1055/s-0042-1742513 PMID:36463867

Theodosiou, A. A., & Read, R. C. (2023). Artificial intelligence, machine learning and deep learning: Potential resources for the infection clinician. *The Journal of Infection*, *87*(4), 287–294. doi:10.1016/j.jinf.2023.07.006 PMID:37468046

Tripathi, S., & Jha, M. (2022). Advancements in Neuroradiology via Artificial Intelligence and Machine Learning. *International Journal of Bioinformatics and Intelligent Computing*, *1*(2), 130–141. doi:10.61797/ijbic.v1i2.159

Uwinama, I., Rusanganwa, J., & Ingabire, E. (2023). The Moral Implications of Big Data and Machine Learning in Healthcare: A Review. *American Journal of Technology*, *2*(1), 45–53.

Wang, C., Liu, S., Yang, H., Guo, J., Wu, Y., & Liu, J. (2023). Ethical considerations of using ChatGPT in health care. *Journal of Medical Internet Research*, *25*, e48009. doi:10.2196/48009 PMID:37566454

Chapter 5
A Comprehensive Analysis of the Health Effects of 5G Radiation

Soumitra Saha
https://orcid.org/0000-0002-0012-9109
Chandigarh University, India

Shubh Kumar
Chandigarh University, India

ABSTRACT

The human race has acquired the peak of sensation today in augmenting development in the technological domain. With gratitude to state-of-the-art technology, humans are now enjoying heavenly delights, and 5G technology with high-wave transmission capability can be considered a universally acknowledged example. However, the radiation from 5G technology, which employs high-frequency waves, puts humankind at unimaginable health risks, which weakens us to further radiation. Seemingly invisible high-power radiations function like dormant volcanoes in the human body. The extremes of radiation in 5G technology have become so widespread that the miscellaneous physical problems provoked by radiation can easily be termed manufactured disorders that humans progressively construct. In addition to causing chronic diseases, this radiation plays a distinct role in suppressing the regular immune system of the human body. This chapter tries to clarify and shed light on the dangers of radiation arriving from 5G technology-supported devices and various types of incurable diseases of the human body.

1. INTRODUCTION

Technology plays the most prominent role in enhancing and improving the quality and standard of living in the 21st century. The technologies we use in 2023 were unimaginable, even twenty-five to thirty years ago. This impossible and unimaginable achievement of technology has significantly impacted our way of life, and in these circumstances, imagining a world without technology is like a dream that may or

DOI: 10.4018/979-8-3693-1822-5.ch005

may not exist in reality. Our daily lives and use of technology are very closely linked together nowadays. The use of technology and, at the same time, the usefulness of technology has increased so much now that there is no area of daily life where we can live comfortably without the smooth use of technology. The promotion and spread of new technologies have increased so much that famous companies constantly strive to innovate new technologies, try to figure out new features of the existing technologies, and constantly update the system to be up-to-date and more user-friendly than ever. In each sector of our routine life, like communication, electrical, energy, manufacturing, medical, transportation, etc., we can quickly notice the broad walk of technology, and the usability rate is increasing tremendously. However, the most important thing about this advancement is that we will continue striving to develop each technology and achieve success to cope with the new generation. The technology used a decade ago seems like a 'neglected, unused, and unnecessary' object or substance. However, as social beings, we naturally let go of the old and embrace the new. In addition, if it is state-of-the-art technology, then there is no question.

The best example is telecommunication technology, where we can use - the largest, most accepted, most talked about, widely used, and expanded network. Even in the 20th century, we used satellite-based telecommunication systems and their associated networks to communicate because the cellular network remains undiscovered and unexplored. The First Generation (1G) cellular network was introduced and became available for the general public to access in 1981 for the first time in the history of telecommunication technology (Liu & Jiang, 2016). As the technology was newly invented and very much new to ordinary people, there were few technological advantages regarding today's technology. In the first part of the 20th century, countries worldwide started incorporating second-generation (2G) technology to build internal communication systems and infrastructures in a wide margin ("Radiolinja's History," 2004). Organizations and technological companies have always worked to improve the existing technology and to invent more modern technology, and in these circumstances, third-generation (3G) technology was introduced at the beginning of the new century's first decade ("DoCoMo Delays 3G Launch," 2001) and fourth-generation (4G) (Li, Wang & Zhang, 2020) at the end of the decade. At the end of the next decade, in 2019, cellular phone companies started using the fifth-generation (5G) network, the next-generation wireless network technology, for the first time worldwide (Hoffman, 2019).

Where earlier technologies could transfer data in kilobits per second (kbps) or Megabits per second (Mbps), the newly developed 5G technology can transfer data in Gigabit per second (Gbps). 5G reached a golden peak of success, technologically speaking or in terms of technological advancement; the newly invented network achieved download speeds of 10 Gbps ("5G explained," n.d.), which is, in short, forever the most incredible and successful invention of the century. 5G can handle more additional devices than any other prior network and also in an energy-efficient manner. The ultimate reason behind these capabilities to handle or work with so many devices once at a time alongside optimal results is that it can transmit high frequencies and handle diverse networks, work with miscellaneous types of devices simultaneously, and have a higher probability of providing the best results. Moreover, it has another excellent capacity to reduce or cut down the quantity of the latency rate for a given network with a top reliability rate (Agiwal, Roy, & Saxena, 2016). Diverse operations like establishing a phone conversation, building up the interconnection between two or more intelligent devices, and sending or receiving data from the server side to the client, or vice versa, are all examples that can occur quickly and remarkably faster.

One of the biggest misconceptions about 5G technology is its high speed, but it is essential to note that certain obstacles are also present. 5G technology has more luxurious bandwidth than any other formerly developed technology. Indeed, the total amount of radiated radiation increases in 5G networks

Figure 1. Transformation of networks from 1G to 5G

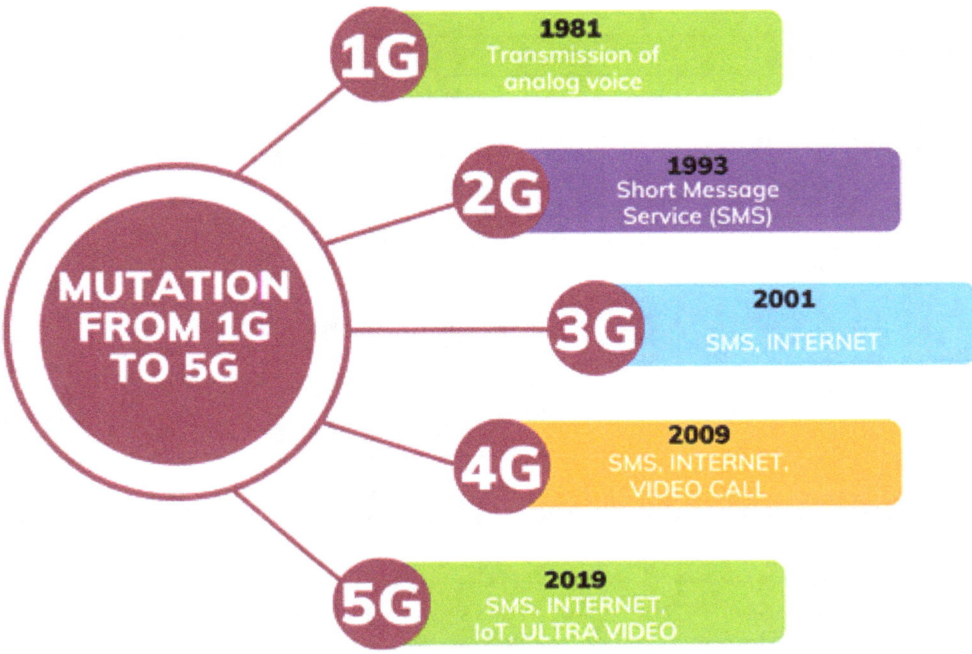

regarding the radiation that comes out from the 2G, 3G, and 4G network sources for the same area or the number of devices (El-Hajj & Naous, 2020). On account of this, in any good, human-beneficial, and newly invented technology or thing, there is a fair amount of evil along with the good; it is a fact of life. Technology courses have become a part of our lives with the constant development of new technologies is a never-ending process, from small to large scale, and much the same has happened with the 5G network. As much as 5G has sped up our lifestyle and made our day-to-day life more accessible than ever; it has also caused various problems and headaches. *Beamforming* is a widely trendy technique in telecommunications, wireless communication, and other relevant fields. These techniques focus on transmitting or receiving specific types of signals in predefined directions and working with the signal quality so that the signal can be transmitted correctly into the right path. At the same time, multiple-input multiple-output (MIMO) technologies are implemented in the 5G network, and extra cells are needed for 5G regarding 3G and 4G (de Vocht & Albers, 2022), as MIMO is also a particular type of communication technology. It will also use multiple antennas for reliability to improve communications performance in transmitting and receiving signals. The expected differences between 1G-5G technologies are shown below.

In these circumstances, required usable frequencies are always in a demanding stage, and these frequencies are injurious and a more significant threat to human health. Furthermore, the most prominent reason for this headache is the waves emitted or received from the high-power 5G technology, harming the human body and race, and children are at more risk (Pölzl, 2011). What kind of damage the frequency of 5G technology can do to an individual, and how much damage it has can be expressed beyond the imagination of ordinary people. Since the frequencies coming from 5G serving technology

Table 1. Cellular Technology Comparison (Sood & Garg, 2014; Eluwole et al., 2018; del Peral-Rosado et al., 2017; Hodara & Skaljo, 2021; Bhandari et al., 2017; Shah, 2022; Goyal et al., 2020; Hollington & Jansen, 2023; Moshood et al., 2020)

Feature name	1G	2G	3G	4G	5G
Development	1981	1993	2001	2009	2019
Speed (max)	Up to 9.6kbps	Up to 384kbps	1.5Mbps - 42Mbps	10Mbps - 1Gbps	150Mbps - 10Gbps
Latency (milliseconds)	> 1000ms	300 - 1000ms	100 - 300ms	Less than 100ms (30ms - 50ms)	<= 1ms
Applications availability	Voice only	Voice and Short Message Service (SMS)	Voice, SMS, Video calling	Voice, SMS, Video calling, Gaming, HD TV, Video streaming	Voice, SMS, Video calling, Gaming, HD TV, Video streaming, Virtual Reality. (VR), Augmented. Reality (AR), Video. streaming (UHD, 4K, 8K),
Switching technology	Circuit	Circuit	Circuit / Packet	Packet	Packet
Bandwidth	Analog	25 Megahertz (MHz.)	25 MHz.	100 MHz.	500 MHz. - 2 Gigahertz
Bandtype (frequency)	Narrow. band	Narrow. band	Wide.. band	Ultra. wide band	Exceedingly Ultra wide band
Spectrum efficiency	Lower	Moderate	Higher	Enormously higher	Exceedingly higher
Communication between devices	N/A	Limited	Enhanced	Advanced	Ultra-fast
Handoff	Horizontal.	Horizontal.	Horizontal.	Horizontal. and Vertical.	Horizontal. and Vertical.
Power consumption	Higher	Moderate	Lower	Enormously lower	Exceedingly lower
Reliability	Lower	Moderate	Higher	Enormously higher	Exceedingly higher
Security	Lower	Moderate	Higher	Enormously higher	Exceedingly higher
Cost	Higher	Moderate	Lower	Enormously lower	Exceedingly lower

are invisible to the naked eye and being a relatively new technology, the all-around public has very little understanding of this technology, so we are not much aware of its harmfulness, disease-causing, and life-destroying aspects. One of the most significant problems with any frequency-induced health concern is the misunderstanding that incoming frequencies can cause physical problems. Highly powerful and newly developed 5G technologies can radiate waves by providing favorable services to the user and building stable connections to stay competitive in the flourishing market. As a result, this cutting-edge technology can cause unimaginable damage to the human body and will remain unknown to those unaware of it. The excessive addition of these 5G base stations because of more instantaneous mobile broadband speeds with acceleration has an impactful question regarding the adverse health effect (Sofri et al., 2021).

The subsequent part of this article is categorized in the following steps: Section II describes the distinct type of health issues that are somehow connected to 5G technology. Section III describes the background studies in 5G technology in healthcare. Section IV describes the conclusion and discus-

Figure 2. The lower and highest limits of various networks' frequencies

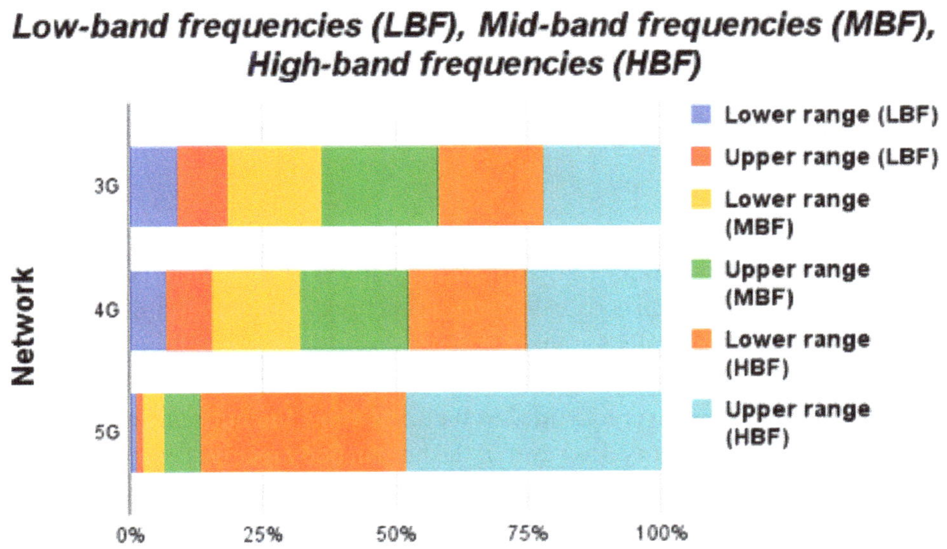

sion of the effects and what kind of work can be done to minimize the damage and for the awareness of unknown persons regarding 5G.

2. 5G RADIATION RELATED ILLNESS

Suppose the topics come into data transmission from one place to another, or build a connection between layers inside a system to communicate. In that case, a maximum number of telecommunication companies often use three types of frequencies in a 5G network: low.-band, mid.-band, and high.-band frequencies. Though low-band and mid-band 5G use frequencies suitably and tolerably, high-band 5G offers ten times better performances and speed regarding the concurrent 4G networks. It often uses more increased frequencies like 20GHz and multiple antennas for short distances (Lin, 2020). Due to globalization, powered base stations frequently employ high frequencies, and in these circumstances, an enormous number of 5G-supported devices is increasing miraculously. The more facilities are increased, and telecommunication technology advances, the more 5G-supported devices are fabricated and spread out worldwide.

In consequence, future planning is proceeding by the wireless telecommunication operators to operate the system in 'millimeter wave,' which has an extensive amount of frequency coverage from 30-300 GHz (Wu, Rappaport, & Collins, 2015). These high waves may be one of the leading causes of damage to the human body. Generally, these technologies' impact affects the human body, especially in the skin, eyes, environment, and plants simultaneously (Hasan et al., 2021).

The high-powered radio frequency used in 5-G technology is one of the most significant contributors to the development of various types of ailments and also the leading cause of various incurable diseases in the human body. This radio frequency is sometimes directly connected to the advancement of some illnesses and indirectly in some circumstances. A considerable amount of widespread body organs will be attacked. There is a high possibility of dominance by these radio frequencies in the body of a human being, and the output of these effects is alarming. We will talk about the various infected organs that might be unhealthy because of this disorder and diseases correlated with radio frequency, what kind of problems the high-level radio frequency used in 5-G technology can cause the human body, which can be the leading or primary cause of some severe diseases, and what kind of problems a person can face.

1.1 Cancer

The human body is composed of millions of cells, and as a result of that, there is a surprising diversity available among the cells in the human body. All kinds of functionalities of the human body are performed by these cells. There are different cells for specific functions to take place. It provides the appropriate structure for a human body so that it can do its regular job quite effortlessly and collect nutrients from the food consumed by the body in everyday life. At the same time, the nutrients consumed by the body from the food are converted into energy. However, if the human body receives excessive radio frequency, the routine functioning of the cells will be disrupted. These radio frequencies play an influential role in disrupting the normal functioning of the cells. It can cause internal and external problems in the human body at the same time. The term mutation is highly familiar in medical science and real-life situations; mutation is one of the leading causes of multiple life-threatening diseases. Nucleic acids are present in the human body and are one of the most vital substances for a human being. The nucleic acids in the genome follow a straightforward sequence, as this acid is susceptible in terms of a human organ, and a slight change in the sequence of nucleic acid may damage the repository system. A change in nucleic acid sequence is a mutation that can easily be caused by receiving too much radio frequency, which might be transmitted from a 5G-supported machine. Mutations can cause changes in deoxyribonucleic acid (DNA) sequences. Affecting DNA or damaging DNA is altogether possible by radiofrequency radiation (Panagopoulos, 2019; ALDelfi et al., 2023) and, most notably, Electromagnetic Fields, primarily generated from mobile phones. At the same time, mutation plays an essential role in forming tumors that can quickly turn into cancer in future and neurodegenerative diseases as well.

1.2 Neurological Disorders

A neurological disorder is a problem that can affect the nerves in different parts of the body, especially the brain and the spinal cord. Signals are transmitted through the nerve cells throughout the different parts of the body, and as a result, abnormality in the nervous system can affect human beings in the long run. Some of the critical physical problems caused by this disease include loss of cognitive ability, which means it is one kind of force to disappear the ability to think correctly and weakness due to various muscle contractions. Neurological Disorders may cause different types of diseases like-

- Brain cells are about to die because of Alzheimer's disease (Krämer, 2023), a neurological disorder. This disorder causes behavioral changes and is so injurious for the brain that the ultimate number of abnormal protein build-ups, called amyloid and tau, is increased inside the brain.
- Parkinson's disease is a developmental or slowly progressive disorder in the human body that significantly impairs normal movement and balance (Gonzalez-Usigli, 2023). As a result, the average speed of movement will slow down, the balance of the body will be changed, and a more

significant margin will also increase in the stiffness of the muscles. This disease is also one of the leading causes of disrupting dopamine production in the brain.

- Nerve cells are affected, which handle the movement of the muscle—for that weakness in the muscle part, facing problems when articulating and devouring may happen because of Amyotrophic lateral sclerosis (ALS) diseases (Yasin, 2023). ALS is a motor neuron progressive neurological disease, meaning that symptoms and the harmfulness of these diseases gradually worsen. Most importantly, no effective treatment or medicine has been invented yet.
- The central nervous system may be long-term affected, and the nerve fibers may be damaged because of Multiple sclerosis (MS) (Cleveland Clinic, n.d.), a chronic disorder. The brain and spinal cord nerve fibers are being harmed and, as a result, cause different problems like balance loss or muscle weaknesses, and fundamental vision problems are one of them.

1.3 Skin Problems

Any number of reasons can cause or might be responsible for skin problems. Nevertheless, skin problems caused by radiation differ from other skin problems. Skin problems caused by radiation, rather than other causes, are exceedingly severe. Since skin is a part of the immune system, skin problems caused by radiation can harm the immune system. Furthermore, the possibility of skin cancer has also increased significantly more than before. That is why, now, it is not a big deal to get skin cancer because of the excessive transmission of radiofrequency. Hair loss from skin problems has become very common now. The radio frequency from 5G technology can cause several types of skin problems, including -

- *Contact dermatitis* is a well-known skin rash problem that might be spread out when anyone comes closer to such things that are allergic and might be caused by exposure to high radio frequency or contact with certain substances and spending a lengthy amount of time (Cleveland Clinic, n.d.).
- *Acne* is a skin problem that may occur because of radiofrequency that can cause various spots and whiteheads on the face, back, and chest. Papules and pustules may be raised in different body parts, which can be very painful to touch.
- *Eczema* is a chronic disease caused by radio frequency, which causes skin rashes, swelling, and itching. *Eczema* is one of those diseases which can affect people of any age; however, children often face problems.
- Itching and burning sensations can occur because of interconnecting with high frequencies, a common skin symptom. Issues in the liver or kidney and diseases like diabetes can occur.

1.4 Immune System Disorders

The immune system plays the most crucial role in protecting the human body from various types of viruses, bacteria, or fungal attacks, and all the organs in the human body function beautifully and smoothly. Therefore, the better and more convenient way the immune system can be managed, the healthier, stronger, and more standard the human body will be. If this system is not functioning correctly, the immunity will be significantly reduced, and as a result, the risk of developing any toxic disease will be enormously increased. When the human body's immune system fails to function correctly, it becomes difficult to distinguish between different types of cells, leading to 80 types of immune system disorders (Orbai, 2022). Nevertheless, the immune system may be severely damaged if the human body is exposed to high

Figure 3. Health effects because of 5G radiation

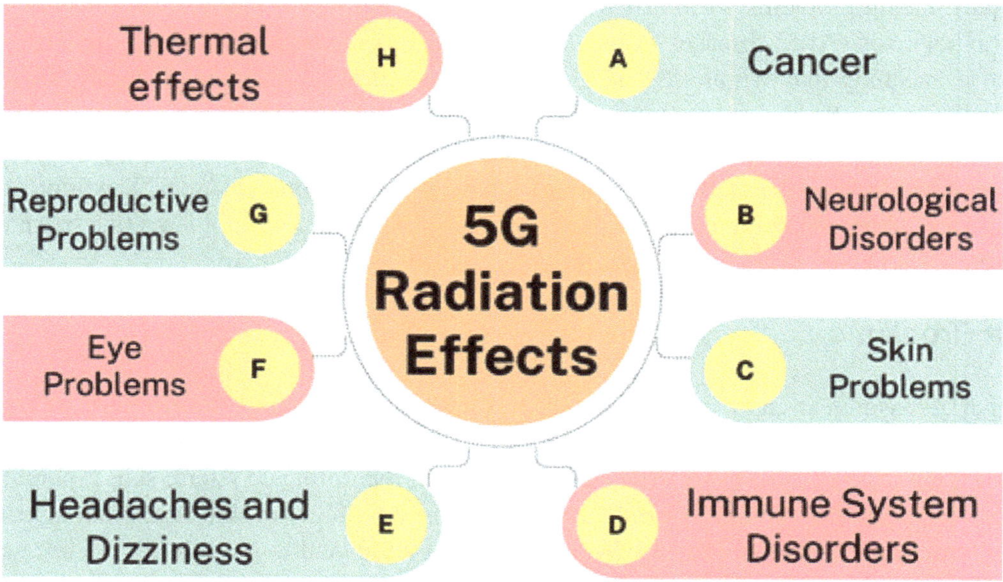

radiation levels for too long, and the ability to resist disease will be broken down. This problem occurs only when high doses of radiation can damage tissues and cells directly related to the immune system. Every cell and tissue is somehow interconnected to the immune system.

- One of the leading causes of white blood cell damage is exposure to high levels of radiation, which in turn increases the susceptibility of the human body to disease and reduces its ability to fight infection. It is the principal symptom of immunodeficiency. If immune system disorders are present in children, average growth and mental development may also be delayed.
- Radiation can sometimes make the immune system overactive. It varies from person to person according to their lifestyle, the variations regarding environmental surroundings, sometimes food habits, and so on. As a result, the immune system attacks healthy and functioning tissues or cells. The result can be much more hazardous if the healthy tissues are affected for some reason, like environmental conditions and any external issues.

- A variety of infections can occur for no specific reason and just because of problems with the immune system, and some of them are infections in the ears, skin, and urinary as well. If this occurs, these issues may last longer than typical diseases and take longer to recover.

- An autoimmune disorder is one of many types of immune system disorder where the immune system attacks the body's tissues by mistake, which may result in various symptoms depending on which types of tissues are affected. These types of infections are uncommon for ordinary people.

1.5 Headaches and Dizziness Because of Sleep Disturbances

A healthy person can experience various physical symptoms if he or she does not get enough sleep or has trouble sleeping correctly. Insufficient and disturbed sleep leads to problems like dizziness and headaches, and one of the leading causes of these problems is exposure to high levels of radiation. High radiation levels can significantly affect a person's sleeping habits and patterns. This problem is more likely to occur if the radiation is directly connected to the brain, spine, or other parts of the nervous system. *Melatonin* is the hormone used to go to sleep and wake up at the right time and in the right way. This hormone, melatonin, regulates the entire cycle. If the melatonin production process is going well and the transfer rate throughout the body works well, we never have to face such problems. However, excessive radiation inhibits the production of this melatonin hormone in moderate amounts and simultaneously inhibits melatonin hormone flow. When melatonin does not flow to its standard rate, then miscellaneous types of issues may arise, and among them, the main symptoms are not being able to sleep at the right time, not being able to sleep for a long time, waking up frequently, and feeling very tired. Lack of proper sleep disrupts the balanced distribution of oxygen to the brain and can also lead to changes in high blood pressure. Other cases may appear, like anxiety, nervousness or panic disturbances, ear infections, low blood sugar, Etc.

1.6 Eye Problems

The eye is considered one of the most significant and vital organs of the human body. Cataracts are one of the most familiar problems when considering eye diseases, and cataracts are one of the leading causes of blindness for everyone. The terrible disease can easily affect people of any age, from young to old. Radiation from 5G devices is thought to be one of the causes of cataracts (McLaughlin, Donoviel & Jones, 2017). Devices with 5G-supported technology are always very close to our eyes. For example, mobile phones are used to be kept very close to the eyes all the time. As a result, the eyes are more likely to be damaged by this radiation than any other part of the body. Exposure to high radiation levels can cause mild eye irritation, eye pain, and even loss of vision. Ultraviolet and ionizing are two unique electromagnetic radiation types, considered the most harmful radiation for any eye problem. High levels of ionizing radiation are usually responsible for severe redness or swelling of the eyes, and this kind of radiation has the ability or is also responsible for extensive damage to the eye tissues. Ultraviolet radiation mainly affects the visible wavelengths of the eye and is capable of causing much more damage when considered in terms of the cornea. The cornea is transparent and can protect the iris and pupil parts of the eye. The most necessary and compulsory work to be done by the cornea is to focus light. Ultraviolet keratitis or snow blindness is a type of disease, mainly known as *Photokeratitis* or *ultraviolet keratitis* and can be provoked by these radiations, which cause severe damage in the cornea section. *Photokeratitis* can cause various problems; sensitivity in front of light, blurred vision, and redness are more common among them.

Figure 4. Consequences of radiation on the human body
(Sciorio et al., 2022).

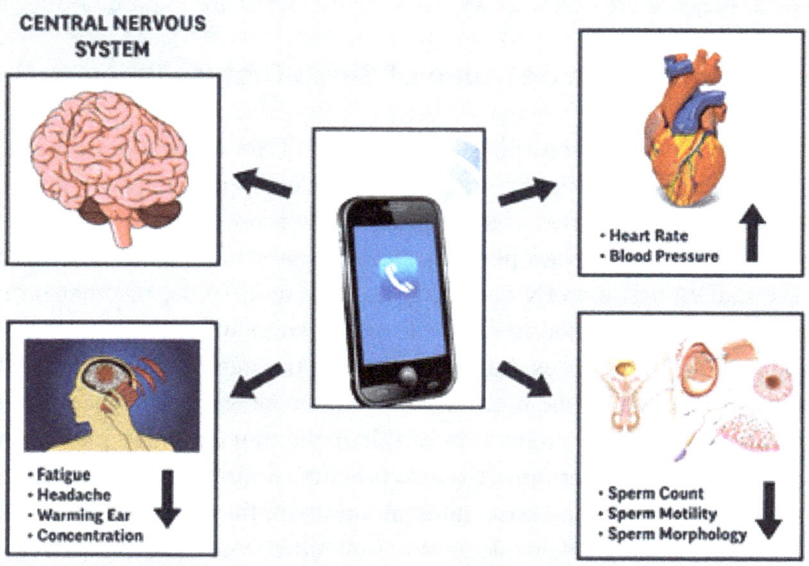

1.7 Reproductive Problems

Among all the problems that high radiation levels can cause the human body, reproductive problems are the most severe and one of the main ones. Reproductive problems arise for several reasons, like biological causes, environmental issues, or lifestyle habits of a human being; however, if they arise because of high radiation levels, they may be life destroyers. High levels of radiation can cause major reproductive problems not only in women but also in men (Wdowiak et al., 2017).

- High radiation exposure can damage women's ovaries and result in subsequent infertility. Also, the effects of these radiations are one of the leading causes of miscarriage and can also cause harmful diseases to any unborn child. At the same time, facing problems during pregnancy is another viral issue for every woman. Hormonal imbalances, uterine abnormalities, or immune system disorders can contribute to these challenges. Hormonal tissue imbalance, urine abnormalities, and immune system disorders may come forward because of infertility issues. Changes in the DNA sequence are the leading cause of fetal loss. The developing fetus may also be infected and is very sensitive regarding pregnancy.
- In men's case, high radiation levels reduce men's ability to produce moderate amounts of sperm and play a significant role in degrading sperm quality. Reproductive organs are crucial in keeping the fertility process active and regular. Structural abnormalities may be highlighted in the reproductive organs when the males are affected by this high-frequency radiation. At the same time, radiation may affect the testicles. Testicles, also known as the testis, are one particular kind of male reproductive organs and are very much responsible for producing sperm in males and testosterone

in the hormone. Radiation may cause them to lose these organs' ability to function, and sperm production may be hampered or stopped permanently.

1.8 Thermal Effects

Thermal effects are another impactful factor in radiofrequency radiation. Thermal effects are primarily used to refer to changes in temperature inside the body. As the body's internal temperature increases significantly because of the thermal effect, it is difficult to estimate the damage this problem can cause. Excessive exposure to radiofrequency energy absorption might heat the body tissues more than usual and hamper the biological process directly connected to immune responses. As the human body absorbs high radiation levels, which different sources can absorb, especially in 5 G-supported devices and technologies, temperature changes can be observed in various tissues in the body. Wave conduction is caused by lengthy exposure to high-frequency waves emitted by 5G devices, and this is also because of thermal effects. The more 5G-supported devices are used, the higher the chance of this radiation entering the human body. Emitted radiation energy is quickly converted into heat energy; heat energy interferes with the body's everyday activities, encourages making wrong decisions, harms tissues, and many more. If the body temperature rises too much, the expected performance of the human body is reduced, attention is diminished, and memory fails to do regular tasks systematically. The safety measures maintained by the 5G technology-supported service provider are not safe enough for the children regarding the thermal effects mentioned by the author (Belpomme et al., 2018). Excessive heat flow can cause kidney problems as kidneys are sensitive to excess heat. The kidney will lose the ability to do its regular duty if it is affected by radiation or any other reason. A patient with heart disease will experience difficulty breathing due to excess radiation exposure and increased hypertension. Dehydration is another problematic situation that may occur because of high temperature and blood volume may be affected. If the radiation is high, this situation can be much more dire. Similarly, asthmatic patients also face problems due to this radiation.

3. BACKGROUND STUDIES IN 5G REGARDING HEALTH ISSUES

Electromagnetic radiation is often responsible for different health issues, most like afflictions in the neurological and has a higher probability of damaging the tissues of the living organs. It is now one of the most concerning issues in the healthcare field because radiation exposure is one of the leading causes of tissue damage and is a more severe cause of tissue damage than any other reason. This type of radiation is emitted from the mobile phone; among this, 5G technology-supported smartphones throw away the maximum amount of radiation (Hemalatha et al., 2020). A review was published after analyzing 94 recent publications by (Simkó & Mattsson, 2019), where they talked about the health consequences of these highly harmful frequencies. Mònica Guxens and company (Guxens et al., 2019) propose the side effects of radiofrequency electromagnetic fields (EMF), which they applied on five-year-old Amsterdam-born Children. The survey was mainly done to show that high radiation levels may also cause extreme harm to children. Though they said the radiation they absorbed from the phone calls was not precisely amalgamated with any emotional and behavioral problems, it is still a thoughtful subject. 5G technology has become so prevalent that children are exposed to it regularly nowadays, and it is hazardous for a growing baby. Different types of radioactive substances are called carcinogens. Different types of chemicals, radiation, viruses, and other factors are available in carcinogen. Human carcinogenicity is much more poisonous to children, is one of the cancer casualties (Moon, 2020), and might injure animals. Expecting high performance and promising results of 5 G-supported devices to meet customer needs, developing high-power radiation-receiving devices is now becoming a trend. At the same time, service providers use more additional radiation in the hope of better results and on-demand devices. The International Commission on Non-Ionizing Radiation Protection (ICNIRP) said it is a worrying sign for human health if the 5G networks often use radiofrequency above 6 GHz (Karipidis et al., 2021).

Hematological parameters (HP) are clinical indicators like other internal cells or organs of a human body. HP includes hemoglobin (Hb), white and red blood cell counts, platelet count, and erythrocyte sedimentation rate (ESR). These parameters are also circuitously affected by the radiation that comes from the smartphone, as shown by (Christopher et al., 2020). Electromagnetic radiation (EMR) in 5G-supported devices directly impacts the human body because this type of appliance uses the 28 GHz band, the world's most tested band. Skin color will be influenced by skin pigmentation disturbance. Skin pigmentation, available in murine melanoma cells (B16F10), might be fatigued because of 5G (28 GHz) EMR (Kim et al., 2022). A murine tumor cell line called B16 melanoma effortlessly may come into contact with 5G (28 GHz) and be affected, which is destructive for the human body because this cell body is usable for research of human skin cancer. One of the most popular histopathology techniques, Fontana-Masson (FM)staining, demonstrated that the extensive availability of 5G radiation could decline the total amount of B16 cells. Ex vivo bovine brain tissue sucks up the radiation from the 5G wireless networks as frequency, and the tissue might transmit the excessively absorbed radiation throughout the body, significantly increasing the body's inner temperature by a considerable margin. For that, a method was proposed by (Gultekin & Siegel, 2020) to measure the temperature rising and frequency absorption rate of the tissue samples. The most significant difference between ionizing and non-ionizing radiation is that ionizing radiation carries a unique energy that makes it able to easily remove electrons from atoms and molecules contained in various material particles (Centers for Disease Control and Prevention, n.d.). However, the matter is more complicated in the case of the non-ionizing radiation to remove those electrons. However, the non-ionizing, non-visible radiation can also cause severe health issues.

The high-intensity radiation affects the internal organs and, at the same time, causes severe damage to the external human organs on a much larger scale. In their proposed paper, evidence was provided by (Kostoff et al., 2020) that the newborn 5G technology impacted external human organs and, more importantly, the adverse system, which will be responsible for diverse types of long-term diseases. Malignant brain cancer is hazardous to the human body and can sometimes be fatal. Different types of tumors can be caused by the abnormal and uncontrolled growth of different types of cells present in the brain. This type of tumor is one of the main symptoms of malignant brain cancer. The International Health Research Agency (IARC) classified radiation as possibly carcinogenic to humans, which raises the chances of glioma, a malignant type of brain cancer (Silva, 2020). The brain and spinal cord have glial cells, and glioma might arise from the glial cell. Prof. Magda Havas says in his article that red blood cells might be contaminated because of the disclosure of mobile phone radiation and mentions that stroke occurrences can be increased (Havas, 2013). In the process of contamination, the blood is not flowing to the brain properly as it is interrupted, and brain cells might be damaged. A correlation exists between radiation density and a person who consumes radiation because a typical receiver boosts the probability of exposure to radiofrequency electromagnetic fields and undergoes unfavorable consequences on human health, like severe headaches, dizziness, anxiety, and nightmares at night (López et al., 2021; Chandan et al., 2021; Hinrikus et al., 2022; Ramakrishnan & Athikary, 2023). Lack of adequate and proper sleep is considered one of the leading causes of any chronic disease, and the author discusses the symptoms of lack of adequate sleep.

EMF disclosure limits must be increased because evaluating the threat in the 5G technology is impartial (Hardell & Carlberg, 2020), and 5G works parallelly with the existing mobile technologies, creating a new point at issue and challenges described by (Pawlak et al., 2019). As the number of 5G-supported devices increases, which must be, and technology promotes daily, the issue and challenges also increase alarmingly. If the electromagnetic field of radio frequency releasing rate is above 6GHz from a mobile phone, then it should be reckoned in terms of incident power density, as said by (Xu et al., 2019). Most telecommunication companies and bandwidth providers are working hard to add extra frequencies to the current network, which might not be a good sign for human beings. Even before the start of the COVID-19 pandemic, more than 500 million people worldwide were living below the poverty line due to healthcare costs alone. Diseases caused by electromagnetic radiation are one of the primary reasons for this poverty (The World Bank, 2020), and one of the main reasons for this is that the symptoms of radiation-induced diseases are not easily understood, and the consequences are far-reaching. The exceeding number of smartphones used is the conclusive point for increasing the radiofrequency(RF) radiation shown by (Carlberg et al., 2019), where the author warns that this amount is not likely to decrease but will only increase day by day. The author also added that, in the era of globalization, we are becoming more modern, trying to simplify and improve the quality of life, and telecommunication technology is gradually improving. The author demonstrates that RF amount was maximum in HayMarket, central parts of Stockholm, Sweden, the most crowded area. Much debate and controversy exists about 5G. However, the author has proven through considerable evidence that radio frequency is capable of causing significant disorders in human health (Frank, 2021), and, considering the environmental aspect, radio frequency is capable of causing substantial damage to plants and animals as well (International EMF Scientists Appeal Statement, 2020; Physicians' Health Initiative for Radiation and Environment, n.d.). Conventional systems like 2G-4Gs dispatch fewer packets of data and use less speed than newly established 5G systems. As they transfer fewer data packets, the efficiency and productivity of these networks are much less than those of 5G networks. Because of these, the possibility of incurable diseases is also much less. Higher amounts of originated radiofrequency from the 5G wireless systems bring health and safety issues addressed by COMAR Technical Information Statement (TIS) (Bushberg et al., 2020).

Substances by any cause may be the reason for cancer, which is called a carcinogen. Whereas radiofrequency (RF) in between the coverage of 30 KHz-30GHz is categorized as a human carcinogen. A human carcinogen is Group 1 stage cancer, narrated by the International Agency for Research on Cancer (IARC) (Hardell & Nyberg, 2020). The horribleness of RF utilized in 5G technologies is smoothly understood from the report published by the author (Hardell & Nilsson, 2023). Because of installing a 5G base station on the roof, dissimilar physiological problems like neurology-related signs, tinnitus, fatigue, insomnia, skin disorders, and blood pressure variability were observed among the residents. However, after leaving the cottage, within a few days, the symptoms vanished utterly. Not specifically, but in precise circumstances, the human body reacts like a dipole, an electromagnetic phenomenon. It happens when the body consumes a baseband beyond 20GHz, automatically heating the tissue (Matthew & Kazaure, 2021). Because of heating the tissues, discrete polarization like dielectric, ionic, and orientational polarization is assembled. Compared to a youngster and an adult, if there is a 5G smartphone near the head, radiation can easily expose the child's brain infrastructure because the bone marrow of the child's skull is very unsubstantial and has a relatively high radiation absorption capacity (Miller et al., 2019). Though the 5G telecommunications technique has become highly controversial because of its two-faced character (Frank, 2021), Moskowitz, J. M. (Moskowitz, 2019) expressed that we have no justification for believing that 5G technology is adequately safe and secure. At the same time, the author highlighted the fact very seriously that if more caution is not taken now, humankind will have to pay a heavy price for it.

4. CONCLUSION

All progressive technological innovations have constructed our lives more comfortably and accessible, simplified living standards, and scientists are constantly developing new and innovative plans. Periodically, they can create entirely new technologies, whereas sometimes, they innovate old technologies by adding new features to the existing ones. Just as these modern and high-powered technologies have enhanced our quality of life, we now encounter diverse problems at different stages of life. High radiation levels from 5G technology are one of the leading causes of this concern in the healthcare industry. Devices that support 5-G technology can absorb high-frequency waves, which can be considered one of the foremost causes of various complex and challenging diseases of the human body. Like every other disease, the symptoms caused by higher waves are not smoothly seen so they can take a severe form shortly. 5G-supported devices utilize at least two to three times more radiofrequency than 4G-supported devices, which is alarming, and the used waves may cause far-reaching and severe diseases. Although research is underway on how much the incoming waves from 5G technology-supported devices can harm the body, humankind must be very cautious. 5G technology should be used sparingly and keeping a safe distance as much as possible. Through this work, we have tried to demonstrate that these waves can cause distinct diseases in the human body, which will be a more significant threat in the healthcare industry, and in the future, we will discuss what kind of work can be done to unravel this problem.

ACKNOWLEDGMENT

I am very grateful to my advisor, Dr. [UMESH KUMAR LILHORE], for all his contributions and recommendations for this project. His expertise and guidance were constructive for me to design my research and write my report. I was also quite thankful to Chandigarh University for allowing me to work with Dr. UMESH. Lastly, I appreciate the opportunity to intercommunicate my analysis judgments to a larger audience.

REFERENCES

Agiwal, M., Roy, A., & Saxena, N. (2016). Next generation 5G wireless networks: A comprehensive survey. *IEEE Communications Surveys and Tutorials*, *18*(3), 1617–1655. doi:10.1109/COMST.2016.2532458

Al Delfi. H. H., Mahdi, Q. S., & Celik, B. (2023, January). The side-effects of 5G between hypothesis and theory. In AIP Conference Proceedings (Vol. 2554, No. 1, p. 080001). AIP Publishing LLC.

Belpomme, D., Hardell, L., Belyaev, I., Burgio, E., & Carpenter, D. O. (2018). Thermal and non-thermal health effects of low intensity non-ionizing radiation: An international perspective. *Environmental Pollution, 242*, 643–658. doi:10.1016/j.envpol.2018.07.019 PMID:30025338

Bhandari, N., Devra, S., & Singh, K. (2017). Evolution of cellular network: from 1G to 5G. *International journal of engineering and techniques, 3*(5), 98-105.

Bushberg, J. T., Chou, C. K., Foster, K. R., Kavet, R., Maxson, D. P., Tell, R. A., & Ziskin, M. C. (2020). IEEE Committee on Man and Radiation—COMAR Technical Information Statement: Health and safety issues concerning exposure of the general public to electromagnetic energy from 5G wireless communications networks. *Health Physics, 119*(2), 236–246. doi:10.1097/HP.0000000000001301 PMID:32576739

Carlberg, M., Hedendahl, L., Koppel, T., & Hardell, L. (2019). High ambient radiofrequency radiation in Stockholm city, Sweden. *Oncology Letters, 17*(2), 1777–1783. PMID:30675237

Chandan, R. K., Suman, P. N., & Sinha, K. (2021). The Environmental Impact of 5G Technology on Humans and Animals. In Handbook of Research on Knowledge and Organization Systems in Library and Information Science (pp. 48-68). IGI Global.

Christopher, B., Mary, Y. S., Khandaker, M. U., Bradley, D. A., Chew, M. T., & Jojo, P. J. (2020). Effects of mobile phone radiation on certain hematological parameters. *Radiation Physics and Chemistry, 166*, 108443. doi:10.1016/j.radphyschem.2019.108443

de Vocht, F., & Albers, P. (2022). The population health effects from 5G: Controlling the narrative. *Frontiers in Public Health, 10*, 1082031. doi:10.3389/fpubh.2022.1082031 PMID:36600933

del Peral-Rosado, J. A., Raulefs, R., López-Salcedo, J. A., & Seco-Granados, G. (2017). Survey of cellular mobile radio localization methods: From 1G to 5G. *IEEE Communications Surveys and Tutorials, 20*(2), 1124–1148. doi:10.1109/COMST.2017.2785181

El-Hajj, A. M., & Naous, T. (2020, September). Radiation Analysis in a Gradual 5G Network Deployment Strategy. In *2020 IEEE 3rd 5G World Forum (5GWF)* (pp. 448-453). IEEE.

Eluwole, O. T., Udoh, N., Ojo, M., Okoro, C., & Akinyoade, A. J. (2018). From 1G to 5G, what next? *IAENG International Journal of Computer Science, 45*(3).

Frank, J. W. (2021). Electromagnetic fields, 5G and health: What about the precautionary principle? *Journal of Epidemiology and Community Health, 75*(6), 562–566. doi:10.1136/jech-2019-213595 PMID:33468601

Gonzalez-Usigli, H. A. (2023, September 29). *Parkinson Disease (PD)*. MSD Manual Consumer Version. https://www.msdmanuals.com/en-in/home/brain,-spinal-cord,-and-nerve-disorders/movement-disorders/parkinson-disease-pd

Goyal, J., & Singla, K., Akashdeep, & Singh, S. (2020). A survey of wireless communication technologies from 1G to 5G. In *Second International Conference on Computer Networks and Communication Technologies: ICCNCT 2019* (pp. 613-624). Springer International Publishing. 10.1007/978-3-030-37051-0_69

Gultekin, D. H., & Siegel, P. H. (2020). Absorption of 5G radiation in brain tissue as a function of frequency, power and time. *IEEE Access : Practical Innovations, Open Solutions*, *8*, 115593–115612. doi:10.1109/ACCESS.2020.3002183

Guxens, M., Vermeulen, R., Steenkamer, I., Beekhuizen, J., Vrijkotte, T. G., Kromhout, H., & Huss, A. (2019). Radiofrequency electromagnetic fields, screen time, and emotional and behavioural problems in 5-year-old children. *International Journal of Hygiene and Environmental Health*, *222*(2), 188–194. doi:10.1016/j.ijheh.2018.09.006 PMID:30314943

Hardell, L., & Carlberg, M. (2020). [Comment] Health risks from radiofrequency radiation, including 5G, should be assessed by experts with no conflicts of interest. *Oncology Letters*, *20*(4), 1–1. doi:10.3892/ol.2020.11876 PMID:32774475

Hardell, L., & Nilsson, M. (2023). Case report: The microwave syndrome after installation of 5g emphasizes the need for protection from radiofrequency radiation. *Annals of Case Reports*, *8*, 1112.

Hardell, L., & Nyberg, R. (2020). [Comment] Appeals that matter or not on a moratorium on the deployment of the fifth generation, 5G, for microwave radiation. *Molecular and Clinical Oncology*, *12*(3), 247–257. doi:10.3892/mco.2020.1984 PMID:32064102

Hasan, H. J., Yosef, A. J., & Hachem, H. A. (2021). 5G radiation and potential risks to the environment and human health. [TURCOMAT]. *Turkish Journal of Computer and Mathematics Education*, *12*(6), 1689–1693.

Havas, M. (2013). Radiation from wireless technology affects the blood, the heart, and the autonomic nervous system1. *Reviews on Environmental Health*, *28*(2-3), 75–84. doi:10.1515/reveh-2013-0004 PMID:24192494

Hemalatha, R., Rhesa, M., & Revathi, S. (2020). 5G mobile communication, challenges: A survey. *Bull Sci Res*, *35*(42), 35.

Hinrikus, H., Koppel, T., Lass, J., Orru, H., Roosipuu, P., & Bachmann, M. (2022). Possible health effects on the human brain by various generations of mobile telecommunication: A review based estimation of 5G impact. *International Journal of Radiation Biology*, *98*(7), 1210–1221. doi:10.1080/09553002.2022.2026516 PMID:34995145

Hodara, H., & Skaljo, E. (2021). From 1G to 5G. *Fiber and Integrated Optics*, *40*(2-3), 85–183. doi:10.1080/01468030.2021.1919358

Hoffman, Chris (2019). *What is 5G, and how fast will it be?* How-To Geek LLC.

Hollington, J., & Jansen, M. (2023, September 9). How fast is 5G? What you need to know about 5G speeds. *Digital Trends*. https://www.digitaltrends.com/mobile/how-fast-is-5g/

Karipidis, K., Mate, R., Urban, D., Tinker, R., & Wood, A. (2021). 5G mobile networks and health—A state-of-the-science review of the research into low-level RF fields above 6 GHz. *Journal of Exposure Science & Environmental Epidemiology*, *31*(4), 585–605. doi:10.1038/s41370-021-00297-6 PMID:33727687

Kim, K., Lee, Y. S., Kim, N., Choi, H. D., & Lim, K. M. (2022). 5G Electromagnetic Radiation Attenuates Skin Melanogenesis In Vitro by Suppressing ROS Generation. *Antioxidants*, *11*(8), 1449. doi:10.3390/antiox11081449 PMID:35892650

Kostoff, R. N., Heroux, P., Aschner, M., & Tsatsakis, A. (2020). Adverse health effects of 5G mobile networking technology under real-life conditions. *Toxicology Letters*, *323*, 35–40. doi:10.1016/j.toxlet.2020.01.020 PMID:31991167

Krämer, K. (2023). Daily briefing: Alzheimer's makes brain cells undergo'cellular suicide'. *Nature*. doi:10.1038/d41586-023-02965-z

Li, Z., Wang, X., & Zhang, T. (2020). *From 5G to 5G+*. Singapore: Springer Singapore., doi:10.1007/978-981-15-6819-0_3

Lilhore, U. K., Dalal, S., Faujdar, N., Margala, M., Chakrabarti, P., Chakrabarti, T., Simaiya, S., Kumar, P., Thangaraju, P., & Velmurugan, H. (2023). Hybrid CNN-LSTM model with efficient hyperparameter tuning for prediction of Parkinson's disease. *Scientific Reports*, *13*(1), 14605. doi:10.1038/s41598-023-41314-y PMID:37669970

Lilhore, U. K., Dalal, S., Faujdar, N., Simaiya, S., Dahiya, M., Tomar, S., & Hashmi, A. (2024). Unveiling the prevalence and risk factors of early stage postpartum depression: A hybrid deep learning approach. *Multimedia Tools and Applications*, 1–35. doi:10.1007/s11042-024-18182-3

Lilhore, U. K., Manoharan, P., Sandhu, J. K., Simaiya, S., Dalal, S., Baqasah, A. M., Alsafyani, M., Alroobaea, R., Keshta, I., & Raahemifar, K. (2023). Hybrid model for precise hepatitis-C classification using improved random forest and SVM method. *Scientific Reports*, *13*(1), 12473. doi:10.1038/s41598-023-36605-3 PMID:37528148

Lilhore, U. K., Poongodi, M., Kaur, A., Simaiya, S., Algarni, A. D., Elmannai, H., Vijayakumar, V., Tunze, G. B., & Hamdi, M. (2022). Hybrid model for detection of cervical cancer using causal analysis and machine learning techniques. *Computational and Mathematical Methods in Medicine*, *2022*, 2022. doi:10.1155/2022/4688327 PMID:35572826

Lilhore, U. K., Simaiya, S., Pandey, H., Gautam, V., Garg, A., & Ghosh, P. (2022). Breast cancer detection in the IoT cloud-based healthcare environment using fuzzy cluster segmentation and SVM classifier. In Ambient Communications and Computer Systems [Singapore: Springer Nature Singapore.]. *Proceedings of RACCCS*, *2021*, 165–179.

Lilhore, U. K., Simaiya, S., Prasad, D., & Guleria, K. (2020). A hybrid tumor detection and classification based on machine learning. *Journal of Computational and Theoretical Nanoscience*, *17*(6), 2539–2544. doi:10.1166/jctn.2020.8927

Lin, J. C. (2020). 5G communication technology and coronavirus disease [Health Matters]. *IEEE Microwave Magazine*, *21*(9), 16–19. doi:10.1109/MMM.2020.2999236

Liu, G., & Jiang, D. (2016). 5G: Vision and requirements for mobile communication system towards the year 2020. *Chinese Journal of Engineering*, *2016*, 8. doi:10.1155/2016/5974586

López, I., Félix, N., Rivera, M., Alonso, A., & Maestú, C. (2021). What is the radiatiotn before 5G? A correlation study between measurements in situ and in real time and epidemiological indicators in Vallecas, Madrid. *Environmental Research, 194*, 110734. doi:10.1016/j.envres.2021.110734 PMID:33434609

Matthew, U. O., & Kazaure, J. S. (2021). Chemical polarization effects of electromagnetic field radiation from the novel 5G network deployment at ultra high frequency. *Health and Technology*, *11*(2), 305–317. doi:10.1007/s12553-020-00501-x PMID:33527067

McLaughlin, M. F., Donoviel, D. B., & Jones, J. A. (2017). Novel indications for commonly used medications as radiation protectants in spaceflight. *Aerospace Medicine and Human Performance*, *88*(7), 665–676. doi:10.3357/AMHP.4735.2017 PMID:28641684

Miller, A. B., Sears, M. E., Morgan, L. L., Davis, D. L., Hardell, L., Oremus, M., & Soskolne, C. L. (2019). Risks to health and well-being from radio-frequency radiation emitted by cell phones and other wireless devices. *Frontiers in Public Health*, *7*, 223. doi:10.3389/fpubh.2019.00223 PMID:31457001

Moon, J. H. (2020). Health effects of electromagnetic fields on children. *Clinical and Experimental Pediatrics*, *63*(11), 422–428. doi:10.3345/cep.2019.01494 PMID:32683815

Moshood, T. D., Shittu, R. A., & Abidin, T. (2020). Covid-19 and 5G radiation are two parallel lines: A systematic review. *Int. J. Innov. Sci. Res. Technol*, *5*(7), 744–751. doi:10.38124/IJISRT20JUL597

Moskowitz, J. M. (2019). We Have No Reason to Believe 5G Is Safe. *Scientific American.*

Orbai, A. M. (2022). *What Are Common Symptoms of Autoimmune Disease? Johns Hopkins Medicine website*. Accessed June.

Panagopoulos, D. J. (2019). Comparing DNA damage induced by mobile telephony and other types of man-made electromagnetic fields. *Mutation Research/Reviews in Mutation Research*, *781*, 53–62. doi:10.1016/j.mrrev.2019.03.003 PMID:31416578

Pawlak, R., Krawiec, P., & Zurek, J. (2019). On measuring electromagnetic fields in 5G technology. *IEEE Access : Practical Innovations, Open Solutions*, *7*, 29826–29835. doi:10.1109/ACCESS.2019.2902481

Physicians' Health Initiative for Radiation and Environment (PHIRE). (n.d.). *Safety limits and political conflicts of interest*. PHIRE. https://phiremedical.org/safety-limits-and-political-conflicts-of-interest/

Pölzl, C. (2011). EMF recommendations specific for children? *Progress in Biophysics and Molecular Biology*, *107*(3), 467–472. doi:10.1016/j.pbiomolbio.2011.09.018 PMID:21967839

Ramakrishnan, M., & Athikary, K. G. (2023). Health effects of fifth-generation technologies. *International Journal of Environmental Health Engineering*, *12*(1), 3. doi:10.4103/ijehe.ijehe_18_22

Sciorio, R., Tramontano, L., & Esteves, S. C. (2022). Effects of mobile phone radiofrequency radiation on sperm quality. *Zygote (Cambridge, England)*, *30*(2), 159–168. doi:10.1017/S096719942100037X PMID:34384508

Shah, A. S. (2022, January). A survey from 1G to 5G including the advent of 6G: architectures, multiple access techniques, and emerging technologies. In *2022 IEEE 12th Annual Computing and Communication Workshop and Conference (CCWC)* (pp. 1117-1123). IEEE.

Simkó, M., & Mattsson, M. O. (2019). 5G wireless communication and health effects—A pragmatic review based on available studies regarding 6 to 100 GHz. *International Journal of Environmental Research and Public Health, 16*(18), 3406. doi:10.3390/ijerph16183406 PMID:31540320

Sofri, T., Rahim, H. A., Abdulmalek, M., Abd Rani, K., Omar, M. H., Yasin, M. N. M., & Soh, P. J. (2021). Health Effects of 5G Base Station Exposure: A Systematic Review. *IEEE Access : Practical Innovations, Open Solutions*, *10*, 41639–41656. doi:10.1109/ACCESS.2021.3139385

Sood, R., & Garg, A. (2014). Digital society from 1G to 5G: A comparative study. [IJAIEM]. *International Journal of Application or Innovation in Engineering & Management*, *3*(2), 186–193.

The World Bank. (2020). *Health Overview: Development News, Research, Data*. World Bank.. https://www. worldbank.org/en/topic/health/overview#1

Wdowiak, A., Mazurek, P. A., Wdowiak, A., & Bojar, I. (2017). Effect of electromagnetic waves on human reproduction. *Annals of Agricultural and Environmental Medicine*, *24*(1), 13–18. doi:10.5604/12321966.1228394 PMID:28378967

Wu, T., Rappaport, T. S., & Collins, C. M. (2015). Safe for generations to come: Considerations of safety for millimeter waves in wireless communications. *IEEE Microwave Magazine*, *16*(2), 65–84. doi:10.1109/MMM.2014.2377587 PMID:26539066

Xu, B., Zhao, K., Ying, Z., Sjöberg, D., He, W., & He, S. (2019). Analysis of impacts of expected RF EMF exposure restrictions on peak EIRP of 5G user equipment at 28 GHz and 39 GHz bands. *IEEE Access : Practical Innovations, Open Solutions*, *7*, 20996–21005. doi:10.1109/ACCESS.2019.2897271

Yasin, N. (2023). *Amyotrophic Lateral Sclerosis (ALS): A Visual Representation*. McMaster Brain Research Society.

Chapter 6
Bridging Bytes and Biology–Advanced Learning and Bioinformatics in Innovative Drug Discovery

Dwijendra Nath Dwivedi
https://orcid.org/0000-0001-7662-415X
Krakow University of Economics, India

Ghanashyama Mahanty
https://orcid.org/0000-0002-6560-2825
Utkal University, India

Varunendra Nath Dwivedi
Vellore Institute of Technology, India

ABSTRACT

In this chapter, the authors explore the synergistic integration of advanced machine learning techniques with bioinformatics to revolutionize drug discovery. This interdisciplinary research is jointed on the premise that combining computational power with biological data can lead to more efficient, precise, and innovative approaches to identifying new therapeutic compounds. The chapter begins by detailing the journey of drug discovery from ancient times to modern. The authors discuss the current challenges in drug discovery and delve into how machine learning algorithms can analyze vast amounts of biological data. A significant portion of the chapter is dedicated to case studies where these techniques have been successfully applied, demonstrating their potential in identifying novel drug candidates and in personalizing treatment. The authors also discuss the various regulations and guidelines that address the challenges of data heterogeneity and the need for robust and interpretable models.

DOI: 10.4018/979-8-3693-1822-5.ch006

1. INTRODUCTION

The history of drug creation is an interesting journey that shows how science and medicine have changed over time. In the past, most drug research was based on empirical methods, which were based on the knowledge of traditional healers who used plants, minerals, and animal parts to treat a wide range of illnesses. This method was based on trying things out and learning from cures that have been passed down from generation to generation. Researchers began to look into the natural world in a more organized way during the Renaissance. This made it easier to find drugs like morphine. The 19th and 20th centuries were the start of modern medicines, as shown in Tables 1 and 2. During this time, big steps forward were made in the creation of drugs, like making aspirin and finding antibiotics like penicillin. The last few years of the 20th century and the start of the 21st have seen the rise of biotechnology and the use of genetic and molecular biology techniques. It made it easier for focused medicines and personalized medicines to get better. Drug development is a difficult and multidisciplinary process that brings together chemistry, biology, computer science, and engineering to create new medicines for a wide range of illnesses.

A new frontier is being created in the field of drug development as a result of the integration of advanced computer learning approaches with bioinformatics. The industry is undergoing a fundamental transformation as a result of the convergence of bytes and biology, which is usually referred to as the "bridging of bytes and biology." Drug development has always been defined by its lengthy duration,

Table 1. The progression of drug discovery

Timeline:	Key Innovations:
Ancient Times	Herbal remedies and traditional medicinal practices
18th-19th Century	Isolation of active ingredients from natural products
Early 20th Century	Development of synthetic drugs (e.g., Aspirin)
Mid-20th Century	Discovery of antibiotics and large-scale drug manufacturing
Late 20th Century	Advancements in molecular biology and genetic engineering
21st Century	Integration of bioinformatics and personalized medicine

Table 2. The progression of drug discovery in 21st century

Timeline:	Key Innovations:
2000s - Genomics Era Begins	Completion of the Human Genome Project, leading to a better understanding of genetic factors in diseases.
2010s - Rise of Personalized Medicine	Development of therapies tailored to individual genetic profiles, targeting specific disease mechanisms.
2010s - Advancements in Biologics	Increased development and approval of biologic drugs, including monoclonal antibodies, for various diseases.
2020s - AI and Machine Learning Integration	Application of AI and ML in drug discovery and development, enhancing efficiency and predictive accuracy.
2020s - CRISPR Gene Editing	CRISPR technology revolutionizing gene editing, with potential for treating genetic disorders.

expensive expenses, and inherent unpredictability. These characteristics have been present throughout the process. Furthermore, the percentage of their failures is an extremely high one. Because of the advancements that have been made in the fields of machine learning, data analytics, and bioinformatics, the paradigm that is now in place is undergoing a tremendous transformation. The purpose of this study is to optimize and improve the process of drug development by introducing an innovative method that makes use of powerful machine learning algorithms and the huge data that is derived from bioinformatics. The utilization of sophisticated machine learning techniques for the purpose of analyzing massive and complex biological datasets is the foundation of this strategy. The algorithms in question are able to identify potential medication candidates and make predictions regarding how they will interact with biological targets. In comparison to the capabilities of traditional methods, they are able to foresee potential bad reactions with a speed and accuracy that is unmatched by other methods.

Classification Methods: High-throughput sequencing technologies have recently made significant advancements, which have enabled scientists to gather huge biological information. Following the incorporation of internet web services, these datasets became readily accessible to scientific audiences in a short amount of time. As a consequence of this, numerous bioinformatics frameworks began to make use of predictive models and a growing dependence on approaches that involve machine learning. The act of allocating data into categories is referred to as classification in the field of bioinformatics. This process encompasses a wide range of biomedical applications, including disease identification, drug response prediction, gene annotation, and exon boundary detection in DNA sequences, amongst many others. Classification that is powered by machine learning has the potential to give solutions that are both powerful and versatile for bioinformatics issues.

Detection Methods: Machine learning models are able to tackle new tasks or classes with only a little amount of labelled data thanks to a technique called Few-Shot Learning. On account of this, this method is particularly useful in situations when the collection of big datasets with labels is either difficult or impossible. Additionally, Few-Shot Learning makes it possible for models to rapidly adapt to new classes in a short amount of time. This is an advantage in bioinformatics applications such as detecting faces on smartphones or identifying rare diseases from radiological data. The pre-training of a deep model that is supervised on a rich support set and then the application of this model for few-shot classification on a query set is one approach that can be taken to overcome this difficulty.

Prediction Methods: Machine Learning (ML) approaches empower researchers to discover intricate patterns within datasets and generate precise statistical predictions regarding similar fresh datasets, thereby revolutionizing various facets of bioinformatics study. The incorporation of machine learning methods into bioinformatics frameworks has significantly enhanced the development of prediction models in the fields of molecular evolution, proteomics, systems biology, and disease genomics. Machine Learning techniques can be used to improve data processing and modeling activities. They can also help find understandable characteristics in biological data for interpretation purposes and assess performance against separate biological samples. Despite the challenges, ML methods have gained significant popularity in the field of biological sequence analysis. Neural networks, in particular, offer visualization and interpretation tools that assist biologists in better understanding the material they are investigating.

The fusion of technology and biology beyond mere effectiveness. It represents a shift towards an individualized approach to medicine. By employing bioinformatics, drugs can be tailored according to individual genetic profiles. However, this nascent field also has challenges and considerations to be mindful of. The primary issues are around data security, the ethical utilization of genetic data, and the necessity for cooperation between technologists and biologists. The study demonstrates that the integration of digital information and biological sciences has the potential to significantly revolutionize the possibilities in drug discovery. This conclusion is drawn from a thorough examination of case studies and future advancements. For leaders in the industry, being abreast of these advancements is not merely an opportunity, but a must in order to sustain competitiveness in a rapidly expanding domain.

2. LITERATURE REVIEW

Walters (2021) presented a comprehensive examination of the various uses of artificial intelligence in the field of drug discovery. The tasks encompassed property prediction, molecule production, picture analysis, and organic synthesis planning. The text explores the development of techniques for representing molecules and the possibility of methods for generating new molecules to explore uncharted chemical territories. Narayanan et. al. (2022) emphasized the significant impact of AI on the pharmaceutical business, particularly in speeding up drug discovery and the creation of novel medications for different ailments. Additionally, it explores the involvement of AI in the fields of nanomedicine and nanorobotics. Blanco-Gonzalez et al. (2022) discussed the advantages, difficulties, and limitations of artificial intelligence (AI) in the field of drug discovery. In their study, Cherney et al. (2023) proposed that the FDA should assess and certify AI technologies used in drug research to promote responsible usage and address concerns related to potential misuse. It underscores the necessity for protective measures in the utilization of AI in the field of medication discovery. Didi & Zecevic (2022) examined the necessity for the advancement of AI in drug discovery, specifically in regard to causality, in order to tackle the difficulties presented by misleading correlations in data. In their study, Yarragunta (2022) emphasized the significance of artificial intelligence (AI) in expediting the drug discovery process, specifically in relation to COVID-19. The text explores the application of artificial intelligence (AI) in the development of pharmaceuticals for different disorders, as well as the concept of drug repurposing. Tripathi et al. (2021) provided a concise overview of the significance of big data and artificial intelligence (AI) in the field of drug discovery. The authors specifically emphasized the creation of novel algorithms and structures for extracting valuable information from Cheminformatics databases. They also highlighted the diverse ways in which these advancements have influenced different stages of the drug discovery process. In their study, Vemula et al. (2022) examined the influence of computer-aided drug design (CADD), artificial intelligence (AI), machine learning (ML), and deep learning (DL) on the drug discovery process. This includes the identification of potential drug candidates and the several stages involved in the drug discovery pipeline. Turon et al. (2023) introduced ZairaChem, an artificial intelligence and machine learning tool designed for quantitative structure-activity relationship/quantitative structure-property relationship modeling. The tool was developed and deployed at the H3D Centre in Africa. The demonstration showcased the utilization of artificial intelligence and machine learning to optimize the process of drug discovery in settings with restricted resources. Philippidis (2021) examined the collaboration between NVIDIA, Schrödinger, AstraZeneca, and UF Health in utilizing artificial intelligence (AI) for drug discovery. The author highlights the significance of computational techniques in the advancement of therapeutic treatments. Cyranoski (2021) examined the increasing utilization of artificial intelligence (AI) in the field of drug research in China. The article emphasized the endeavors of Chinese enterprises and prominent information technology (IT) corporations in harnessing AI to facilitate groundbreaking medication development. Li, L. et al. (2022) presented ImDrug, a thorough benchmark and Python library designed to tackle the issues posed by uneven data distribution in AI-assisted drug discovery.

3. INTEGRATION OF COMPUTATIONAL METHODS AND CELLULAR SCIENCE

3.1 Machine Learning for Drug Design and Discovery

Machine learning (ML) is a strong tool that has changed the way medicines are made and found (Table 3). It is changing the way that pharmaceutical study is done. Using machine learning techniques in this area is changing the old way of finding new drugs, which takes a long time and costs a lot of money. That's because neural networks, more specifically deep learning models, can predict molecular action, toxicity, and pharmacokinetics. This helps scientists find new drugs. To get the best results from pharmaceutical substances, reinforcement learning is used to improve their structures over and over again until they reach certain goals. Drug repurposing is the process of changing existing medicines so that they can be used for different medical reasons. Machine learning (ML) could be an important part of this process. Through finding the right chemicals to test for new pharmacological properties and interesting targets for further research. The author of Cassidy's (2020) study looked into how machine learning (ML) methods can be used in drug discovery and development. In this study, they focused on using computer methods to find pharmacological targets and make lead molecules work better, with a main focus on cancer. In 2022, Priya et al. looked at how machine learning and e-learning are used in chemoinformatics to help make new drugs. This included guessing what drug targets would be, molecular docking, and building QSAR models to guess biological action. In 2022, Shoaib et al. looked into how machine learning and deep learning could be used in the area of drug discovery. The study was mostly about how to use

Table 3. Machine learning case studies for drug discovery (Source: Author)

#	Case Study	Description
1	Deep Learning for Drug Discovery	Use convolutional neural networks (CNNs) to predict how compounds will link to target proteins, which speeds up the identification of compounds.
2	Generative Models for Novel Molecule Design	Generative Adversarial Networks (GANs) and Variational Autoencoders (VAEs) are used to create new molecules with the qualities that are wanted.
3	Predicting Drug-Drug Interactions	Natural language processing (NLP) models are used to guess how drugs might interact with each other based on medical books and databases.
4	Biomarker Discovery for Precision Medicine	using machine learning techniques to find disease biomarkers in genetic and clinical data, which helps the development of targeted therapies.
5	Drug Repurposing Using ML	ML models can find new uses for drugs that are already on the market, like quickly testing drugs to see how well they work against COVID-19.
6	Predictive Models for Clinical Trial Outcomes	ML models are being made to guess how clinical trials will turn out by looking at data from past trials.
7	AI-Driven Drug Formulation	Predicting how drug formulations will behave using AI and ML models to improve drug delivery and effectiveness.

image-based profiling to find traits that are linked to disease and guess how well a treatment will work. In 2022, Jain looked into how machine learning and deep learning algorithms could be used in different stages of drug creation, such as validating targets, running clinical trials, and keeping an eye on drugs after they have been made.

3.2 Genomics and Next-Generation Sequencing Analysis

Artificial intelligence algorithms scrutinize genomic sequences to pinpoint potential targets for novel pharmaceuticals. Machine learning models are utilized to forecast the impact of various genetic variations on drug response, facilitating the customization of treatments for particular patients, a practice known as precision medicine.

AI algorithms necessitate extensive datasets of genetic sequences, which can be acquired from diverse sources, such as public genomic databases and scientific research. Next, a preprocessing phase is performed to clean and prepare the data. This step may comprise aligning sequences, addressing missing data, and normalizing genetic variants. AI systems scrutinize genomic sequences to detect particular characteristics or patterns that are pertinent to diseases. These traits may encompass gene expressions, mutations, epigenetic alterations, and other genomic markers. Methods such as Principal Component Analysis (PCA) are employed to decrease the intricacy of the data while preserving the most significant information. Support vector machines (SVM), random forests, and neural networks are trained using labeled genomic data to detect patterns linked to particular diseases or therapeutic responses.AI systems scrutinize genomic data to pinpoint certain genes or genetic pathways that exhibit a substantial correlation with specific diseases. AI models have the ability to forecast the impact of genetic variants on the development of diseases or the effectiveness of drugs, which helps in identifying possible targets for drug development. Potential targets are genes or proteins that have robust connections with diseases and are

Table 4. Machine learning case studies for genomics and next-generation sequencing analysis (Source: Author)

#	Case Study	Description
1	Cancer Genomics and Personalized Medicine	AI can look at genomic data in cancer cases to make personalized medicine.
2	Predictive Models for Genetic Diseases	Deep Genomics uses machine learning to guess what genetic differences will do, which helps us understand and treat genetic illnesses.
3	Gene Editing and CRISPR-Cas9	Using AI tools to find successful gene targets and predict off-target effects will help make CRISPR-Cas9 gene editing more accurate.
4	Microbiome Analysis for Health and Disease	Using AI to look at the human microbiome and figure out what part it plays in health and illness.
5	Non-Invasive Prenatal Testing (NIPT)	AI programs look at cell-free DNA in a mother's blood to find genetic problems early on in the pregnancy process.
6	Drug Discovery and Development	Using AI in genetics to find drug targets and learn more about how diseases work, like how GNS Healthcare and Biogen work together.
7	Neurogenetics and Brain Disorders	Using AI tools to look into the genetic causes of neurodevelopmental and neurodegenerative diseases.
8	Agricultural Genomics for Crop Improvement	Using AI to study crop genomics to boost output, resistance to disease, and ability to handle stress, like what Benson Hill Biosystems is doing.

projected to be amenable to modification by medicines.Subsequently, these targets undergo additional biological and clinical investigations in order to facilitate the development of novel pharmaceuticals.

Sharma (2023) examined the contributions of different types of AI, such as machine learning, neural-based learning, and network-based learning, in the process of discovering drugs and identifying targets for cancer treatment. The focus lies on the utilization of artificial intelligence technologies, web services, and databases for the purpose of finding and validating targets. In Smith's (2004) study, the author examined the difficulties associated with discovering vulnerabilities in possible therapeutic targets based on whole genome sequences. It emphasizes the initial phase of identifying targets for drug development. Manchado (2017) presented a method for identifying pharmacological targets using RNA interference (RNAi) tools and inducible RNAi platforms. This approach involves combining these techniques with genetically modified animal models based on embryonic stem cells to validate the targets in vivo. Hu (2019) conducts a comprehensive analysis of recent research on the identification of pharmacological targets, with a specific emphasis on the data and methodologies employed. This encompasses databases, biological experiments, and machine learning methodologies, encompassing feature selection and model optimization.

3.3 Proteomics and Protein Structure Prediction

Artificial intelligence computers search genomic sequences for novel drug targets. Precision medicine uses machine learning models to anticipate how genetic variations affect drug responsiveness.

AI algorithms need large genomic sequence datasets from public genomic databases and scientific study. Preparation involves data cleaning and formatting. This may involve sequencing alignment, missing data, and genetic variant normalization. AI algorithms search genetic sequences for disease-related patterns. These features include gene expressions, mutations, epigenetic changes, and genomic markers. Principal Component Analysis (PCA) reduces data complexity while keeping key information. Support vector machines (SVM), random forests, and neural networks train on labeled genomic data to find disease or therapeutic response patterns.AI algorithms analyze genomic data to identify genes or genetic pathways linked to certain disorders. AI models can predict how genetic polymorphisms affect disease development and treatment responsiveness, helping find therapeutic targets. Targets are genes or proteins having strong disease associations that may be modified by drugs.Further research is conducted on these targets to develop new pharmaceuticals. Sharma (2023) explored the role of AI, including machine learning, neural-based learning, and network-based learning, in discovering cancer treatment targets. It uses AI to identify and qualify targets using web services and databases. Smith (2004) studied the challenges of discovering therapeutic target vulnerabilities using entire genome sequencing. It emphasizes therapeutic target identification first. Pharmacological targets can be identified utilizing RNA interference (RNAi) methods and inducible RNAi platforms, according to Manchado (2017). In vivo validation of targets was achieved using genetically engineered animal models based on embryonic stem cells. Hu (2019) analyzes recent pharmaceutical target identification research, focusing on data and methods. Databases, biological experiments, and machine learning methods include feature selection and model optimization.

Artificial intelligence has improved protein architecture prediction by examining amino acid sequences. This is called protein structure prediction. Understanding protein function and drug discovery are crucial.

The input data is protein structure. AI algorithms look for patterns or qualities in the sequence that may indicate structural attributes. AI models called convolutional neural networks (CNNs) predict amino acid sequences' alpha helices and beta sheets. The models learn how amino acid order affects secondary

Table 5. Machine learning case studies for proteomics and protein structure prediction (Source: Author)

Case Study	Application	Description
AlphaFold by DeepMind	Protein Structure Prediction	Utilizes deep learning to accurately predict protein structures, enhancing understanding of protein folding and disease mechanisms.
Proteomics in Cancer Research	Cancer Biomarker Identification	AI algorithms analyze proteomic data for patterns and biomarkers, aiding in cancer detection and personalized treatment.
Drug Discovery and Development	Pharmaceutical Research	AI models predict protein-drug interactions, speeding up drug development processes, as seen in COVID-19 treatment research.
Understanding Rare Diseases	Disease Mechanism Analysis	AI is used to predict protein structures associated with rare diseases, providing insights into disease mechanisms for targeted therapies.
Biomarker Discovery for Neurodegenerative Diseases	Biomarker Identification	AI analyzes proteomic data to identify biomarkers for diseases like Alzheimer's and Parkinson's, contributing to early detection and new treatments.
Personalized Medicine	Tailored Treatment Strategies	AI-driven analysis of individual proteomes assists in predicting disease risk and drug responses, leading to personalized healthcare.
Enzyme Engineering	Biotechnological Applications	AI in the design and prediction of enzyme structures for specific industrial applications, such as biofuels and waste degradation.
Antibody Design	Therapeutic Antibody Development	AI aids in designing antibodies with high specificity and affinity for therapeutic use, particularly in targeting specific disease markers.

structure likelihood. Advanced models can predict protein three-dimensionality. Prediction of amino acid residue distances and angles, which are crucial to the three-dimensional structure, is used. AlphaFold, developed by DeepMind, advances protein structure prediction. The algorithm accurately predicts protein 3D configuration using advanced deep learning. For instance, RosettaFold uses sequencing data and physical and biological knowledge to predict protein forms using deep learning. Additional computational methods are often used to improve structural predictions. The predicted configurations are compared to X-ray crystallography or NMR spectroscopy-derived configurations for verification. Raha et al. (2000) developed a method for predicting amino acid sequences that match a protein's three-dimensional structure. The system generates sequences that closely resemble natural protein family members and uses statistical profile scores to evaluate protein design strategies. Guan et al. combined closest neighbor algorithms, neural networks, heuristic rules, and evolutionary algorithms to predict protein structure in 1994. The methods were applied to several protein sequences, yielding structures that approximated crystallography. Senior et al. (2020) created AlphaFold, a highly accurate protein structure prediction method. AlphaFold uses a neural network to estimate residue distances, providing a mean force potential that describes the protein's structure and can be improved using gradient descent. Dorn (2011) investigated how machine learning and optimization can solve the Protein Structure Prediction (PSP) problem. This article describes how AI can predict three-dimensional protein structures from amino acid sequences.

The application of AI techniques has significantly transformed the field of protein structure prediction by offering very accurate predictive tools. The progress has substantial ramifications for biological investigation and pharmaceutical exploration(Table 5).

3.4 Chemoinformatics

Genomic sequences are analyzed by AI systems to find drug targets. Precision medicine uses machine learning models to predict how genetic variants affect drug response.

AI algorithms need large genomic sequence datasets from public genomic databases and scientific study. Preparing the data comprises cleaning and formatting. It may involve aligning sequences, correcting missing data, and normalizing genetic variations. AI algorithms search genetic sequences for disease-related patterns. These features include gene expressions, mutations, epigenetic changes, and genomic markers. Principal Component Analysis (PCA) reduces data complexity while keeping key information. Support vector machines (SVM), random forests, and neural networks train on labeled genomic data to find disease or therapeutic response patterns.AI algorithms analyze genomic data to identify genes or genetic pathways linked to certain disorders. AI algorithms can predict how genetic polymorphisms affect disease progression and medication response, helping find therapeutic targets. Genes or proteins linked to diseases and potentially changed by drugs are potential targets.To create new drugs, these targets undergo further biological and clinical studies.

Sharma (2023) studied how machine learning, neural-based learning, and network-based learning can find cancer treatments and targets. Artificial intelligence, web services, and databases are used to discover and validate targets. Smith investigated whole genome sequencing challenges in identifying therapeutic target vulnerabilities in 2004. Early drug development target identification is emphasized. Manchado reported a strategy for discovering pharmaceutical targets utilizing RNA interference (RNAi) tools and inducible RNAi platforms in 2017. In vivo validation of targets was achieved using genetically engineered animal models based on embryonic stem cells. Hu (2019) analyzes recent pharmaceutical target identification research, focusing on data and methods. This course covers databases, biological experiments, and machine learning, including feature selection and model optimization. Artificial intelligence has improved protein structure prediction from amino acid sequences. This is called protein structure prediction. Understanding protein function and drug discovery are crucial.

Sharma & Sharma (2018) presented a comprehensive examination of artificial intelligence algorithms employed in chemoinformatics, with a specific focus on their application in the field of drug development. The text explores the significance of artificial intelligence (AI) in predicting bioactivity, designing new molecules, predicting synthesis outcomes, and analyzing biological images. It also provides a comprehensive overview of the tools, frameworks, and platforms employed in these specific applications. Baldi (2007) examined the difficulties and potential advantages of utilizing AI and machine learning in the field of chemoinformatics. The text outlines the creation of a comprehensive public database called ChemDB, which contains information about various chemicals and reactions. It also discusses the utilization of machine learning techniques to forecast molecular features and their practical use in drug screening and design. Navathe (2007) discussed the utilization of machine learning in the fields of bioinformatics and Geographic Information Systems (GIS), with a specific emphasis on text mining and ontology applications. Bajorath (2021) conducted a comprehensive evaluation of the most advanced techniques in artificial intelligence applied to medicinal chemistry. The primary emphasis was on deep learning, specifically utilizing neural network designs with several layers. Jiménez-Luna et al. (2021) provided an overview of the present state of artificial intelligence (AI) in chemoinformatics. This includes the application of AI in quantitative structure-activity/property relationship and structure-based modeling, as well as de novo molecular design and chemical synthesis prediction. In this study, Fang et al. (2014) examined various methods for drug target prediction using chemoinformatics. These methods

Table 6. Machine learning case studies for Chemoinformatics (Source: Author)

Case Study	Area of Application	Description
Drug Discovery and Molecular Docking	Pharmaceutical Research	AI predicts how chemical compounds bind to target proteins, aiding in rapid drug discovery and development.
Predicting Chemical Properties and Toxicity	Safety Analysis	AI is used to anticipate the properties of new compounds, including toxicity, improving drug safety and efficacy.
Optimizing Chemical Synthesis	Chemical Engineering	AI models design efficient synthesis pathways for chemical compounds, reducing resource expenditure in chemical synthesis.
Material Science and Discovery	Material Development	AI aids in discovering and designing new materials with specific properties, applicable in various industries.
Personalized Medicine	Healthcare	AI analyzes chemical data for tailoring drug regimens to individual patients, enhancing treatment effectiveness and reducing side effects.
Environmental Chemoinformatics	Environmental Analysis	AI assesses the environmental impact of chemical compounds, predicting their behavior and effects on ecosystems.
Virtual Screening for Drug Repurposing	Drug Development	AI conducts virtual screenings of existing drugs to identify new therapeutic uses, speeding up the drug repurposing process.
Chemical Image Analysis	Data Analysis	AI is applied in the analysis of chemical images from high-throughput screening to identify complex patterns.

encompassed ligand-based, receptor-based, and data mining-based predictions. Moeller (2011) presented a comprehensive overview of chemoinformatics, covering its fundamental principles and practical uses.

3.5 Parmacogenomics and Personalized Medicine

AI models combine genomic data and clinical outcomes to analyze the effects of genetic differences on the effectiveness and safety of drugs, therefore facilitating the development of personalized medicine. AI models are essential for merging genomic data with clinical results. This integration facilitates the comprehension of how genetic variants influence the effectiveness and safety of drugs. Data on clinical outcomes, patient health records, and therapeutic response are gathered. AI models combine genomic data and clinical data to generate a comprehensive dataset that establishes a connection between genetic information and health outcomes. Artificial intelligence systems detect genetic markers (biomarkers) that are linked to pharmacological responses or negative drug reactions. Machine learning methods, particularly those employing supervised learning, identify patterns in the data that associate certain genetic differences with clinical outcomes. Artificial intelligence algorithms utilize genetic profiles to forecast the response of individuals to particular medications. Genetic factors are examined in Safety and Efficacy models to see how they impact the safety and effectiveness of drugs. These models help identify potential risks of adverse responses. Tailored Therapies: By utilizing the predictions, healthcare providers can create individualized treatment strategies that are more probable to be successful and result in fewer adverse reactions. Dosage Optimization: AI models can propose the most advantageous medicine dosages by considering a patient's genetic composition.

Parekh et al. (2023) examined the impact of artificial intelligence (AI) on personalized medicine, namely in the development of customized treatment plans that take into account an individual's genetic makeup and medical background. The study conducted by Chafai et al. (2023) assessed the efficacy of machine learning algorithms employed in the field of genetic medicine and healthcare. The text pres-

Table 7. Machine learning case studies for Parmacogenomics and personalized medicine (Source: Author)

Case Study	Medical Condition	Description
Tailoring Cancer Treatment	Breast Cancer	Genetic testing identifies patients who will benefit from Herceptin (effective only in HER2-positive individuals).
Warfarin Dosing	Blood Clot Prevention	Pharmacogenomic testing guides precise dosing of Warfarin, reducing risks of bleeding or clotting, based on genetic factors.
Abacavir Hypersensitivity in HIV Treatment	HIV/AIDS	Identifies HIV patients at risk of hypersensitivity to Abacavir, preventing adverse reactions in those with HLA-B*5701 variant.
Cystic Fibrosis Treatment	Cystic Fibrosis	Ivacaftor treatment is tailored to patients with specific CFTR gene mutations, targeting the underlying cause of the disease.
Depression and Psychiatric Medications	Mental Health Disorders	Genetic testing predicts responses to antidepressants and psychiatric medications, aiding in effective, side-effect-minimized treatment selection.
Statins and Cholesterol Management	Cholesterol Management	Identifies individuals likely to experience side effects from statins, facilitating alternative treatments or dosage adjustments.
Asthma Treatment Personalization	Asthma	Pharmacogenomic testing enables tailored treatment, matching patients with the asthma medication they respond to best.
Pediatric Acute Lymphoblastic Leukemia (ALL)	Childhood Leukemia	Genetic testing informs chemotherapy intensity, reducing relapse risk and long-term side effects in pediatric ALL treatment.

ents an overview of the uses of artificial intelligence (AI) in clinical medicine, drug development, and genomic medicine. It also discusses the limitations and challenges of integrating AI in the healthcare industry. Johnson et al. (2020) investigated the merging of AI and precision medicine, emphasizing how AI utilizes computation and inference to produce individualized insights for diagnosis and treatment. In their study, Vadapalli et al. (2022) conducted a comparison of artificial intelligence (AI) and machine learning techniques employed in the fields of genomics and precision medicine. Tsai et al. (2016) outlines a bioinformatics approach to handle and transmit genomic data for clinical analysis in whole genome sequencing (WGS). Kumar & Gandhi (2023) presented a comprehensive examination of the utilization of artificial intelligence (AI) and machine learning (ML) in the field of breast cancer care. This encompasses several aspects such as early detection, diagnosis, prognosis, medication development, and individualized therapy planning. In their study, Restrepo et al. (2023) examined the progress made in genetic data and biomarkers in non-small cell lung cancer (NSCLC), emphasizing the contribution of artificial intelligence (AI) in improving customized medicine by analyzing biomarkers. In their study, Bai et al. (2019) assessed the influence of clinical genomic sequencing on Chinese patients diagnosed with stage IV cancer.

3.6 Biological Network Analysis

AI systems uncover therapeutic targets and paths in complex biological networks. To understand drug polypharmacology, network pharmacy models are used. AI systems to evaluate complex biological networks require numerous key methods to identify treatment targets and pathways. AI methods first process large amounts of genomic, proteomic, and metabolomic data. This stage ensures data is error-free, standardized, and ready for analysis. Deep learning algorithms excel at finding patterns in complex datasets. AI can develop and analyze biological networks like protein-protein interactions and

Table 8. Machine learning case studies biological network analysis (Source: Author)

Case Study	Application	Description
Predicting Protein-Protein Interactions (PPIs)	Molecular Biology	AI models predict protein-protein interactions, crucial for understanding cellular functions and disease mechanisms.
Gene Regulatory Network Analysis	Genomics	AI is used to decipher gene regulatory networks, aiding in understanding gene expression patterns and their impact on diseases.
Neural Network Analysis in Neuroscience	Neuroscience	AI helps map and analyze complex neural networks, aiding in understanding brain functions and disorders.
Metabolic Network Modeling	Metabolomics	AI is employed in metabolic network modeling to predict how changes in metabolic pathways affect cellular processes.
Microbiome Network Analysis	Microbiology	AI techniques analyze interactions within microbiomes, helping to understand their impact on human health and the environment.
Drug-Target Interaction Networks	Pharmacology	AI models predict interactions between drugs and their targets, aiding in drug discovery and repurposing.
Pathway Analysis in Disease Research	Disease Mechanism Study	AI is utilized for pathway analysis in disease research, identifying key biological pathways involved in diseases.
Ecological Network Analysis	Ecology	AI assists in analyzing complex ecological networks, key for conservation biology and understanding ecosystem dynamics.

gene-regulation networks. AI can find genes, proteins, and relationships that affect disease processes using graph theory and network analysis. AI algorithms can model how gene or protein changes affect the biological network. This predicts the effects of planned treatment actions, such as a medicine on a disease pathway. Artificial intelligence can identify illness biomarkers by analysing biological data patterns. AI can analyze biological networks to find drug targets. They can also evaluate large chemical libraries to find substances that may affect these targets. Artificial intelligence can analyze patient data to recommend personalized treatments. AI can use an individual's biological network to recommend the most effective therapies with the fewest side effects.

In their study, Chai et al. (2019) introduced an innovative algorithm that integrates self-organizing and genetic evolutionary methods to enhance the effectiveness and structural efficiency of neural networks. Silva & Saraiva (2015) presented a methodology for the application of Social Network Analysis (SNA) metrics to the study of biological interaction networks. In their study, Xia et al. (2014) developed a tool that effectively tackles the difficulties associated with researching extensive networks. This tool offers a comprehensive solution to researchers in this field. In their study, Alhasson et al. (2016) presented a new method for creating and measuring biological networks from 2D images. They achieved this by employing an adaptive rectangular mesh refinement technique.

3.7 Image Analysis in Cellular and Molecular Biology

Artificial intelligence, specifically deep learning models, can greatly improve the processing of cell pictures for high-throughput screening (HTS) of medication effects. Cell pictures can be analyzed using deep learning algorithms, enabling efficient screening of medication effects on a large scale. Artificial intelligence (AI) is employed to analyze images in order to comprehend the connections between molecules and the behaviors of cells when exposed to various medication treatments(Table 9). Deep learning methods, particularly convolutional neural networks (CNNs), excel at autonomously identifying and

Table 9. Machine learning case studies image analysis in cellular and molecular biology (Source: Author)

Case Study	Application	Description
Single-Cell Image Analysis	Cell Classification	AI algorithms analyze single-cell images for cell type identification, development tracking, and understanding cellular dynamics.
High-Throughput Screening in Drug Discovery	Pharmaceutical Research	AI in high-throughput screening for analyzing cellular responses to chemical compounds, aiding in drug candidate identification.
3D Reconstruction of Cellular Structures	Structural Biology	AI models assist in 3D reconstruction of cellular structures from 2D images, enhancing understanding of cellular architecture.
Automated Analysis of Fluorescence Microscopy Images	Cell Biology & Genetics	AI automates fluorescence microscopy image analysis, identifying and quantifying cellular components and markers.
Diagnosis of Diseases from Cellular Images	Medical Diagnostics	AI algorithms detect abnormalities in cellular images for disease diagnosis, such as cancer detection.
Studying Protein Localization and Movement	Proteomics	AI tools analyze images to understand protein localization and movements within cells, important for cell signaling studies.
Synthetic Biology and Genetic Engineering	Synthetic Biology	AI image analysis monitors effects of genetic modifications at the cellular level, aiding in synthetic biology applications.
Morphological Profiling of Cells	Cell Morphology	AI is used for morphological profiling to understand the impact of genetic or chemical changes on cell morphology.

extracting pertinent characteristics from cell pictures. Being able to discern tiny changes in cell shape, texture, or other phenotypic traits is essential for determining the impact of a medicine. Grafton et al. (2021) showcased the application of deep learning to swiftly identify patterns of cardiotoxicity in high-content picture analysis utilizing induced pluripotent stem cell-derived cardiomyocytes (iPSC-CMs). Hashemzadeh et al. (2021) created a Computer-Aided Diagnosis (CAD) system that use deep learning to classify photos of lung cancer cell lines. Li et al. (2020) propose a technique that combines high content screening (HCS) with U-Net based deep learning image analysis models to accurately measure the distribution of drugs in tissues. Haslum et al. (2023) investigated the utilization of deep learning in conjunction with Cell Painting, an image-based assay, for the prediction of chemical bioactivity in the initial stages of drug screening.

4. AI RISK AND REGULATIONS

The benefits of using AI, such as accelerated discovery processes, cost reductionandmore precise targeting of diseases, are significant. However, ongoing vigilance and responsible use of AI are essential to mitigate these risks.

1. **Data Privacy Breaches:** AI systems in drug discovery often handle sensitive patient data. Any breach or leak of this data can have serious privacy implications.
2. **Biased or Inaccurate Predictions**: If AI models are trained on biased or incomplete datasets, they can make inaccurate predictions. This could lead to overlooking potential drug candidates or pursuing ineffective ones.
3. **Intellectual Property Issues**: AI-generated molecules or compounds might raise complex intellectual property issues, including questions about data ownership and patentability.

Table 3. Key guidelines from various countries (Source: Author)

Region / Country	Key Regulations / Guidelines:
United States (FDA):	FDA's Digital Health Innovation Action Plan
	Guidelines for Software as a Medical Device (SaMD)
	Regulatory Framework for AI/ML-based Software as Medical Device
European Union (EMA):	EU's Medical Device Regulation (MDR)
	Guidelines on Good Machine Learning Practice
	Data Protection Regulation (GDPR) implications on AI technologies
United Kingdom (MHRA):	UK's Medicines and Medical Devices Act 2021
	AI and Machine Learning guidance by NICE
China (NMPA):	Regulations for the Supervision and Administration of Medical Devices
	Guidelines on AI medical software product registration
Japan (PMDA):	Pharmaceuticals and Medical Devices Agency's guidelines on AI in healthcare
	Regulatory framework for AI-based medical devices
Canada (Health Canada):	Medical Devices Regulations under Canada's Food and Drugs Act
	Digital Health Technology assessment guidelines

4. **Ethical Concerns:** AI's role in decision-making processes might raise ethical concerns, especially if AI systems are tasked with making critical decisions about drug development priorities(Dwivedi, D. N. et.al(2022)).
5. **Regulatory Compliance:** AI applications in drug discovery must comply with stringent regulatory standards. Failure to meet these standards could lead to delays in drug approval or potential legal issues.
6. **Cybersecurity Threats:** AI systems, being heavily reliant on digital data, are susceptible to cybersecurity threats. Hacking or tampering with these systems could have severe consequences.
7. **Over-reliance on AI:** There's a risk of over-reliance on AI predictions without sufficient human oversight, which could lead to missed opportunities or misinterpreted results.
8. **Unforeseen Side Effects:** AI might identify drug candidates that perform well in simulated environments but have unforeseen side effects when used in real-world clinical settings.

There had been multiple guidelines issued by various government agencies around the world on ris of usinging AI in drug discovery (Table 3).

5. SUMMARY AND CONCLUSION

Adding AI to the process of making new drugs marks the start of a new era in pharmaceutical research. Handling and analyzing large datasets, making predictions about the safety and usefulness of drugs, and speeding up the discovery process all help solve many of the industry's usual problems. Still, AI's success in this field relies on having access to better data, using AI technologies in an ethical way, and AI experts and biological researchers working together all the time. The future of AI in drug finding looks bright, and it could make huge steps forward in the field of medicine. As AI technologies continue to improve, they are likely to become an important part of the drug development process. This will lead to better efficiency, lower costs, and new solutions in medicine and healthcare.

- AI greatly speeds up the process of finding new drugs, cutting down on the time it takes from the start of study to clinical trials. It can take over ten years to find new drugs the old way, but AI can cut that time down to a few years.
- Artificial intelligence (AI) is very good at guessing how well and how dangerous chemicals will be, which helps narrow down the list of possible drugs early on in the finding process.
- There is a lot of data in the life sciences, and AI helps handle and analyze huge amounts of it, finding patterns and insights that people would not be able to find quickly.
- Show AI's skill in modeling mathematical relationships between structure, activity, and property, allowing for effective virtual screening of compounds
- AI helps find new uses for old drugs. This is called "drug repurposing," and it can be faster and cheaper than making a whole new drug.
- AI can be used to find new drugs in any part of the world, so people from all over the world can work together and make the best use of resources, even in places where resources are limited.

In spite of its benefits, AI in drug discovery meets problems such as poor data quality and moral issues, especially when it comes to data privacy and the possibility of AI technologies being used for two different purposes.

REFERENCES

Alhasson, H. F., & Obara, B. (2016). *2D Adaptive Grid-Based Image Analysis Approach for Biological Networks*. IEEE. doi:10.1109/BIBE.2016.17

Bai, Y., Wang, G., Wei, J., Dai, C., Xu, X., Cai, X., Wu, B., Sun, W., Xu, Q., & Jiao, S. (2019). *Using clinical genomic sequencing to guide personalized cancer therapy in China*. PME. doi:10.2217/pme-2018-0056

Bajorath, J. (2021). State-of-the-art of artificial intelligence in medicinal chemistry. doi:10.2144/fsoa-2021-0030

Baldi, P. (2007). *Machine Learning Challenges in Chemoinformatics and Drug Screening and Design*. Future Design. doi:10.1109/ICMLA.2007.125

Blanco-Gonzalez, A., Cabezón, A., Seco-Gonzalez, A., Conde-Torres, D., Antelo-Riveiro, P., Piñeiro, Á., & García-Fandiño, R. (2022). The Role of AI in Drug Discovery: Challenges, Opportunitiesand-Strategies. *Pharmaceuticals (Basel, Switzerland)*, *16*(6), 891. doi:10.3390/ph16060891 PMID:37375838

Cassidy, J. (2020). *Applications of Machine Learning in Drug Discovery I: Target Discovery and Small Molecule Drug Design*. In Tech Open. doi:10.5772/intechopen.93159

Chafai, N., Bonizzi, L., Botti, S., & Badaoui, B. (2023). *Emerging applications of machine learning in genomic medicine and healthcare*. Taylor & Francis. doi:10.1080/10408363.2023.2259466

Chai, Z., Yang, X., Liu, Z., Lei, Y., Zheng, W., Ji, M., & Zhao, J. (2019). *Correlation Analysis-Based Neural Network Self-Organizing Genetic Evolutionary Algorithm*. IEEE. doi:10.1109/ACCESS.2019.2942035

Cherney, R., Major, R., & Fitzpatrick, T. (2023). Qualify AI Drug Discovery Tools through FDA ISTAND Program to Model Responsible Drug Discovery AI and Mitigate Dual Use Concerns. *Journal of Science Policy & Governance*, *22*(3). http://www.sciencepolicyjournal.org/uploads/5/4/3/4/5434385/cherney_etal_jspg_22-3.pdf. doi:10.38126/JSPG220302

Cyranoski, D. (2021). AI drug discovery booms in China. *Nature Biotechnology*, *39*(8), 785–786. doi:10.1038/s41587-021-01016-0 PMID:34345035

Didi, K., & Zecevic, M. (2022). On How AI Needs to Change to Advance the Science of Drug Discovery. https://dx.doi.org//arXiv.2212.12560 doi:10.48550

Dorn, M., Buriol, L., & Lamb, L. (2011). *Combining Machine Learning and Optimization Techniques to Determine 3-D Structures of Polypeptides*. IJCAI. doi:10.5591/978-1-57735-516-8/IJCAI11-469

Dwivedi, D. N. (2024). The Use of Artificial Intelligence in Supply Chain Management and Logistics. In D. Sharma, B. Bhardwaj, & M. Dhiman (Eds.), *Leveraging AI and Emotional Intelligence in Contemporary Business Organizations* (pp. 306–313). IGI Global.

Dwivedi, D. N., & Mahanty, G. (2024). AI-Powered Employee Experience: Strategies and Best Practices. In M. Rafiq, M. Farrukh, R. Mushtaq, & O. Dastane (Eds.), *Exploring the Intersection of AI and Human Resources Management* (pp. 166–181). IGI Global. doi:10.4018/979-8-3693-0039-8.ch009

Dwivedi, D. N., Mahanty, G., & Pathak, Y. K. (2023). AI Applications for Financial Risk Management. In M. Irfan, M. Elmogy, M. Shabri Abd. Majid, & S. El-Sappagh (Eds.), The Impact of AI Innovation on Financial Sectors in the Era of Industry 5.0 (pp. 17-31). IGI Global. doi:10.4018/979-8-3693-0082-4.ch002

Dwivedi, D. N., Mahanty, G., & Vemareddy, A. (2022). How Responsible Is AI?: Identification of Key Public Concerns Using Sentiment Analysis and Topic Modeling. [IJIRR]. *International Journal of Information Retrieval Research*, *12*(1), 1–14. doi:10.4018/IJIRR.298646

Dwivedi, D. N., Pandey, A. K., & Dwivedi, A. D. (2023). Examining the emotional tone in politically polarized Speeches in India: An In-Depth analysis of two contrasting perspectives. *SOUTH INDIA JOURNAL OF SOCIAL SCIENCES, 21*(2), 125-136. https://journal.sijss.com/index.php/home/article/view/65

Grafton, F., Ho, J. J., Ranjbarvaziri, S., Farshidfar, F., Budan, A., Steltzer, S. S., Maddah, M., Loewke, K., Green, K., Patel, S., Hoey, T., & Mandegar, M. (2021). Deep learning detects cardiotoxicity in a high-content screen with induced pluripotent stem cell-derived cardiomyocytes. *eLife*, *10*, e68714. Advance online publication. doi:10.7554/eLife.68714 PMID:34338636

Guan, X., Mural, R., & Uberbacher, E. (1994). *Protein structure prediction using hybrid AI methods*. IEEE. doi:10.1109/CAIA.1994.323633

Gupta, A., Dwivedi, D. N., & Shah, J. (2023). Overview of Money Laundering. In: Artificial Intelligence Applications in Banking and Financial Services. Springer, Singapore. doi:10.1007/978-981-99-2571-1_1

Gupta, A., Dwivedi, D. N., & Shah, J. (2023). Financial Crimes Management and Control in Financial Institutions. In: Artificial Intelligence Applications in Banking and Financial Services. Future of Business and Finance. Springer, Singapore. doi:10.1007/978-981-99-2571-1_2

Gupta, A., Dwivedi, D. N., & Shah, J. (2023). Overview of Technology Solutions. In: Artificial Intelligence Applications in Banking and Financial Services. Future of Business and Finance. Springer, Singapore. doi:10.1007/978-981-99-2571-1_3

Gupta, A., Dwivedi, D. N., & Shah, J. (2023). Applying Machine Learning for Effective Customer Risk Assessment. In: Artificial Intelligence Applications in Banking and Financial Services. Future of Business and Finance. Springer, Singapore. doi:10.1007/978-981-99-2571-1_6

Gupta, A., Dwivedi, D. N., & Shah, J. (2023). Artificial Intelligence-Driven Effective Financial Transaction Monitoring. In: Artificial Intelligence Applications in Banking and Financial Services. Future of Business and Finance. Springer, Singapore. doi:10.1007/978-981-99-2571-1_7

Gupta, A., Dwivedi, D. N., & Shah, J. (2023). Machine Learning-Driven Alert Optimization. In: Artificial Intelligence Applications in Banking and Financial Services. Future of Business and Finance. Springer, Singapore. doi:10.1007/978-981-99-2571-1_8

Gupta, A., Dwivedi, D. N., & Shah, J. (2023). Setting up a Best-In-Class AI-Driven Financial Crime Control Unit (FCCU). In: Artificial Intelligence Applications in Banking and Financial Services. Future of Business and Finance. Springer, Singapore. doi:10.1007/978-981-99-2571-1_11

Hashemzadeh, H., Shojaeilangari, S., Allahverdi, A., Rothbauer, M., Ertl, P., & Naderi-manesh, H. (2021). A combined microfluidic deep learning approach for lung cancer cell high throughput screening toward automatic cancer screening applications. *Scientific Reports*, *11*(1), 9804. doi:10.1038/s41598-021-89352-8 PMID:33963232

HaslumJ. F.LardeauC.KarlssonJ.TurkkiR.LeuchowiusK. J.SmithK.MüllersE. (2023). Cell Painting-based bioactivity prediction boosts high-throughput screening hit-rates and compound diversity. bioRxiv doi:10.1101/2023.04.03.535328

Hu, Y., Zhao, T., Zhang, N., Zhang, Y., & Cheng, L. (2000). *A Review of Recent Advances and Research on Drug Target Identification Methods*. Bentham Science. doi:10.2174/1389200219666180925091851

Jain, G. (n.d.). *Application of Machine Learning in Drug Discovery and Development Lifecycle. AI Publications*. doi:10.22161/ijmpd.6.6.4

Jiménez-Luna, J., Grisoni, F., Weskamp, N., & Schneider, G. (2021). *Artificial intelligence in drug discovery: recent advances and future perspectives*. Taylor & Francis. doi:10.1080/17460441.2021.1909567

Johnson, K. B., Wei, W.-Q., Weeraratne, D., Frisse, M., Misulis, K., Rhee, K., Zhao, J., & Snowdon, J. (2020). *Precision Medicine, AI and the Future of Personalized Health Care*. ASCPT. doi:10.1111/cts.12884

Kumar, K., & Gandhi, H. (2023). *Artificial Intelligence for the Management of Breast Cancer: An Overview*. Bentham Science. doi:10.2174/0115701638262066231030052520

Li, L., Zeng, L., Gao, Z.-C., Yuan, S., Bian, Y., Wu, B., Zhang, H., Lu, C., Yu, Y., Liu, W., Xu, H., Li, J., Zhao, P., & Heng, P. (2022). ImDrug: A Benchmark for Deep Imbalanced Learning in AI-aided Drug Discovery. https://dx.doi.org//arXiv.2209.07921 doi:10.48550

Li, Z., Xiao, Y., Peng, J., Locke, D., Holmes, D., Li, L., Hamilton, S., Cook, E., Myer, L., Vanderwall, D. E., Cloutier, N., Siddiqui, A. M., Whitehead, P., Bishop, R., Zhao, L., & Cvijic, M. (2020). Quantifying drug tissue biodistribution by integrating high content screening with deep-learning analysis. *Scientific Reports*, *10*(1), 14408. doi:10.1038/s41598-020-71347-6 PMID:32873881

Lilhore, U. K., Manoharan, P., Sandhu, J. K., Simaiya, S., Dalal, S., Baqasah, A. M., Alsafyani, M., Alroobaea, R., Keshta, I., & Raahemifar, K. (2023). Hybrid model for precise hepatitis-C classification using improved random forest and SVM method. *Scientific Reports*, *13*(1), 12473. doi:10.1038/s41598-023-36605-3 PMID:37528148

Lilhore, U. K., Poongodi, M., Kaur, A., Simaiya, S., Algarni, A. D., Elmannai, H., Vijayakumar, V., Tunze, G. B., & Hamdi, M. (2022). Hybrid model for detection of cervical cancer using causal analysis and machine learning techniques. *Computational and Mathematical Methods in Medicine*, *2022*, 2022. doi:10.1155/2022/4688327 PMID:35572826

Lilhore, U. K., Simaiya, S., Pandey, H., Gautam, V., Garg, A., & Ghosh, P. (2022). Breast cancer detection in the IoT cloud-based healthcare environment using fuzzy cluster segmentation and SVM classifier. In Ambient Communications and Computer Systems [Singapore: Springer Nature Singapore.]. *Proceedings of RACCCS*, *2021*, 165–179.

ManchadoE.HuangC.-H.TasdemirN.TschaharganehD. F.WilkinsonJ.LoweS. (2017). A Pipeline for Drug Target Identification and Validation. https://dx.doi.org/ doi:10.1101/sqb.2016.81.031096

Narayanan, R. R., Durga, N., & Nagalakshmi, S. (2022). Impact of Artificial Intelligence (AI) on Drug Discovery and Product Development. *Indian Journal of Pharmaceutical Education and Research*, *56*(3s), s387–s397. https://ijper.org/sites/default/files/IndJPhaEdRes-56-3s-s387.pdf. doi:10.5530/ijper.56.3s.146

Navathe, S. (2007). *Text Mining and Ontology Applications in Bioinformatics and GIS*. IEEE. doi:10.1109/ICMLA.2007.122

Parekh, A. E., Shaikh, O. A., Simran, F., Manan, S., & Hasibuzzaman, M. A. (2023). *Artificial intelligence (AI) in personalized medicine: AI-generated personalized therapy regimens based on genetic and medical history: short communication*. doi:10.1097/MS9.0000000000001320

Philippidis, A. (2021). From Model to Molecule: NVIDIA Doubles Down on AI Drug Discovery. *GEN Edge*, *3*(1), 238–246. doi:10.1089/genedge.3.1.038

Priya, S., Tripathi, G., Singh, D., Jain, P., & Kumar, A. (2022). Machine learning approaches and their applications in drug discovery and design. https://dx.doi.org/ doi:10.1111/cbdd.14057

Raha, K., Wollacott, A., Italia, M. J., & Desjarlais, J. (n.d.). *Prediction of amino acid sequence from structure*. Wiley. doi:10.1110/ps.9.6.1106

Restrepo, J. C., Dueñas, D., Corredor, Z., & Liscano, Y. (2023). *Advances in Genomic Data and Biomarkers: Revolutionizing NSCLC Diagnosis and Treatment*. MDPI. doi:10.3390/cancers15133474

Senior, A., Evans, R., Jumper, J., Kirkpatrick, J., Sifre, L., Green, T., Qin, C., Zídek, A., Nelson, A. W. R., Bridgland, A., & Hassabis, D. (2020). *Improved protein structure prediction using potentials from deep learning*. doi:10.1038/s41586-019-1923-7

Sharma, S., & Sharma, D. (2018). *Intelligently Applying Artificial Intelligence in Chemoinformatics*. doi:10.2174/1568026619666181120150938

Sharma, V., Singh, A., Chauhan, S., Sharma, P. K., Chaudhary, S., Sharma, A., Porwal, O., & Fuloria, N. (2023). *Role of Artificial Intelligence in Drug Discovery and Target Identification in Cancer*. doi:10.2174/1567201821666230905090621

Shoaib, T., & Parveen, D. S. (2022). Deep Learning Driven Drug Discovery and Use of Machine Learning Strategies. *RE:view*, *13*(SO3). doi:10.47750/pnr.2022.13.S03.112

Silva, J. S., & Saraiva, A. (2015). *A methodology for applying social network analysis metrics to biological interaction networks*. ACM., doi:10.1145/2808797.2808824

Tripathi, M., Nath, A., Singh, T., Ethayathulla, A. S., & Kaur, P. (2021). Evolving scenario of big data and Artificial Intelligence (AI) in drug discovery. *Molecular Diversity*, *25*(3), 1827–1844. doi:10.1007/s11030-021-10256-w PMID:34159484

Tsai, E., Shakbatyan, R., Evans, J., Rossetti, P., Graham, C., Sharma, H., Lin, C.-F., & Lebo, M. S. (2016). Bioinformatics Workflow for Clinical Whole Genome Sequencing at Partners HealthCare. *Personalized Medicine*, *6*(1), 12. doi:10.3390/jpm6010012 PMID:26927186

Turon, G., Hlozek, J., Woodland, J. G., Chibale, K., & Duran-Frigola, M. (2023). First fully-automated AI/ML virtual screening cascade implemented at a drug discovery centre in Africa. *Nature Communications*, *14*(1), 1–10. doi:10.1038/s41467-023-41512-2 PMID:37714843

Vadapalli, S., Abdelhalim, H., Zeeshan, S., & Ahmed, Z. (2022). *Artificial intelligence and machine learning approaches using gene expression and variant data for personalized medicine*. doi:10.1093/bib/bbac191

Vemula, D., Jayasurya, P., Sushmitha, V., Kumar, Y. N., & Bhandari, V. (2022). CADD, AI and ML in Drug Discovery: A Comprehensive Review. *European Journal of Pharmaceutical Sciences*, *176*, 106324. doi:10.1016/j.ejps.2022.106324 PMID:36347444

Walters, W., & Barzilay, R. (2021). Critical assessment of AI in drug discovery. *Expert Opinion on Drug Discovery*, *16*(5), 497–507. doi:10.1080/17460441.2021.1915982 PMID:33870801

Xia, J., Benner, M. J., & Hancock, R. (2014). NetworkAnalyst - integrative approaches for protein–protein interaction network analysis and visual exploration. *Nucleic Acids Research*, *42*(W1), W167–W174. Advance online publication. doi:10.1093/nar/gku443 PMID:24861621

Yarragunta, A. (2022). The Future of Pharmaceuticals: AI Drug Discovery & COVID-19 Treatments. *Journal of Student Research*, *11*(3). doi:10.47611/jsrhs.v11i3.3046

Ziska Fields, I. G. I. (Ed.). (2023). Human Creativity vs. Machine Creativity: Innovations and Challenges. Multidisciplinary Approaches in AI, Creativity, Innovation, and Green Collaboration (pp. 19–28). Global. doi:10.4018/978-1-6684-6366-6.ch002

Chapter 7
Challenges and Limitations of Few-Shot and Zero-Shot Learning

V. Dankan Gowda
https://orcid.org/0000-0003-0724-0333
BMS Institute of Technology and Management, Bangalore, India

Sajja Suneel
https://orcid.org/0009-0007-7759-7924
Institute of Aeronautical Engineering, Dundigal, India

P. Ramesh Naidu
https://orcid.org/0000-0003-4941-9569
Nitte Meenakshi Institute of Technology, Bangalore, India

S. V. Ramanan
PPG Institute of Technology, Coimbatore, India

Sampathirao Suneetha
https://orcid.org/0009-0005-2714-6442
Andhra University College of Engineering, India

ABSTRACT

Essential to the development of AI and machine learning, this chapter explores the complex areas of few-shot and zero-shot learning. There have been great advancements towards more efficient and adaptive AI systems with few-shot learning and zero-shot learning, respectively, which can learn from minimal data and infer from particular data instances without previous exposure. Nevertheless, there are several limits and difficulties associated with these procedures. This chapter delves deeply into the theoretical foundations of both techniques, explaining how they work and what problems they solve in different ways. It examines the semantic gap, domain adaptation problems, and model bias, as well as the computational restrictions, overfitting, and model generalizability that are intrinsic to few-shot learning and zero-shot learning, respectively. We may better understand the ideas' potential use in different real-world contexts by comparing and contrasting them.

DOI: 10.4018/979-8-3693-1822-5.ch007

1. INTRODUCTION

In an era of increasing interest in church the fast developing fields of machine learning and artificial intelligence (AI), one must find a proper solution to missing data. Today, few-shot learning as well as zero shot have become two hot topics. Few-Shot Learning means to used, as little labelled data as possible in training a model. Accuracy with traditional machine learning models must require a huge number of labelled data. Collecting this can be expensive and time consumording. On the other hand, few-shot learning aims to improve efficiency and adaptability by enabling AI models to provide reliable predictions with very few examples (J. Chen, 2023). These are achieved by using contemporary techniques like as meta-learning, embedding learning, and transfer learning that help the model generalize from sparse datasets to new data. On the other hand, Zero-Shot Learning exten ds enormously the frontiers of learning with few data. In the case of zero-shot learning, however, the model has to make predictions on classes it didn't see during training. This is achieved by, during training, learning something from attending one set of classes, and then using it to figure out something about another that was missing (i.e., a completely different class of things). To the end of making inferences based on what it understands from descriptions or attributes, the model usually uses this method of understanding and use. It does not need to have any direct experience.

1.1 Importance in the Field of AI and Machine Learning

Few-shot and zero-shot learning are two important steps on the road to developing AI systems that are more intelligent, flexibleand economical with resources (J. Guan, Z. Lu, T. Xiang 2021). Their importance in the field of AI and machine learning cannot be overstated for several reasons:

Data Efficiency: It overcomes one of the chief limitations common to existing AI models, namely that they must use large sets of labelled data. In applications which lack or whose data is expensive to gather, these methods open up the possibility of learning with few or no direct examples. The degree to which this is true determines how applicable and scalable any given AI application will be.

Generalization Capabilities: These learning paradigms will improve the ability of AI models to deal with missing information, and bring their performance closer to our own. This is particularly important in changing environments where the model must quickly adapt to solve a new task or change its environment.

Broadening AI Applications: Few-shot and zero-shot learning, by giving computers the ability to learn from limited or incomplete data, open up a wide range of applications for fields which possesses data that are either lacking quantity or dynamically changing--for example, medical diagnosis and robotics. They also have many applications in natural language processing.

Ethical and Practical Implications: Besides raising ethical and practical questions about AI, they also stimulate better ways to develop responsible and sustainable AI.

Recognition by human eyesight is a complex interaction of the eyes, retina and brain. First, the human eyes and retina work together to capture the visual scene, forming an image that the human brain subsequently interprets for various recognition tasks (S. Rahman, S. Khan and F. Porikli, 2018). These tasks encompass identifying objects within the scene, pinpointing their locations, and acquiring knowledge about object relationships, among others.

In the realm of technology, cameras serve as the functional equivalent of human eyes and retina, while Machine Learning (ML) and Deep Learning (DL) algorithms tackle the diverse spectrum of visual recognition tasks. A typical ML-DL framework revolves around two key components: the algorithm itself and the training data it relies upon. Traditional ML algorithms have historically been engineered to perform adequately with limited training data. Nevertheless, DL models have recently risen to prominence because to the availability of large training datasets. Convolutional Neural Networks (CNNs) are the backbone of many DL models in the image recognition domain because of their remarkable performance and adaptability(F. Pourpanah, 2023). There are two primary ways to solve visual recognition tasks: supervised learning and unsupervised learning. Unsupervised learning takes use of the structure of unlabeled data, in contrast to supervised learning that depends on labelled training data. The standard supervised learning framework for deep learning (DL) models consists of three stages: training, validation, and testing. Consider the problem of recognizing cats in photographs to demonstrate this technique. A diversified dataset including multiple cat photos is fed into a Convolutional Neural Network (CNN) model during the training phase. During training, the main goal is for the CNN model to fully grasp the unique characteristics of cats. The next step is to test the model by using the trained CNN to extract features from fresh, unknown pictures.

The presence or absence of a cat in a given picture may be determined by comparing the trained model of the cat's holistic features with the model of the test image. Both supervised and unsupervised learning techniques play crucial roles in addressing visual recognition challenges, allowing for a broad spectrum of applications in computer vision and image analysis. Despite the remarkable success of Convolutional Neural Networks (CNNs) and deep learning (DL) models in addressing a wide array of visual recognition tasks, they heavily depend on extensive annotated data for effective training(R. V. Mailapur and M. K, 2023). Take the example of recognizing cat images mentioned earlier: a typical DL method necessitates the availability of hundreds of annotated cat images for successful training. While collecting and annotating images for a few object classes may be manageable, the challenge escalates as the number of object classes increases. In real-world scenarios aiming to develop efficient AI systems, dealing with thousands of object classes becomes a formidable task. This labeling endeavor is exemplified by the ImageNet dataset, which encompasses more than 20,000 object classes and a staggering 14 million images. Furthermore, collecting a substantial volume of labeled data can be unattainable for certain object classes. For instance, gathering images of rare bird and animal species presents significant challenges(N. Bendre, K. Desai and P. Najafirad, 2021). Additionally, the continuous introduction of new object classes, such as newly launched car models or consumer products, further complicates the situation.

To adapt AI systems to accommodate these ever-expanding object classes, it becomes necessary to make repeated modifications to the DL network architecture and initiate retraining processes from scratch. This dynamic nature of the data landscape underscores the ongoing challenges and complexities associated with scaling up visual recognition systems. In contrast, humans possess a remarkable ability to recognize unfamiliar objects with just a handful of training images. In most cases, it doesn't require hundreds of images to identify new object classes(B. Jiang, Y. Lu, B. Zhang and G. Lu, 2023). In some extraordinary instances, humans can even recognize completely new object classes without any prior exposure to visual training data related to those classes. Images and high-level semantic data, including textual descriptions, are the backbone of this remarkable power. Figure 1 might be used as an example to demonstrate this phenomena. Think about your acquaintance with dogs, cats, tigers, and zebras, as well as your understanding of their textual high-level semantic descriptions. A zebra, for example, is often thought of as a horse-like animal with distinctive black and white stripes. Present the

Figure 1. Illustrative example serves to elucidate the concept of zero-shot learning (ZSL)

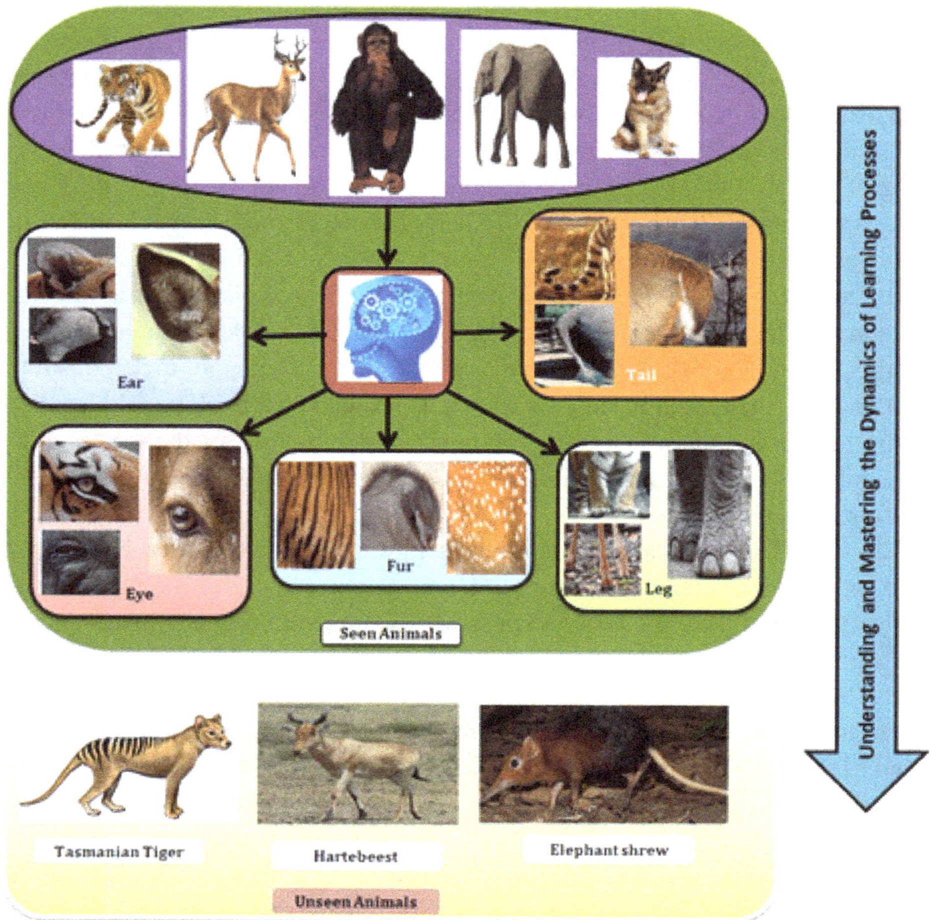

written accounts of three endangered species: the hartebeest, the elephant shrew, and the Tasmanian tiger. Think of the Tasmanian tiger as a dog-like creature with tiger-like stripes and a stiff tail to make your point. When encountering images of these previously unseen animals for the first time, individuals demonstrate an uncanny ability to correctly identify each of these three novel creatures. Proficiency in recognizing previously unfamiliar objects can be attributed to the capacity for transferring knowledge acquired from familiar objects to unfamiliar ones. This transfer of learning is made possible by leveraging semantic information provided in textual descriptions. This learning scenario is formally known as Zero-shot Learning, where "zero" signifies the absence of any visual training data when recognizing new object classes(W. Cho and E. Kim, 2022). A typical Zero-shot Learning (ZSL) model undergoes evaluation on standard benchmark datasets, where object classes are divided into two distinct sets: "seen classes" and "unseen classes." Each data sample within this dataset consists of a pair of visual-semantic features, where visual features can be images or videos, and semantic features are typically represented as high-level descriptions, such as textual descriptions. It's important to note that samples belonging to

Figure 2. Training and testing phases of ZSL

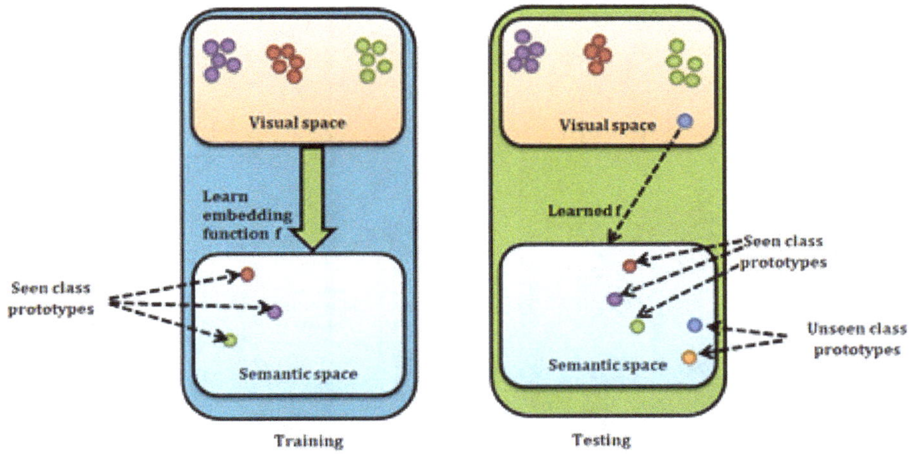

the same object class share a common semantic feature, often referred to as a semantic prototype, class prototype, or simply, a prototype.

In this context, "seen classes" pertain to those classes available for training the ZSL models, while "unseen classes" are introduced during the testing phase. The training and testing phases are illustrated in Figure 2. During training, the primary objective is to learn an embedding or mapping function that associates visual features with their corresponding semantic features for the seen class data. This semantic space effectively serves as an embedding space.

When the testing phase commences, the visual feature of a test sample from a previously unseen class is presented as input to the trained ZSL model, specifically the embedding function. The model then predicts the semantic feature associated with the test sample(S. Venkatakiran and N. S. Reddy, 2023). This projected semantic feature, or embedding, is subsequently compared with the ground truth semantic prototypes of the unseen classes, utilizing a nearest neighbor criterion. This comparison aids in inferring the label for the unseen class based on its closest match.

Since the ultimate objective of ZSL is to accurately label unseen class samples, it becomes imperative to compare the test embedding with the ground truth semantic descriptions of the unseen classes. Consequently, during the testing phase, the true prototypes of the unseen classes are revealed (Sadashiva V. Chakrasali, 2023). It's worth noting that the semantic space can be any form of embedding space, whether it's visual, semantic, or another latent space where nearest neighbor search can be performed for ZSL tasks.

The visual and semantic domains inherently differ from each other, leading to a fundamental distinction in the class-neighborhood structure within these domains. This contrast is effectively demonstrated in Figure 3. Notably, classes that are visually proximate may not necessarily share similar relationships in the semantic space. Moreover, during the learning of the embedding function, the model can take advantage of relationships among seen classes in the visual domain, but it is expected to remain indifferent to relationships between seen and unseen classes.

Consistency in the relationships among seen classes is crucial in both visual and semantic spaces, as it directly impacts the relationships between seen and unseen classes during testing. Ensuring these

Figure 3. Class-neighborhood structures between visual and semantic spaces

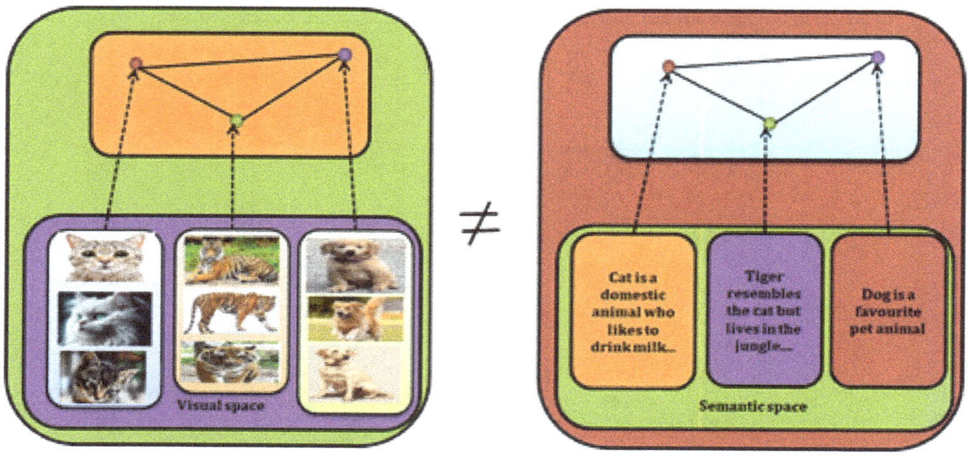

factors are incorporated into the process of learning an embedding function is essential. However, it's important to note that existing (Generalized) Zero-shot Learning ((G)ZSL) methods often do not consider this aspect while training the embedding function.

2. CONCEPTUAL FOUNDATIONS

2.1 Brief Overview of Machine Learning Concepts

A subfield of artificial intelligence, Machine Learning (ML) focuses on the development of systems that recognize patterns in data, learn from them, and make decisions with minimal human intervention. Machine learning comprises an extensive array of methodologies and algorithms(Hombalimath, and N. Patwari, 2023). It progresses from supervised learning, which involves training models on annotated data, to unsupervised learning, which requires models to deduce patterns from unlabeled data. Key concepts include:

Training and Testing: In ML, models are trained on a dataset and then tested on a separate dataset to evaluate their performance.

Overfitting and Underfitting: Models may be overfit (i.e., capturing noise in the data) or underfit (i.e., failing to identify underlying patterns) depending on their complexity.

Generalization: The ability of a model to perform well on new, unseen data.

Algorithms and Models: Models for tasks like grouping, regression, and classification are built using a variety of approaches. Among them, you may find neural networks, decision trees, and support vector machines.

2.2 Distinction Between Traditional Machine Learning, Few-Shot Learning, and Zero-Shot Learning

Traditional Machine Learning: For conventional ML to work, models need to be trained on massive datasets, with thousands—if not millions—of instances. The availability of abundant data is assumed by this strategy, which is not always the case in many real-world settings.

Few-Shot Learning: When training models, few-shot learning is used. This is because it requires extremely little data, usually just a few samples per class(Namitha A R, 2022). Methods that allow a model to generalize from a limited dataset to unlabeled data are crucial to this process. When gathering data is difficult or costly, few-shot learning is a lifesaver.

Zero-Shot Learning: The ultimate case of this is zero-shot learning, which entails teaching models to do tasks they have never seen before. The model must be able to generalize from well-defined categories to less-defined ones, often making use of semantic links or features(P. Ramesh Naidu and N. Guruprasad, 2021). When it is not feasible to collect data that is representative of every category, zero-shot learning becomes highly important.

2.3 Key Technologies Enabling These Learning Methods

Neural Networks: The backbone of many modern ML systems, neural networks, particularly deep learning networks, have the ability to learn complex patterns and representations from data, making them suitable for few-shot and zero-shot learning.

Transfer Learning: this is all about taking what you've learned and applying it to a new, related problem. As another example, when doing few-shot learning it is common to fine-tune a model learned on a big dataset with one understood better in relation to some particular task.

Meta-Learning: Often referred to as "learning to learn," meta-learning involves designing models that can quickly adapt to new tasks with minimal data(G. A, A. B. Naik and N. HG, 2021). It is a critical strategy in few-shot learning.

Embedding Learning: This involves learning a space in which data points are represented such that similar points are closer together. This is particularly important in zero-shot learning, where these embeddings can represent high-level attributes of classes.

Semantic Attribute Learning: In zero-shot learning, models often use attributes (like 'has wings' or 'can swim') to make inferences about unseen classes based on their semantic relationships with seen classes.

While seen and unseen classes in (Generalized) Zero-shot Learning ((G)ZSL) models are separate, the semantic space enables knowledge transfer from seen to unseen classes during testing. However, since these models are exclusively trained on seen classes, they exhibit a bias towards those classes. As an example, consider the seen class "cat" and unseen classes "tiger" and "zebra." The model, trained on "cat," may misclassify unseen class "tiger" samples as "cat" due to their semantic similarity(K. Jeevan and B. M. Sathisha, 2020). This model bias is more pronounced in GZSL compared to ZSL, as depicted in the upper section of Figure 4. In ZSL, test sample embeddings are compared to prototypes of unseen classes, reducing the risk of misclassification into seen classes.

In Generalized Zero-shot Learning (GZSL), the test embedding is compared to prototypes of both seen and unseen classes, leading to misclassification of unseen test samples into seen classes due to

Figure 4. The figure highlights the pronounced model bias towards seen classes in GZSL testing

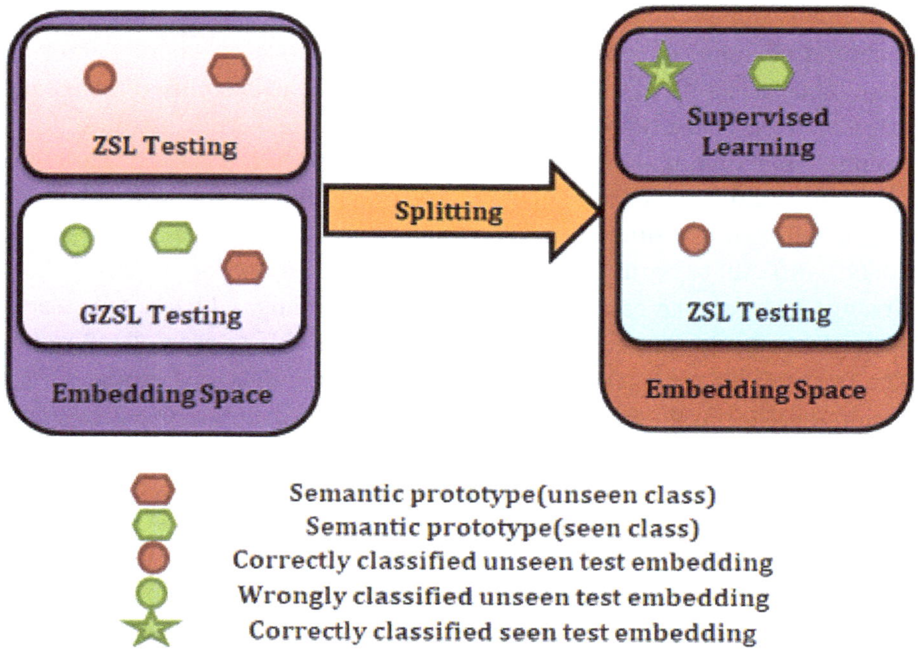

model bias(U. K N and R. V M, 2022). Although the training process is consistent in both Zero-shot Learning (ZSL) and GZSL, the difference in testing protocols results in distinct model behaviors for the same unseen test samples.

However, a closer examination reveals that model bias can be effectively mitigated by splitting the challenging GZSL problem into more manageable supervised learning and ZSL problems, as illustrated in the lower section of Figure 4. By classifying unseen class test samples using the ZSL framework and seen class test samples using a supervised learning framework, model bias can be significantly reduced(M. R. G. and H. Anandaram, 2022). In the experiments conducted on the datasets mentioned earlier, the standard splits provided with these datasets were initially used. However, it's important to note that these standard data splits may not adhere to the Zero-shot Learning (ZSL) assumption of having separate train and test classes. This discrepancy arises due to the fact that ZSL models are trained on visual features extracted using Deep Neural Networks (DNNs) pretrained on the extensive ImageNet dataset, which includes some of the test classes from the (G)ZSL datasets mentioned(Revanna C and Parismita Sarma, 2022). As a result, certain test classes are exposed during training due to the utilization of these pretrained DNNs for visual feature extraction. To resolve this challenge, a new data split proposed **in** (S. Reddy P, P. S. Patwal, 2022) has been introduced to ensure that the ZSL assumption of disjoint train and test classes is not violated. In the experiments conducted, both the standard splits of the datasets and the new splits from (S. R. Kawale, S. P. Diwan, 2022) were examined. Figure 5 provides a visual representation of sample images from these datasets. By leveraging these technologies, few-shot and zero-shot learning aim to overcome the limitations of traditional machine learning, enabling broader and more flexible AI applications. The following sections will delve deeper into these advanced learning paradigms, exploring their intricacies, applications, and the challenges they face. It

Figure 5. Sample images from various (generalized) zero-shot learning

is common practice to train mapping or embedding functions using multivariate regression, but there is little focus on guaranteeing that the embeddings are discriminative. It could be difficult to account for differences inside classes when just one semantic prototype is used for each class(B. Kameswara Rao and Naziya Hussain, 2022). Therefore, it's possible to have less-than-ideal embedding functions when visual characteristics have large intra-class fluctuations mapped to a single semantic prototype. In order to address this problem, earlier studies have proposed learning embedding functions using triplets, as detailed in (M. Swathi Pai and B. Naveen K, 2020). The complexity of multi-modal distribution in visual space as an embedding space is the target of this technique(K. B. Naveen and G. N. Pai, 2020). By using triplet-based embedding learning, we can draw in samples from the same class and separate them from data from other classes. It is important to remember that there may be a huge number of possible triplets, which makes the process of selecting them difficult, even though this objective function based on triplets helps with learning a discriminative embedding space.

The model for (Generalized) Zero-shot Learning ((G)ZSL) that has been suggested uses neural networks, as shown in Figure 6. Two separate embedding functions, fv(·) for visual to latent space and fs(·) for projecting semantic features into the same latent space, are learned concurrently by the model using

Figure 6. Neural network model designed for (generalized) zero-shot learning ((G)ZSL)

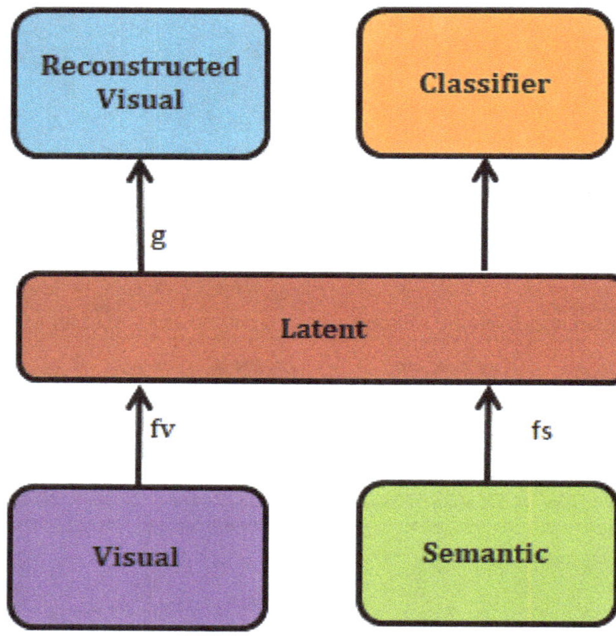

an autoencoder (AE) encoder. Further, the AE decoder (g(·)) uses its latent representation to recreate the input visual sample(P. Ramesh Naidu, N. Guruprasad, 2020). The classifier is fed these latent representations after they have been discriminatively enhanced using cross-domain triplets. In the testing step, zero-shot classification is carried out using the visual test sample and prototypes of the unseen classes projected into latent space.

3. FEW-SHOT LEARNING

Few-Shot Learning relates to a machine learning model's capacity to acquire knowledge with a sparse set of labelled data. While few-shot learning models may generalize from little datasets—as few as one or two samples per class—rather than the massive datasets needed by traditional machine learning models.

3.1 Examples

Image Classification: In a scenario where a model needs to recognize rare animal species from a handful of images.
Medical Diagnosis: Diagnosing rare diseases using limited case studies or imaging data.
Robotics: Teaching a robot new tasks with minimal demonstrations.

3.2 Techniques and Algorithms

Few-shot learning relies on specialized techniques and algorithms that enable models to make the most out of limited data:

Matching Networks: These use an attention mechanism to weigh the importance of support (training) examples in classifying new instances(K. Prasad and G. Poornima, 2022). They create a weighted nearest neighbor classifier for each new instance based on the learned similarities.
Prototypical Networks: These networks compute a prototype (a mean vector) for each class based on the available examples. New instances are classified based on their proximity to these prototypes in the embedding space.
Model-Agnostic Meta-Learning (MAML): Training a model to solve novel learning challenges with a minimal amount of training samples is the goal of MAML. This model is trained on a range of learning tasks.
Bayesian Approaches: These approaches incorporate prior knowledge and update the model's understanding as new data comes in, which is particularly useful when dealing with sparse data.
Transfer Learning: Using a pre-trained model on a large dataset and fine-tuning it on the few-shot task. This leverages the general features learned from the larger dataset.

3.3 Case Studies and Successful Applications

Natural Language Processing (NLP): Few-shot learning has been instrumental in NLP tasks like sentiment analysis or named entity recognition, where labeled data can be scarce for certain languages or domains.
Computer Vision: Few-shot learning is used in facial recognition systems to correctly identify faces even when the amount of training data is small, improving privacy and efficiency.
Autonomous Vehicles: Training models to recognize rare road signs or unusual pedestrian behaviors with limited examples.
Personalized Recommendations: E-commerce platforms often use few-shot learning to make product recommendations based on minimal user interaction data.
Agriculture: Identifying plant diseases or pest infestations from a small number of field images.

The interconnection of billions of devices across a wide range of IoT-enabled application areas has been expedited by recent developments in IT and hardware(H. Anandaram, and B. K. Verma, 2023). Healthcare, transportation, manufacturing, environmental monitoring, and home automation are just a few examples of the vital infrastructure industries that have recently seen the widespread use of smart and adaptable technologies(K. Prasad and B. Ashreetha, 2023). The development of IoE frameworks is warranted by this tendency.

Datasets from specific devices or sceneries are sometimes severely lacking in sample size, even if these many terminal devices provide a vast quantity of data. Conventional data-driven and domain-specific algorithms fail miserably in these situations(R. Kishore Kumar and M. S. C. Senthil Kamalesh, 2023). As a result, a lot of work has gone into finding effective learning algorithms that can work with small samples and in many domains. Two new methods that show promise are meta-learning and few-shot

Figure 7. Comprehensive conceptual framework of few-shot learning (FSL)

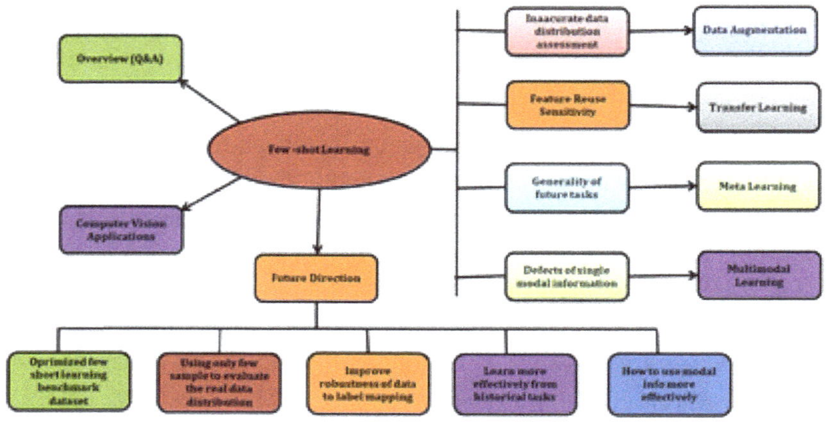

learning (FSL). One of the biggest problems in FSL nowadays is how to get useful information out of limited datasets, especially when applied to different domains.

It is also difficult to generalize a single model across different domains since long-tail effects are common in real-world data distribution. Consider smart manufacturing industrial inspection as an example; intelligent models in this domain have a substantial barrier due to this lack of generality.

Modern technology often need precise illumination conditions when used for industrial quality testing. Artificial intelligence algorithms "generalize" poorly to different illumination scenarios after being trained under only one. Additionally, AI models usually depend on supervised learning and need a large number of flawed samples for training, which might be problematic in industrial quality inspection settings where high accuracy is required(B. Ram Vishal and K. A. M, 2023). However, due to the low percentage of faulty goods in real production scenarios, it is tough to gather a sufficient number of acceptable samples. Impracticality also arises while trying to transmit knowledge across different fields. For example, it would be a huge mistake to use a model that was developed for PC appearance defect detection to identify flaws in mobile phone displays, fridges, washing machines, or even other PC models. Similarly, it is not feasible to gather enough examples for each kind in situations where character pictures on components and circuit boards need to be recognized due to the large number of suppliers, different kinds of devices, and different character styles. For certain kinds, this means there is either no sample pool at all or very little.

A new approach using data, algorithms, and models known as Few-shot Learning (FSL) has emerged to tackle these problems more efficiently(D. Palanikkumar, A. Y. Begum, 2023). There are two main traditional schools of thought when it comes to FSL: meta-learning and metric-based learning. In this study, nevertheless, we classify FSL into four separate domains data augmentation, transfer learning, meta-learning, and multimodal learning taking into account the current difficulties. In Figure 7, we can see the whole conceptual framework of Few-shot Learning (FSL).

Data Augmentation: In this area of FSL, metric or generative approaches are used to simulate data in different situations. The objective is to increase the spread of accessible data as much as feasible.

Transfer Learning: To retrieve previous information from massive auxiliary datasets, transfer learning in FSL mostly makes use of pretraining and fine-tuning(A. Nareshkumar and B. Gururaj, 2022). There may be limits to transfer learning when domain relevance is scarce or when large auxiliary datasets are unavailable.

Meta-learning: At this time, meta-learning is the most popular strategy for fixing the FSL issue. It entails teaching models to learn efficiently, especially when given little data.

These various fields represent a number of different attempts FSL is making to solve the problem data scarcity and domain discrepancy.

These applications deal with few-shot learning, because of insufficient data at the very label level. This method not only lets machine learning potentially solve more tasks, but also saves huge sums of time and money spent collecting data to train the system.

4. CHALLENGES IN FEW-SHOT LEARNING

4.1 Data Scarcity and Overfitting

Data Scarcity: December 19, 2017 And the model only has a partial understanding of the variety and complexity that each class encompasses, because it looks at just examples from few classes. Therefore, because of this scarcity, it is difficult to get a model which accurately reflects the true underlying distribution of data.

Overfitting: But with so few data available, there's no shortage of overfitting in these models. In the same way, in training data where they may have been trained noise or irrelevant details, and performs badly on new unseen data (R. Kavitha & V. Hariram 2023). Balancing complexity and data is one of the most difficult problems.

4.2 Generalizability of Models

Narrow Learning Scope: So models based on few examples could always be too specialized to those cases, and not broad enough in applicability.

Variability Across Tasks: Few-shot learning models are supposed to quickly learn new tasks. But every task can be very different, making this a major challenge for the model to maintain consistent performance across all tasks.

4.3 Dependence on Feature Similarity

Feature Representation Limitations: The main assumption employed by few-shot learning models is that the examples provided are representative of the entire class (S. Archana Shreee, G. Jeevitha Sai, 2023). Even if such examples do not encompass the class's heterogeneity, however, the model's generalization power becomes weak.

Similarity Measure Challenges: Such models rely on measuring feature similarity. But it is very difficult to define an effective and stable similarity measure that can span different classes and domains.

4.4 Computational Complexity

Resource-Intensive Training Processes: One such technique is meta-learning, which can be computationally expensive and time-consuming.

Algorithmic Complexity: To avoid overfitting and learn efficiently from sparse data, algorithms must have complex designs that require high computing power.

Optimization Challenges: High risk of overfitting and the necessity for thorough adjustment of regularization and hyper-parameters make few-shot learning a more intractable optimization environment.

Few shot learning lacks problems unique to machine learning, such as over-fitting and the omnipresent similarity of features. Generalization can also be considered a problem in many cases. However, this type of learning that has the potential for being so helpful may at once teach us how to learn from very few data. These problems require new modeling and algorithm design, as well as data descriptions.

5. ZERO-SHOT LEARNING

finding itself identifying some things it has never seen before There is a state-of-the art method in machine learning which they call zero shot (or if you prefer, infinite shot) learning. Here the idea or object exists but hasn't been defined and the model learns to identify it on its own. In order to do this, you have got to know what generalizable representations or attributes are.

5.1 Examples

Object Recognition: A model trained on an animal dataset can analogous use learned properties (such as fur, four legs) to link with and find rare species or other unfamiliar animals.

Language Translation: Translation of sentences into languages which are outside the training dataset, relying on the knowledge about language structures and semantics.

Content Categorization: Understanding keywords and context, Within the frame of excluded classifications in training set.

5.2 Techniques and Algorithms

Zero-shot learning relies on several key techniques and algorithms:

Embedding Learning: This is to learn a domain with similar concepts near each other. for instance, encoding words or phrases in high-dimensional space according to meaning.

Attribute Learning: During training, the model comes to learn common attributes or features among these different classes. For zero-shot tasks it predicts new items using similarity to learned attributes as a basis.

Semantic Output Codes: The procedure is as follows. By associating a semantic descriptor with each class (i.e., like a set of attributes), the model is trained to predict these descriptors for an image or paragraph.

Hybrid Models: By integrating different types of information such as class attributes and the relationship between classes, it is hoped that good generalization to new classes will be attained.

Generative Models: These models can output new data examples, which then serve to imitate unseen classes when the model is being trained.

5.3 Case Studies and Successful Applications

Natural Language Processing: Applied to tasks such text categorization or sentiment analysis, this zero-shot learning function permits models to process categories and sentiments unseen in training.

Image Classification: Recognizing new objects in images based on learned attributes from a different set of objects, such as identifying specific breeds of dogs or types of plants that the model has not been explicitly trained on.

Environmental Monitoring: Identifying species or environmental conditions not previously recorded in the training data but inferred through learned characteristics.

Healthcare: Diagnosing rare medical conditions based on symptoms and patterns learned from more common diseases.

Retail: Systems that can propose other items based on user likes and characteristics, even if that particular product wasn't part of the training set.

In order to make the more difficult Generalized Zero-shot Learning (GZSL) inference easier to handle, we will be splitting it into smaller, more manageable tasks, such as ZSL and ZSL-supervised learning. This simplification may be achieved if there is previous knowledge about the domain (visible or unseen) of a test sample(M. Kranthi, and R. C. Tanguturi, 2023). Using a domain classifier that can predict the domain of a test sample is one easy way to get this previous knowledge. In a standard Open Set Recognition (OSR) architecture, the visual data from classes that are visible may be used to train the domain classifier, while the visual data from classes that are not visible can be used to represent the open set. In an inductive GZSL context, however, data from the unseen classes (semantic and visual) is unavailable, making GZSL a formidable obstacle during training. To circumvent this problem, we provide visual examples of Prototype Unseen Classes (PUCs) by referencing the open set representations in the prototypes of related PUCs. For each observed class, we create PUCs that look somewhat like the observed class data while yet retaining certain semantic distinctions. In order to do this, we suggest creating PUC visual samples that belong to a certain region beyond the visual feature space's support for the observed class. There is a connection between every visible class and its corresponding PUC; taken together, all of these PUCs stand in for the open set.

The visual-semantic data of PUCs is derived entirely from the visible classes, which is an important point. Using a Generative Adversarial Network (GAN), an adversarial training technique is used to generate visual samples of both observed classes and PUCs. One prototype for each observed class is used to produce visual samples, whereas several prototypes for each PUC are used to generate visual examples of their adaptable prototypes.

The first part of Figure 8 shows a visual example of a Prototype Unseen Class (PUC) enclosing its equivalent visible class. (B) When an end-to-end model is trained, it learns to link semantic information to visual information, as shown in an example figure. (C) Shows the suggested model, which uses their separate adaptable prototypes to create visual examples of PUCs in an adversarial framework during training (R. C. Tanguturi, and V. C. H, 2023). To control the production of visual PUC samples, the

Figure 8. Visual summary ZSL

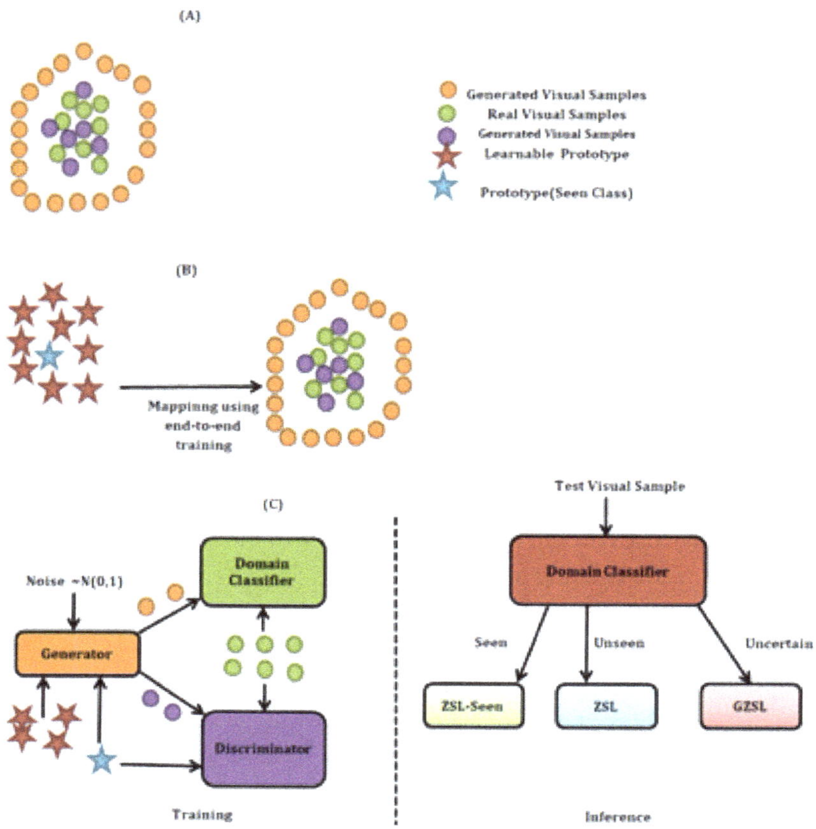

model additionally makes use of the domain classifier and the Generative Adversarial Network (GAN) via cooperative training. To begin with, the test sample's domain is labeled using the learnt domain classifier as a pretreatment step. Then, the class label is inferred using the ZSL-Seen, ZSL, or GZSL framework depending on the domain label, which may be classified as seen, unseen, or uncertain. See the colorful version for the clearest understanding.

Visual examples, including seen classes and Prototype Unseen Classes (PUCs), are generated by applying an end-to-end training strategy to the domain classifier and the Generative Adversarial Network (GAN). The construction of the model is shown in figure 8 (C). The inference step requires using the learned domain classifier to decide whether or not test samples are seen, unseen, and unsure (N. Madapana and J. Wachs 2021). To prevent such debatable offshore domains from being mistakenly classified as either visible or invisible, the doubtful domain category was established. When the test sample has its class label flagged as domain uncertain, the commonly used Generalized Zero-shot Learning (GZSL) method is applied. On the other hand, when answering to inference In this case, if domain classifier assumes that The ZSL-Seen or ZSL framework is used for prediction of matchingclass label. This new work represents a start Y. Feng and X. Huang (202). Two crucial training processes are included in the Open Set Recognition (OSR) approach for performing GZSL: The first step is generating PUC samples, and the second step is learning an end-to-end model using a domain classifier, GANs. Based on the results of the domain classifier, test samples are given domain labels during the inference step.

Zero-shot learning's ability to generalize beyond the trained data makes it an invaluable tool in fields where data collection is challenging or infeasible. Its applications showcase the potential of AI to learn and adapt in dynamic, real-world environments, pushing the boundaries of traditional machine learning.

6. CHALLENGES IN ZERO-SHOT LEARNING

6.1 Semantic Gap Problem

Definition: A semantic gap is an expression used to describe the distance between high-level, abstract terms familiar in human minds and low- level features which machine learning models require for space allocation. In zero-shot learning, this gap emerges when models fail to properly grasp and generalize abstract concepts or attributes such as color across classes.

Complexity in Abstract Representation: That semantic gap is hard to cross, though. Given that the model must infer about unseen classes on the basis of their relationship with seen ones or even just a written description of an attribute in text form-not something it can sense for itself--many stumble at this point and fall out.

6.2 Domain Adaptation Issues

Generalization Across Domains: A problem plaguing zero-shot learning models, for instance that they can only be used on data from the same domain as it was trained at. Because the model was exposed to a problem in that particular environment, there is no guaranteeing that features or attributes will migrate into other areas.

Variability and Shift in Data: This indeed adds to problems in domain adaptation, because data distribution between the training and target domains may be different.

6.3 Attribute Selection and Model Bias

Attribute Relevance: properly choosing attributes with descriptive and differentiating effects upon classes are essential to zero-shot learning. If attributes are not correct or sufficient, the model may perform poorly.

Bias in Attribute Selection: Therefore if the attributes are chosen manually, there is a chance of introducing bias. Biased model prediction can arise from such biases.

6.4 Reliance on Auxiliary Information

Dependence on External Data: With zero-shot learning, one needs auxiliary information like class descriptions and attribute annotations or semantic relationships. The quality and quantity of this auxiliary information directly influence the performance of any model.

Inconsistency and Inaccuracy: The supplementary information might be inconsistent, incomplete or incorrect. That is, it gives the model few chances to make reliable inferences about unseen classes.

Integration Challenges: In fact, the correct incorporation of auxiliary information into learning is a challenging problem and even requires advanced modeling techniques.

In summary, the zero-shot learning problem is especially thorny. The semantic gap Some obstacles are trying to narrow the gap between different domains; some problems prohibit cross-domain generalization. Others still concern when choosing proper attributes, avoiding biases and how much auxiliary information should be used. It is especially important that zero-shot learning be able to overcome these obstacles in order for its capabilities as a practical tool to keep pace withdevelopments, and extending it into different areas of application.

7. COMPARATIVE ANALYSIS: FEW-SHOT VS. ZERO-SHOT LEARNING

In the world of machine learning, when it comes to working with scarce data few-shot and zero- shot techniques both really shine. However they do so in very different ways. For noisy labels, it outstrips itself in effectiveness when some specific data has already been obtained. Because of this, its relatively low exposure allows for greater adaptiveness to specific tasks and less dependence upon auxiliary or external information. However, few-shot learning method has its problems. One concerns overfitting, the other generalization beyond the examples given.

On the other hand, zero-shot learning (models making inferences about classes that have never been seen during training) is shown in many ways to be very capable. It is particularly helpful where data collection is problematic. But zero-shot learning actually has to bridge a semantic gap between conceptual level and feature representation, so it is critical dependent on the quality of any auxiliary information that you can give it (class descriptions or attributes).

Both methods have their uses and are suited to different applications. Few-shot learning is helpful when large amounts of data are not available. One area where this method can find application, for instance, lies in medical diagnostics as wells as personalized applications that adapt to users 'preferences. On the other hand, zero-shot learning works well in fields like natural language processing--tasks on languages not included in training data and content categorization to classify into new genres.

One traditional method that faces this difficult problem of Few-shot Learning (FSL) with few or no labelled data is Transfer learning. Transfer learning's key idea of feature reuse can help ease FSL's data sparseness problem. The basic procedure involves training the model using a large and diverse dataset, then fine-tuning on the small support set available for the task at hand.

But when the source and destination domains are very different, the effectiveness of such knowledge transfer is greatly impaired. And this new difficulty becomes a new problem for FSL in cross-domain situations.

As a foundational strategy, FSL's transfer learning consists of two phases: pre-training and fine-tuning. The entire process is depicted in Figure 9. Recent advances in Few-shot Learning (FSL) require labeled data to be used during training. But it's generally impossible to gather varied datasets designed for certain jobs in many real-world applications. Problems arise in FSL as a result of this restriction, especially when working with very diverse domains. By integrating FSL with domain adaptation challenges, Cross-domain Few-shot Learning generates a thorough and rigorous environment. The conventional dataset format and extreme resemblance to real-world scenarios have long been problems for the benchmark datasets used in FSL. Models do well on benchmark datasets but fail miserably when put to

Figure 9. Illustration of transfer learning

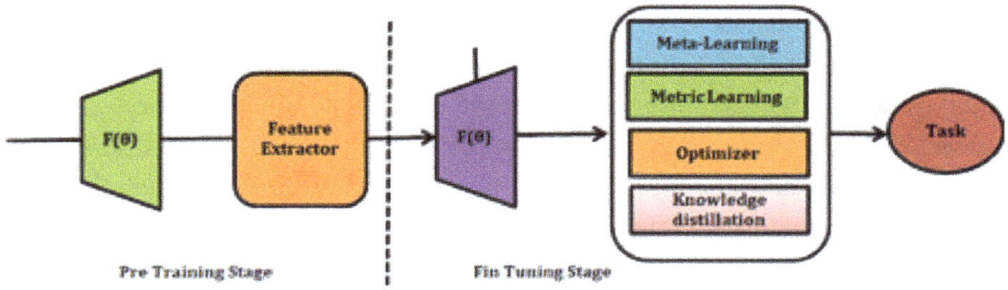

Figure 10. Framework for end-to-end learning

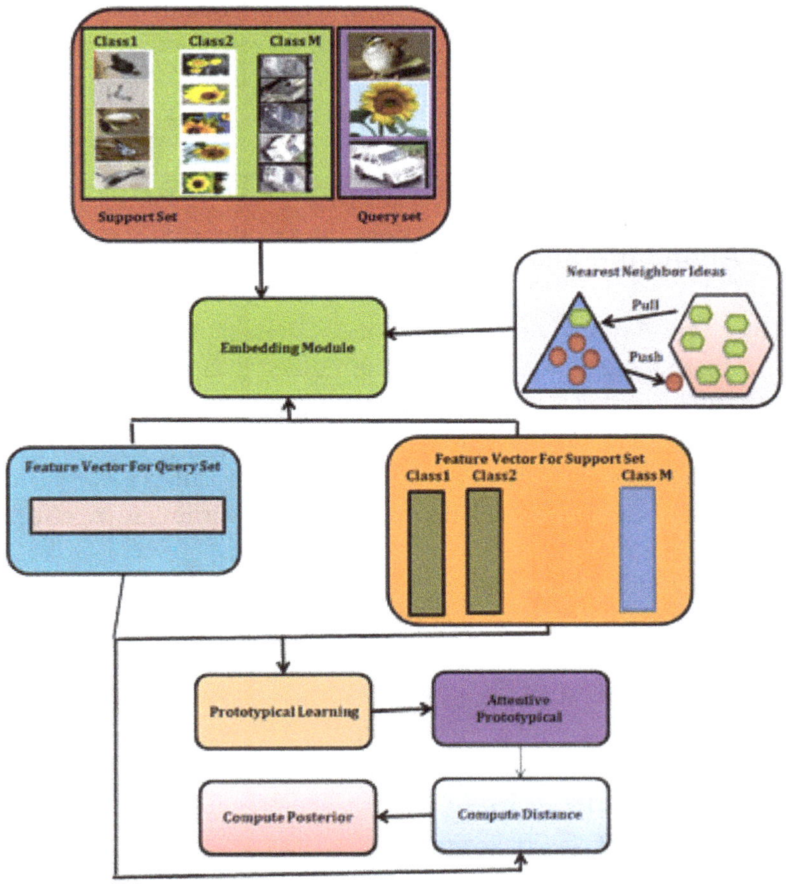

use in the real world because of this. Figure 10 shows an example of a prominent learning approach; it is based on a network prototype that has been fine-tuned to perform very well on a classification benchmark dataset. Considered a binary classification issue, the Siamese neural network was one of the first models in metric learning. In the inference phase, it evaluates picture similarity using a collection of positive or negative sample pairs as input. An other method for dealing with more than two input pairs in FSL metric learning is triple loss. Triple loss, in contrast to SNN, requires the simultaneous availability of positive, negative, and anchor samples. To convert few-shot and unlabeled information into embedding space vectors, Matching Networks provide a more generic network architecture. Modeling the sample distance distribution by learning the embedding representation, they combine the strengths of parametric and nonparametric models, similar to the closest neighbor technique. A more refined embedding manifold is the outcome of embedding propagation, as shown experimentally. To improve the model's accuracy, it is essential to learn high-quality embedding representations efficiently within a time constraint. However, by combining visual embeddings, semantic embeddings, and gating measures, GVSE automatically balances the relative relevance of each metric dimension.

In essence, the choice between few-shot and zero-shot learning depends on the specific requirements of the task at hand, the nature and availability of data, and the desired outcome. As advancements in AI continue, both learning methods are evolving, addressing their respective challenges, and broadening their applicability in various fields.

8. FUTURE DIRECTIONS AND POTENTIAL SOLUTIONS

8.1 Advances in AI and Machine Learning

Improved Algorithms and Architectures: Few-shot and zero-shot learning rely on existing models, so the ability of these models to generalize from limited data is affected by ongoing research in developing more sophisticated algorithms and neural network architectures. Self-supervised learning and advanced embedding methods look promising.

Enhanced Meta-Learning: Another area where few-shot learning shines is meta-learning, or the ability to learn to learn. Discoveries here can help make few-shot learning more effective and even faster than before.

Integrating Generative Models: Generative adversarial networks (GANs) and other generative models can be used to synthesize new training examples, which could overcome the problem of data scarcity in few-shot learning.

Semantic Understanding and Natural Language Processing: NLP and semantic analysis could help close the semantic gap in zero-shot learning, aiding better, more robust models.

8.2 Cross-Disciplinary Approaches

Cognitive Science Insights: Knowledge about how humans can learn with imperfect information derived from cognitive science and psychology provides valuable material for the design of faster-learning algorithms.

Collaboration with Domain Experts: With experts in their respective fields (say healthcare, linguistics or ecology), such partnerships are likely to offer an insightful contribution about the selection of features, data quality and practical applications in these models.

Data Science and Big Data Analytics: Feature extraction and preprocessing the relevant raw dataset by means of big data also helps to improve model performance in few-shot learning. In extreme cases, it may even result in no need for any kind of extra step at all.

8.3 Ethical Considerations

Bias and Fairness: In light of these learning methods dealing typically with thin data; the probability of bias is heightened. A key component of ethical AI development has to be methods for identifying and countering bias in the models.

Transparency and Explainability: The understanding of few-shot and zero-shot learning models is very limited due to their profound complexity, so development needs to go in the direction of more interpretable models with a transparent decision process, and such efforts are especially vital for tasks important to people's lives like healthcare.

Data Privacy and Security: Given the growing volume of sensitive data, privacy and security is particularly important. Various privacy-preserving techniques, including differential privacy and federated learning, could be incorporated in the system to protect user data.

Regulatory Compliance: It is important to follow and help shape the polices and regulations governing AI, machine learning applications, which in turn make these technologies increasingly pervasive in high-stakes areas.

These advances in artificial intelligence and machine learning, cross-disciplinary cooperation as well as an insistence on ethical considerations will all be elements of the future of few-shot and zero-shot learning. As AI technology advances, these learning paradigms can become ever stronger and more flexible, and morally compatible and closer to real life. The ability for a less abstract AI to better solve complex, practical problems will continue to get greater. In recent research on Few-shot Learning (FSL), progress has been made in a number of task contexts. The problem, of course, is that in the more difficult situations there is a severe lack of validation and training datasets. This explains why there is not enough data for training or validation to do successful transfer learning, and if that's the case, there simply isn't enough data that can be used during testing of real samples. Further, there are few options to initiate the parameters of the model for meta-learning. Few-shot Learning is interesting in that its methods are more flexible and adaptive than traditional Deep learning problems, which require great quantities of data for learning. It is modular in structure and solving many problems, due to its smooth compatibility with techniques such as data augmentation, transfer learning and meta-learning.

9. CONCLUSION

In summary, this chapter examined the complex domains of few-shot and zero- shot learning--two new approaches in machine learning that allow one to learn from a small amount of data nonf Bun Out as well as make predictions. But when there are some classes with many instances or even just a few but scarce, the technique of simple to handle examples--few-shot learning has shown amazing potential around. One step beyond that is zero shot learning. A model can generalize from attributes alone, or based on both the class's buffer of attributes and semantic relationships learned between all classes. Despite their potential, both methods are hampered by limitations. In few-shot learning, the pernicious problem of overfitting limits its strength. In zero shot learning there is a semantic gap. from data and attribute choice to computational complexity, there are all kinds of other potential obstacles. Naturally, as IA continues to progress in complexity of algorithms and enhanced manners of learning (beginning with meta-learning methods), new ideas that can only be born from interdisciplinary cooperation will create even larger waves. Still, the development and application of these technologies are clouded by many ethical considerations--particularly with regard to bias on grounds not related to originality; transparency as well data privacy. Heading into the future, few-shot and zero shot learning will be a key for developing AI. With regard to the means of teaching, both are more flexible and effective than previous methods. Spreading of this approach will not only solve today's technical problems with the use of A1, it can also provide a path that allows people to enjoy technology while protecting themselves from harm.

REFERENCES

Anandaram, H. N. B. A., Gupta, N., & Verma, B. (2023). *IoT Wearable Breast Temperature Assessment System.* 2023 7th International Conference on Computing Methodologies and Communication (ICCMC), Erode, India.

Archana Shreee, S., Maheshwari, B., & Jeevitha Sai, G. (2023). A Novel Method of Identification of Delirium in Patients from Electronic Health Records Using Machine Learning. 2023 World Conference on Communication & Computing (WCONF), RAIPUR, India.

Bendre, N., Desai, K., & Najafirad, P. (2021). Generalized Zero-Shot Learning Using Multimodal Variational Auto-Encoder With Semantic Concepts. *2021 IEEE International Conference on Image Processing (ICIP)*, Anchorage, AK, USA. 10.1109/ICIP42928.2021.9506108

Chakrasali, S. V., Kumar, C., Chaturvedi, A., & Jaisudhan Pazhani, A. A. (2023). Computer vision based healthcare system for identification of diabetes & its types using AI, Measurement. *Sensors (Basel)*, *27*, 10075.

Chen, J., Geng, Y., Chen, Z., Pan, J. Z., He, Y., Zhang, W., Horrocks, I., & Chen, H. (2023). Zero-Shot and Few-Shot Learning With Knowledge Graphs: A Comprehensive Survey. *Proceedings of the IEEE*, *111*(6), 653–685. doi:10.1109/JPROC.2023.3279374

Cho, W., & Kim, E. (2022). Improving Augmentation Efficiency for Few-Shot Learning. *IEEE Access : Practical Innovations, Open Solutions*, *10*, 17697–17706. doi:10.1109/ACCESS.2022.3151057

Feng, Y., Huang, X., Yang, P., Yu, J., & Sang, J. (2022). Non-generative Generalized Zero-shot Learning via Task-correlated Disentanglement and Controllable Samples Synthesis. *2022 IEEE/CVF Conference on Computer Vision and Pattern Recognition (CVPR)*, New Orleans, LA, USA. 10.1109/CVPR52688.2022.00913

Naik, G. (2021). Covid-19 Prevention Kit Based on an Infrared Touchless Thermometer and Distance Detector. *2021 5th International Conference on Electronics, Communication and Aerospace Technology (ICECA)*, (pp. 358-362). IEEE.

Guan, J., Lu, Z., Xiang, T., Li, A., Zhao, A., & Wen, J.-R. (2021). Zero and Few Shot Learning With Semantic Feature Synthesis and Competitive Learning. *IEEE Transactions on Pattern Analysis and Machine Intelligence*, *43*(7), 2510–2523. doi:10.1109/TPAMI.2020.2965534 PMID:31940521

Hombalimath, D. Palanikkumar, & Patwari, N. (2023). *Symmetrized Feature Selection with Stacked Generalization based Machine Learning Algorithm for the Early Diagnosis of Chronic Diseases.* 2023 5th International Conference on Smart Systems and Inventive Technology (ICSSIT), Tirunelveli, India.

Jeevan, K., & Sathisha, B. M. (2020). Implementation of IoT Based Wireless Electronic Stethoscope. *2020 Third International Conference on Multimedia Processing, Communication & Information Technology (MPCIT)*, (pp. 103-106). IEEE.

Jiang, B., Lu, Y., Zhang, B., & Lu, G. (2023). Few-Shot Learning for Image Denoising. *IEEE Transactions on Circuits and Systems for Video Technology*, *33*(9), 4741–4753. doi:10.1109/TCSVT.2023.3248585

Kavitha, R., Kumar, A., Kalpana, V., & Hariram, V. (2023). Artificial Intelligence based Health Monitoring System on IoTH platform. *2023 Second International Conference on Augmented Intelligence and Sustainable Systems (ICAISS)*, Trichy, India.

Kawale, S. R., & Diwan, S. P. (2022). Intelligent Breast Abnormality Framework for Detection and Evaluation of Breast Abnormal Parameters. *2022 International Conference on Edge Computing and Applications (ICECAA)*, (pp. 1503-1508). IEEE.

Kishore Kumar, R., Pandidurai, M., & Senthil Kamalesh, M. S. C. (2023). *Design of IoT based Rural Health Helper using Natural Language Processing.* 2023 4th International Conference on Electronics and Sustainable Communication Systems (ICESC), Coimbatore, India.

Kranthi, M., & Tanguturi, R. C. (2023). *Design of Intelligent Medical Integrity Authentication and Secure Information for Public Cloud in Hospital Administration.* 2023 2nd International Conference on Edge Computing and Applications (ICECAA), Namakkal, India.

Lilhore, U. K., Manoharan, P., Sandhu, J. K., Simaiya, S., Dalal, S., Baqasah, A. M., Alsafyani, M., Alroobaea, R., Keshta, I., & Raahemifar, K. (2023). Hybrid model for precise hepatitis-C classification using improved random forest and SVM method. *Scientific Reports*, *13*(1), 12473. doi:10.1038/s41598-023-36605-3 PMID:37528148

Lilhore, U. K., Poongodi, M., Kaur, A., Simaiya, S., Algarni, A. D., Elmannai, H., Vijayakumar, V., Tunze, G. B., & Hamdi, M. (2022). Hybrid model for detection of cervical cancer using causal analysis and machine learning techniques. *Computational and Mathematical Methods in Medicine*, *2022*, 2022. doi:10.1155/2022/4688327 PMID:35572826

Lilhore, U. K., Simaiya, S., Pandey, H., Gautam, V., Garg, A., & Ghosh, P. (2022). Breast cancer detection in the IoT cloud-based healthcare environment using fuzzy cluster segmentation and SVM classifier. In Ambient Communications and Computer Systems [Singapore: Springer Nature Singapore.]. *Proceedings of RACCCS*, *2021*, 165–179.

Lilhore, U. K., Simaiya, S., Prasad, D., & Guleria, K. (2020). A hybrid tumor detection and classification based on machine learning. *Journal of Computational and Theoretical Nanoscience*, *17*(6), 2539–2544. doi:10.1166/jctn.2020.8927

Madapana, N., & Wachs, J. (2021). *ZF-SSE: A Unified Sequential Semantic Encoder for Zero-Few-Shot Learning*. 2021 16th IEEE International Conference on Automatic Face and Gesture Recognition (FG 2021), Jodhpur, India. 10.1109/FG52635.2021.9667025

Mailapur, R. (2023). *Implementation of GUI based Vital Track Ambulance for Patient Health Monitoring*. 2023 8th International Conference on Communication and Electronics Systems (ICCES), Coimbatore, India.

Namitha, A. R., Manu, Y. M., & Rashmi, G. R. (2022), IOT Based Smart Health Care System to Monitor Covid-19 Patients. *IJEER, 10*(1), 36-40.

Nareshkumar, A., & Gururaj, B. (2022). An Integrated IoT Technology for Health and Traffic Monitoring System with Smart Ambulance. *2022 IEEE North Karnataka Subsection Flagship International Conference (NKCon)*, Vijaypur, India.

Naveen, K. B., Ramesha, M., & Pai, G. N. (2020). Internet of things: Internet revolution, impact, technology road map and features. *Adv. Math. Sci. J.*, *9*(7), 4405–4414. doi:10.37418/amsj.9.7.11

Onyema, E. M., Lilhore, U. K., Saurabh, P., Dalal, S., Nwaeze, A. S., Chijindu, A. T., Ndufeiya-Kumasi, L. C., & Simaiya, S. (2023). Evaluation of IoT-Enabled hybrid model for genome sequence analysis of patients in healthcare 4.0. *Measurement. Sensors*, *26*, 100679. doi:10.1016/j.measen.2023.100679

Palanikkumar, D., Mary, P. A., & Begum, A. Y. (2023). *A Novel IoT Framework and Device Architecture for Efficient Smart city Implementation*. 2023 7th International Conference on Trends in Electronics and Informatics (ICOEI), Tirunelveli, India.

Pourpanah, F., Abdar, M., Luo, Y., Zhou, X., Wang, R., Lim, C. P., Wang, X.-Z., & Wu, Q. M. J. (2023). A Review of Generalized Zero-Shot Learning Methods. *IEEE Transactions on Pattern Analysis and Machine Intelligence*, *45*(4), 4051–4070. doi:10.1109/TPAMI.2022.3191696 PMID:35849673

Prasad, K., Anil Kumar, N., Reddy, N. S., & Ashreetha, B. (2023). Technologies for Comprehensive Information Security in the IoT. *2023 International Conference for Advancement in Technology (ICONAT)*, Goa, India.

Prasad, K., Dekka, S., Tanguturi, R. c., & Poornima, G. (2022). *An Intelligent System for Remote Monitoring of Patients Health and the Early Detection of Coronary Artery Disease*. 2022 International Conference on Smart Generation Computing, Communication and Networking (SMART GENCON), Bangalore, India.

Rahman, S., Khan, S., & Porikli, F. (2018). A Unified Approach for Conventional Zero-Shot, Generalized Zero-Shot, and Few-Shot Learning. *IEEE Transactions on Image Processing*, *27*(11), 5652–5667. doi:10.1109/TIP.2018.2861573 PMID:30072324

Ram Vishal, B. & Shankar, M. (2023). *Cardiovascular Disease Prediction Using LSTM Algorithm based On Cytokines*. 2023 4th International Conference for Emerging Technology (INCET), Belgaum, India.

Ramesh, T. R., Lilhore, U. K., Poongodi, M., Simaiya, S., Kaur, A., & Hamdi, M. (2022). Predictive analysis of heart diseases with machine learning approaches. *Malaysian Journal of Computer Science*, 132–148.

Ramesh Naidu, P., Guruprasad, N., & Dankan Gowda, V. (2020). Design and implementation of crypt-cloud system for securing files in cloud. *Adv. Math. Sci. J.*, *9*(7), 4485–4493. doi:10.37418/amsj.9.7.17

Ramesh Naidu, P., Guruprasad, N., & Dankan Gowda, V. (2021). A High-Availability and Integrity Layer for Cloud Storage, Cloud Computing Security: From Single to Multi-Clouds. *Journal of Physics: Conference Series*, *1921*(1), 012072. doi:10.1088/1742-6596/1921/1/012072

Rao, B. K., Chaturvedi, A., & Hussain, N. (2022). Industrial quality healthcare services using Internet of Things and fog computing approach, Measurement. *Sensors (Basel)*, *24*, 100517.

Rathore, M. S., Poongodi, M., Saurabh, P., Lilhore, U. K., Bourouis, S., Alhakami, W., Osamor, J., & Hamdi, M. (2022). A novel trust-based security and privacy model for internet of vehicles using encryption and steganography. *Computers & Electrical Engineering*, *102*, 108205. doi:10.1016/j.compeleceng.2022.108205

Reddy, S., & Patwal, P. P. S. (2022). Data Analytics and Cloud-Based Platform for Internet of Things Applications in Smart Cities. *2022 International Conference on Industry 4.0 Technology (I4Tech)*, (pp. 1-6). IEEE.

Revanna, C. R., Kameswara Rao, B., & Sarma, P. (2022). Enhanced Diagnostic Methods for Identifying Anomalies in Imaging of Skin Lesions. *IJEER 10*(4), 1077-1085.

Sharma, S. K., Lilhore, U. K., Simaiya, S., & Trivedi, N. K. (2021). An improved random forest algorithm for predicting the COVID-19 pandemic patient health. *Annals of the Romanian Society for Cell Biology*, 67–75.

Simaiya, S., Lilhore, U. K., Prasad, D., & Verma, D. K. (2021). MRI brain tumor detection & image segmentation by hybrid hierarchical K-means clustering with FCM based machine learning model. *Annals of the Romanian Society for Cell Biology*, 88–94.

Swathi Pai, M. Shruthi, M., & Naveen, K. (2020). Internet of Things: A Survey on Devices, Ecosystem, Components and Communication Protocols. *2020 4th International Conference on Electronics, Communication and Aerospace Technology (ICECA)*, (pp. 611-616). IEEE.

Tanguturi, R. C. (2023). Implementation of Machine Learning Approach for Detecting Cardiovascular Diseases. *2023 3rd International Conference on Intelligent Technologies (CONIT)*, Hubli, India.

U. K N & R. V M. (2022). *Arduino based COVID-19 Suspect Detection Device*. 2022 6th International Conference on Electronics, Communication and Aerospace Technology, Coimbatore, India.

Venkatakiran, S., Ashreetha, B., & Reddy, N. S. (2023). Implementation of a Machine Learning-based Model for Cardiovascular Disease Post Exposure prophylaxis. *2023 International Conference for Advancement in Technology (ICONAT)*, Goa, India.

Chapter 8
Unveiling the Potential: A Comprehensive Exploration of Deep Learning and Transfer Learning Techniques in Bioinformatics

Umesh Kumar Lilhore
University of Louisiana, USA

Sarita Simaiya
University of Louisiana, USA

ABSTRACT

In the rapidly evolving field of bioinformatics, the surge in biological data necessitates innovative approaches for practical analysis and comprehension. This chapter presents an in-depth exploration of the potential applications of deep learning and transfer learning methods, showcasing their transformative impact on various facets of bioinformatics research. Deep learning methodologies significantly enhance the capacity to extract intricate patterns and features from diverse biological datasets, fostering a more profound understanding of genomic, proteomic, and metabolomic dynamics. Additionally, transfer learning assumes a pivotal role in optimizing model performance, particularly in scenarios where labeled data is scarce, by leveraging knowledge acquired from pre-trained models. The integration of deep learning and transfer learning techniques in bioinformatics yields valuable insights, serving as a guiding framework for researchers and practitioners aiming to harness the full potential of these methodologies in deciphering the complexities of biological systems.

1. INTRODUCTION

Bioinformatics is a discipline that evolved from the intersection of the academic and biological domains. Although it is a relatively new field, it is of critical importance. Both the study of intricate molecular structures and the rapid expansion of biological data have significantly benefited from the confluence of the fields of computing and biology, which has resulted in the development of a number of innova-

DOI: 10.4018/979-8-3693-1822-5.ch008

Figure 1. Structural bioinformatics and proteomics

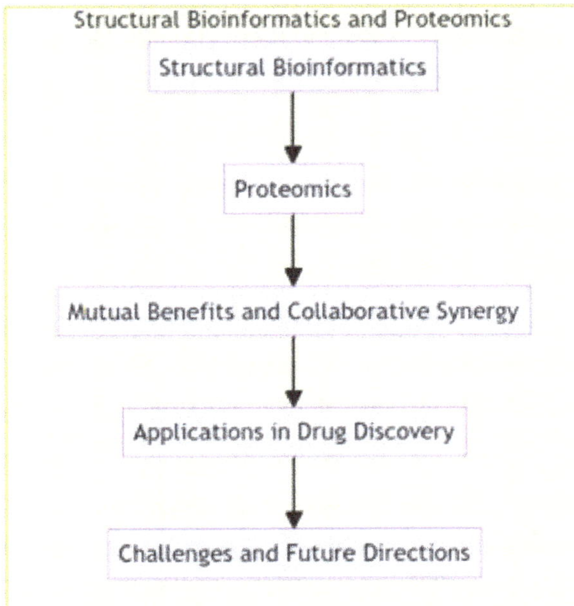

tive approaches and instruments. Within the context of our data-driven era, these creative aspects bring about a profound transformation in the manner in which we see, evaluate, and comprehend information pertaining to biology (Abdar et al., 2021).

The purpose of this article is to provide insights that are intended to stimulate new ideas and collaborative efforts in the arena where biological and computing technology intersect. This may lead to breakthroughs that will have a significant influence on how we comprehend and manage biological processes. The ever-increasing diversity and amount of biological data are posing a number of complex difficulties that call for the most cutting-edge statistical methodologies. DL and TL techniques, which are examples of modern approaches, are primarily responsible for the resolution of these situations.

An interdisciplinary effort in the large area of bioinformatics, which is defined by the examination of proteomics, for example, a method identified as the arrangement of genomics (Cao et al., 2020), is being made in order to discover the secrets that are hidden in the fundamental principles of life. As shown in Figure 1, we are able to see the structure of proteomics incorporating bioinformatics.

The introduction of high-throughput technology resulted in an explosion of biological data, which was a consequence of the opening of this technology. Not only did this result in the manifestation of new possibilities, but it also resulted in the manifestation of new problems. Despite the fact that conventional computational methods are essential, they have a difficult time evaluating, analysing, and drawing conclusions from the vast and complicated datasets that are used in current biology. Even if these tools are essential, this is the situation that has arisen. On the other hand, in order for bioinformatics to fulfil the revolutionary promise that it holds of comprehending intricate biological processes, these issues must be resolved (Gu et al., 2021).

In order to address the challenges that were brought about by the availability of extensive data sets, a subfield of machine learning, which is known as deep learning, was developed. In response to these issues, deep learning was created as a technological solution. In the realm of machine learning, deep learning is a subfield that has been shown to be one of the most effective uses of this notion. The topologies of deep learning models have a substantial effect on the design of neural networks that are present in the human brain. This is because the topology of the models influences the design of neural networks. As a result of the fact that neural networks can be seen in the human brain, this phenomenon may be explained at great length.

This impact is significant enough to be considered a quantity of equal or higher importance in relation to the overall significance with regard to the fact that it is substantial with respect to the matter of the whole. It is possible to apply the design of this neural network to a tremendous number of different kinds of neural networks at the same time. Both convolutional neural networks (CNNs) and recurrent neural networks (RNNs) are examples of neurons that are employed in neural networks. Both convolutional neural networks (CNNs) and recurrent neural networks (RNNs) are examples of neural networks. A further widespread use of CNNs may be found in neural networks. There is no more basic component of neural networks than the fact that they are nothing more than hierarchies of neuronal connections that are associated with one another.

There are two distinct varieties of neural networks, and each of these neural networks has a structure that is distinct from the other kinds of neural networks. An analysis of both of these particular varieties of neural networks is going to be presented in the paragraphs that are going to come after this one. The fact that these neural networks are organised in a hierarchical structure is something that must be taken into mind at all times. The highly minute patterns that these networks detected across the various datasets that they gathered made it feasible for them to attract the attention of each person on an individual basis. This was made possible by the fact that these networks were able to do so independently. To be more specific, with regard to this particular aspect of the issue, it is really beneficial for us to own this specific aspect of the circumstance that is being taken into account.

In accordance with the results that were presented by Karmamit et al. (2019), bioinformaticians possess the ability to swiftly uncover pathways, carry out gene expression analysis, and make predictions about the structures of proteins while they are in the process of functioning. Furthermore, this is consistent with the results of the researchers that were discussed earlier in this paragraph. Despite the fact that this enormous reservoir of biological knowledge is constantly being updated with new data, there is a growing need for the transmission of information on a daily basis. This demand is expected to continue growing. The fact that the repository is being updated on a consistent basis does not change the fact that this is continuing to occur. A component that is presently becoming increasingly crucial is the process of sharing information. There is a shortfall in the quantity of data that has been appropriately labelled, which is one of the reasons and may be one of the causes entirely.

Everything is related to this in some fashion or another. I believe that it has an impact, in addition to the fact that we are entirely in agreement with each one of the topics discussed. In this instance, the idea that has been provided is a credible one that might be taken into account. Without a doubt, the validity of such a remark is one that no reasonable individual could ever question! There is no differentiation between the most basic and the most intricate variants of transfer learning applications. This is because there is no difference between the two different types. To be able to effectively adapt one's expertise in one subject to another, transfer learning abilities are essential.

Transfer learning skills are learned via experience. The ability of a person to educate other people is one example of anything that may be stated in this fashion. Because it is so easy to express, this medium is the most efficient mode of information transmission because it is so easy to comprehend. To this day, the vast majority of our knowledge is founded on information that was acquired in the past. This is essential for the dissemination of new information since it allows for the accumulation of knowledge. There is a possibility that taking this into account might provide more insight into the current situation. Lastly, but certainly not least, the outcome is shown here.

There is a significant quantity of data and information that must be gathered in order to complete this project, making the first stage a considerable endeavour. When this action is taken, it creates the conditions for such behaviour to take place. These activities served as the impetus for the creation of risky patterns of behaviour, which were subsequently put into practice. When it comes to dealing with bioinformatics datasets that include annotations of variable density, one of the most successful techniques is to make use of transfer learning. At least for the time being, it seems to be contributing to the growth of this area of study. As a result of the characteristics of the files, it is essential to keep this in mind while working with datasets that include annotations.

One of the reasons for this predicament is that the records include shockingly inaccurate information. Transfer learning is highly advantageous for this particular reason, which further illustrates the efficacy of the strategy that was previously described. The issue is only going to become worse from this point forward, despite the fact that transfer learning does have some good impacts despite the fact that it is now functioning. Similar knowledge is transferable and thus a vital component of the whole learning process. This is of the utmost importance, mainly when one takes into consideration the fact that the annotation of bioinformatics data needs a considerable amount of time and effort to complete. Lee was a participant in the framework of a project that was being carried out in the year 2023.

At this point, we are presented with a wide range of fresh and challenging difficulties, which makes it tough for us to interpret the subtle patterns that may be seen in biological data. What would be the most efficient way for bioinformatics to make use of the most cutting-edge computational methods if it were to apply them? What would be the most effective way for it to employ these approaches? What are some of the ways that computational biology may be of aid to them in resolving the particular problems that they are now dealing with? If you could describe the several ways in which they might make the data obtained from biological sources more accurate and complete, it would be of great assistance. As a result of the fact that the field of bioinformatics is currently undergoing a transition into the era of big data, which requires the management of a substantial quantity of datasets, it is of the utmost importance to address ethical concerns as well as the potential limitations of the methodologies that are currently being utilised.

Article by Li H. 2018 et al. The complexity of molecular systems is expected to continue to expand, which will result in the emergence of new problems that will need more collaboration between computational approaches and the biological sciences. For these problems to be overcome, further contact between the two disciplines will be required. Increasing the amount of cooperation that exists between the two sectors will be necessary in order to be successful in overcoming these issues.

Our discussion extends upon the prior evaluation of the bioinformatics potential of DL and TL that was presented in the paper. This is an additional point of interest that might be of interest to you. Taking into account the most current developments in the area of applications for deep learning is something that is done at every stage. Several instances of such advancements include the power to evaluate gene expression, the identification of novel pathways, and the prediction of structures. These are only a few examples.

Additionally, we explore the ways in which the principles of TL impact the reliability and consistency of predictions provided by biological research. This is something that we do in addition to the previous information. Specifically, this is something that we do. Let me be more precise. We are doing research into a wide range of deep learning architectures and transfer learning approaches with the goal of, among other things, improving the adaptability of these systems and their ability to cope with various biological environments that are difficult to manage.

The objective of this article is to provide an overview of the many ways in which contemporary computational techniques have influenced the area of bioinformatics research for the purpose of presenting the findings of this work. If anybody happens to come across this, it would be much appreciated if they could provide some light on the topic that is now being discussed. The primary objective of this paper is to provide a case for the broad use of deep learning and transfer learning by a diverse group of individuals, including academics from universities and executives from businesses. Specifically, the application of deep learning will be the subject of discussion over the course of this discussion. The purpose of this investigation is to provide evidence that can be used to support the implementation of either of the two educational strategies that were mentioned before.

The achievement of this goal is within the scope of this study. At this very moment, we are going to conduct this investigation. For the purpose of establishing the foundation for this study, the information that was derived from the research conducted by Li and his colleagues in the year 2020 was used. This particular article served as the primary motivation for carrying out this study. During the time that I was working on this article, I conducted research, which ultimately served as the foundation for the article that I actually produced. According to the circumstances that now exist, one of the effects that cannot be avoided is the fact that we will invariably acquire further knowledge about the biological systems that are functioning. It is one of the challenges that absolutely cannot be ignored, and it is necessary to take action to address it. This specific consequence is simply one of many that may occur, and it is not even close to being feasible to stop it from happening. There are many other potentially occurring outcomes as well. Because of the vast amount of information that is taken into consideration, it is possible to achieve this level of comprehension. Not only do these data include the current situation, but they also include all of the historical patterns. In addition to that, it takes into consideration potential events in the future and incorporates hypothetical causes and consequences.

Deep learning is discussed in depth in the second portion of this chapter, which focuses on its application in the area of bioinformatics. The whole chapter is included in this section. Additionally, the accompanying PDF has a detailed explanation of the chapter that may be found throughout the document. Moving forward, we will investigate the potential benefits of transfer learning in relation to the process of addressing data availability issues. The third part of this series established these ideas, and in this section, we will elaborate on them.

This is based on the information that is presented in Part 3. In this article's fourth section, we will describe the several applications that are accomplished via the use of genomic sequencing. Section 5, which is devoted to the study of structural bioinformatics and proteomics, is occupied by a significant portion of the research that is conducted in these areas. Within the sixth segment, the method of merging data from a number of different omics research trials is presented for the first time. We discuss the problems that arise in the seventh section, as well as the possible remedies to those problems, and we also discuss the prospective solutions. It is common practice to refer to the eighth and final portion of the chapter as "Part 8."

2. BIOINFORMATICS DEEP LEARNING FUNDAMENTALS

The biotechnology revolution will have far-reaching effects. This phenomenon demands deep learning's scientific data analysis. Deep learning powers this transformation. Deep learning paradigm shift affects biological sciences, claim Lilhore et al. (2023). Change is needed now. This innovation involves neural network research. It may also imply complex brain architecture. Numerous arguments support this approach.

Artificial neural networks will thrive and matter. Modification should include artificial neural networks. CNNs prioritise genetic data for understanding. Many call these "CNN." Lilhore and colleagues discovered in 2022 that CNNs understand gene sequences. Study results confirmed that. DNA gene sequences might reveal its physiological functions.

RNNs can find sequence correlations for transcript expression monitoring. Because RNNs excel in sequence correlations. A major benefit. Consider fast neural networks. They help us comprehend gene expression patterns, critical sequences, and biological processes. This research hypothesised that RNNs may learn from gene expression patterns. Lindorff-Larsen et al. (2021) suggest deep learning, CNNs, and recurrent neural networks may help genetically driven fields like biology. These topologies are "deep learning." IT may manage data better. After that, the industry can examine data faster and seize possibilities. Improved data management would help the sector capture new opportunities. They're right. Lindorff-Larsen & Partners.

Deep learning helps bioinformatics. Deep understanding significantly improves protein structure prediction. The AlphaFold programme visualises two-dimensional chemical structures. This is unprecedented. This helps us comprehend pharmaceutical composition synthesis. Deep learning helps functional genomics analyse complex gene regulation networks and find event-associated genes. This has significantly advanced functional genomics. Deep learning allows this. Practical genomics researchers may understand cells by finding gene areas that behave (Liu et al., 2021). The industry would grow if considered. Researchers presented their input and tested the proposal.

Thus, "transfer learning" is applying what you've learned in one job to another similar one. The term describes informatics processes. Transfer learning is one of its many meanings. Labelling issues may benefit from the approach. With few labels, deep learning solves new problems. It uses model training data to succeed. Fact-grabbing aids learning. Transfer learning improves sparse annotations, Luo and colleagues found in 2019. The study supported this. Transferring a model's skills to another area may increase its performance. This method looks plausible.

Deep learning has improved, but bioinformatics still has problems. Though deep learning has changed, this remains true. Large labelled datasets, model complexity, and ethical concerns about results are essential. No one addressed numerous concerns. This field requires understanding outcomes explanations. Thus, models must be easy to use. In 2017, Min and colleagues studied. Researching rare diseases or biological processes requires plenty of labelled data. Study participation requires this. Table 2 shows bioinformatics deep learning papers. Below, we'll compare and contrast these journals' articles and publications.

The bioinformatics paradigm shift to deep learning may help researchers discover new biological systems. Since innovation is available, this is possible. There were new physiological systems. We've found and investigated many bodily systems. Researchers may use this finding to create new biological systems. Because of this, there is a greater possibility that new biological systems may be located. These previously unimaginable avenues of inquiry into biological systems are now officially within the realm

Table 1. Comparative analysis of existing research of deep learning in bioinformatics

Reference	Technique Used	Key Contribution	Applications	Future Scope
Abdar et al., 2021	Predicting human viral protein-protein interactions using CNN layers of various sizes and complexity.	Human–virus PPI prediction improves with multiscale CNN.	Multiscale CNN predicts human–virus PPI significantly.	Significant correlation between multiscale CNN and human–virus PPI. This conversation matters.
Cao et al., 2020	Brain-computer interfaces that utilise electroencephalography signals.	Interfaces between brains and computers that employ EEG signals.	Brain-computer interfaces	Progress in signal sensor technology and intelligence-based computational methods.
Gu et al., 2020	Application of transfer learning to brain operational connectives'.	Novel transfer learning approach to enhance deep neural network classification of brain functional connectomes.	Brain functional connectomes	Advances in transfer learning for deep neural networks of brain functioning connectome analysis.
Karimi et al., 2020	Transfer learning and genome-scale metabolic modelling	Utilising genome-scale metabolism modelling and transfer learning for reconstructing the gene regulatory structure in humans.	Reconstruction of the regulation network of human genes.	Investigation of transfer learning within various biological networks.
Lee, M. (2023)	Compound-protein affinity studied using deep learning techniques.	Construct a comprehensible DeepAffinity framework for compound–protein connection deep learning.	Prediction of the affinity between compounds and proteins.	Improvements in interpretable deep learning algorithms for drug discovery continue to progress.
Li, et al., 2020	Analysis of recent advancements in Protein-Protein Interaction research	An exhaustive examination of recent progress in deep learning for the study of protein-protein interactions.	PPI analysis	Investigation of novel deep learning frameworks and their applications in protein-protein interaction analysis.
Li et al., 2020	Quantifying unpredictability in deep learning	Analysis of uncertainty assessment methods in deep learning, focusing on their applications and the problems they present.	Quantifying unpredictability in deep machine learning.	Improvement of methods for handling uncertainty in diverse deep learning contexts.
Lilhore et al., 2023	Enhancement of techniques for Bio-informatics using deep learning scenarios.	Creation of a mixed deep neural network through the utilisation of transfer learning for the purpose of interpreting EEG motor imagery.	Decoding movement images using EEG	Advancements in composite deep neural networks are being investigated for electroencephalogram applications.
Lindorff-Larsen, K., & Kragelund, B. B. (2021)	DeepTorrent computational tool to accurately predict the locations of DNA N4-methylcytosine sequences.	DeepTorrent developed to forecast the positions of genetic material N4-methylcytosine patterns precisely.	Prediction of N4-methylcytosine sites in DNA	Progress in forecasting DNA methylation events through the utilisation of deep learning techniques.
Liu et al., 2021	Advancements in predicting DNA methylation occurrences using deep learning methodologies.	Creating an improved convolutional neural network with an extended short-term memory model to identify diseases in cassava leaves.	Identification of cassava leaf-related illnesses.	Optimising plant disease recognition using hyperparameter tuning and deep learning model improvements.
Luo et al., 2019	Enhancing the efficiency of plant disease identification through hyperparameters and algorithm enhancements.	Enhancing plant pathogen recognition with hyperparameters and algorithms.	Classification of apicomplexan parasites	Considering deep cycle-based transfer learning towards parasite identification.
Min et al., 2017	Exploring the application of deep cycle transfer learning for the purpose of identifying parasites.	An investigation into methods for incorporation of multi-omics information used for the framework of machine learning.	Analysis of multiple omics data sets	Enhancing artificial intelligence with multi-omics fusion.
Min et al., 2016	A hybrid CNN and LSTM model for Parkinson's disease.	Parkinson's disease prediction uses a CNN-LSTM hybrid approach.	PPD	Advancement in neuro-degeneration disease identification employing hybrid architectures.
Pio et al., 2022	PAS recognition with DeeReCT-PolyA	DeeReCT-PolyA is an effective deep-learning way to identify polymerization signals.	PAS detection in Bioinformatics dataset	DeeReCT-PolyA is a highly efficient deep-learning method for detecting polymerization patterns.

of possibilities as a result of this. Only this fresh information is responsible for the paradigm shift that has taken place. A change in paradigm has occurred as a result of the newly discovered knowledge and the realisation. Both our attempts to interpret the intricacies of proteome, metabolomic, genomic, and deep learning data, as well as our research into the interaction between bioinformatics and deep learning, are going hand in hand.

At the exact moment that we are working, all of this takes place. This project is now receiving our attention. We are now concentrating on a number of topics, including this particular one. In addition to that, we are also engaged in other activities at the same time. It is impossible to differentiate this operation from any of the different conceivable functions since it takes place at the same time as those other activities. This particular combination has the potential to provide ground-breaking discoveries, in addition to giving insight into the complicated molecular dance that life may sometimes play. Each of these two outcomes is a distinct possibility that this combination may bring about. Considering that this combination has the potential to cause both of these outcomes, it is not entirely out of the question that both of them may occur.

There is a good possibility that this model will be able to profit from both of these outcomes. If this were included in the strategy as a whole, it would be very beneficial. It is quite feasible to take into consideration any of these two possibilities for this model. According to Picard et al. (2022), the scientific community is on the verge of a future in which the computational power of computers and the biological understanding of scientists will be brought together, which will result in ground-breaking discoveries. This position receives more credibility as a result of the projected shift in the profession. It is anticipated that a similar kind of event will take place in the near future. There is a possibility that you may arrive at a prognosis that is consistent with the predictions made by other experts in the field. Every person is looking forward to its arrival with great excitement. Everyone has high hopes and prayers that it will take place in the near future.

On the other hand, this is only one of many predictions that have been made about the course of the issue. A significant amount of forecasting has been done. We will get a peek at the actual future of every one of them. Now that the challenges are out in the open, they may be investigated more thoroughly, worked on together, and effectively remedied by innovative means.

3. TRANSFER LEARNING: BRIDGING THE DATA GAP

There is a persistent issue of data that is not appropriately tagged, and the word "Transfer Learning: Bridging the Data Gap" offers a proactive solution to this problem. When it comes to the intricate field of machine learning, this book is an essential resource. For a lengthy period, this issue has been present in the world. Transfer learning is based on the core idea that it is possible to increase performance in a task that is similar to but different from the one that is being studied by transferring the knowledge that is received from pre-trained models to the work that was initially known. Figure 2 illustrates the subchapter in its component components, which are divided down there.

In many cases, the lack of annotated datasets is a significant roadblock to creating and using good machine-learning models. This issue is tackled by using this forward-thinking technique, therefore solving the problem. that year (2016) In (2017), Wang et al. One way to successfully adapt models to new tasks is via transfer learning, which helps to bridge the data gap. A bridge is used to do this. The term used to describe this strategy in the realm of education is "transfer learning." In order to achieve this goal, data collected from a specific location is first used. When discussing the feature extraction step of transfer learning, an equation known as Equation 1 is used. The letter I represent input, the sign AS

Figure 2. Transfer learning in bio-informatics

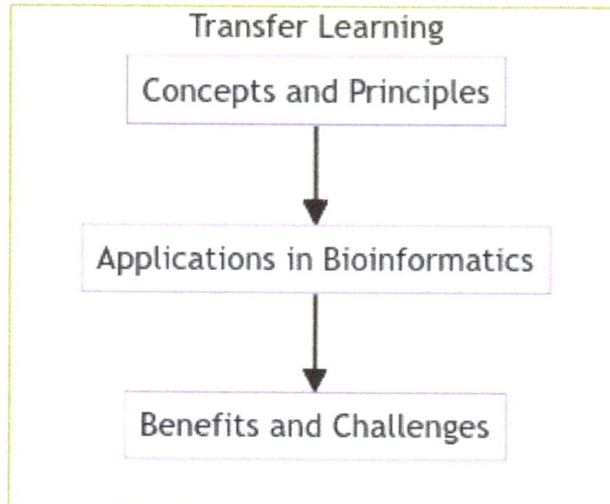

represents absolute characteristics, the letter BV represents the bias vector within the context of these specific circumstances, and Wg represents the weight parameter.

$$AS = BV + Wg \times I \tag{1}$$

In the field of bioinformatics, the most compelling evidence of the effectiveness of transfer learning may be found. This is a possibility. In spite of the fact that there are many potential applications for transfer learning, this is the situation. In the course of doing research in the domains of genetics and biomedicine, obtaining data that has been tagged might likely be a complex process that reduces the amount of resources that are available to researchers. In spite of this, the implementation of transfer learning has resulted in a complete and entire shift in this respect. Annotated datasets may be used by academics to complete their research. Such cases use little labelled data. Your success is greatly appreciated. The pharmaceutical industry, studies of biological interactions, and medical diagnostics could all stand to gain from this, say Xia et al. (2019). Most likely, it's because of how quickly it can do the task. Improving transfer learning is undoubtedly possible using Equation 2.

Annotated datasets allow academics to do research even when labelled data is not available. Your achievement has garnered admiration from everybody. Pharmaceutical manufacturing, biological connection comprehension, and medical problem diagnostics could all profit from this, as stated by Xia et al. (2019). Its ability to speed up the process is probably the main reason. Using Equation 2 to improve transfer learning is a sure bet.

$$Wg_{new} = \left[Wg_{old} \left[- \left[\beta \times \frac{\varphi f_{tuneloss}}{\varphi \times W} \right. \right. \right] \tag{2}$$

Where, Wg_{new} : New weight results, Wg_{old} : old weight results, $f_{tuneloss}$: Fine-tuning loss

Machine learning has been able to accomplish tremendous success in several domains, including computer vision and natural language processing, because of the adaptability of transfer learning, which goes beyond the constraints of bioinformatics. The reason is that transfer learning is applicable outside of bioinformatics as well. This is driving the rapid expansion of the industry. An example of a specialised task that may be easily added to a model trained on a large dataset of generic photographs for picture recognition is medical image analysis. The reason for this is that a massive amount of photos were used to train the model. This is happening because different sorts of pictures were used to train the model. This problem originated when the model was introduced in stock photographs in the past. Although labelled datasets are essential for precise diagnosis, in this specific context, there are often just a few datasets available.

Transfer learning is not simple, yet it might lead to change. This holds irrespective of the reality of the possibility. Important considerations include the model's adaptability to the issue domain, the usefulness of the pre-trained model, and the presence or absence of biases introduced by the source domain. Each one needs your undivided attention. The degree to which the disciplines of origin and destination are congruent determines the efficacy of any information transfer process. In order for the treatment to be successful, this is crucial. To ensure fair and responsible use of transfer learning across disciplines, Yang et al. (2019) argue that we must resolve the ethical issues raised by the inherent biases in existing models.

It is crucial to employ transfer learning appropriately and fairly. Hone in on this; it's vital. Objective transfer learning is essential, unlike what most people think.

Thanks to transfer learning's abundance of new ideas and opportunities, machine learning algorithms have come a long way. This resulted from course-to-course learning transfer. Being able to express oneself vocally is a necessary skill for personal development. Because transfer learning is so popular, we got ourselves into this jam. These young guys achieved this. Since our model-generation approach reduces the requirement for enormous datasets that are labelled at each iteration, efficiency and sustainability are also improved. Because it works and doesn't harm the environment, this technique is doable. There will be no delay in resolving the data lack or gap; everything is proceeding as expected. The most striking example of the flexibility made possible by modern technology is this fresh take on machine learning. The pace of technological advancement is dizzying. By combining data from apparently unconnected sectors, this technology simplifies operations. The door has been opened by technological development.

Data scarcity and "Transfer Learning: Bridging the Data Gap" are both brought up by Zhang et al. (2021). This is indeed the case, as shown by the statistics. Thanks to this shift in perspective, collaboration, research, and machine learning are now more accessible than ever before. Now they're all less difficult. To begin, it upends the status quo, simplifies app development, and promotes collaboration. There are a number of benefits to this upgrade. This enhancement offers several advantages. Its popularity may be attributed, in part, to the fact that it facilitates teamwork, shortens development cycles, and simplifies application development. What makes it so attractive? Due to this. I can do this. Anyone can. Academic research and cooperation are the two main goals of this organisation. In this article, we provide some basic guidelines for developing AI apps.

4. BIOINFORMATICS USE CASES

Consuming "Applications in Genomic Sequencing" is equivalent to diving headfirst into the intricate realm of genetics. Looking at genomics through the lens of "Applications in Genomic Sequencing," its many practical uses and far-reaching impact may become apparent. These tasks are completed by using contemporary methods and tools. Genomic sequencing entails reading an organism's whole DNA sequence. This technique is described by genomic analysis. Many more types of application classes may be traced back to this method.

The term "genome sequencing" appears in a few of these titles, although it isn't included in others. Still, others use other words to describe the same kind of DNA sequencing. The intricate mechanisms that lead to the development of inherited disorders may only be better understood with the help of this method. The one we're discussing here is only one of several potential applications for this technology. And it doesn't even begin to cover all the other significant cases where this strategy may be beneficial. The development of individualised treatment programmes that take into consideration a patient's specific genetic composition could be made more accessible by this. The first achievement is genomic sequencing. Thanks to DNA sequencing, we can now find genetic variations and mutations, which makes this conceivable. We can accomplish this goal now that DNA sequencing is a reality. This hope is probably going to materialise as it opens the door for DNA sequencing.

Genome sequencing has the potential to revolutionise healthcare in several ways, according to Abdar et al. (2021). What is needed in these areas includes, but is not limited to, more precise diagnosis, improved outcome prediction, and the creation of personalised treatments. The study they carried out led directly to this conclusion. A great deal of research went into arriving at this conclusion. Because of how important it is, you shouldn't let your mind wander while making this choice.

On several occasions, studies performed by cancer experts have shown the great value of genome sequencing. This is because academic understanding of the genetic elements that encourage cancer development has grown substantially in recent years. The scenario is based on this reasoning. Thus, everything about the present situation has not changed. The fundamental genetic abnormalities that cause cancer may be better understood with a deeper understanding of the genetic makeup of cancer cells. It is thought that these abnormalities lead to the development of cancerous tumours. For malignant tumours, these abnormalities mark the beginning of the subsequent growth.

There has been a dramatic shift in the cancer therapy landscape due to the most recent discoveries. Based on the data given here, early cancer detection and prediction are within reach, and patient-specific treatments may be possible in the future. Using this data also makes it simpler to make personalised drugs. The field of pharmacogenomics investigates the effects of drug dose and a person's genetic makeup on health outcomes. Genome sequencing is expected to have a significant impact on pharmacogenomics, the study of how to determine an individual's optimal medication dosage and frequency of administration based on their genetic makeup. It is critical to consider the individual's biology while deciding on a treatment plan, including dosage and medicine. Reducing the severity of side effects without sacrificing therapeutic effectiveness is the goal of this approach.

Contrary to popular belief, this all-encompassing approach has caused a paradigm change in cancer patient care and treatment. The way chemotherapy is administered causes this transformation. As a result, a novel method of administering chemotherapy has been developed.

Research in genetics, sometimes known as "Applications in Genomic Sequencing," is often believed to be a huge undertaking that may have far-reaching consequences. The success of this trip depends on navigating the complex world of genetics. This is crucial to your success. Anything having to do with genomics falls under the umbrella of "Applications in Genomic Sequencing". This is because of the close connection it has with heredity. We have decided to focus our investigation on genetics since it is a highly creative field that is both practically applicable and has far-reaching implications.

Figure 3. Application of deep learning in genomics

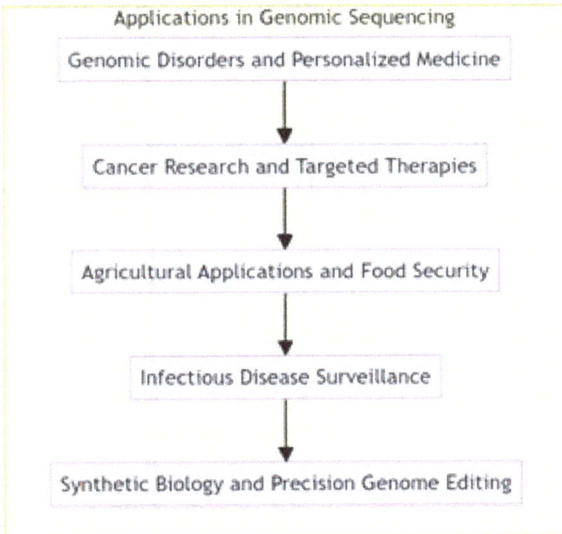

Research on genomics is the primary focus of this investigation. Through the process of genome sequencing, it is now feasible to read any DNA sequences that are present in an organism. In the event that you adhere to this strategy, you should be able to complete the work at hand successfully. The phrase "genomic sequencing" is often used by scientists to speak about this technique. This technology's essential capabilities are augmented by a multitude of applications, which are all available to users. Due to the fact that these applications are built on top of it, it is necessary for the manufacturing of these applications. The multi-stage method that is referred to as genomic sequencing and genetic sequencing belongs to the same category. A good illustration of this would be the sequencing of the genome.

One of the primary goals of this investigation is to advance the area of genomics research, which is the primary target of this examination. Because of genome sequencing, we are now able to grasp the whole of the genetic code that is housed inside one creature. This had previously been impossible. It is expected that you will be able to complete the work that you are now working on if you adhere to this method. This technique is sometimes referred to as "genomic sequencing" by researchers when they are presenting it from their point of view. The fundamental capabilities of this technology are enhanced by a large number of applications that may be used by people from a variety of educational and cultural backgrounds. Considering that it serves as the basis upon which these apps are built, it is an essential component in the process of their development. Due to the fact that both genetic sequencing and genomic sequencing entail a significant number of stages, the two methodologies may be deemed interchangeable. One excellent illustration of this is the sequencing of the genome, which demonstrates this point.

As a result of our recent discoveries on the operation of genes, we are now in a position to modify pharmaceuticals to correspond with the unique genetic profile of each individual. When viewed from a theoretical perspective, it is conceivable to improve these techniques in such a way that they take into consideration the genetic makeup of an individual. In order to successfully use this method, the first step is to undergo the sequencing of the genome. The sequencing of the genome is the first stage in the process that will be carried out.

DNA sequencing has made it feasible to find genetic variations and mutations which were previously unattainable. This possibility was previously unavailable. This alternate approach is worthy of further consideration by experts since it prepares the way for DNA sequencing, which is a process that involves sequencing DNA. Do you think there is anything that we need to take into consideration? According to Abdar et al. (2021), the sequencing of the genomes of humans has the potential to revolutionise the diagnosis, prognosis, and treatment regimens of persons. In spite of the fact that these are only a few of the probable outcomes that the implementation of precision healthcare treatment might bring about, there is a probability that there will be a great deal more. There is a high probability that the use of this particular strategy will lead to the accomplishment of this specific aim, which is the culmination of the process. The findings of this investigation, which were published in the journal Scientific Reports, which is subject to peer review, were consistent with the hypothesis.

Regarding this specific subject, the writers have expressed their wish that the viewers would keep it in mind. They have found data that lends credence to an inevitable conclusion as a consequence of the combined efforts of the researchers who have been working together. They arrived at this interpretation as a result of the study that they conducted. It is crucial to keep this choice in mind going forward because of its importance, as it merits more attention than it now gets. This focus is currently being given to it.

In multiple instances, it has been shown that sequencing the genome is a vital component of cancer research. This has been demonstrated on numerous occasions. Currently, this particular field of study is more critical than it has ever been before. Our understanding of the genetic variables that contribute to the development of cancer has considerably expanded over the course of the last few years, which is the basis for this discovery. This is the reason why everything is in the condition that it is in at present. This is the meaning behind everything. To put it in the most straightforward words imaginable, that is the fundamental reason why we are now going through this crisis.

This scenario may be responsible for the outcome. Except for this underlying effect, nothing has happened; it is the single source from which all that has occurred has emanated. Nothing else has happened. If the genetic code that is found within cancer cells is decoded, it has the potential to shed information about the underlying congenital disabilities that are responsible for the development of tumours. This is because cancer cells carry a genetic code that is housed inside them. These anomalies will inevitably lead to the growth of malignant tumours, and this outcome is unavoidable. When it comes to describing this method of evaluating the DNA sequence of cancer cells, the word "genetic analysis" is the most acceptable technique to use. There is a higher likelihood that malignant tumors may develop when specific abnormalities are present in the body.

Given that these abnormalities are the primary factors that lead to the development of cancer, the probability of having the disease is greatly enhanced. If these defects are not addressed over a lengthy period, there is a possibility that they may develop into malignant tumors. These abnormalities were the first spark that set the cancer on the route to development and led to its progression. The advancement of malignant tumours through the phases of development is facilitated by these anomalies, which make it simpler for the tumours to move to other areas of the body and to spread to other locations.

This abnormality, which is essential for the development of malignant growths, is caused by the production of malignant growths, which is the thing that is accountable for this abnormality. As a consequence of these anomalies, the formation of malignant tumours is a potential outcome. Inevitably, these abnormalities will ultimately lead to the formation of cancerous tumours if they continue to rise in frequency. A significant number of anomalies have surfaced as a consequence of the fact that some abnormalities have led to the appearance of even other monsters. It is directly accountable for the significant shift that has taken place in the many cancer treatment alternatives that are accessible today, in that the march towards information democratisation has taken place.

Every single one of these developments is a direct consequence of the information that has been obtained about the circumstance. The information presented here not only contributes to the creation of individualised therapies but also sheds light on the possibility of early cancer detection and prognosis in the context of cancer. The significance of this particular facet of the material cannot be overstated. Both of these aspects are taken into account, which increases the value of this information. In addition, the field of pharmacogenomics might potentially gain a great deal from the sequencing of the genome. The practice of pharmacogenomics refers to the process of adapting the medication regimen of a patient to their specific genetic composition in order to get the best possible health results. The sequencing of

Table 2. Comparative analysis of deep learning applications in bioinformatics

Reference	Key Application	Method Used	Outcomes	Future Directions
Min et al., 2017	Identification of Genetic Disorders	Deep learning (e.g., CNNs, RNNs)	Enabled precise detection of genetic variations and mutations.	Explore advanced deep-learning architectures for more accuracy.
Min et al., 2016	Cancer Research	Genomic Sequencing, Deep Learning (e.g., CNNs).	Revolutionized understanding of the genetic basis of tumours.	Investigate the potential of ensemble deep learning approaches.
Pio et al., 2022	Pharmacogenomics	Genomic Sequencing, Deep Learning (e.g., CNNs, RNNs).	Personalized drug selection based on genetic profiles.	Address challenges in data interpretation and ethical concerns.
Lee, M. (2023)	Agricultural Crop Improvement.	Genomic Sequencing, Deep Learning (e.g., CNNs, RNNs).	Accelerated development of crops with enhanced traits.	Optimize bioinformatics for handling diverse datasets.
Karimi et al., 2019	Infectious Disease Surveillance.	Genomic Sequencing, Deep Learning (e.g., CNNs, RNNs).	Real-time monitoring of pathogen spread during outbreaks.	Enhance the robustness of deep learning for real-time applications.
Gu et al., 2021	Synthetic Biology	CRISPR-Cas9, Deep Learning (e.g., CNNs)	Precision genome editing and creation of genetically modified organisms.	Investigate interpretability and ethical considerations in synthetic biology.
Luo et al., 2019	Variant Calling in Genomics	Deep learning (e.g., CNNs)	Improved accuracy in identifying genetic variants.	Develop deep learning models for non-coding genome regions.
Lindorff-Larsen et al., 2021	Non-Coding Genome Interpretation	Deep learning (e.g., CNNs, RNNs).	Enhanced understanding of regulatory elements.	Explore novel deep-learning architectures for gene expression.
Min et al., 2016	Deep Learning in Genomic Data Analysis.	Deep learning (e.g., CNNs, RNNs).	Contribution to tasks like variant calling and pathway analysis.	Investigate the integration of multi-omics data in deep learning.
Picard et al., 2021	Predicting DNA Methylation Sites	Deep learning (e.g., CNNs)	Development of DeepTorrent for predicting DNA methylation sites	Address challenges in scalable and cost-effective sequencing technologies
Wang et al., 2019	Plant Disease Identification	Deep Learning (e.g., CNN-LSTM)	Enhanced model for identification of cassava leaf diseases	Investigate hybrid models for improved plant disease identification
Li et al., 2018	Parasite Classification in Apicomplexan Parasites	Deep Learning (e.g., DCTL)	Achieved parasitologist-level classification of apicomplexan parasites.	Explore the potential of deep cycle transfer learning for broader parasite classification.
Cao et al., 2020	Integration of Multi-Omics Data for ML Analysis	Machine Learning, Multi-Omics Data	Exploration of integration strategies for multi-omics data in ML analysis.	Further developments in multi-omics integration for machine learning applications.
Lilhore et al., 2022	Prediction of Parkinson's Disease	Hybrid CNN-LSTM, Deep Learning (e.g., CNN-LSTM)	Application of a hybrid CNN-LSTM model for the prediction of Parkinson's disease.	Improve predictive modelling for neurodegenerative diseases using hybrid architectures.
Lilhore et al., 2023	Prediction of Polyadenylation Signals	Deep Learning (e.g., DeeReCT-PolyA)	Introduction of a robust deep learning method for Polyadenylation Signal identification.	Advanced techniques for predicting Polyadenylation Signals.
Luo et al., 2019	Examining Relationship in Intrinsically Disordered Proteins.	Machine Learning	Evaluation of machine learning potential in examining IDP relationships	Investigate the relationship between sequence, structure, dynamics, and function in IDPs

the genome is an essential piece of news in this context. Before making any judgements on therapeutic drugs and dosages, it is necessary to pay close attention to the biochemical composition of the patient.

In terms of treatment, this is, without a doubt, the most advantageous choice that can be made for the individual being treated. The purpose of this medical strategy is to achieve the maximum feasible degree of therapeutic effectiveness while simultaneously minimising the adverse effects of the treatment to the greatest extent possible. Utilising this tactic offers the opportunity to accomplish the goal that has been set for oneself. You will find below a comprehensive description of the protocol that is currently in place. Over the course of the past few years, there has been a substantial change in the manner in which chemotherapy is provided to patients who are recovering from cancer.

The conventional and tried-and-true approach of administering chemotherapy is being phased out in favour of this comprehensive procedure, which constitutes a significant departure from the standard practice. Modifications to the management and administration of chemotherapy are required to accommodate this parallel development. These alterations are necessary. As a direct result of this advancement, the process by which chemotherapy is administered has been subjected to a number of modifications that significantly impact its effectiveness. For your convenience, the following table presents a comparison of different uses of deep learning which are employed in the field of biological information technology. Table 2 shows a Comparative analysis of deep learning applications in Bioinformatics.

The fact that these apps are fantastic does not change the reality that there are still problems that need to be fixed. Some of the topics that are being explored in this article are the need for sequencing technologies that are both scalable and cost-effective, the storage and interpretation of data, and the ethical issues that are associated with genetic privacy. Genome sequencing technologies are finding more and more uses, which is in line with the growing capabilities of our technological infrastructure.

This has the potential to significantly improve several fields, including agriculture, healthcare, and our knowledge of the environment. Thus, the investigation into "Applications in Genomic Sequencing" develops into a plot that is constantly changing. According to Luo et al. (2019), additional businesses have been motivated to rethink and grow due to this story's deep comprehension of the genetic code. Rewriting this language is one of its other responsibilities.

5. PROTEOMICS, BIOINFORMATICS, STRUCTURAL ANALYSIS

One of the best ways to determine how persistent an issue is is to look at the number of times that individuals try to find a solution to the problem that they are seeking. No matter how much optimism is expressed, the realities of the situation will not alter in any manner, regardless of how much optimism is expressed. Some of the topics that are discussed in this lengthy book include the storage and interpretation of data, the significance of sequencing technology that is both scalable and economical, and the ethical considerations that are associated with genetic privacy. All of these topics are discussed in detail. In addition to these subjects, the relevance of the sequencing approach is also brought up and explored. When it comes to these topics, a sizeable percentage of individuals have shown a desire to acquire further knowledge.

There is a high probability that this will include a significant amount of the land that is now under evaluation. It is continually discovered that there are new applications for genome sequencing, and these applications may either be beneficial or detrimental to the scientific community. With regard to a significant number of things, the fulfilment of these expectations is necessary. This is something that may be regarded a possibility due to the fact that the capabilities of the technological system are constantly developing. It is directly related to the fact that technical expertise has been steadily increasing over the course of the years that this accomplishment has been accomplished.

The change that was brought about by the one that came before it was directly responsible for bringing about this adjustment. In the event that we are successful in our endeavour, we may be able to get further information on environmental issues and sectors that are linked to them, such as agriculture and medicine. Due to the fact that this is a possibility, this may take place. On the other hand, this is something that a considerable number of people are very positive will take place in the not-too-distant future. The inquiry into "Applications in Genomic Sequencing" that takes place during this phase of the research gives birth to a narrative that is not only dynamic but also continually under development. This narrative is a result of the investigation. The reason for this is because the story is becoming better and better all the time.

Because the advancement of the story is focused on this issue and the way in which it is responded to, this is the reason why this is the case. An earlier body of research that arrived at the same result has made a direct contribution to the problem that people are now facing. In accordance with Luo et al. (2019), this story affected a wide range of different areas, which led to a number of ground-breaking discoveries. Due to the fact that we have a comprehensive grasp of DNA, which is frequently referred to as the "language of life," we were able to accomplish this breakthrough effectively. This was the primary motivating factor in achieving our goals. Throughout the whole of the book, the material is presented in a manner that is not only respectful but also maintains consistency throughout.

There is still another essential aspect to take into consideration, and that is the fact that this narrative has the potential to give rise to new points of view. The act of rewriting text is only one of the many things that this language is used for; it is just one of the numerous applications that this language has. Despite the fact that it is one of its tasks, there is no relationship between that fact and the controversy that is now being discussed.

6. UNIFIED ANALYSIS OF MULTI-OMICS DATA

The fact that studies are being conducted does not change anything, regardless of the severity of the concerns. To fix some problems, you have to act quickly. The occasional glitch in these applications is annoying, but it doesn't detract from the greatness of the series themselves. While we are actively seeking answers, it is essential to note that the issues are still there and being addressed. It is easily observable that there are problems. This article intends to cover a lot of terrain, including topics like the administration of data storage and analysis, the need for inexpensive and scalable DNA sequencing equipment, and the ethical implications surrounding genomic privacy.

These are just a few examples of what people are talking about at the moment. The many topics now under discussion are vast, and this is only a tiny sampling. There are a lot of other choices. We are also thinking about adding more. All of these distinct matters are now the subject of much debate. The fact that we are now discussing a more comprehensive range of issues is encouraging. But this is among the most cutting-edge technologies available today. There have been other remarkable occurrences lately, and these are only a few of them. Still, genome sequencing has an overwhelming amount of possible applications. This trend is expected to persist, according to forecasts. One indication of this condition is a propensity to worsen continuously. The occurrence of this occurrence has led to advancement.

This is possible because technological systems are constantly becoming better. Recent circumstances have led to this choice. The latest technological developments, which will allow for its ultimate implementation, make this even more accurate. To be more specific, this change might be associated with the product and increasing power of technology. This improvement, which is a result of technological development, is responsible for responsibility. This latest development, which has contributed to technology's ongoing expansion, is a vital byproduct of that progress. Evolution, or this transformation, is driven by the addition of technological capabilities. Some have proposed the phrase "new product development" to include this change. These changes are a direct result of the advances that have been made.

This goal is within reach since technical systems are constantly improving. This decision is crucial in light of recent developments. This will come to fruition as a result of the widespread use of new technology. Perhaps a more immediate factor influencing this shift is the rate of technical advancement. Advancements in technology are causing this shift. Because of that advancement, which has driven the exponential growth of technology, this new creation is a direct outcome. The primary factor propelling these changes is the continuous improvement of technical capacities. Some have suggested broadening the definition of "new product development" to include this updated idea. Something has changed recently, and this is an indication of it.

There is, without a doubt, utterly no space for uncertainty on this. This one study is turning the so-called "Applications in Genomic Sequencing" probe into a tale that is constantly developing and changing. This is because, for now, the inquiry is yielding valuable findings. This result has materialised since this particular aspect of the research served as the impetus for the story's growth, which ultimately led to its creation. This is due to the tale being prefabricated. Something concrete has come to light because of the research that has been carried out so far. So far, inquiries have led to the discovery of this specific situation as a direct result of the probe. The authors Luo et al. (2019) assert that this narrative has had the most influence and is accountable for the expansion of innovation as well as the rethinking of several sectors. In particular, we will be delving further into this narrative. The authors have made an effort to draw attention to this for the sake of their readers. This study and this accomplishment could not have been possible without a fundamental grasp of DNA, the language of life. A complete comprehension of DNA was required to discover the answer to this problem. The realisation that this was true pushed me to complete the finding.

Being suddenly aware that this was real was the epiphany. Since the reader is aware of this truth throughout the novel, it sticks out. It is my belief as a storyteller that this is what makes this narrative unique. Remember that this story serves as a springboard for your creativity. This particular object demands careful examination. To be more precise, this is not only an extra factor but an integral part of the whole.

To put it another way, things will become worse. This language also has a strong emphasis on text rewriting. This language also takes care of another crucial aspect. Also, it's not just accountable for the many things it may influence; it's responsible for the whole process. Its intriguing feature and one of its purposes is that. Remembering this will enhance the event's overall appeal.

7. CHALLENGES AND OPPORTUNITIES

With the intention of providing a critical analysis of the challenges and constraints that are connected with the application of deep learning and transfer learning in the field of bioinformatics at the current moment, the objective of this article is to provide an overview of the situation. To provide a greater level of specificity, the research will focus on the challenges and limitations that were encountered. In addition to this, it assesses potential future paths, which may include ground-breaking developments in the interpretability of models, ethical concerns, and collaborative initiatives with the objective of developing standards that are widely acknowledged (Min et al., 2017). The establishment of standards is the way by which it is envisioned that these goals will be accomplished.

7.1 Challenges

Listed below is a compilation of the significant difficulties that are currently being faced in bioinformatics research.

- Diversity and integration: Bioinformatics combines several data types. This area comprises genomes, proteomics, and metabolomics. Harmonizing this data requires seamless integration and interoperability.
- Insufficient Labeled Statistics: This article briefly highlights bioinformatics difficulties. The probe will concentrate on concerns and constraints for more information. Future methods include collaborative, ethical problem-solving, open-source standards, and simpler models (Min et al., 2017).
- Deep Learning Model Interpretation: Reliable methods provide interpretation issues for sickness treatment programs. The next generation should construct models that predict and explain future occurrences.
- Computing Resources: Interpretability difficulties make public deep-learning algorithms unsuitable for disease diagnosis. Research should concentrate on predictive and explanatory models.

7.2 Opportunities

The key opportunities are as follows.

- Integration of Multi-Omics Data: Integrating multi-omics data smoothly with deep learning methods is vital for future developments. Deep patterns and interactions at all stages of life will be uncovered via intensive analysis, which is intended to expand our understanding of biological processes.
- TL for Smaller Statistics: Health diagnosis is not suitable for deep learning methods owing to difficulties in understanding their results. The next phase of the study ought to concentrate on constructing models capable of correctly forecasting the future along
- Edge Computing for Point-of-Care Applications: Modern computational solutions for real-time applications, as well as point-of-care solutions, should be the primary emphasis of bioinformatics research. More accessible and faster-decentralised analyses are possible with framework installations on lightweight devices, which is especially useful in healthcare contexts.
- Collaborative Research Initiatives: The primary focus of bioinformatics should be the development of state-of-the-art computational techniques for use in real-time and point-of-care purposes. When healthcare organisations implement frameworks on mobile devices, decentralised analysis becomes more straightforward and faster.

8. CONCLUSION

Research into bioinformatics's deep learning and transfer learning has shown both the field's great potential and its formidable obstacles. The interpretation of large biological datasets has significantly been improved because of deep learning's capacity to recognise patterns at a more sophisticated level. Genomic, proteomics, and metabolomics datasets are all part of this. Transfer learning remains a viable bioinformatics method, especially when faced with little labelled data, despite the obstacles. Several issues, such as data heterogeneity, interpretability challenges, and ethical limits, limit the revolutionary potential of these approaches to change biological research. We need to simplify deep learning models so they are easy to work with and comprehend if we want people to believe their decisions. For bioinformatics applications to remain viable in the future, it is essential to address data scarcity with improved transfer learning algorithms and resolve ethical concerns.

For the sake of a stable and consistent foundation for bioinformatics research, researchers should work together to develop defined criteria. To speed up the transition to a community-driven approach and guarantee the validity of results, it may be helpful to promote open dialogue, make dataset sharing more effortless, and stick to established standards. Bioinformatics is a discipline that bridges computational biology and computational science; as such, it is well-positioned to reap the benefits of DL and TL. Where healthcare, drug development, and fundamental understanding of life go from here depends on collaboration, ethics, and constant progress. These technologies will significantly enhance our knowledge of the intricate workings of biological systems.

REFERENCES

Abdar, M., Pourpanah, F., Hussain, S., Rezazadegan, D., Liu, L., Ghavamzadeh, M., Fieguth, P., Cao, X., Khosravi, A., Acharya, U. R., Makarenkov, V., & Nahavandi, S. (2021). A review of uncertainty quantification in deep Learning: Techniques, applications and challenges. *Information Fusion*, *76*, 243–297. doi:10.1016/j.inffus.2021.05.008

Cao, Y., Geddes, T. A., Yang, J. Y. H., & Yang, P. (2020). Ensemble deep Learning in bioinformatics. *Nature Machine Intelligence*, *2*(9), 500–508. doi:10.1038/s42256-020-0217-y

Gu, X., Cao, Z., Jolfaei, A., Xu, P., Wu, D., Jung, T. P., & Lin, C. T. (2021). EEG-based brain-computer interfaces (BCIs): A survey of recent studies on signal sensing technologies and computational intelligence approaches and their applications. *IEEE/ACM Transactions on Computational Biology and Bioinformatics*, *18*(5), 1645–1666. doi:10.1109/TCBB.2021.3052811 PMID:33465029

Karimi, M., Wu, D., Wang, Z., & Shen, Y. (2019). DeepAffinity: Interpretable deep learning of compound–protein affinity through unified recurrent and convolutional neural networks. *Bioinformatics (Oxford, England)*, *35*(18), 3329–3338. doi:10.1093/bioinformatics/btz111 PMID:30768156

Lee, M. (2023). Recent Advances in Deep Learning for Protein-Protein Interaction Analysis: A Comprehensive Review. *Molecules (Basel, Switzerland)*, *28*(13), 5169. doi:10.3390/molecules28135169 PMID:37446831

Li, H., Parikh, N. A., & He, L. (2018). A novel transfer learning approach to enhance deep neural network classification of brain functional connectomes. *Frontiers in Neuroscience*, *12*, 491. doi:10.3389/fnins.2018.00491 PMID:30087587

Li, S., Yang, Q., Jiang, H., Cortés-Vecino, J. A., & Zhang, Y. (2020). Parasitologist-level classification of apicomplexan parasites and host cell with deep cycle transfer learning (DCTL). *Bioinformatics (Oxford, England)*, *36*(16), 4498–4505. doi:10.1093/bioinformatics/btaa513 PMID:32413103

Lilhore, U. K., Dalal, S., Faujdar, N., Margala, M., Chakrabarti, P., Chakrabarti, T., Simaiya, S., Kumar, P., Thangaraju, P., & Velmurugan, H. (2023). Hybrid CNN-LSTM model with efficient hyperparameter tuning for prediction of Parkinson's disease. *Scientific Reports*, *13*(1), 14605. doi:10.1038/s41598-023-41314-y PMID:37669970

Lilhore, U. K., Imoize, A. L., Lee, C. C., Simaiya, S., Pani, S. K., Goyal, N., Kumar, A., & Li, C. T. (2022). Enhanced convolutional neural network model for cassava leaf disease identification and classification. *Mathematics*, *10*(4), 580. doi:10.3390/math10040580

Lindorff-Larsen, K., & Kragelund, B. B. (2021). On the potential of machine learning to examine the relationship between sequence, structure, dynamics and function of intrinsically disordered proteins. *Journal of Molecular Biology*, *433*(20), 167196. doi:10.1016/j.jmb.2021.167196 PMID:34390736

Liu, Q., Chen, J., Wang, Y., Li, S., Jia, C., Song, J., & Li, F. (2021). DeepTorrent: A deep learning-based approach for predicting DNA N4-methylcytosine sites. *Briefings in Bioinformatics*, *22*(3), bbaa124. doi:10.1093/bib/bbaa124 PMID:32608476

Luo, F., Wang, M., Liu, Y., Zhao, X. M., & Li, A. (2019). DeepPhos: Prediction of protein phosphorylation sites with deep learning. *Bioinformatics (Oxford, England)*, *35*(16), 2766–2773. doi:10.1093/bioinformatics/bty1051 PMID:30601936

Min, S., Lee, B., & Yoon, S. (2017). Deep learning in bioinformatics. *Briefings in Bioinformatics*, *18*(5), 851–869. PMID:27473064

Min, X., Chen, N., Chen, T., & Jiang, R. (2016, December). DeepEnhancer: Predicting enhancers by convolutional neural networks. *2016 IEEE International Conference on Bioinformatics and Biomedicine (BIBM)* (pp. 637-644). IEEE. 10.1109/BIBM.2016.7822593

Picard, M., Scott-Boyer, M. P., Bodein, A., Périn, O., & Droit, A. (2021). Integration strategies of multi-omics data for machine learning analysis. *Computational and Structural Biotechnology Journal*, *19*, 3735–3746. doi:10.1016/j.csbj.2021.06.030 PMID:34285775

Pio, G., Mignone, P., Magazzù, G., Zampieri, G., Ceci, M., & Angione, C. (2022). Integrating genome-scale metabolic modelling and transfer learning for human gene regulatory network reconstruction. *Bioinformatics (Oxford, England)*, *38*(2), 487–493. doi:10.1093/bioinformatics/btab647 PMID:34499112

Wang, D., Zeng, S., Xu, C., Qiu, W., Liang, Y., Joshi, T., & Xu, D. (2017). MusiteDeep: A deep-learning framework for general and kinase-specific phosphorylation site prediction. *Bioinformatics (Oxford, England)*, *33*(24), 3909–3916. doi:10.1093/bioinformatics/btx496 PMID:29036382

Xia, Z., Li, Y., Zhang, B., Li, Z., Hu, Y., Chen, W., & Gao, X. (2019). DeeReCT-PolyA: A robust and generic deep learning method for PAS identification. *Bioinformatics (Oxford, England)*, *35*(14), 2371–2379. doi:10.1093/bioinformatics/bty991 PMID:30500881

Yang, X., Yang, S., Lian, X., Wuchty, S., & Zhang, Z. (2021). Transfer Learning via multiscale convolutional neural layers for human–virus protein–protein interaction prediction. *Bioinformatics (Oxford, England)*, *37*(24), 4771–4778. doi:10.1093/bioinformatics/btab533 PMID:34273146

Zhang, R., Zong, Q., Dou, L., Zhao, X., Tang, Y., & Li, Z. (2021). Hybrid deep neural network using transfer learning for EEG motor imagery decoding. *Biomedical Signal Processing and Control*, *63*, 102144. doi:10.1016/j.bspc.2020.102144

Chapter 9
Unlocking the Future of Healthcare:
Biomarkers and Personalized Medicine

Samiksha Garse
School of Biotechnology and Bioinformatics, D.Y. Patil University, India

Divya Dalal
School of Biotechnology and Bioinformatics, D.Y. Patil University, India

Sneha Dokhale
B.K. Birla College, India

Shine Devarajan
School of Biotechnology and Bioinformatics, D.Y. Patil University, India

ABSTRACT

Modernization and advancement in technology have developed a new field called Health Informatics. It is an interdisciplinary that accommodates bioinformatics, artificial intelligence (AI), and machine learning (ML) applications. Additionally, a large amount of data that travels around the global network of cloud computing has coined the term internet of things (IoT), which has fostered a new wave of innovation in the healthcare sector. This chapter provides insights into how recognition systems, self-monitoring systems, big data analysis and interpretation, etc. use AI and IoT-based models to accelerate the capabilities of providing advanced healthcare outcomes. It not only unlocks futuristic smart healthcare, but also platforms a revolutionary milestone in the medical world. Moreover, the chapter addresses the challenges of security, scalability, cost-effectiveness, standardization, and quality care encountered by the healthcare sector.

DOI: 10.4018/979-8-3693-1822-5.ch009

1. INTRODUCTION

Healthcare is one of the basic needs of life since the good health of an individual is the primary objective for survival. Like any other industry, healthcare is also evolving to what we call smart healthcare which is said to have several billion-dollar turnover in the coming few years. Smart healthcare not only comprises of edge-cutting technologies like Artificial Intelligence (AI) and the Internet of Things (IoT) but it also makes use of the Internet of Medical Things (IoMT), Cloud Computing, Wireless Communication Technology (Fatima Alshehri, 2020), Telemedicine, Web-Technologies and Big Data (Asoke K. Talukder, 2021). It is proving to be a promising field, both for the patients and the healthcare workers. The early identification of forthcoming health concerns, planning of suitable future monitoring treatments, and creation of new evaluations can all be facilitated by the applications of new advances in precautionary laws and behavioral systems (Fatima Alshehri, 2020). Intelligent healthcare covers all aspects of prevention, identification, assessment and treatment of a disease. The goal is to overcome the barriers of awareness, accessibility and affordability in traditional healthcare (Asoke K. Talukder, 2021). The illustration of Smart Healthcare is shown in **Figure 1.**

Pharmaceuticals and Life Sciences organizations use a variety of AI technologies nowadays, which brings us to the idea of Personalized Healthcare. It lays down solutions for the patients based on their genomic data which provides the most adequate information about how a person will respond to a particular drug and how much chance a person is prone to a particular disease in the future. The biomarker technology is also a promising field for identifying certain conditions in the body. A definition of biomarkers has been established based on International Chemical Safety by the World Health Organisation (WHO) as a collaborative initiative with the United Nations and the International Labour Organisation. According to this definition, a biomarker is described as any compound, its structure or its biological process, detectable in the body or its derivatives that has the capacity to impact or forecast the occurrence of outcomes for any disease (Kyle Strimbu, 2011).

Despite the fact that AI can perform healthcare tasks in many instances just as well as or better than humans, the difficulties of implementation will prevent the roles of medical practitioners from being extensively automated for a considerable amount of time (Thomas Davenport, 2019). In this chapter, we shall discuss how AI and IoT play a major role in the advancement of healthcare and assist health

Figure 1. Aspects of Smart Healthcare (Shuo Tian, 2019)

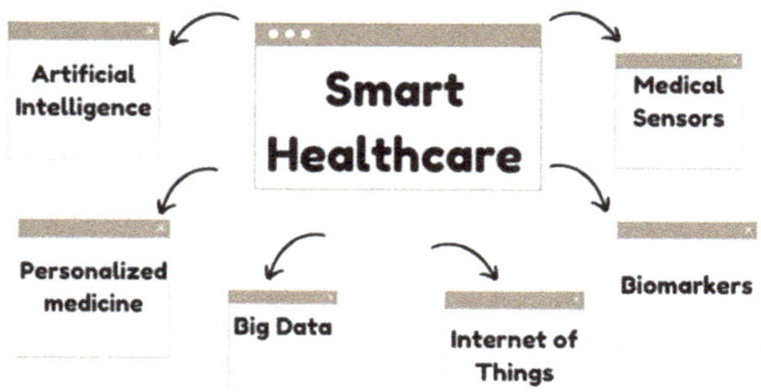

professionals in making wiser decisions in less time to prevent adversity. With all their potential and capabilities, a few societal and ethical complexities that they possess, have also been discussed here (Pavel Hamet, 2017).

2. Link Between Bioinformatics, Artificial Intelligence (AI) and Internet of Things (IoT)

2.1 Bioinformatics

The combination of biology and information technology resulted in the creation of a new name and an entirely new branch of study known as "Bioinformatics." In the early 1970s, Ben Hesper and Paulien Hogeweg introduced the term "*Bioinformatics*" to describe the research they envisioned, defining it as "*the study of informatics processes in biotic systems*" (Paulien Hogeweg, 2011). It is a multi-disciplinary that integrates Biology and Computer Science with Statistical Analysis and balances the methods to analyze "Big Data" notably Genomics. Bioinformatics is the combination of health information, data and knowledge that works on computational techniques and tools to move one step ahead in the field of research (Vinayak Majhi, 2019). It has paved its way in almost all areas from designing new drugs, and studying how drugs bind to the protein (docking), to comparing various sequences for evolutionary insights, i.e. sequence alignment. There are various softwares and online tools which help in the pipeline of the same. Bioinformatics techniques can be employed to detect genetic variations that serve as indicators for a particular disease or disorder.

This has further led to the emergence of the field called Health Informatics. According to Michael Imhoff et. al. 2001, Health informatics involves the creation and evaluation of approaches and systems to gather, process, and interpret patient data, utilizing insights from scientific research. This definition indicates that Health Informatics extends beyond computer applications, encompassing the comprehensive management of information in healthcare where the emphasis is on the patient and the care process (Michael Imhoff, 2001).

The main difference is that Bioinformatics primarily focuses on biological data, utilizing computational methods to analyze genetic and molecular information whereas, Health Informatics encompasses a broader scope, integrating information and data across various medical domains, including clinical, administrative, and healthcare management. The major applications of Bioinformatics are illustrated in **Figure 2.**

2.2 AI and its history

AI is the discipline of inducing intelligence in machines to carry out tasks that conventionally need a human brain (Ashish Ghosh, 2018). Such machines can be seen in an enormous number in every field today. It is the data from all the interdisciplinary fields that are being fed into the machines to make them artificially intelligent (Ashish Ghosh, 2018). It comprises domains from philosophy, mathematics, sociology, biology and statistics. AI formally came into the picture in 1956. The term was coined by John McCarthy (Yoav Mintz, 2019) and is associated with the creation of robots which is a Czech origin word. (Pavel Hamet, 2017). AI is not just one technology but a combination of several technologies allied to make the field prosper. In 1983, Elaine Rich defined AI as "*the study of how to make computers do*

Figure 2. Applications of Bioinformatics on the Broader Term

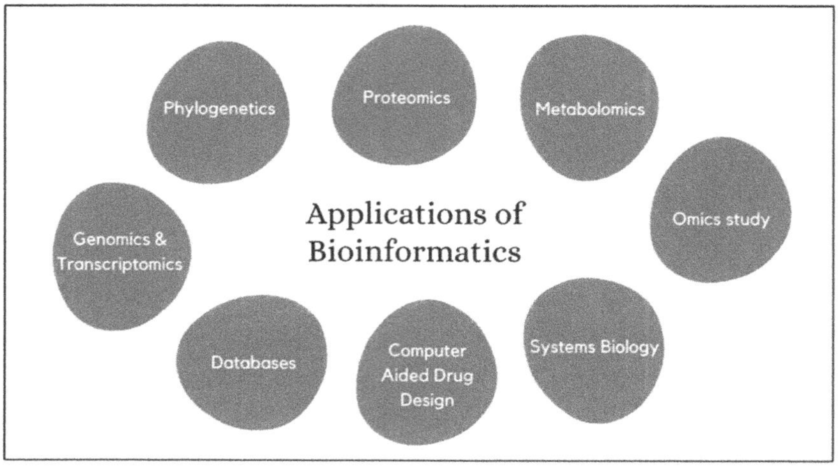

things at which, at the moment, people are better" (Ian Mackie, 2017). A machine is said to be artificially intelligent when it can act logically, solve problems, plan, and learn from its environment (Stuart J. Russell, 2010). Critical difficulties in sustainable manufacturing can also be resolved with the help of AI.

2.3 What is the Internet of Things (IoT)?

IoT is the network of objects made up of technology that helps to exchange data over the internet. Anything that has the ability to connect to the internet, collect and share data, comes under the domain of IoT. From smart agriculture to smart cities and smart healthcare IoT penetrates all aspects of daily life. For example, things are interconnected when one finds ads for gyms and protein shakes on their Facebook or Instagram feed, only hours after a Google search about Pilates. The data is stored in the form of cookies by websites that further get sold to companies that provide ads, catering to one's interests. Therefore, IoT has increased the pervasiveness of the Internet by integrating interactive devices with embedded systems that shall eventually result in a widely dispersed ecosystem of devices to interact with humans and other devices (Feng Xia, 2012).

2.4 AI and IoT in Healthcare

The Biological Sciences, Automation, and Healthcare Diagnosis including "Omics" uses AI and IoT together to make smart healthcare. There are two primary sub-fields of AI in medicine: virtual and physical (Pavel Hamet, 2017). In the virtual one, by the provision of Machine Learning (ML) algorithms and knowledge management, AI has hastened and continues to enhance discoveries in genetics and molecular medicine. Moreover, innovative computational techniques are been developed that can detect single nucleotide variations (SNPs), using innovative evolutionary embedding algorithms that are more resilient and less prone to the problems associated with over-fitting, which happens when a model has too many parameters concerning the quantity of observations.

The physical branch includes medical devices, daily sensors or trackers and increasingly sophisticated robots taking part in the delivery of healthcare (Pavel Hamet, 2017). These are IoT-powered devices that use semiconductor and communication technology. Wearable gadgets are recommended by medical practitioners who want to monitor their patients for a longer period of time to collect data about their medical statistics. These devices can measure glucose levels, physical activity, heart rate, and other metrics (Shaftab Ahmed, 2018). This application of AI and IoT is so extensive that today probably everyone has the tools for smart healthcare but only the semantics of it have to be understood more deeply where the knowledge needs to be extracted from the data. A variety of patient data can be processed and evaluated using AI to more accurately forecast a patient's prognosis, survival time, and disease progression (Shigao Huang, 2020). AI plays a crucial role in advancing personalized medicine and biomarker discovery by leveraging large datasets, analyzing complex biological information, and providing more precise and tailored treatment approaches. One of the major examples of AI and IoT in medicine is that of a company called PathAI which is using AI-powered pathology to advance the diagnostic phase. This revolution is discussed further in the sections below with some examples of some other phases of medicine that are also automated.

3. Stone Age of Healthcare

From IBM's "Smart Planet" concept, emerged the notion of Smart Healthcare (Armonk, NY, USA) in 2009 (Shuo Tian, 2019). There were some challenges before technology was integrated with healthcare, i.e. medical treatment and diagnosis largely relied only on the expertise of medical professionals, their experience, and knowledge. Earlier healthcare providers used paper-based methods to collect, and maintain patient records, to communicate with patients and other healthcare providers. This manual approach often led to inefficiencies and errors, as medical records were easily misplaced or misinterpreted. Patients had limited access to their medical records, which made it challenging to keep track of their health history and treatment progress. In addition, medical professionals had limited access to real-time patient data, which made it challenging to provide timely accurate diagnoses and treatment recommendations. This often resulted in delays in treatment, increased healthcare costs, and reduced patient outcomes.

The process of drug discovery was completely based on wet lab experiments and paper-based management systems during clinical trials which took a lot of time for a drug to reach the market. Due to the huge amount of chemical compounds present, the lack of technology made the process of drug development time-consuming and expensive (Debleena, 2021). Eventually, the process of a drug reaching the market from its initial concept stage required more than a decade and billions of dollars.

Misdiagnosing or late diagnosis of a disease was also a major challenge before AI was introduced in healthcare, some of the challenges are highlighted in **Figure 3**. Even in the clinical trials for a particular drug, patients find it difficult to understand and determine their eligibility due to the trial eligibility criteria's complexity in terms of number and medical language. Here, it also becomes difficult to monitor the patient (Stephen Harrer, 2019).

AI has proven to be as good as or even better than human professionals in disease diagnosis. For example, algorithms are outperforming radiologists in differentiating between benign and malignant tumors (Thomas Davenport, 2019). The healthcare industry was characterized by a lack of integration and coordination between various healthcare providers, leading to fragmented and inefficient care. However, with the emergence of smart healthcare technologies, the industry has seen significant improvements in patient outcomes, efficiency, and overall quality of care with reduced cost and automated processes.

Figure 3. Challenges Before Smart Healthcare

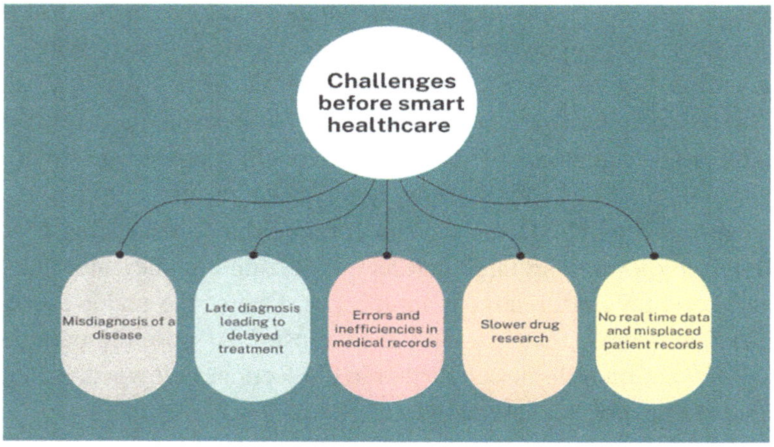

With all this, it is still important to note that AI will always assist medical professionals or healthcare workers and never replace them as there are social issues like accountability.

4. Trendy Big Data: Fusion of Large Datasets

4.1 What is Big data?

"*A large dataset which comprises a good amount of data, impossible for humans to analyze manually or even for a traditional local computer is called Big Data*". When studying the bubonic plague, which was ravaging Europe in 1663, John Graunt dealt with a huge volume of information for the first time (Adilin Beatris, 2021). This was the first time Big Data came into the picture which often comes from multiple sources, and arrives in multiple formats (Anurag Agrahari, 2017). Big Data is like the ore waiting to be mined and vital minerals to be extracted from it. It is a field valuable to all sectors of the economy as we can obtain patterns or extract the necessary information related to each field by computationally resolving the data. Industries that realize its potential, are continuously transforming this raw data into a wealth of information (P. Zikopoulos, 2011), making their business more efficient, and boosting their revenue by optimizing products and services based on the insights that the data provides (Ronald Schmelzer, 2023). But only a small chunk of companies can truly achieve the meaningful information that they are looking for. The growth of population and their connectivity with the internet is increasing the raw data exponentially which needs to be managed. The management of data can be done through databases and cloud computing (for storing), AI and ML (for analyzing), and smart devices (for delivering the service).

Big Data is growing quickly these days, causing individuals to progressively understand the immense worth and numerous applications concealed in data. Hence, developing a thorough understanding of the Big Data industry is essential. (Jinxuan Tang, 2020). Given these characteristics, there are new types of architecture and processes for collecting, processing and analyzing Big Data are emerging. (Ronald Schmelzer, 2023). The National Institute of Standards and Technology (NIST), USA has defined the characteristics of Big Data in four v's i.e. "Volume", "Variety", "Velocity" and "Value". But as it got

Figure 4. Characteristics of Big Data

incorporated into various domains, Big Data is said to have 42 V's as its characteristics (Anurag Agrahari, 2017) as shown in **Figure 4**.

Following are the top 4 Vs discussed:

VOLUME: Volume refers to the amount of data that exists. It includes the massive volume of data created every second. The volume of data is said to increase twice every two years. From 800,000 petabytes (PB) in the year 2000, it is now stored in zettabytes (P. Zikopoulos, 2011).

VARIETY: The different forms in which data is present are called its Variety. It can be unstructured, semi-structured or structured. Variety also means the format in which the data is present like texts, tables, images, excel files, email, etc.

VELOCITY: Velocity refers to the inundation of data which is always going to show exponential growth due to the internet and connectivity. It also considers how quickly it is being stored and retrieved (P. Zikopoulos, 2011).

VALUE: Value is the measure of how crucial the data is.

4.2 Categories of Big Data

Big data can be divided into three basic groups:

Structured: Data that can be maintained, retrieved, and manipulated in a predetermined manner is referred to as "Structured" data. For example; students' related data can be stored in a row and column format similar to that of a matrix data structure (Anurag Agrahari, 2017).

Unstructured: The collected, unprocessed, raw data in its unknown form is termed as unstructured data. Having a huge size, unstructured data poses various challenges in Big Data analysis. In reality, 80% of the world's data is still unstructured (Anurag Agrahari, 2017).

Semi-Structured Data: Data that consists of both the aforementioned types of information, are referred to as semi-structured data, which is structured but not precisely defined (Anurag Agrahari, 2017).

4.3 Big Data and AI (Fusion of Large Datasets)

Big Data and AI work in a symbiotic manner. AI is equipped for Big Data analysis, in turn, AI requires large data to learn and improve decision-making processes. Today's concerns are the hidden insights of the data, that are essential for building a futuristic model or a machine that would behave like humans as closely as possible. To achieve this, there is a breakthrough in Big Data analysis integrated with ML (Ashish Ghosh, 2018), that puts the data together from all the interdisciplinary fields to be fed into the machines to make them artificially intelligent (Ashish Ghosh, 2018).

The fusion of large datasets is a key component of the Big Data movement which involves the integration of multiple sources of data to create a comprehensive viewpoint within the data. This approach allows organizations to analyze data from a variety of sources. For example, in the healthcare industry, the fusion of large datasets aids clinical providers, in analyzing the pattern drifts within the patient data that may not be evident when analyzing individual data sets. By integrating data from electronic health records, medical imaging, and other sources, these providers can gain a more comprehensive view of patient health and identify potential health risks. Similarly, in the financial industry, the fusion of large datasets can help organizations identify fraudulent activity by analyzing transactional data from multiple sources, including credit card transactions, bank transfers, and other financial data. Overall, the trend toward the fusion of large datasets represents a significant opportunity for such organizations to gain new insights and drive innovation across a wide range of industries.

4.4 Big Data in Healthcare

Big Data helps in managing patient and hospital records in the healthcare sector which the otherwise normal technologies can't. It keeps the record of the patients and helps doctors to understand as much as possible and as early as possible about a person to make better decisions. It is the digitized form of data that can be retrieved and kept for future analysis. The first step of collecting the data is done by smart wearable devices and sensors that feed real-time data to electronic health records. After the COVID-19 pandemic, organizations are using Big Data for the prediction of epidemic outbreaks and telemedicine. Early detection of diseases and proficient analysis of medical images is also possible due to the fusion of large datasets from various sources. These are some well-known fields of research that demonstrate how Big Data analytics is currently used in medicine:

Telemedicine: Remote monitoring systems used for communication between the patient and the healthcare worker, rely on Big Data analysis of the patient's prior history to estimate the future hazards of the patient's condition. In addition, for individuals from remote places with no clinics and doctor's reach, telehealth is like a blessing. Telemedicine is also becoming widely used in disaster aid. "Ambulance Drones" are used as an assistance to paramedics with medicines and medical kits for the out-of-reach patients. These are used for various delivery purposes like transportation of blood products, medicines, etc. proving to be faster and more economical (Mackle, 2020; Zailani, 2021).

Image processing: The data produced by imaging techniques like MRI, CT and ultrasound requires multiple terabytes of storage. This data is then used to train AI models for image recognition in diseases like melanoma. Evaluating these images increases the accuracy in the detection of a disease. There are many medical image datasets for several diseases that have large volumes of images with varying resolution, quality and modality (Ashwin Belle, 2015).

Signal processing: Clinical Decision Support Systems (CDSSs) based on Big Data are being developed with the help of increasing amounts of signal processing research. The electrophysiologic monitoring of patients has generated interest in neurology research as a way to not only investigate complex disorders in a novel way but also to create diagnostic and treatment tools for the future. (Ashwin Belle, 2015)

Omics study: Databases like GEO, SwissProt, PDB are major pools of data in genomics and proteomics, which further help in high throughput screening and discovery of drugs and drug targets. The Human Genome Project is a pioneering endeavor that aims to create an extensive repository of human genetic variants by routinely sequencing the whole genomes of a large number of people (Tao Zeng, 2016).

Natural Language Processing (NLP): To identify the context and avoid the medical jargon with subtle nuances created in the communication between the patient and doctors, NPL aids to bring clarity in the diagnosis and the treatment. In the same context, the National Library of Medicine is creating the Specialist System, which will serve as an information-extraction tool for the biomedical knowledge base in general and the Medline abstracts in particular (P. Spyns, 1996). Some detailed examples along with certain algorithms are mentioned in **section 4.**

4.5 Big Data for Personalized Medicine

The high-throughput data analysis methods have revolutionized the medicine industry. The rapidly generated data and its volume from genomic sequencing mechanisms are augmenting the capacity of each genetic profile to forecast risk and tailor therapeutic approaches for any diseases (MR Naqvi, 2020). Personalized medicine entails adapting medical treatments and therapies to each patient's unique traits. This customization is based on genetic information, specifically on pharmacogenetic and pharmacogenomic data, to determine the most effective and safe treatments for an individual. By considering a person's genetic makeup and other relevant factors, healthcare professionals aim to optimize treatment outcomes and minimize adverse effects. The genetic makeup is obtained by various sequencing techniques which are aided with big data technologies to obtain faster results. There are various projects like the AACR Project GENIE, a global initiative for sharing data with the primary aim to advance Precision Healthcare. It seeks to catalyze collaboration by involving numerous National and International Cancer Centers in the collective effort to share information. JINTEL (https://www.deepthinkhealth.com) is a pioneering company that employs advanced analytics and AI in the realm of Precision Medicine. The company utilizes large-scale structured data, encompassing clinical and genetic datasets, and applies advanced analytics and ML techniques to revolutionize new studies and enhance Precision Medicine for patients.

4.6 Big Data for Biomarkers

Numerous publicly accessible data sources in the fields of biology and medicine provide opportunities for investigating biomarkers. The notable databases in this context encompass various categories: (i) repositories storing high-throughput data, such as TCGA, NCBI GEO, EMBL-EBI, or GigaDB; (ii)

Table 1. Biomarker Classification Based on Network Analysis (Tao Zeng, 2016).

Processes of Biomarker Classification		Mechanism
Big-Data-Based Edge Biomarkers	Node or Molecular Biomarkers	Utilizes differentially expressed data from gene network
	Network-Based Biomarkers	
	Edge Biomarkers/ Network Biomarkers	Utilizes correlational data of biomolecules in learning and predicting models.
	Dynamical Edge Biomarker	Utilizes and characterizes dynamics of Big Data in higher dimensions

databases dedicated to biological sequences and elements, exemplified by ENCODE; (iii) databases illustrating physical or interconnected networks, like KEGG or STRING; and (iv) repositories archiving pre-existing functional annotations of biological elements, such as Gene Ontology (Tao Zeng, 2016). The biomarker identification has two evident phases (i) Finding biomarkers using the information that is already available about the samples that were taken from various people. (ii) Once identified, utilizing these markers onto a completely new set of samples for novel applications. Apart from their physical molecular properties, biomarkers can be classified based on their therapeutic implications and methods used for their identification (Samia Afzal,2022). Considering the Big Data, associated databases and based on Network Analysis, **Table 1** discusses some major types of Biomarkers based on their network information (Malik Yousef, 2009).

5. Revolutionized Healthcare

5.1 AI and ML in Oncology

The preliminary step of curing any disease is detecting it correctly at the right time. Many patients with chronic diseases have a low disease-curing rate due to late diagnosis of their condition. One such disease is cancer. Numerous scientists have applied ML to create prediction models that extract logical patterns from a large amount of historical data to forecast a patient's chances of survival (Shigao Huang, 2020). AI-based algorithms have a great deal of potential and have opened the door to the early detection of genetic alterations and abnormal protein interactions leading to cancer (Muhammad Javed Iqbal, 2021).

For example, to improve patient care, breast cancer prognosis includes assessing the chances of the disease recurrence and projecting a patient's survival (Shigao Huang, 2020). Thereafter, researchers typically use multimodal deep neural networks (DNNs) to investigate the receiver operating characteristic (ROC) curve and AUC values while integrating multidimensional data. The results demonstrate the efficacy of combining several data types and ensemble DNN approaches for enhancing human breast cancer prognostic forecasts (Shigao Huang, 2020).

Digital image analysis and processing, which involves extracting relevant information from images using segmentation (differentiation in the image variables) and classification (description of labels) algorithms, have drawn researchers' attention specifically for cancer diagnoses (Hanya Mahmood, 2021). There are AI-based apps developed to capture the images and compare them to a vast number of images in various datasets to predict whether the image lesion is a tumor or not. Dermatologists have yet to confirm the accuracy of their outcomes. Deep Learning (DL) Systems for image detection with proper pathological diagnosis can alleviate the burdens of radiologists and at the same time increase the efficiency of patient care.

For example in breast cancer, DL Systems are used for image processing and detecting the symptoms of the cancer. Out of all the algorithms used or the traditional ways of image processing used, Support Vector Machines (SVM) showed the highest accuracy (Farahnaz Sadoughi, 2022).

5.2 Pre-clinical Symptom Checkers and Chatbots

Many online symptom checkers like Symptomate by Infermedica and Mediktor by IBM are being used by people for pre-clinical online diagnosis. These AI technologies are based on NLP, Bayesian Networks, Hidden Markov Model (HMM) and SVM at the backend to accurately predict the diseased condition based on the symptoms entered by the patients. Such models help people in making decisions about visiting a doctor at the right time (Lekha Athota, 2020).

NLP: It focuses on how computers and human language interact, especially when it comes to programming computers to handle and analyze huge amounts of human language written data. Algorithms like SVM and HMM are being used to infer the patterns and interpret the handwritten data. Speech recognition technology is also developed on such models (Prakash M Nadkarni, 2011). NLP is categorized into high-resource languages (HRLs) and low-resource languages (LRLs), wherein nowadays the systems are generated for all the HRLs. Taking into account the world's population with its diversity, millions of native languages with different dialects lack these NLP systems and resources. Creating information and tools for 1000+ languages is a challenge for the language community in the future (Julia Hirschberg, 2015).

Moreover, the AI Chatbots that assist in disease diagnosis before visiting a doctor, are completely based on the same algorithms (Lekha Athota, 2020) and can be involved in almost all healthcare facilities. In the upcoming years, they are targeted to provide better communication between the clinical teams and patients, line up appointments, fetch and analyze medical data (Tom Nadarzynski, 2019) at a personal and community research level. These can also monitor regular sleep, nutrition habits, etc. and give preventive measures based on the daily data fed into the system. These systems have shown a huge scope due to the COVID-19 pandemic, in assisting with the treatment and prevention of the viral disease. In similar terms, the WHO has also developed a chatbot that can be accessed through WhatsApp and Facebook (Manal Almalki, 2020).

5.3 Smart Monitoring Systems

Using IoT, many devices can monitor heart rate, body temperature, and spO2 levels in the body of a person. Through sensors built into smartphones and other smart portable devices, doctors can remotely check the wellness of their diabetic patients. Amine Rghioui et. al., 2020 have created and discussed how a wireless, smartphone, glucose monitoring device, etc. can communicate via the Wi-Fi protocol. It is a tool for automatically detecting and categorizing the glucose level data. The prediction of data was done by several algorithms out of which the J48 algorithm gave the highest accuracy of 99.17%. When atypical behavior is found, the system can also send out a notification to the nearby medics as a precautionary measure (Amine Rghioui, 2020).

5.4 Early Diagnosis of Fatal Diseases

A few diseases might not show noticeable symptoms or even if they do, the disease state might have advanced to an extent that the treatment becomes complicated. One such disease is Thyrotoxic Periodic Paralysis (TPP) characterized by acute weakness, with low potassium levels and overactive thyroid. Chin Lin et al., 2021 used a Deep Learning Model (DLM) to ascertain hypokalemia and hyperthyroidism. This study evaluated the use of regular laboratory data and AI-assisted Electrocardiography (ECG) for the premature identification of TPP. The outputs showed that when combined with normal blood chemistries, an AI-ECG system precisely identified hypokalemia in paralytic patients and offered helpful decision-making characteristics for primary TPP symptoms (Chin Lin, 2021).

AI tools are used to interpret and augment lab investigations by combining data from numerous inputs. While human clinicians rely on the standard ranges, AI can make blood reports more precise by identifying patterns that are unique to each individual's medical history. For example, PMcardio (Powerful Medical; Bratislava, Slovakia), can digitize an ECG by AI processing and instantly screens for 38 cardiovascular diseases (Chris Brown, 2023). Many different AI models have been developed to detect pathology on chest X-rays.

Another example is Sepsis, a detrimental organ dysfunction caused by an uncontrolled host response to an infection. The obstacle arises in distinguishing sepsis from other acute conditions because of the similarities of symptoms between them (Ivana Pepic, 2021). M. Schinkel et al., 2019 studied 25 patients with sepsis for the onset or prediction of sepsis using 11 different algorithms. Gradient-boosted tree model based on supervised learning (by Desautels, 2016) was found to be most efficient in detecting sepsis with an AUROC of 0.952. AI models can detect sepsis hours before, wherein clinicians can further prescribe antibiotics for its reversal (M. Schinkel, 2019). Sepsis detecting systems have also been set up in the ICU and emergency departments. These models were found to decrease mortality or shock duration in patients.

DL algorithms are used by the Viz LVO app to automatically identify strokes on CT images and notify on-call stroke specialists about individuals who may be eligible for treatment. This device is a game changer because it reduces the amount of time it takes for patients to obtain the care they require, which is crucial. After all, stroke is one of the most time-sensitive medical disorders. The largest health AI dataset to date was used in a study where Viz LVO obtained 96% sensitivity and 94% specificity. This application's quicker triage allows for the detection and treatment of more patients, improving patient outcomes and lowering the likelihood of long-term incapacity. ML methods like Random Forest Learning and CNNs are used to build similar types of apps that can detect strokes (Nick M Murray, 2020). Arterys' imaging platform detects cardiac diseases and has been privileged to be the first DL platform approved by the FDA (B Marr, 2017).

5.5 Drug Discovery

A promising strategy for speeding up the development of pharmaceutical products is AI-based formulation development (Junhuang Jiang, 2022). In the traditional process, a drug progresses from its conceptual stage to the market stage in more than 15 years with an investment of more than a billion dollars. Using AI, these time and cost variables can be reduced to a larger extent. Under the domain of Bioinformatics, some software use ML and DL algorithms for processes like Virtual Screening, Toxicity Prediction, Drug Monitoring, etc. (Rohan Gupta, 2021), saving millions of dollars spent on wet lab experiments for the same process. AI is capable of identifying the hits and lead compounds, further accelerating the validation of medicinal targets and optimizing their structural design (Debleena Paul, 2021). The drug design process such as Peptide Synthesis, Protein-Protein Interaction and Quantitative Structure–Activity Relationship (QSAR), all use AI-based algorithms like Artificial Neural Networks (ANN), DNN and Meta-Learning (Rohan Gupta, 2021). To find potential therapeutic candidates, QSAR modeling tools have incorporated AI-based algorithms like Linear Discriminant Analysis (LDA), SVMs, Random Forest (RF), and Decision Trees to hasten the QSAR analysis (Debleena Paul, 2021).

5.6 Robots in Surgery

Robotic surgery offers the potential to increase technical accuracy while reducing surgeon fatigue (Jonathan J. Rasouli, 2021). Clay M. Elswick et al.,2020 found that 28 patients who underwent spinal surgery for the placement of pedicle screws showed 97.6% accuracy with the Excelsius GPS robot (Clay M. Elswick, 2020). The technology has not only helped the patients but has decreased the surgeons' and operating room staff's exposure to harmful radiation through these robotic-assisted instruments (Jonathan J. Rasouli, 2021).

5.7 AI in Clinical Trials

Clinical trials are the most time and cost-consuming part after the discovery of a potential drug. Regarding eligibility, suitability, motivation, and empowerment to enroll, each clinical trial imposes specific conditions on participating patients (Stephen Harrer, 2019). AI helps in predicting the likeliness of a subject backing out from the trial based on his/her behavioral data, interpreted by the machine. By enabling the automation of formerly manual processes like organ, tissue, and structural segmentation, AI and its implementation through ML, and particularly DL techniques, bring several advantages to medical image processing, that lead to understanding the difference in the patient's behavioral patterns (Irene Mayorga-Ruiz, 2019).

5.8 Personalized Medicine

It is also called Precision Medicine, Personalized Medicine is a revolutionary field in itself that aims to curb the hurdles of detection and accurate disease diagnosis (Jamilu Awwalu, 2015), and provide targeted treatment to each patient based on his/her unique genetic makeup, medical history, age and lifestyle. With accuracy and decreased false positive predictions, AI uses algorithms like ANN, SVM, Naïve Bayes, and fuzzy logic for predicting several diseases with a personalized touch. (Jamilu Awwalu, 2015).

Naïve Bayes is based on a probabilistic model that enables the systematic capture of uncertainty in a model by calculating the odds of various events. It is used to predict relationships between diseases and treatments (Shashikant U. Ghumbre, 2011) and is based on Bayers' Theorem:

$$P(A|B) = P(B|A) . P(A) / P(B)$$

Where A and B are two events related to one another just like Disease and its Treatment; P(A|B) is a posterior probability of A with the occurrence of B; P(B|A) is a conditional probability of B with the occurrence of A; P(A) is a probability of A; P(B) is a probability of B.

Heart attacks are one of the main yet unpredictable causes of death in the world. Hence, its diagnosis at the right time is crucial, but due to a lot of uncertainties and risk factors, specialists sometimes have trouble with its diagnosis. Powerful reasoning techniques like fuzzy logic excel at coping with data ambiguity (Tanmay Kasbe, 2017). Fuzzy expert systems are used to analyze datasets and predict heart attacks on new data. Mittal et al., 2023 discussed various fuzzy expert systems with 94% accuracy for heart disease diagnosis. A proactive approach to providing individualized preventative treatments can serve as new predictors for the early onset of diseases (David Thesmar, 2019).

5.9 Biomarkers

Biomarkers are defined as "an attribute that can be used to test and assess scientifically to determine if it is a sign of pathological processes, normal biological processes, or pharmacological reactions to a therapeutics". AI algorithms can identify potential biomarkers by analyzing patterns and relationships within biological data. Biomarkers are very objective attributes of biological processes. It is conceivable that measurable biological features exist without aligning with a patient's clinical state; and variations in these markers may be both undetectable and inconsequential to health (Kyle Strimbu, 2011). Several types of biomarkers have recently emerged as potential detectors for diseases and are discussed in **Table 2.**

5.9.1 Voice biomarkers

Voice modulation and pronunciations act as crucial indicators of diseases like Alzheimer's and Parkinson's. Voice biomarkers are a way to potentially identify the symptoms of these diseases. John M. Tracy et al., 2020 collected control and voice data of patients with Parkinson's Disease. They used three machine learning models out of which the gradient-boosted model outperformed RF and Logistic Regression models in detecting the disease. This helped in the accuracy and cost management of the diseases (John M. Tracy, 2020).

5.9.2 Image Biomarkers

The field of imaging biomarkers has been highly dynamic, particularly with the success rate of AI-based algorithms on larger datasets (Peter Savadjiev, 2019). The proliferation of large digital datasets in a variety of scientific and technical fields has led to an upsurge in interest in data-derived and hypothesis-free analytic methods (Y LeCun, 2015). These methods prioritize the exploration of large datasets to uncover new patterns and assist in formulating novel hypotheses that led to the generation of a new field called Deep Radiomics. (WL Kraus, 2015).

5.9.3 Tissue biomarkers

Tissue biomarkers serve as data embedded in tissues and play a crucial role in Pathology by identifying distinct patient subsets for diagnostic, prognostic, or predictive purposes. (VN Harry, 2010). These systems are capable of monitoring and predicting the changes in the treatment routine, with the possibility to modify the course of action leading toward a personalized treatment regime (JPB O'Connor,

Table 2. Revolutionized Healthcare

Sr. No.	Application	Use	Mechanism
1.	Oncology	Detecting the symptoms of the cancer and prognosis	Deep learning systems like Support Vector Machines are used for image processing (Farahnaz Sadoughi, 2022)
2.	Preclinical symptom checker and chatbots	Symptom checkers like symptomate	Natural Language Processing and Bayesian Networks with Hidden Markov Model and Support Vector Machine at the backend (Prakash M Nadkarni, 2011)
3.	Smart monitoring systems	Gadgets that track the body's spO2 levels, body temperature, and heart rate	J48 algorithm with the highest accuracy of 99.17% (Amine Rghioui, 2020)
4.	Early diagnosis of fatal diseases	Thyrotoxic periodic paralysis (TPP) Interpret and augment lab investigations	Chin Lin et. al. used a deep learning model (DLM) to ascertain hypokalemia and hyperthyroidism AI ML models (Chris Brown, 2023) and data processing using IoT and big data
5.	Drug discovery	Development of pharmaceutical products	Artificial neural networks, deep neural networks and meta-learning (Rohan Gupta, 2021)
6.	Robots in surgery	Technical accuracy (Jonathan J. Rasouli, 2021)	Excelsius GPS robot (Clay M. Elswick, 2020) based on computer vision, natural language processing
7.	Clinical trials	Patient Recruitment and Eligibility (Stephen Harrer, 2019) Protocol Design and Optimization Safety Monitoring	AI algorithms can analyze electronic health records (EHRs) and other relevant data to identify potential participants analyzing large datasets, including historical trial data, published literature, and real-world evidence algorithms can detect adverse events, identify potential safety concerns, and provide real-time alerts
8.	Personalized medicine	Accuracy and reduces the false positive rates of disease prediction	SVM, Naïve Bayes, Fuzzy Logic (Jamilu Awwalu, 2015)
9.	Biomarkers	early detection and monitoring	Pattern recognition, RNN, Feature extraction, Gradient-boosted ML model
10.	Digital health devices	Medical-grade ECG recorder	Kardia algorithm analyzes readings in real-time

2017). The anticipation is that AI-based devices will streamline the assessment of tissue biomarkers and uncover novel insights that might be overlooked through human examination. These findings can then be employed to formulate precise predictions (Cesare Lancellotti, 2021). AI-based biomarkers forecast genetic variations, therapeutic response, or longevity of patients that can be used to establish a Breast Digital Pathology Lab (Cesare Lancellotti, 2021). Osareh et al. 2010 introduced a model to reliably differentiate between benign and malignant tissues based on 10 different physiological characteristics of a cell (A. Osareh, 2010). In similar terms, the Prostate Pathology Lab can be established to study and classify prostatic adenocarcinoma tissues (Cesare Lancellotti, 2021).

5.10 Digital health devices

The disease known as Atrial Fibrillation, or aFib, can make people more susceptible to heart problems such as heart failure and stroke. Yet up until recently, managing it has been extremely challenging because it necessitates ongoing ECG monitoring to provide information about heart rate and rhythm. And with the advent of digital health gadgets, AI has already altered that. For instance, AliveCor's Kardia, a medical-grade ECG recorder approved by FDA has the Kardia algorithm that analyzes readings in real-time while running virtually undetectable in the background, shifting the point of care from hospitals to the patient's location (Giebel, 2019). In the US, epilepsy ranks fourth among neurological conditions, behind migraine, stroke, and Alzheimer's disease. However, Wearable technology such as the AI-powered Empatica Wristband, is intended to alert users and carers of a possible seizure. In clinical trials, the generalized tonic-clonic seizure detection accuracy of Embrace was 98%. Of course, the long-term goal is to also be able to predict seizures.

6. Healthcare Sector: A Million Dollar Opportunity for AI and IoT

Healthcare systems could benefit from the application of AI by addressing issues with resource allocation, streamlining hospital operations, and enhancing the infrastructure of healthcare as a whole (C Wang, 2021). Furthermore, the interconnectedness of IoT devices makes it possible to build an ecosystem for health data, which will facilitate data exchange and improve the quality of future research (D Chen, 2018). The integration of IoT, AI, and bioinformatics in healthcare presents several problems despite the positive outlook. The expense of implementation and upkeep for this is a big worry. Healthcare organizations may face financial difficulties as a result of the initial investment in AI and IoT infrastructure as well as continuing operating costs, especially in regions with limited resources. Integrating AI into healthcare involves interactions between AI systems and a diverse user base, potentially leading to biases from users who may perceive AI as "dangerous" based on prevalent cultural narratives (Oksana Iliashenko, 2019). Yet another obstacle for AI in healthcare pertains to the issue of privacy and data security. Patient data, including medical histories, are utilized to train AI systems. This requires the technicians to take measures to prevent unauthorized access by third parties (Oksana Iliashenko, 2019). Safeguarding patient data and ensuring the responsible use of AI technologies are critical aspects that require attention to build and maintain public trust in these innovations. The results of a single diagnostic technique do not determine a diagnosis; instead, a comprehensive approach combines information from other sources, such as physical findings and major complaints. In the future, there will be a need for a system that effectively integrates various specific diagnostic data to enhance diagnostic accuracy (Ayumu Asai, 2021). The evolution of AI and IoT in healthcare has enormous potential to improve patient outcomes, maximize resource use, and elevate the administration of healthcare as a whole. They will also contribute to the growth of the biomarkers and personalized medicine fields. The social and ethical ramifications of technology must be considered as it develops wherein the ethical code of conduct should not be broken by AI. With the progression in technology where AI and IoT greatly contribute to the improvement of global healthcare, cooperation between healthcare experts, legislators, technology developers, etc., human-race shall dive into a "Million Dollar" fortune to be unlocked in the AI and IoT-based Ecosystem.

REFERENCES

Adeli, A., & Neshat, M. (2010, March). *A fuzzy expert system for heart disease diagnosis*. In *Proceedings of international multi conference of engineers and computer scientists,* Hong Kong.

Adilin Beatris. (2019). *All About the Basics of Big Data: History*. Types and Applications.

Afzal, S., Hassan, M., Ullah, S., Abbas, H., Tawakkal, F., & Khan, M. A. (2022). Breast cancer; discovery of novel diagnostic biomarkers, drug resistance, and therapeutic implications. *Frontiers in Molecular Biosciences*, *9*, 783450. doi:10.3389/fmolb.2022.783450 PMID:35265667

Agrahari, A., & Rao, D. T. V. D. (2017). A review paper on Big Data: Technologies, tools and trends. *Int Res J Eng Technol*, *4*(10), 10.

Ahmed, S., Ilyas, M., & Raja, M. Y. A. (2018, March). Internet of Things: application in smart healthcare. In *Proceedings of the 9th International Conference on Society and Information Technology (IICSIT 2018)* (pp. 19-24). IEEE.

Alshehri, F., & Muhammad, G. (2020). A comprehensive survey of the Internet of Things (IoT) and AI-based smart healthcare. *IEEE Access : Practical Innovations, Open Solutions*, *9*, 3660–3678. doi:10.1109/ACCESS.2020.3047960

Asai, A., Konno, M., Taniguchi, M., Vecchione, A., & Ishii, H. (2021). Computational healthcare: Present and future perspectives. *Experimental and Therapeutic Medicine*, *22*(6), 1–13. doi:10.3892/etm.2021.10786 PMID:34659497

Athota, L., Shukla, V. K., Pandey, N., & Rana, A. (2020, June). Chatbot for healthcare system using artificial intelligence. In *2020 8th International conference on reliability, infocom technologies and optimization (trends and future directions)(ICRITO)* (pp. 619-622). IEEE. 10.1109/ICRITO48877.2020.9197833

Awwalu, J., Garba, A. G., Ghazvini, A., & Atuah, R. (2015). Artificial intelligence in personalized medicine application of AI algorithms in solving personalized medicine problems. *International Journal of Computer Theory and Engineering*, *7*(6), 439–443. doi:10.7763/IJCTE.2015.V7.999

Belle, A., Thiagarajan, R., Soroushmehr, S. M., Navidi, F., Beard, D. A., & Najarian, K. (2015). Big data analytics in healthcare. *BioMed Research International*. PMID:26229957

Brown, C., Nazeer, R., Gibbs, A., Le Page, P., Mitchell, A. R., & Mitchell, A. R. (2023). Breaking bias: The role of artificial intelligence in improving clinical decision-making. *Cureus*, *15*(3). doi:10.7759/cureus.36415 PMID:37090406

Carbonell, J. G., Michalski, R. S., & Mitchell, T. M. (1983). An overview of machine learning. *Machine Learning*, 3–23.

Davenport, T., & Kalakota, R. (2019). The potential for artificial intelligence in healthcare. *Future Healthcare Journal*, *6*(2), 94–98. doi:10.7861/futurehosp.6-2-94 PMID:31363513

DeepThink Health. 2020. Deepthinkhealth.com. https://www.deepthinkhealth.com

Elswick, C. M., Strong, M. J., Joseph, J. R., Saadeh, Y., Oppenlander, M., & Park, P. (2020). Robotic-assisted spinal surgery: Current generation instrumentation and new applications. *Neurosurgery Clinics*, *31*(1), 103–110. PMID:31739920

Food and Drug Administration & National Institutes of Health. (n.d.). *BEST (biomarkers, endpoints, and other tools) resource.* NCBI. https://www.ncbi.nlm.nih.gov/books/NBK326791

Ghosh, A., Chakraborty, D., & Law, A. (2018). Artificial intelligence in Internet of things. *CAAI Transactions on Intelligence Technology*, *3*(4), 208–218. doi:10.1049/trit.2018.1008

Ghumbre, S., Patil, C., & Ghatol, A. (2011, December). Heart disease diagnosis using support vector machine. In *International conference on computer science and information technology (ICCSIT') Pattaya* (pp. 84-88). IEEE.

Giebel, G. D., & Gissel, C. (2019). Accuracy of mHealth devices for atrial fibrillation screening: Systematic review. *JMIR mHealth and uHealth*, *7*(6), e13641. doi:10.2196/13641 PMID:31199337

Hamet, P., & Tremblay, J. (2017). Artificial intelligence in medicine. *Metabolism: Clinical and Experimental*, *69*, S36–S40. doi:10.1016/j.metabol.2017.01.011 PMID:28126242

Harrer, S., Shah, P., Antony, B., & Hu, J. (2019). Artificial intelligence for clinical trial design. *Trends in Pharmacological Sciences*, *40*(8), 577–591. doi:10.1016/j.tips.2019.05.005 PMID:31326235

Harry, V. N., Semple, S. I., Parkin, D. E., & Gilbert, F. J. (2010). Use of new imaging techniques to predict tumour response to therapy. *The Lancet. Oncology*, *11*(1), 92–102. doi:10.1016/S1470-2045(09)70190-1 PMID:20129132

Hirschberg, J., & Manning, C. D. (2015). Advances in natural language processing. *Science*, *349*(6245), 261–266. doi:10.1126/science.aaa8685 PMID:26185244

Hogeweg, P. (2011). The roots of bioinformatics in theoretical biology. *PLoS Computational Biology*, *7*(3), e1002021. doi:10.1371/journal.pcbi.1002021 PMID:21483479

Huang, S., Yang, J., Fong, S., & Zhao, Q. (2020). Artificial intelligence in cancer diagnosis and prognosis: Opportunities and challenges. *Cancer Letters*, *471*, 61–71. doi:10.1016/j.canlet.2019.12.007 PMID:31830558

Iliashenko, O., Bikkulova, Z., & Dubgorn, A. (2019). Opportunities and challenges of artificial intelligence in healthcare. In *E3S Web of Conferences* (Vol. 110, p. 02028). EDP Sciences. doi:10.1051/e3sconf/201911002028

Imhoff, M., Webb, A., & Goldschmidt, A. (2001). On behalf of the ESICM. *Intensive Care Medicine*, *27*(1), 179–186. doi:10.1007/PL00020869

Iqbal, M. J., Javed, Z., Sadia, H., Qureshi, I. A., Irshad, A., Ahmed, R., Malik, K., Raza, S., Abbas, A., Pezzani, R., & Sharifi-Rad, J. (2021). Clinical applications of artificial intelligence and machine learning in cancer diagnosis: Looking into the future. *Cancer Cell International*, *21*(1), 1–11. doi:10.1186/s12935-021-01981-1 PMID:34020642

Iqbal, M. S., Abd-Alrazaq, A., & Househ, M. (2022). Artificial Intelligence Solutions to Detect Fraud in Healthcare Settings: A Scoping Review. *Advances in Informatics, Management and Technology in Healthcare*, 20-23.

Iroju, O. G., & Olaleke, J. O. (2015). A systematic review of natural language processing in healthcare. *International Journal of Information Technology and Computer Science*, *8*(8), 44–50. doi:10.5815/ijitcs.2015.08.07

Jiang, J., Ma, X., Ouyang, D., & Williams, R. O. III. (2022). Emerging artificial intelligence (ai) technologies used in the development of solid dosage forms. *Pharmaceutics*, *14*(11), 2257. doi:10.3390/pharmaceutics14112257 PMID:36365076

Kasbe, T., & Pippal, R. S. (2017, August). Design of heart disease diagnosis system using fuzzy logic. In *2017 International Conference on Energy, Communication, Data Analytics and Soft Computing (ICECDS)* (pp. 3183-3187). IEEE. 10.1109/ICECDS.2017.8390044

Kraus, W. L. (2015). Would you like a hypothesis with those data? Omics and the age of discovery science. *Molecular Endocrinology (Baltimore, Md.)*, *29*(11), 1531–1534. doi:10.1210/me.2015-1253 PMID:26524008

Lancellotti, C., Cancian, P., Savevski, V., Kotha, S. R. R., Fraggetta, F., Graziano, P., & Di Tommaso, L. (2021). Artificial intelligence & tissue biomarkers: Advantages, risks and perspectives for pathology. *Cells*, *10*(4), 787. doi:10.3390/cells10040787 PMID:33918173

LeCun, Y., Bengio, Y., & Hinton, G. (2015). Deep learning. *nature, 521*(7553), 436-444.

Lin, C., Lin, C. S., Lee, D. J., Lee, C. C., Chen, S. J., Tsai, S. H., Kuo, F.-C., Chau, T., & Lin, S. H. (2021). Artificial Intelligence–assisted electrocardiography for early diagnosis of thyrotoxic periodic paralysis. *Journal of the Endocrine Society*, *5*(9), bvab120. doi:10.1210/jendso/bvab120 PMID:34308091

Mackle, C., Bond, R., Torney, H., McBride, R., McLaughlin, J., Finlay, D., ... McEneaney, D. (2020). A data-driven simulator for the strategic positioning of aerial ambulance drones reaching out-of-hospital cardiac arrests: A genetic algorithmic approach. *IEEE Journal of Translational Engineering in Health and Medicine*, *8*, 1–10.

Mahmood, H., Shaban, M., Rajpoot, N., & Khurram, S. A. (2021). Artificial Intelligence-based methods in head and neck cancer diagnosis: An overview. *British Journal of Cancer*, *124*(12), 1934–1940. doi:10.1038/s41416-021-01386-x PMID:33875821

Majhi, V., Paul, S., & Jain, R. (2019, February). Bioinformatics for healthcare applications. In 2019 Amity international conference on artificial intelligence (AICAI) (pp. 204-207). IEEE. doi:10.1109/AICAI.2019.8701277

Marr, B. (2017). *First FDA approval for clinical cloud-based deep learning in healthcare. Forbes.* Forbes Publishing Company.

Marsland S. (2015). *Machine learning: an algorithmic perspective*. CRC press.

Mayorga-Ruiz, I., Jiménez-Pastor, A., Fos-Guarinos, B., López-González, R., García-Castro, F., & Alberich-Bayarri, Á. (2019). The role of AI in clinical trials. *Artificial Intelligence in Medical Imaging: Opportunities, applications and risks*, 231-243.

Mittal, V., Malik, P., Purohit, K. C., & Ram, M. (2023). Fuzzy logic based decision systems in the healthcare sector. In *Advances in Soft Computing Applications* (pp. 157–184). River Publishers.

Murray, N. M., Unberath, M., Hager, G. D., & Hui, F. K. (2020). Artificial intelligence to diagnose ischemic stroke and identify large vessel occlusions: A systematic review. *Journal of Neurointerventional Surgery*, *12*(2), 156–164. doi:10.1136/neurintsurg-2019-015135 PMID:31594798

Nadarzynski, T., Miles, O., Cowie, A., & Ridge, D. (2019). Acceptability of artificial intelligence (AI)-led chatbot services in healthcare: A mixed-methods study. *Digital Health*, *5*, 2055207619871808. doi:10.1177/2055207619871808 PMID:31467682

Nadkarni, P. M., Ohno-Machado, L., & Chapman, W. W. (2011). Natural language processing: An introduction. *Journal of the American Medical Informatics Association : JAMIA*, *18*(5), 544–551. doi:10.1136/amiajnl-2011-000464 PMID:21846786

Naqvi, M. R., Jaffar, M. A., Aslam, M., Shahzad, S. K., Iqbal, M. W., & Farooq, A. (2020, June). Importance of big data in precision and personalized medicine. In *2020 International Congress on Human-Computer Interaction, Optimization and Robotic Applications (HORA)* (pp. 1-6). IEEE. 10.1109/HORA49412.2020.9152842

O'connor, J. P., Aboagye, E. O., Adams, J. E., Aerts, H. J., Barrington, S. F., Beer, A. J., Boellaard, R., Bohndiek, S. E., Brady, M., Brown, G., Buckley, D. L., Chenevert, T. L., Clarke, L. P., Collette, S., Cook, G. J., deSouza, N. M., Dickson, J. C., Dive, C., Evelhoch, J. L., & Waterton, J. C. (2017, March). Imaging biomarker roadmap for cancer studies. *Nature Reviews. Clinical Oncology*, *14*(3), 169–186. doi:10.1038/nrclinonc.2016.162 PMID:27725679

Osareh, A., & Shadgar, B. (2010). Machine learning techniques to diagnose breast cancer. *5th international symposium on health informatics and bioinformatics 2010 Apr 20* (pp. 114-120). IEEE. 10.1109/HIBIT.2010.5478895

Paul, D., Sanap, G., Shenoy, S., Kalyane, D., Kalia, K., & Tekade, R. K. (2021). Artificial intelligence in drug discovery and development. *Drug Discovery Today*, *26*(1), 80–93. doi:10.1016/j.drudis.2020.10.010 PMID:33099022

Pepic, I., Feldt, R., Ljungström, L., Torkar, R., Dalevi, D., Maurin Söderholm, H., Andersson, L.-M., Axelson-Fisk, M., Bohm, K., Sjöqvist, B. A., & Candefjord, S. (2021). Early detection of sepsis using artificial intelligence: A scoping review protocol. *Systematic Reviews*, *10*(1), 1–7. doi:10.1186/s13643-020-01561-w PMID:33453724

Qian, L., Zheng, H., Zhou, H., Qin, R., & Li, J. (2013). Classification of time series gene expression in clinical studies via integration of biological network. *PLoS One*, *8*(3), e58383. doi:10.1371/journal.pone.0058383 PMID:23516469

Rasouli, J. J., Shao, J., Neifert, S., Gibbs, W. N., Habboub, G., Steinmetz, M. P., Benzel, E., & Mroz, T. E. (2021). Artificial intelligence and robotics in spine surgery. *Global Spine Journal, 11*(4), 556–564. doi:10.1177/2192568220915718 PMID:32875928

Rghioui, A., Lloret, J., Harane, M., & Oumnad, A. (2020). A smart glucose monitoring system for diabetic patient. *Electronics (Basel), 9*(4), 678. doi:10.3390/electronics9040678

Russell, S. J., & Norvig, P. (2010). *Artificial intelligence a modern approach.*

Sadoughi, F., Kazemy, Z., Hamedan, F., Owji, L., Rahmanikatigari, M., & Azadboni, T. T. (2018). Artificial intelligence methods for the diagnosis of breast cancer by image processing: a review. *Breast Cancer: Targets and Therapy*, 219-230.

Schinkel, M., Paranjape, K., Panday, R. N., Skyttberg, N., & Nanayakkara, P. W. (2019). Clinical applications of artificial intelligence in sepsis: A narrative review. *Computers in Biology and Medicine, 115*, 103488. doi:10.1016/j.compbiomed.2019.103488 PMID:31634699

Spyns, P. (1996). Natural language processing in medicine: an overview. *Methods of information in medicine, 35*(04/05), 285-301.

Strimbu, K., & Tavel, J. A. (2010). What are biomarkers? *Current Opinion in HIV and AIDS, 5*(6), 463–466. doi:10.1097/COH.0b013e32833ed177 PMID:20978388

Talukder, A., & Haas, R. (2021, June). AIoT: AI meets IoT and web in smart healthcare. In *Companion Publication of the 13th ACM Web Science Conference 2021* (pp. 92-98). ACM.

Tang, J., Ma, T., & Luo, Q. (2020). Trends prediction of big data: A case study based on fusion data. *Procedia Computer Science, 174*, 181–190. doi:10.1016/j.procs.2020.06.073

Tian, S., Yang, W., Le Grange, J. M., Wang, P., Huang, W., & Ye, Z. (2019). Smart healthcare: Making medical care more intelligent. *Global Health Journal (Amsterdam, Netherlands), 3*(3), 62–65. doi:10.1016/j.glohj.2019.07.001

Tracy, J. M., Özkanca, Y., Atkins, D. C., & Ghomi, R. H. (2020). Investigating voice as a biomarker: Deep phenotyping methods for early detection of Parkinson's disease. *Journal of Biomedical Informatics, 104*, 103362. doi:10.1016/j.jbi.2019.103362 PMID:31866434

Wolfgang, E. (2017). *Introduction to artificial intelligence*. Springer.

Xia, F., Yang, L. T., Wang, L., & Vinel, A. (2012). Internet of things. *International Journal of Communication Systems, 25*(9), 1101–1102. doi:10.1002/dac.2417

Yousef, M., Ketany, M., Manevitz, L., Showe, L. C., & Showe, M. K. (2009). Classification and biomarker identification using gene network modules and support vector machines. *BMC Bioinformatics, 10*(1), 1–7. doi:10.1186/1471-2105-10-337 PMID:19832995

Zailani, M. A., Azma, R. Z., Aniza, I., Rahana, A. R., Ismail, M. S., Shahnaz, I. S., ... Mahdy, Z. A. (2021). Drone versus ambulance for blood products transportation: An economic evaluation study. *BMC Health Services Research, 21*, 1–10.

Zeng, T., Zhang, W., Yu, X., Liu, X., Li, M., & Chen, L. (2016). Big-data-based edge biomarkers: Study on dynamical drug sensitivity and resistance in individuals. *Briefings in Bioinformatics*, *17*(4), 576–592. doi:10.1093/bib/bbv078 PMID:26411472

Zikopoulos, P., & Eaton, C. (2011). *Understanding big data: Analytics for enterprise class hadoop and streaming data*. McGraw-Hill Osborne Media.

Chapter 10
Unveiling the World of Bioinformatics

Khushboo Dhiman
Kurukshetra University, India

Hardik Dhiman
Uniwersytet Łódzki, Poland

ABSTRACT

This chapter take a fascinating voyage across the complex bioinformatics landscapes. The authors begin by defining bioinformatics as the essential connection that connects unprocessed biological data to useful knowledge, laying the groundwork for an investigation into the core fields of proteomics, genomics, and data mining. The story flows naturally, showing how feature refinement techniques, machine learning's predictive power, and deep learning's revolutionary ability to understand intricate biological processes all work together in harmony. Applications in the real world emerge, demonstrating the palpable effects of bioinformatics on drug discovery, the prediction of protein functionality, and the rapidly expanding field of personalised medicine. But this journey is not without its difficulties; we must negotiate the nuances of data integration, contend with the interpretability of machine learning models, fight with moral dilemmas, and face the computing demands of big data. Every obstacle turns into a doorway for creativity.

1. INTRODUCTION

Fundamentally, bioinformatics is the fusion of biology and information technology; it is a field that uses computational methods and tools to extract valuable information from biological data. This discipline emerged from the realisation that advanced computational tools are required for the interpretation and analysis of biological data due to its sheer bulk and complexity, which includes protein structures, functional annotations, and genetic sequences. Essentially, bioinformatics serves as a translator between the biological and data languages, allowing scientists to understand the complexities of life at the molecular level.

DOI: 10.4018/979-8-3693-1822-5.ch010

It is difficult to overestimate the importance of bioinformatics in contemporary biology. In a time of unparalleled technological progress and a deluge of biological data, bioinformatics becomes an indispensable cornerstone, enabling the conversion of unprocessed data into knowledge(Wang, N., & He, Q. (2023)). Large datasets produced by systems biology techniques, high-throughput technologies, and genome sequencing initiatives are too big for conventional analytical techniques to handle. In addition to managing and storing these enormous datasets, bioinformatics plays a crucial role in identifying biological markers, identifying meaningful patterns, and revealing the functional links that support the complexity of living things. Furthermore, bioinformatics is essential to precision medicine, which customises treatment plans based on each patient's unique genetic profile. Bioinformatics applications include understanding the genetic basis of disease(Guo, Q. (2023)), finding biomarkers, and forecasting treatment responses. Bioinformatics' capabilities are further enhanced by its junction with machine learning, which makes it possible to create predictive models, pattern recognition algorithms, and advanced analytics that go beyond the scope of conventional statistical methods.

As we set off on the journey covered in this chapter, the breadth includes a thorough examination of the principles of bioinformatics. We will go into structural biology, predicting protein structures and using structural databases for insights(Helmick, H., Rodriguez, N., & Kokini, J. L. (2023)), as well as sequence analysis, delving into the nuances of DNA and RNA sequencing. This chapter will introduce

Figure 1. Scope of bioinformatics

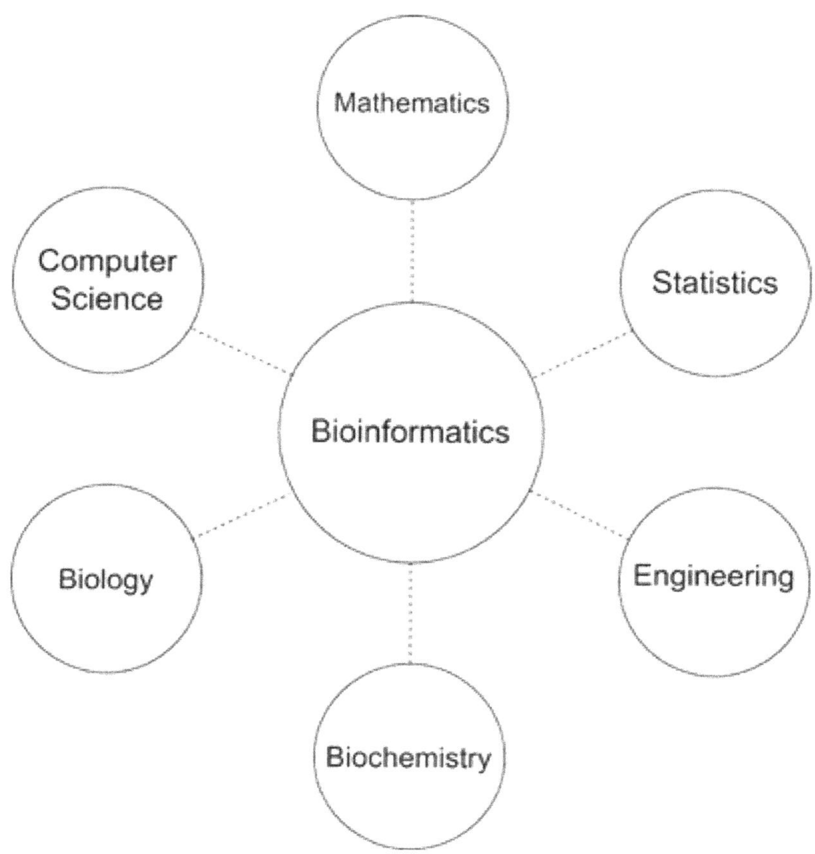

the field of data mining in bioinformatics, covering techniques such as clustering, classification, and association rule mining that are crucial for extracting valuable information from large and intricate datasets(Mey, F., Clauwaert, J., Van Huffel, K., Waegeman, W., & De Mey, M. (2021)). The meeting point of bioinformatics and machine learning will be explored, revealing the ways in which these two dynamic domains work together to transform the processing of biological data. We will examine the nuances of feature engineering and selection, the use of regression and classification models, and the development of deep learning techniques in the interpretation of intricate biological events. We'll talk openly about the benefits and challenges this convergence presents, giving readers a comprehensive picture of the scene(Tang, Y. J., & Aristilde, L. (2020)).

In conclusion, this chapter lays the groundwork for a thorough investigation into bioinformatics, encompassing its foundational ideas and revolutionary uses in contemporary biology and the vast scope of Bioinformatics as shown in figure 1. The integration of biological knowledge with computer methods becomes increasingly important as we navigate the complex network of biological data, leading the field of bioinformatics into a new era of innovation and discovery.

2. BIOINFORMATICS FUNDAMENTALS: DECIPHERING THE MOLECULAR CODE

Through the examination of "Bioinformatics Fundamentals," we travel through the complex fields of data mining, structural biology, and sequence analysis. These principles form the cornerstone of bioinformatics, allowing scientists to unravel the molecular makeup of life. This part functions as a thorough guide, explaining the intricacies of protein structure prediction, DNA and RNA sequencing, and data mining approaches(Riza, L. S., Zain, M. I., Izzuddin, A., Prasetyo, Y., Hidayat, T., & Abu Samah, K. A. F. (2023)).

2.1 Sequence Analysis: Unraveling the Genetic Tapestry

The delicate art of sequence analysis, which is similar to deciphering the genetic code encoded in DNA and RNA, is at the core of bioinformatics. A fundamental aspect of genomics is DNA sequencing, which reveals the nucleotide sequence of a DNA molecule and offers a blueprint for an organism's genetic composition. Thanks to its insights into inherited features, illness susceptibility, and possible treatment targets, this discovery has revolutionised sectors such as genetics and personalised medicine. Simultaneously, RNA sequencing explores the dynamic field of transcriptomics, revealing the variety and quantity of RNA molecules present in a cell. Understanding gene expression patterns, alternative splicing processes, and the intricate workings of biological responses has become greatly aided by this technology.

2.2 Structural Biology: Bridging the Molecular Gulf

Structural biology explores the three-dimensional details of biomolecules, especially proteins, going beyond the simple nucleotide code. One of bioinformatics' greatest challenges is predicting protein structures, which has significant effects on medication development and our comprehension of cellular processes(Muzyukina, P., & Soutourina, O. (2023)). Researchers navigate the wide conformational space to predict how a protein folds into its functional structure by using computational techniques. In addition

Figure 2. Process of data mining

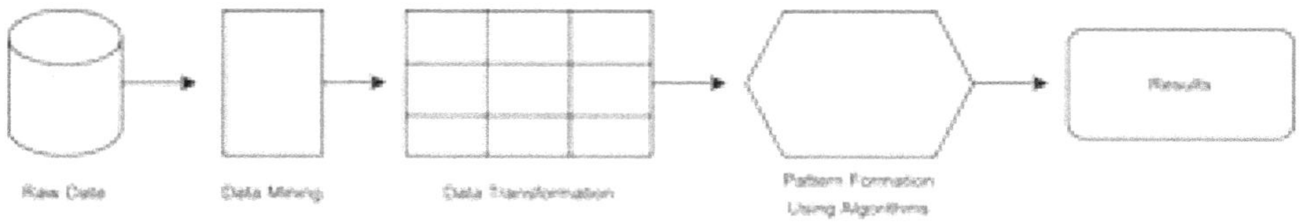

to being a matter of great intellectual interest, structural determination experiments can be guided by this topic, which also provides a window into the molecular machinery that controls life's operations. Structural databases also contain a multitude of data; they are a storehouse for structures that have been found through experimentation and a useful tool for comparing and confirming computer forecasts.

2.3 Data Mining in Bioinformatics: Extracting Knowledge From Complexity

In the age of massive biological data, it is critical to extract useful information efficiently. As the metaphorical compass, data mining techniques in bioinformatics help researchers navigate the wide fields of proteomics, genomics, and other related fields(Kuang, Z., Zhao, Y., & Yang, X. (2023)). Figure 2 Shows the process of Data Mining, How raw data converted into useful information. Algorithms that use clustering to group biological elements together reveal hidden patterns and relationships within datasets. In contrast, classification models are predictive tools that use learned patterns to place biological items into predetermined groups. As this is going on, association rule mining reveals complicated relationships between many biological variables, providing insight into the intricate interactions that occur within cellular systems. Together, these methods enable researchers to extract insights from complexity, facilitating the discovery of biomarkers, comprehension of disease processes, and deciphering the intricate workings of biological networks.

The convergence of biological understanding and computational approaches becomes clear as we work through these bioinformatics principles. Data, structures, and sequences together constitute the molecular symphony of life; they are not discrete entities(Patra, P., B.R., D., Kundu, P., Das, M., & Ghosh, A. (2023)). These foundations have applications in a wide range of sectors, including environmental science, agriculture, and medical, in addition to research laboratories. We will examine in the following sections of this chapter how these foundations pave the way for the use of machine learning techniques, bringing in a new phase of research and development at the nexus of computational biology and bioinformatics.

3. INTERSECTION WITH MACHINE LEARNING: UNVEILING THE SYNERGIES

The symbiotic relationship between bioinformatics and machine learning opens up new possibilities in the field, revolutionising biological data analysis and interpretation. The smooth merging of bioinformatics and machine learning is examined in this part, where computer models and algorithms help us better understand the intricacies of life.

3.1 Machine Learning in Bioinformatics: Elevating Analytical Capacities

In the field of bioinformatics, machine learning (ML) has become a disruptive force that is changing the way scientists evaluate and understand intricate biological data. Machine learning algorithms are particularly good at finding patterns, predicting outcomes, and deriving valuable insights from large datasets in the fields of proteomics, genomics, and other related fields. Here, we examine the main uses and contributions of machine learning in bioinformatics, delving into its enormous influence.

1. Data Analysis and Pattern Recognition:
   - Genomic Sequencing: Machine learning algorithms examine DNA and RNA sequences to identify patterns linked to functional, regulatory, and genetic variants(Anderson, B., Ouyang, D., D'Agostino, A., Bonin, B., Smith, E., Kraushaar, V., Rudman, S. L., & Ho, D. E. (2023)).
   - Proteomic Data: ML models help with protein identification, quantification, and post-translational modification analysis by identifying patterns in mass spectrometry data(Park, J., Wilkins, C., Avtonomov, D., Hong, J., Back, S., Kim, H., Shulman, N., MacLean, B. X., Lee, S.-W., & Kim, S. (2023)).
2. Feature Selection and Engineering:
   - To improve model performance and interpretability, machine learning (ML) techniques refine the selection of pertinent features (genetic or molecular properties) from high-dimensional datasets.
   - Feature engineering is the process of adding new features or changing already-existing ones in order to find hidden relationships and increase a model's capacity for prediction.
3. Classification and Regression Models:
   - Illness Prediction: Using genetic or clinical data, ML models categorise people into illness groups to enable early diagnosis and individualised treatment.
   - Regression models aid in the selection of specialised therapeutic interventions by forecasting patient reactions to particular medications.
4. Deep Learning Approaches:
   - Neural Networks: Convolutional neural networks (CNNs) and recurrent neural networks (RNNs), two examples of deep learning architectures, perform exceptionally well in tasks like image analysis and sequence data processing(Dou, H., Shen, F., Zhao, J., & Mu, X. (2023)).
   - Omics Data Integration: In order to provide comprehensive insights, deep learning models use multi-omics data and capture the complex interactions between transcriptomics, proteomics, and genomes(Zheng, X., Wang, M., Huang, K., & Zhu, E. (2024)).
5. Challenges and Opportunities:
   - Data Interpretability: Although ML models are quite good at predicting outcomes, it can be difficult to understand their choices in a biological context. This calls for the creation of explainable AI techniques.
   - Integration with Biological Knowledge: Linking machine learning predictions to known biological concepts guarantees the data' biological significance and facilitates the formulation of hypotheses.

6. Future Directions:
   - Multi-Modal Data Integration: In order to conduct thorough studies, advances in machine learning will concentrate on integrating heterogeneous datasets and merging genomes, imaging, and clinical data.
   - Transfer Learning: By fine-tuning ML models trained on a single biological dataset for related tasks, one can increase efficiency and decrease the requirement for large amounts of labelled data.

Bioinformatics machine learning is at the vanguard of data-driven biological research, providing unmatched chances for creativity and discovery. The symbiotic relationship between machine learning and bioinformatics will continue to reveal the secrets of life stored in molecular data as technology develops and interdisciplinary collaborations deepen.

3.2 Feature Selection and Engineering: Refining Biological Insights

Within the complex field of bioinformatics, feature engineering and selection are essential steps in the larger machine learning process. By improving the relevance and interpretability of data, these methods significantly contribute to the refinement of biological insights. This is especially true when dealing with high-dimensional biological datasets such as proteomics and genomics.

3.2.1 Feature Selection: Navigating High-Dimensional Spaces

Numerous features, or variables, are frequently included in high-dimensional datasets, as those produced in proteomics or genomics research, and they might not all contribute equally to the underlying biological processes. In feature selection, a subset of the most informative features is systematically chosen, and redundant or unnecessary features are eliminated. This is aided by machine learning methods, which can be as basic as statistical testing or as complex as recursive feature elimination(Smith, C. M., Weathers, A. L., & Lewis, S. L. (2023)). Feature selection reduces the dimensionality of the data, which improves model performance and reveals the essential chemical elements underlying biological processes.

3.2.2 Feature Engineering: Crafting Insights from Data

In order to extract more meaningful information from the data, feature engineering entails generating new features or modifying existing ones in addition to picking only the most relevant ones. This could involve creating new molecular descriptors, combining features to depict biological pathways, or modifying data to identify non-linear interactions in the field of bioinformatics. In order to find hidden patterns and relationships in the data that may be essential for comprehending intricate biological processes, feature engineering is a skill that demands a thorough understanding of the biological context.

3.2.3 Applications in Genomics: Unmasking Genetic Signatures

The fields of genomics and feature engineering are vital because of the large datasets generated by the investigation of DNA and RNA sequences. In the field of cancer genomics, for example, feature selection is used to pick significant genetic markers, which in turn helps find prospective biomarkers for disease diagnosis and prognosis. Feature engineering encompasses the development of composite features that represent gene interactions or the integration of genomic rearrangement data. By using these methods, the genomic landscape is fine-tuned, revealing minute differences and patterns that may be suggestive of certain biological circumstances(Salehin, I., Islam, Md. S., Saha, P., Noman, S. M., Tuni, A., Hasan, Md. M., & Baten, Md. A. (2023)).

3.2.4 Applications in Proteomics: Enhancing Protein Insights

The study of proteins and their functions, known as proteomics, greatly benefits from feature engineering and selection. These methods become essential in mass spectrometry-based proteomics, where protein identification and quantification produce complex datasets(Reilly, L., Lara, E., Ramos, D., Li, Z., Pantazis, C. B., Stadler, J., Santiana, M., Roberts, J., Faghri, F., Hao, Y., Nalls, M. A., Narayan, P., Liu, Y., Singleton, A. B., Cookson, M. R., Ward, M. E., & Qi, Y. A. (2023)). In order to direct further investigations, feature selection aids in identifying the most discriminative peptides or protein fragments. Feature engineering could entail combining data from several peptides to more precisely describe protein abundance or developing new features based on post-translational changes.

3.2.5 Challenges and Considerations

Overfitting: Overfitting occurs when a model performs well on training data but has difficulty generalising to new, unseen data. This can be caused by aggressive feature selection or engineering.

Biological Interpretability: Since the ultimate objective is to produce insights that advance our comprehension of biological processes, it is imperative to make sure that chosen or created features are in line with biological knowledge.

To sum up, in the fields of bioinformatics and machine learning, feature selection and engineering serve as compass points that lead researchers over the wide stretches of high-dimensional biological data. These methods provide new features that capture complex relationships and refine datasets to only include the most pertinent information, which paves the way for more accurate models and ultimately deeper biological insights.

3.3 Classification and Regression Models: Decoding Biological Phenomena

In the field of bioinformatics, classification and regression models—essential elements of the machine learning toolbox—are essential for deciphering intricate biological events. These models are useful tools in the fields of genomics, proteomics, and systems biology because they help predict and comprehend complex connections between biological variables.

3.3.1 Classification Models: Categorizing Biological Entities

Classification models are highly effective in bioinformatics for classifying biological items into predetermined classes or groupings. These models, for instance, are able to differentiate between states of health and illness based on proteome or genetic profiles. In the field of diagnostics, the capacity to categorise samples into discrete groups is very useful as it facilitates the identification of putative disease markers and the creation of patient outcome prediction models(Lin, L., Liu, T., Zhang, H., Xiong, N., Leng, J., Wei, L., & Liu, Q. (2023)). Among the algorithms frequently used for classification tasks are Support Vector Machines (SVMs), Random Forests, and Neural Networks; each has a distinct advantage based on the intricacy of the biological problem at hand.

3.3.2 Regression Models: Predicting Numerical Outcomes

Conversely, regression models are highly effective at forecasting numerical results, which makes them indispensable for understanding quantitative elements of biological processes. Regression models in genomics can forecast levels of gene expression or protein abundance depending on a number of variables, offering insights into the regulatory processes that underlie biological phenomena. These models help researchers comprehend the quantitative effects of environmental or genetic factors on biological systems by helping to unravel the complex interactions between molecular variables. Regression problems in bioinformatics are commonly solved using ensemble techniques such as Gradient Boosting and Ridge regression in addition to linear regression(Fernández-Durán, J. J., & Gregorio-Domínguez, M. M. (2024)).

3.3.3 Applications and Implications: Precision in Biological Understanding

Bioinformatics uses regression and classification models in a variety of contexts, including as disease diagnosis, treatment response prediction, and biomarker discovery. Through the use of these models, scientists are able to find patterns in high-dimensional information and identify minute signals that may have biological significance. These models' precision helps to develop hypotheses and experimental designs, leading to a more sophisticated knowledge of biological events. The interplay between classification and regression models is a keystone of bioinformatics' ongoing evolution, allowing researchers to decipher the intricate workings of living systems and successfully negotiate the complexities of biological data.

3.4 Deep Learning Approaches: Navigating Complex Biological Landscape

A type of machine learning called deep learning has been a game-changer in the field of bioinformatics because of its unmatched ability to navigate the complex and diverse biological terrain. Deep learning techniques use multi-layered artificial neural networks, which are modelled after the structure of the human brain. This allows for the extraction of complex patterns and representations from large and detailed biological information.

3.4.1 Convolutional Neural Networks (CNNs): Decoding Visual Data

The use of Convolutional Neural Networks (CNNs) to decode visual data is one of the most notable uses of deep learning in bioinformatics, especially in areas like structural biology and medical imaging. CNNs are very good in image recognition, which enables the recognition of patterns in biological pictures such as protein structures or microscopy slides. CNNs can be trained to examine spatial patterns in DNA sequences, providing information about genome architecture, gene expression, and regulatory areas in genomics(Yuan, F., Zhang, Z., & Fang, Z. (2023)).

3.4.2 Recurrent Neural Networks (RNNs) and Long Short-Term Memory Networks (LSTMs): Modeling Sequential Data

In genomics, where information is frequently given in a sequential fashion, such as time-series gene expression data, deep learning extends its capabilities to the modelling of sequential data, an important area of study. Long Short-Term Memory Networks (LSTMs) and Recurrent Neural Networks (RNNs) are powerful tools for extracting dependencies and temporal correlations from sequential data(Perez Cortes, V. M., Chatterjee, A., & Khovalyg, D. (2024)). Because of this, they are very useful for tasks like identifying dynamic interactions within biological systems or forecasting patterns of gene expression across time.

3.4.3 Applications in Drug Discovery and Multi-Omics Integration

Deep learning techniques are being used in drug development, where the ability of deep models to identify intricate correlations is advantageous for both designing new compounds and predicting drug-target interactions. Furthermore, deep learning architectures make the integration of multi-omics data—the simultaneous analysis of transcriptomics, proteomics, genomes, and other layers—more practical. These models have the ability to automatically learn hierarchical representations of data from various omics datasets, providing thorough insights into the molecular causes of various diseases and their reactions to treatment.

Although deep learning has tremendous potential in bioinformatics, there are a number of obstacles to overcome, including interpretability issues, the requirement for big labelled datasets, and ethical issues with healthcare applications. These factors make a careful and deliberate approach crucial. As the field develops, the incorporation of deep learning techniques into bioinformatics holds the potential to reveal previously undiscovered levels of biological system complexity. This could lead to revolutionary discoveries and applications in personalised medicine, drug development, and our overall comprehension of life at the molecular level.

3.5 Challenges and Opportunities: Navigating the Frontier

The combination of machine learning and bioinformatics creates previously unheard-of opportunities, but it is not without difficulties. Obstacles that require careful consideration include the intrinsic complexity of biological systems, the requirement for robust validation, and the need for interpretability in models. A careful approach is even more important in light of ethical considerations, data protection, and the correct application of machine learning in healthcare. Opportunities are there, though. There is much to be discovered in the fields of integrating multi-omics data, improving predictive models, and investigating unsupervised learning for new insights. As the discipline develops, it becomes increasingly important for biologists, data scientists, and machine learning specialists to work with transdisciplinary to ensure that advances are both technologically advanced and based on biological understanding.

The convergence of machine learning and bioinformatics represents the ideal combination of computation and biology. This combination not only strengthens our analytical abilities but also changes the way we perceive and investigate the complexities of life. This is the beginning of a paradigm shift in the field of bioinformatics and computational biology. The next sections of this chapter will go into practical applications, demonstrating how these machine learning paradigms are translating into groundbreaking discoveries in drug discovery, proteomics, and genomics.

4. REAL-WORLD APPLICATIONS: BRIDGING THE GAP BETWEEN DATA AND DISCOVERY

The ultimate litmus test in the symbiotic dance of bioinformatics and machine learning is how well computational discoveries are translated into practical applications. This section highlights the significant influence of this dynamic partnership in the fields of drug development, proteomics, genomics, and personalised medicine. We demonstrate the revolutionary potential of these applications in transforming the field of biology and medicine through in-depth investigation and compelling case examples.

4.1 Genomics and Genetic Variation: Decoding the Blueprint of Life

At the nexus of molecular biology and bioinformatics, genomics is poised to become a groundbreaking tool in deciphering the complex life story recorded in our DNA. Within the roughly 3 billion nucleotide base pairs that make up the human genome are the instructions needed to create and sustain an organism(Tran, D. M., Nguyen, T. H., & Nguyen, A. D. (2023)). Genomic analysis and interpretation of this massive amount of genomic data, within the bioinformatics environment, aims to unlock the secrets of life's blueprint and comprehend the genetic variants that underlie population diversity and play a role in human health and disease.

4.1.1 Sequencing the Genome: From Nucleotides to Insights

The expedition starts with DNA sequencing, a novel tool that enables scientists to ascertain the nucleotide sequence—adenine (A), thymine (T), cytosine (C), and guanine (G)—that comprises an organism's DNA. Next-generation sequencing (NGS) and other sequencing technology advancements have revolutionised genomics by making it possible to analyse entire genomes at low cost and with high throughput(Singh, S., Sarma, D. K., Verma, V., Nagpal, R., & Kumar, M. (2023)). These technologies have played a key role in understanding the genomes of many different creatures, including humans, and have opened up a wealth of information on their evolutionary history, genetic diversity, and functional components.

4.1.2 Identifying Genetic Variations: Unraveling the Tapestry

Variations at the individual and population levels can be found within the genome sequence, providing the foundation for diversity and adding to the mosaic of life. The various genetic changes that mould the genetic landscape include structural differences, insertions, deletions, and single nucleotide polymorphisms (SNPs). The identification and characterization of these variants, as well as their possible correlations with illnesses and functional implications, are greatly aided by bioinformatics. By comparing the genomes of individuals or populations, genomic studies use statistical techniques and computational algorithms to analyse large datasets and identify genetic variants linked to certain traits or disorders.

4.1.3 Predicting Phenotypes: From Genotype to Function

Knowing genetic variations involves more than just identifying them; it also involves forecasting the phenotypic effects of those changes. In order to identify patterns in genomic data and connect genetic variants to observable features or disease predispositions, machine learning methods are utilised. Researchers can deduce the possible effects of particular genetic variants on cellular functions, protein function, or general health thanks to predictive modelling(Anandan, J., Dwivedi, D. P., & Govindaraj, V. (2023)). Classification models, for instance, can be trained to forecast the likelihood that a certain genetic mutation would result in a particular disease or affect how a medicine will work. Predictions like these direct experiments, rank variations according to their functional validity, and advance precision medicine as a whole.

4.1.4 Applications in Disease Research and Personalized Medicine: A Glimpse into the Future

The field of genomics, driven by bioinformatics and machine learning, holds significant potential for personalised treatment and disease research. Finding genetic variants linked to an illness makes it easier to comprehend the underlying molecular pathways, which in turn opens up new possibilities for the creation of drugs and targeted therapies. Furthermore, genetic data plays a critical role in treatment decision-making, patient group stratification, and drug response prediction. The era of personalised medicine is emerging, wherein treatments are customised based on a person's genetic composition, maximising benefits and reducing side effects.

4.2 Proteomics and Protein Function Prediction: Unraveling the Proteomic Tapestry

Entering the field of proteomics at the nexus of bioinformatics, we look at the complex and dynamic world of proteins—the molecular builders that orchestrate life's intricate details. Within the field of bioinformatics, proteomics aims to decipher the global patterns of protein interactions, functions, and structures by analysing the proteome tapestry. The knowledge that the proteome—the entire collection of proteins—executes the instructions and expresses the phenotypic diversity seen in living creatures, while the genome serves as the blueprint, propels this discipline.

4.2.1 Deciphering Protein Structures: A Computational Odyssey

The goal of determining the three-dimensional structures of proteins is at the core of proteomics. These structures, which resemble the folds and twists of a molecular origami, control how proteins interact and perform specific tasks in cells. Protein structure prediction is a challenging task because of the large conformational space. Computational methods and bioinformatics methodologies are used to facilitate this process. In this endeavour, machine learning in particular has shown to be a powerful tool, using training data from protein structures that have been discovered experimentally to predict the shapes of unknown proteins. This ability to predict outcomes not only helps to comprehend the basic biology of proteins but also has great potential for drug development, as protein architectures inform the creation of medicines that target certain molecular weaknesses.

4.2.2 Functional Annotation: Illuminating the Biological Significance

Beyond the analysis of their structures, proteins are fully understood in terms of their roles in biological activities. This is where the prediction of protein function becomes important. Sophisticated computational techniques are necessary because to the immense diversity of proteins, which includes structural proteins that sustain cells and enzymes that catalyse biological activities. Protein functions that are not yet known are predicted by machine learning algorithms that are trained on datasets that contain experimentally verified protein functions. These predictions are extremely helpful in explaining how proteins act in biological pathways, providing understanding of the mechanisms underlying disease, and guiding the development of experimental plans intended to decipher the intricate workings of cells.

4.2.3 Applications in Drug Discovery and Systems Biology: Bridging the Bench to Bedside Gap

Proteomics and protein function prediction applications have a significant impact on drug development and systems biology, among other disciplines. Understanding protein structures and functions serves as a compass for drug development, helping scientists create compounds with the best binding affinities and find possible therapeutic targets. Furthermore, systems biology's comprehensive grasp of cellular networks is aided by the prediction of protein functions. It makes it easier to identify important participants in metabolic pathways, regulatory networks, and signalling cascades, illuminating the connections between various biological processes.

Proteomics is a key field in the rapidly developing field of bioinformatics that helps to solve the puzzles of the proteome tapestry. The promise of proteomics lies not only in deepening our understanding of basic biology but also in spurring ground-breaking discoveries with far-reaching consequences, ranging from personalised medicine to novel drug therapies, as we navigate this dynamic field by utilising computational methodologies and machine learning. Proteomics is a guiding thread in the tapestry of life, connecting the complex stories of molecular dynamics and cellular function.

4.3 Drug Discovery and Repurposing: Accelerating Therapeutic Innovation

The combination of machine learning and computational approaches has revolutionised drug discovery and repurposing in the field of bioinformatics, driving unprecedented levels of therapeutic innovation. Previously a lengthy and resource-intensive process, drug discovery today benefits from the accuracy and efficiency provided by bioinformatics technologies.

4.3.1 Virtual Screening and Drug-Target Prediction: Navigating the Chemical Space

A key component of bioinformatics-driven drug development is virtual screening, which uses computational techniques to sort through enormous libraries of chemical compounds and forecast how they will interact with biological targets. This process is accelerated by machine learning models that anticipate possible interactions between medications and certain proteins. These models are trained on databases of known drug-target interactions. This streamlines the early phases of drug development by hastening the identification of prospective drug candidates. Drug-target prediction models also help in the logical design of treatments with improved efficacy and fewer adverse effects by providing insights into the intricate relationships between pharmaceuticals and their intended targets.

4.3.2 Drug Repurposing: Uncovering Hidden Potentials

Bioinformatics is essential for drug repurposing, which is the process of finding new therapeutic applications for already-approved medications. This goes beyond the conventional paradigm of drug discovery. Researchers find previously undiscovered links between medications and illnesses by using machine learning algorithms and mining large databases of data pertaining to pharmaceuticals. This method reduces the risks involved in creating brand-new medications while simultaneously hastening the identification of possible treatments. The process of translating discoveries from bench to bedside is sped up by repurposing existing medications for new indications, which capitalises on their established safety profiles.

4.3.3 Case Studies in Drug Discovery and Repurposing: A Glimpse Into Success

1. **Virtual Screening for Antiviral Agents:**
   - Scenario: A new viral outbreak prompts the urgent need for antiviral drugs.
   - Application: Virtual screening employs machine learning models to identify existing compounds with potential antiviral activity.
   - Outcome: The rapid identification of drug candidates accelerates the development of treatments for the emerging viral threat.
2. **Drug Repurposing for Neurological Disorders:**
   - Scenario: Confronting the challenges of developing novel therapeutics for neurological conditions.
   - Application: Bioinformatics-driven approaches reveal existing drugs with potential efficacy for neurological disorders.
   - Outcome: Repurposed drugs show promise in preclinical studies, providing a shortcut to clinical trials and potential new treatments.

In conclusion, bioinformatics is revolutionising the conventional drug development environment by acting as a catalyst for drug discovery and repurposing. The combination of machine learning and computational approaches accelerates the discovery of promising treatments, leading to a paradigm change in precision medicine and innovative medicines. Bioinformatics emerges as a guiding light in the search for breakthrough medicines, speeding up the rate of discovery and giving patients in need new hope.

4.4 Personalized Medicine: Tailoring Treatments to Individual Variability

One-size-fits-all approaches to medical care are being replaced by personalised medicine, a novel paradigm at the nexus of bioinformatics and healthcare.[23] It recognises that each person is unique and that variations in genetics, chemistry, and health can have a significant impact on how well a treatment works. With its ability to analyse large amounts of data, bioinformatics supports personalised medicine by combining predictive modelling with individualised data.

The foundation is genomic data, which is obtained using cutting-edge sequencing technology. Large genomic databases are analysed by bioinformatics algorithms, which find genetic differences that affect treatment outcomes, prognosis, and disease susceptibility. Trained on a variety of datasets, machine learning models anticipate unique patient outcomes according to genetic profiles, allowing physicians to customise interventions.

Personalised medicine is further enhanced by the incorporation of electronic health records (EHRs), which incorporate clinical data to provide a holistic picture of patient health. Bioinformatics-guided personalised medicine maximises therapeutic efficacy, minimises side effects, and optimises medication selection by taking both genetic and clinical aspects into account. With medicines that are not only prescribed but also precisely tailored to each person's unique genetic composition, this innovative technique heralds a new age in healthcare that promises more accurate, efficient, and patient-centered medical interventions.

4.5 Case Studies: Illustrating Real-World Impact

1. Genomics and Cancer Therapy:
   a. Scenario: A patient with cancer is given a genetic profiling diagnosis.
   b. Use: By analyzing the genomic data, machine learning models forecast how the patient would react to various chemotherapy treatments.
   c. Outcome: By reducing side effects and maximizing therapeutic efficacy, the oncologist chooses a customized treatment strategy depending on the anticipated medication sensitivity.
2. Proteomics in Drug Development:
   a. Scenario: The goal of research is to find new medication targets for a complicated illness.
   b. Use: By analyzing proteomic data, machine learning algorithms make predictions about the interactions and functions of unidentified proteins.
   c. The result is the prioritization of possible drug targets, which speeds up the drug discovery process and opens up new options for therapeutic intervention.
3. Drug Repurposing for Neurological Disorders:
   a. Situation: Looking for cutting-edge remedies for a neurological illness.
   b. Use: Machine learning algorithms search through current drug databases to find compounds that may be effective since they share mechanisms with well-established treatments.
   c. Outcome: Drugs that were repurposed but were initially intended for different indications perform well in preclinical models, hastening the transition to clinical trials.
4. Personalized Medicine in Cardiology:
   a. Hypothesis: Individuals with heart-related disorders go through genetic testing and have their medical histories examined.
   b. Use: Using clinical and genetic data, machine learning algorithms forecast each person's unique reaction to anticoagulant drugs.
   c. Outcome: Individualized treatment regimens minimize the possibility of negative reactions and improve the handling of cardiovascular conditions.
5. Multi-Omics Integration for Disease Subtyping:
   a. Scenario: In order to provide specialized treatments, researchers want to categorize individuals with a heterogeneous disease into subtypes.

Table 1. Real-world applications of bioinformatics and machine learning

Application Area	Key Techniques	Illustrative Case Studies
Genomics and Genetic Variation	- SNP Prediction - Genomic Profiling	- Predicting Drug Responses in Cancer Patients
Proteomics and Protein Function Prediction	- SVMs, Deep Learning - Functional Annotation	- Identifying Novel Drug Targets Through Protein Function Prediction
Drug Discovery and Repurposing	- Virtual Screening - Drug-Target Prediction	- Repurposing Existing Drugs for Neurological Disorders
Personalized Medicine	- Predictive Modeling - Integration of EHRs	- Tailoring Anticoagulant Therapies for Individual Patients in Cardiology
Multi-Omics Integration	- Integrated Analysis of Genomics, Transcriptomics, and Proteomics Data	- Disease Subtyping for Precision Medicine Strategies

b. Use: To find unique molecular subtypes, machine learning combines data from transcriptomics, proteomics, and genomes.
c. Result: Each subtype is given a precision medicine strategy, opening the door to more individualized and successful therapeutic interventions.

In conclusion, these case studies demonstrate the real-world effects of machine learning and bioinformatics on several facets of healthcare and biomedical research. Applications such as customizing cancer medicines, forecasting medication reactions, and repurposing current treatments highlight how revolutionary this dynamic synergy may be.

These illustrations, along with the Table 1 that goes with them, show the range and complexity of practical uses and how the combined forces of bioinformatics and machine learning are changing patient care and scientific research.

5. CHALLENGES AND FUTURE DIRECTIONS IN BIOINFORMATICS: NAVIGATING THE FRONTIER OF DISCOVERY

Although bioinformatics is transforming our knowledge of biological systems, it also presents a number of difficulties and is at the cusp of exciting new opportunities. Unlocking new frontiers in biomedical research and customized therapy requires tackling these issues as biological data grows more complicated and technology advances.

5.1. Data Integration and Interoperability

Challenge: Biological data is heterogeneous and comes from a variety of sources and experimental platforms, making it difficult to integrate and interoperate seamlessly. Effective data sharing and cooperation are hampered by disparate data formats, standards, and nomenclatures.

Future Direction: Ontologies, standardized data formats, and interoperable systems are being developed. Researchers may more successfully integrate and analyze heterogeneous datasets because to the adoption of FAIR (Findable, Accessible, Interoperable, and Reusable) principles, which are encouraging a culture of data sharing and collaboration.

5.2. Interpretability of Machine Learning Models

Challenge: There are questions regarding the interpretability of certain machine learning models in bioinformatics due to their black-box nature. It is important to comprehend the reasoning behind forecasts, particularly in biological applications where useful information is essential.

Future Direction: Improving the interpretability of machine learning models is the goal of the development of explainable AI (XAI) techniques. Models are made more comprehensible and reliable by utilizing model-agnostic techniques, integrating domain expertise, and adding transparency features to algorithms.

5.3. Privacy and Ethical Concerns

Challenge: As bioinformatics explores personal genetics and health data, consent, data ownership, and privacy become more complicated ethical issues. Finding a balance between personal privacy and data utility is a constant struggle.

Future Direction: It is imperative to create strong ethical frameworks, open data-sharing guidelines, and safe data anonymization methods. Sensitive biological information can be handled responsibly and ethically if researchers, institutions, and legislators work together.

5.4. Handling Big Data and Computational Resources

Challenge: High-throughput technologies are driving the extraordinary rise of biological data, which creates storage, processing, and analytical issues. In bioinformatics, computational resources need to grow to satisfy the needs of big data analytics.

Future Direction: Scalable solutions are provided by ongoing developments in distributed computing systems, cloud computing, and parallel processing. Overcoming computational issues is facilitated by collaborative platforms that allow for resource sharing and access to high-performance computing facilities.

5.5. Translational Gap between Research and Clinical Applications

Challenge: There is still much work to be done to close the gap between bioinformatics research and therapeutic applications. Overcoming obstacles related to clinical validation and regulatory complexity is necessary to translate findings from experimental settings into practical interventions in the field of medicine.

Future Direction: It is imperative that bioinformaticians, physicians, and regulatory agencies work together more closely. The seamless integration of research findings into clinical practice is made possible by initiatives such as the development of real-world evidence (RWE) and the heightened focus on clinical bioinformatics education.

Figure 3. Key insights

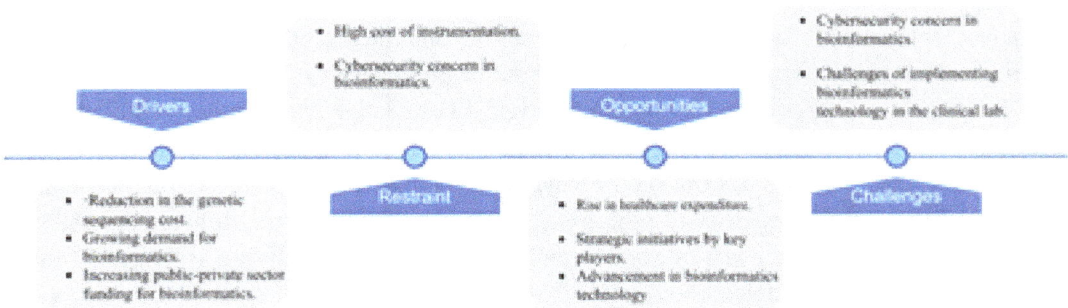

5.6. Progressing Personalized Health Care

Challenge: Overcoming obstacles pertaining to patient classification, data heterogeneity, and the integration of multi-omics data for targeted treatment methods are necessary to fully realize the promise of personalized medicine.

Future Direction: A more comprehensive picture of an individual's health is made possible by developments in wearable technology, single-cell omics, and the integration of patient-reported outcomes. Including multi-omics data and improving prediction models will help the field move closer to customizing care according to each patient's own biological profile.

In conclusion, a wave of creativity and teamwork are used to meet the problems in bioinformatics. By tackling these issues, we not only improve our analytical techniques but also move bioinformatics toward a day when the tiny intricacies of biological systems are precisely uncovered, enabling revolutionary breakthroughs in biology and medicine.

6. CONCLUSION

Our journey has revealed the many levels of biological complexity within the enormous field of bioinformatics, which is bound together by scientific research and computing skill. we covered the key insights as depicted in figure 3. Starting from the basic knowledge of bioinformatics as the keystone that links unprocessed biological data to useful insights, we explored the domains of proteomics, genomics, and data mining to shed light on the dynamic interactions between molecular information. The combination of deep learning and machine learning proved to be a game-changer, enabling us to explore the future, improve features, and identify trends. Practical uses transformed the ideas into significant realities, directing medication development, forecasting the roles of proteins, and opening the door for customized healthcare. But this journey was not without its difficulties, which gave us the chance to innovate and improve our processes. Complexities in data integration, the interpretability of machine learning models, moral conundrums, and the computing needs of big data all encourage us to go beyond the bounds of what is now considered acceptable. The future of bioinformatics beckons with the promise of improved collaboration, standardized procedures, and responsible innovation as we traverse these obstacles. The complex and dynamic field of bioinformatics encourages both practitioners and researchers to start their own knowledge journeys.

To sum up, this journey through bioinformatics goes beyond the lines that separate scientific research from technical advancement. It challenges us to solve the enigmas surrounding life, explore the intricacies of cellular functions, and imagine a time when biological knowledge is not only dynamic but also acts as a driving force behind revolutionary developments in the fields of healthcare and other fields. As the journey continues, the changing field of bioinformatics is a monument to the relentless human spirit of discovery, where the coming together of biology, computation, and data not only shapes the stories of our past but also moves us closer to a future in which the complexities of living systems are understood with great insight, purpose, and accuracy.

REFERENCES

Anandan, J., Dwivedi, D. P., & Govindaraj, V. (2023). Clinical phenotypes of COPD and their impact on quality of life: A cross-sectional study. In *Respiratory Medicine* (Vol. 220, p. 107452). Elsevier BV., doi:10.1016/j.rmed.2023.107452

Anderson, B., Ouyang, D., D'Agostino, A., Bonin, B., Smith, E., Kraushaar, V., Rudman, S. L., & Ho, D. E. (2023). *Leveraging Genomic Sequencing Data to Evaluate Disease Surveillance Strategies. In iScience*. Elsevier BV. doi:10.1016/j.isci.2023.108488

Dou, H., Shen, F., Zhao, J., & Mu, X. (2023). Understanding neural network through neuron level visualization. In *Neural Networks* (Vol. 168, pp. 484–495). Elsevier BV. doi:10.1016/j.neunet.2023.09.030

Fernández-Durán, J. J., & Gregorio-Domínguez, M. M. (2024). Regression models for circular data based on nonnegative trigonometric sums. In *Journal of Statistical Planning and Inference* (Vol. 230, p. 106114). Elsevier BV. doi:10.1016/j.jspi.2023.106114

Guo, Q. (2023). Bioinformatics analysis of the diversity of gut microbiota and different microbiota on insulin resistance in diabetes mellitus patients. In *Heliyon* (p. e22117). Elsevier BV. doi:10.1016/j.heliyon.2023.e22117

Helmick, H., Rodriguez, N., & Kokini, J. L. (2023). Utilization of creep ringing and bioinformatic modelling in study of cold denatured pea protein emulsions. In Innovative Food Science & Emerging Technologies (Vol. 88, p. 103420). Elsevier BV. https://doi.org/ doi:10.1016/j.ifset.2023.103420

Kuang, Z., Zhao, Y., & Yang, X. (2023). Machine learning approaches for plant miRNA prediction: Challenges, advancements, and future directions. In Agriculture Communications, 1(2). Elsevier BV. doi:10.1016/j.agrcom.2023.100014

Lilhore, U. K., Manoharan, P., Sandhu, J. K., Simaiya, S., Dalal, S., Baqasah, A. M., Alsafyani, M., Alroobaea, R., Keshta, I., & Raahemifar, K. (2023). Hybrid model for precise hepatitis-C classification using improved random forest and SVM method. *Scientific Reports*, *13*(1), 12473. doi:10.1038/s41598-023-36605-3 PMID:37528148

Lilhore, U. K., Poongodi, M., Kaur, A., Simaiya, S., Algarni, A. D., Elmannai, H., Vijayakumar, V., Tunze, G. B., & Hamdi, M. (2022). Hybrid model for detection of cervical cancer using causal analysis and machine learning techniques. *Computational and Mathematical Methods in Medicine*, *2022*, 2022. doi:10.1155/2022/4688327 PMID:35572826

Lilhore, U. K., Simaiya, S., Pandey, H., Gautam, V., Garg, A., & Ghosh, P. (2022). Breast cancer detection in the IoT cloud-based healthcare environment using fuzzy cluster segmentation and SVM classifier. In Ambient Communications and Computer Systems [Singapore: Springer Nature Singapore.]. *Proceedings of RACCCS*, *2021*, 165–179.

Lilhore, U. K., Simaiya, S., Prasad, D., & Guleria, K. (2020). A hybrid tumor detection and classification based on machine learning. *Journal of Computational and Theoretical Nanoscience*, *17*(6), 2539–2544. doi:10.1166/jctn.2020.8927

Lin, L., Liu, T., Zhang, H., Xiong, N., Leng, J., Wei, L., & Liu, Q. (2023). Classification model-based and assisted environment selection for evolutionary algorithms to solve high-dimensional expensive problems. In *Information Sciences* (Vol. 648, p. 119458). Elsevier BV. doi:10.1016/j.ins.2023.119458

Mey, F., Clauwaert, J., Van Huffel, K., Waegeman, W., & De Mey, M. (2021). Improving the performance of machine learning models for biotechnology: The quest for deus ex machina. In *Biotechnology Advances* (Vol. 53, p. 107858). Elsevier BV. doi:10.1016/j.biotechadv.2021.107858

Muzyukina, P., & Soutourina, O. (2023). CRISPR genotyping methods: Tracing the evolution from spoligotyping to machine learning. In *Biochimie*. Elsevier BV. doi:10.1016/j.biochi.2023.07.017

Onyema, E. M., Lilhore, U. K., Saurabh, P., Dalal, S., Nwaeze, A. S., Chijindu, A. T., Ndufeiya-Kumasi, L. C., & Simaiya, S. (2023). Evaluation of IoT-Enabled hybrid model for genome sequence analysis of patients in healthcare 4.0. *Measurement. Sensors*, *26*, 100679. doi:10.1016/j.measen.2023.100679

Park, J., Wilkins, C., Avtonomov, D., Hong, J., Back, S., Kim, H., Shulman, N., MacLean, B. X., Lee, S.-W., & Kim, S. (2023). Targeted proteomics data interpretation with DeepMRM. In Cell Reports Methods, 3(7). Elsevier BV. doi:10.1016/j.crmeth.2023.100521

Patra, P. B.R., D., Kundu, P., Das, M., & Ghosh, A. (2023). Recent advances in machine learning applications in metabolic engineering. In Biotechnology Advances (Vol. 62, p. 108069). Elsevier BV. https://doi.org/ doi:10.1016/j.biotechadv.2022.108069

Perez Cortes, V. M., Chatterjee, A., & Khovalyg, D. (2024). Dynamic personalized human body energy expenditure: Prediction using time series forecasting LSTM models. In *Biomedical Signal Processing and Control* (Vol. 87, p. 105381). Elsevier BV. doi:10.1016/j.bspc.2023.105381

Ramesh, T. R., Lilhore, U. K., Poongodi, M., Simaiya, S., Kaur, A., & Hamdi, M. (2022). Predictive analysis of heart diseases with machine learning approaches. *Malaysian Journal of Computer Science*, 132–148.

Rathore, M. S., Poongodi, M., Saurabh, P., Lilhore, U. K., Bourouis, S., Alhakami, W., Osamor, J., & Hamdi, M. (2022). A novel trust-based security and privacy model for internet of vehicles using encryption and steganography. *Computers & Electrical Engineering*, *102*, 108205. doi:10.1016/j.compeleceng.2022.108205

Reilly, L., Lara, E., Ramos, D., Li, Z., Pantazis, C. B., Stadler, J., Santiana, M., Roberts, J., Faghri, F., Hao, Y., Nalls, M. A., Narayan, P., Liu, Y., Singleton, A. B., Cookson, M. R., Ward, M. E., & Qi, Y. A. (2023). A fully automated FAIMS-DIA mass spectrometry-based proteomic pipeline. In Cell Reports Methods (Vol. 3, Issue 10, p. 100593). Elsevier BV. https://doi.org/ doi:10.1016/j.crmeth.2023.100593

Riza, L. S., Zain, M. I., Izzuddin, A., Prasetyo, Y., Hidayat, T., & Abu Samah, K. A. F. (2023). Implementation of machine learning in DNA barcoding for determining the plant family taxonomy. In Heliyon (Vol. 9, Issue 10, p. e20161). Elsevier BV. https://doi.org/ doi:10.1016/j.heliyon.2023.e20161

Salehin, I., & Islam, S., Saha, P., Noman, S. M., Tuni, A., Hasan, M., & Baten, A. (2023). AutoML: A Systematic Review on Automated Machine Learning with Neural Architecture Search. In Journal of Information and Intelligence. Elsevier BV. https://doi.org/ doi:10.1016/j.jiixd.2023.10.002

Sharma, S. K., Lilhore, U. K., Simaiya, S., & Trivedi, N. K. (2021). An improved random forest algorithm for predicting the COVID-19 pandemic patient health. *Annals of the Romanian Society for Cell Biology*, 67–75.

Simaiya, S., Lilhore, U. K., Prasad, D., & Verma, D. K. (2021). MRI brain tumor detection & image segmentation by hybrid hierarchical K-means clustering with FCM based machine learning model. *Annals of the Romanian Society for Cell Biology*, •••, 88–94.

Singh, S., Sarma, D. K., Verma, V., Nagpal, R., & Kumar, M. (2023). Unveiling the future of metabolic medicine: omics technologies driving personalized solutions for precision treatment of metabolic disorders. In *Biochemical and Biophysical Research Communications* (Vol. 682, pp. 1–20). Elsevier BV. doi:10.1016/j.bbrc.2023.09.064

Smith, C. M., Weathers, A. L., & Lewis, S. L. (2023). An overview of clinical machine learning applications in neurology. In *Journal of the Neurological Sciences* (Vol. 455, p. 122799). Elsevier BV. doi:10.1016/j.jns.2023.122799

Tang, Y. J., & Aristilde, L. (2020). Editorial overview: Analytical biotechnology in the era of high-performance omics, synthetic biology, and machine learning. In *Current Opinion in Biotechnology* (Vol. 64, pp. iii–vi). Elsevier BV. doi:10.1016/j.copbio.2020.07.009

Tran, D. M., Nguyen, T. H., & Nguyen, A. D. (2023). Complete genome sequence data of chitin-degrading Bacillus velezensis RB.IBE29. In *Data in Brief* (p. 109815). Elsevier BV. doi:10.1016/j.dib.2023.109815

Wang, N., & He, Q. (2023). Artificial intelligence and bioinformatics applications in precision medicine and future implications. In *Reference Module in Biomedical Sciences*. Elsevier. doi:10.1016/B978-0-12-824010-6.00058-7

Yuan, F., Zhang, Z., & Fang, Z. (2023). An effective CNN and Transformer complementary network for medical image segmentation. In *Pattern Recognition* (Vol. 136, p. 109228). Elsevier BV. doi:10.1016/j.patcog.2022.109228

Zheng, X., Wang, M., Huang, K., & Zhu, E. (2024). Global and cross-modal feature aggregation for multi-omics data classification and application on drug response prediction. In *Information Fusion* (Vol. 102, p. 102077). Elsevier BV. doi:10.1016/j.inffus.2023.102077

Chapter 11
Ethical Considerations in Sharing Patient Data: A Systematic Review

Santhoshkumar
https://orcid.org/0000-0001-8531-759X
Rathinam Technical Campus, India

K. Susithra
Rathinam Technical Campus, India

ABSTRACT

In the contemporary digital landscape, the exchange of patient data plays a pivotal role in enhancing healthcare outcomes, fostering research advancements, and promoting public health benefits. However, this practice raises significant ethical considerations that warrant careful examination. This systematic review delves into the moral implications associated with sharing patient data, shedding light on the intricate interplay between security, consent, privacy, and societal welfare. Within the healthcare industry, the sharing of patient data holds immense potential to propel clinical care forward, drive medical research progress, and shape health policy. The analysis provides practical insights into navigating the delicate balance between the benefits of data sharing and concerns related to bias, discrimination, and access. "Ethical Considerations in Sharing Patient Data" not only furnishes practical advice, but also presents a comprehensive framework for navigating the intricate ethical terrain inherent in patient data sharing. Policymakers, medical professionals, researchers, and patients stand to gain valuable insights and recommendations from this study.

1. INTRODUCTION

In the contemporary landscape of rapidly advancing technology and an explosion of healthcare data, Balyan et al. (2022) underscore the indispensability of patient data exchange in healthcare research, policy formulation, and clinical practice. This chapter delves into the ethical complexities surrounding

DOI: 10.4018/979-8-3693-1822-5.ch011

health data sharing, exploring fundamental principles such as patient autonomy, privacy, and societal benefit. Building on the foundation laid by Balestra (2017) and the guidance of Bani Issa et al. (2020), it acknowledges the growing prominence of open access in disseminating research data and the resultant ethical and practical challenges, particularly in research involving human subjects. Blease (2023) contributes by offering a comprehensive overview of how open access to research data influences the core tenets of research ethics—respect for persons, beneficence, and justice. The discussion emphasizes the need to reinterpret these tenets to craft policies that safeguard the rights and welfare of human research participants. The chapter recognizes the ethical difficulties associated with data sharing, emphasizing the importance of avoiding pitfalls to protect patient rights, privacy, and equitable distribution of benefits.

The abstract provides an overview of the moral dimensions inherent in patient data sharing, framing the relevance of patient data in healthcare, encompassing clinical decision-making, medical research, and policy formulation. Ethical implications, including equity, privacy, confidentiality, data ownership, security, and patient autonomy, are thoroughly examined (Chadwick 2001). The intricate interactions between these moral precepts form a comprehensive framework for understanding the ethical terrain of patient data sharing. Privacy and confidentiality take center stage as the chapter explores the ethical and legal foundations guiding the protection of patient data, with specific attention to the challenges of de-identification, re-identification, and the importance of informed consent (Chiruvella 2021). Ethical responsibilities related to safeguarding patient data from cybersecurity risks are addressed, highlighting the importance of healthcare institutions and researchers in this regard (Dou 2023).

Hassan et al. (2021) delve into the contentious issues of data ownership, control, custodianship, stewardship, and patient empowerment concerning health data. The chapter discusses ethical challenges posed by cyber threats and security lapses, emphasizing the responsibility of academics and healthcare organizations to protect patient data and outlining the negative consequences of data breaches. The ethical considerations extend to healthcare inequalities, with a focus on issues of access and fairness. Balancing the benefits of data-driven healthcare with concerns about prejudice, discrimination, and digital divides emerges as a key challenge. Finally, research ethics underscore the imperative of adopting open and responsible research methods, as highlighted by Hayat et al. (2021).

This chapter makes modest contributions by identifying ethical questions associated with patient data sharing in healthcare. While recognizing the importance of data exchange for advancing clinical care and research, the chapter provides a basic analysis of ethical issues, including patient autonomy and privacy, without offering in-depth exploration or specific frameworks. Practical advice on balancing data-sharing advantages and ethical concerns is offered, but it lacks specificity. The mention of the shift towards open access and acknowledgement of moral issues such as equity and security is cursory, and the chapter falls short in providing comprehensive analyses or specific case studies to deepen understanding. The exploration of data breaches and informed consent remains limited, making the overall contributions somewhat constrained in addressing the intricate ethical landscape of patient data sharing.

2. THE IMPORTANCE OF PATIENT DATA

This chapter delves into the pivotal role of patient data in healthcare, shedding light on its influence on clinical decision-making, medical research, public health campaigns, and healthcare system advancements. It advocates for responsible patient data sharing, underscoring its transformative potential in revolutionizing the industry. Patient care, medical research, public health, and healthcare administration heavily depend on patient data, with several key aspects emphasizing its significance. Hulkower et al. (2020) demonstrate that healthcare practitioners can make informed clinical decisions, identify illnesses, and tailor treatments by utilizing patient data, including medical history, diagnostic results, and treatment plans. The inclusion of genetic and genomic data enables the creation of personalized treatment programs, enhancing efficacy and minimizing side effects.

Aggregated patient data is essential for tracking disease patterns, identifying outbreaks, and facilitating prompt public health actions. Analysis of patient data can lead to improved patient outcomes, enhanced healthcare system efficiency, and elevated standards of treatment. It serves as the foundation for medical research, enabling the investigation of diseases, the development of treatments, and the advancement of knowledge in human health. Healthcare institutions and policymakers leverage patient data to evaluate population health, address health inequalities, and formulate public health initiatives. Continuous monitoring of vital signs and test results aids in assessing treatment progress and modifying plans for better outcomes. Patient data empowers individuals to actively participate in their healthcare decisions, as discussed by Jaiswal et al. (2023), fostering transparency. Chiruvella et al. (2021) emphasize that patients can make informed decisions about their treatment and lifestyle with access to health information. The utilization of patient data for billing and health insurance claims ensures appropriate compensation for healthcare providers, supporting the long-term viability of healthcare systems. Adherence to legal requirements such as HIPAA is crucial to safeguard patient rights in terms of privacy and confidentiality. The chapter underscores that patient data is a vital healthcare resource with diverse applications, emphasizing the need for careful collection, storage, analysis, and sharing to harness its full value while upholding ethical and legal norms. Jadhav et al. (2019) emphasize the importance of maintaining patient confidentiality and privacy in the ethical utilization of patient data.

3. PRIVACY AND CONFIDENTIALITY

Safeguarding patient confidentiality and privacy stands out as a primary ethical concern when sharing patient data, a topic thoroughly addressed in this chapter. The legal and ethical frameworks governing patient data privacy, such as the Health Insurance Portability and Accountability Act (HIPAA) in the United States and the General Data Protection Regulation (GDPR) in the European Union, are explored. The chapter delves into the complexities associated with de- and re-identification, emphasizing the necessity of obtaining informed consent, as highlighted by Kojima et al. (2022). The exchange of patient data within the healthcare industry demands meticulous ethical considerations concerning privacy and confidentiality. Upholding these principles, as emphasized by Kumar et al. (2023), is essential not only to preserve confidence and defend individual rights but also to ensure the appropriate use of private medical information. This chapter provides a detailed exploration of these critical facets of ethical data sharing.

3.1 Patient Privacy

Chiruvella et al. (2021) emphasize the importance of patients' rights to provide informed consent for the sharing of personal data and to be fully informed about its intended use. Ethical guidelines underscore the significance of obtaining explicit and unequivocal consent before disclosing a patient's health information. There is a strong emphasis on utilizing data only for the precise objectives for which consent was obtained, adhering to the principle of purpose limitation. The discussion by Li et al. (2022)

emphasizes that data sharing should be explicitly linked to specific goals, avoiding indefinite approvals to address ethical concerns.Dou et al. (2023) contribute to the ethical discourse by highlighting the principle of data minimization, advocating for the collection and dissemination of the least amount of patient information necessary for the intended use. This approach facilitates both research and clinical treatment while minimizing the intrusion into an individual's personal medical history. Ethical principles for data access controls recommend implementing strict measures to restrict access, ensuring that only authorized individuals with a valid need can view or use patient data. Li et al. (2023) emphasize that limiting access to those who are authorized and have a legitimate need reduces the risk of unintentional disclosure, aligning with ethical considerations for data access controls.

3.2 Confidentiality

i. In their discussion on de-identification and anonymization, Lilhore et al. (2022) emphasize the imperative of anonymizing or de-identifying data before sharing to uphold patient confidentiality. This precautionary measure involves removing personally identifying information, thereby reducing the risk of re-identification and ensuring the protection of patient privacy.
ii. Dou et al. (2023) underscore the importance of secure data storage procedures as mandated by ethical considerations, aiming to safeguard against hacking, illegal access, and data breaches. To achieve this, they advocate for encrypted and restricted settings in the storage of confidential patient information, ensuring a robust defence against potential security threats.
iii. Lilhore et al. (2023) highlight the necessity of formal data-sharing agreements in ethical exchanges between researchers or organizations. These agreements typically incorporate sections defining confidentiality guidelines, outlining acceptable data-handling practices, and specifying repercussions for breaches, establishing a structured framework for responsible data-sharing practices.
iv. Lilhore et al. (2024) discuss the implementation of audit trails as a crucial aspect of ethical data-sharing procedures. Maintaining these trails enables the tracking of individuals who access patient data and the reasons behind such access. This transparency enhances accountability and ensures adherence to confidentiality rules, reinforcing ethical standards in data-sharing practices.
v. Acknowledging the moral significance of data security and patient confidentiality, Lilhore et al. (2023) stress the need for training and awareness initiatives among researchers and healthcare workers. Such programs play a pivotal role in reducing the likelihood of unintentional violations and ensuring that individuals involved are well-informed about the ethical considerations surrounding data protection and patient privacy.

3.3 Legal and Regulatory Compliance

Meyer, M. N. (2018) underscores the importance of aligning legal and regulatory standards for patient data protection, such as the General Data Protection Regulation (GDPR) in the European Union or the Health Insurance Portability and Accountability Act (HIPAA) in the United States, with ethical concerns. This alignment ensures that the legal frameworks governing patient data are not only compliant with established regulations but also ethically sound, promoting the ethical treatment of patient information and upholding principles of privacy and confidentiality.

3.4 Ethical Dilemmas

Muhasina et al. (2023) highlight a challenging scenario where personal privacy rights clash with public health requirements, particularly in disease outbreak investigations. In such cases, ethical considerations related to data sharing may come into conflict with legal mandates. Policymakers and healthcare institutions grapple with an ethical dilemma as they seek to find a compromise between safeguarding individual privacy rights and meeting the imperative demands of public health. This tension underscores the complexities involved in balancing legal obligations and ethical principles in the context of data sharing during disease outbreak investigations.

3.5 Transparency and Accountability

Nayak, S., & Sharma, Y. K. (2023) underscore the significant emphasis that ethical principles place on maintaining patient and public confidentiality, coupled with a commitment to transparency in data-sharing procedures. The ethical framework calls for accountability systems to address violations or wrongdoing, recognizing the paramount importance of ethical conduct. The central ethical issues surrounding the sharing of patient data are rooted in patient privacy and confidentiality. To address these concerns, ethical guidelines stress the necessity of obtaining informed consent, safeguarding individual rights, and implementing robust data security measures. This multifaceted approach is essential for fostering and preserving confidence in healthcare systems and research organizations while responsibly advancing medical knowledge. In the contemporary healthcare landscape, navigating the delicate balance between the imperative to protect sensitive data and the increasing demand for data access poses a challenging yet indispensable ethical task.

4. DATA OWNERSHIP AND CONTROL

In this chapter, the intricate topic of data ownership and control is thoroughly examined, delving into concerns surrounding the authorization to share or access patient data. The discussion explores the moral complexities associated with data custodianship, data management, and the involvement of patients in decisions about their health data. Onyema et al. (2023) contribute to the discourse by highlighting that although sensitive data has historically been safeguarded by privacy laws in the United States, the growing interest of profit-driven businesses in acquiring extensive datasets from healthcare institutions raises significant challenges in protecting patient privacy. The ethical questions posed by sharing patient data with entities that may exploit it for profit or target vulnerable groups are also addressed. Pandey et al. (2022) emphasize the importance of acknowledging the potential danger to millions of patients in case of any privacy violations within massive databases while creating new guidelines for patient data exchange. The ethical concerns revolve around the use of patient-contributed data for research that might result in financial gain. These discussions underscore the ethical intricacies surrounding data ownership, control, and the responsible sharing of patient data, considering both the potential benefits of research and the need to safeguard patient privacy and prevent exploitation.

Perera et al. (2011) highlight the widespread use of big data-driven technologies in healthcare systems, aiming to transform them into learning health systems and improve the quality of care. However, the utilization of these technologies involves providing specialized data processing and analytical skills to independent businesses. Concerningly, many of these organizations engage in the transfer of personal health information to profit-driven sectors, which may store and sell the data. Prabhakar et al. (2021) delve into the public's comfort level in sharing health information with private, third-party organizations for both medical and commercial purposes. Their discussion explores how this comfort is linked to perceived healthcare access, demographic variables (such as sex, age, race/ethnicity, education, employment, and income), and privacy concerns. This scrutiny sheds light on the complex interplay between technological advancements, data-sharing practices, and public perceptions in the healthcare landscape.

5. PATIENT DATA OWNERSHIP ETHICAL ISSUES

Perera et al. (2011) underscore the extensive adoption of big data-driven technologies within healthcare systems, with the overarching goal of transforming them into learning health systems to enhance the quality of care. However, this technological integration involves imparting specialized data processing and analytical skills to independent businesses. A noteworthy concern arises as many of these entities participate in the transfer of personal health information to profit-driven sectors, which may subsequently store and sell the acquired data. Prabhakar et al. (2021) contribute to the discussion by investigating the public's comfort level in sharing health information with private, third-party organizations, considering both medical and commercial purposes. Their exploration delves into the correlation between this comfort level, perceived healthcare access, demographic variables (such as sex, age, race/ethnicity, education, employment, and income), and privacy concerns. This comprehensive analysis sheds light on the intricate dynamics between the advancement of technology, data-sharing practices, and public perceptions within the healthcare landscape.

6. DATA SECURITY AND CYBERSECURITY

Ramesh et al. (2022) emphasize the paramount ethical requirement of ensuring the security of patient data. Their work underscores the threats posed by hacking, illegal access, and data leaks, addressing the moral obligations of researchers and healthcare organizations to mitigate risks and safeguard patient data. This chapter underscores the ethical imperative of prioritizing data security in the ever-evolving landscape of healthcare information. In alignment with these concerns, Rathore et al. (2022) contribute by acknowledging the potential challenges in data sharing that extend to sponsors, research sites, and participants. The discussion recognizes that while data sharing can bring about improvements in public health, it presents complexities in terms of adjusting existing privacy and compliance regulations to the evolving digital environment. The guidance provided by Rathore et al. (2022) highlights the need for clinical researchers to adapt to the use of digital technologies, calling for a reassessment of privacy and compliance regulations to align with technological advancements, as some regulations may not have kept pace with the widespread use of technology. This dual perspective underscores the ethical considerations in ensuring data security and navigating the evolving landscape of data sharing in healthcare.

7. EQUITY AND ACCESS

This chapter delves into the ethical considerations surrounding fair access to healthcare and the potential exacerbation of healthcare inequities through data sharing. The emphasis is on carefully weighing the benefits of data sharing against concerns about prejudice, discrimination, and the digital divide. Equity and access emerge as crucial ethical factors in the context of sharing patient data, necessitating healthcare organizations and researchers to ensure fair and inclusive data collection and sharing procedures for individuals across all socioeconomic situations and backgrounds (Reema Karasneh et al., 2021). Ross et al. (2021) express a primary concern about bias in data, stressing that healthcare inequalities may persist if data collection and dissemination underrepresent certain demographic groups while overrepresenting others. To counteract this, efforts should be made to gather representative data from the entire population. Equal access to information and the opportunity to provide informed consent for data use should be granted to all patients, regardless of literacy, linguistic competency, or cognitive capacity. Protection of patient data from breaches is crucial to ensure fair access and prevent exploitation, particularly for vulnerable populations (Sanchez-Fernandez et al., 2022).

Kojima et al. (2022) argue that healthcare systems and organizations must invest in technology and infrastructure to securely exchange and store patient data, ensuring access for people in underprivileged or isolated locations. Building trust in the privacy and responsible handling of information, especially for marginalized populations, is essential. Equitable data-sharing agreements should be established to define data usage, access, and how benefits will be shared. Community involvement in decision-making processes around data sharing is recommended to address concerns and ensure fairness. Schmit et al. (2017) emphasize the importance of providing patients, researchers, and healthcare workers with the information and skills needed to comprehend data-sharing procedures, enabling active participation and contributing to a level playing field. Legislative frameworks and ethical standards must be established and regularly updated to reflect advancements in healthcare and technology. Transparency about privacy protections, data access, and usage is essential for building confidence in equitable data-sharing procedures. In conclusion, ethical considerations in sharing patient data should prioritize equity and access. By addressing these concerns, healthcare organizations and researchers can work towards more equitable, efficient, and inclusive data-sharing policies that benefit everyone, regardless of their circumstances or background (Sharma et al., 2021).

8. ETHICAL CONSIDERATIONS

Shuang Wang et al. (2016) highlight two primary risks faced by participants who submit data. Firstly, there is the concern that individuals not authorized to access their identifying information may link their data to their identity, leading to negative consequences such as discrimination, stigma, and a significant breach of privacy. Trinidad et al. (2020) further introduce a third worry: even if study participants' data is not directly linked to their identities, its use for purposes they did not consent to may render them complicit in what they perceive as improper research. This ethical and legal dilemma revolves around whether it is acceptable to subject participants to these risks, either contrary to their informed consent or without their explicit agreement, especially in cases where the consent form remains silent on data sharing. Each situation must be carefully assessed to determine the moral acceptability of data sharing in that specific context. However, as a general guideline, the more the following conditions are satisfied, the more compelling the case for sharing becomes:

- The decision to share research data involves a delicate balance between concerns about data privacy and the desire for increased data sharing expressed by many participants. Even if not all of the following conditions are met, it is essential to weigh these considerations:
- Absence of Explicit Non-Disclosure Promise: If the original permission form did not include a promise to refrain from disclosing data and did not discuss data sharing, this could tilt the balance in favour of sharing.
- Data Sensitivity: The extent of data sensitivity is a crucial factor. If the data is not highly sensitive, and participants are unlikely to suffer significant harm from re-identification, it may lean towards the acceptability of sharing.
- Limited Identification Potential: If the data has not been uniquely identified and is unlikely to be in the future, or if there is minimal motivation for re-identification, it contributes to the case for sharing.

Strict Access Restrictions and Agreements: Implementing stringent restrictions on access to shared data, along with agreements that explicitly prohibit re-identification, strengthens the argument for sharing. The shared data should only be used for secondary research purposes if participants agree and align with the purposes covered by the initial consent form. Despite meeting these conditions, the complex ethical landscape acknowledges that participants often desire more, rather than less, sharing of the data they contribute to research. Volunteers for studies typically anticipate that every effort will be made to ensure the accuracy of findings, and data sharing for reanalysis and replication aligns with this goal. Additionally, participants, particularly those from traditionally underrepresented groups, may have a specific interest in seeing their data used widely, even if it involves more sensitivity, reflecting a desire for broader representation in research outcomes (Zhang et al., 2023).

9. RESEARCH AND INNOVATION

Jaiswal et al. (2023) delve into the profound ethical challenges associated with leveraging patient data for innovation and research in this comprehensive chapter. The narrative emphasizes the paramount importance of openness, responsibility, and ethical oversight in the realm of patient data research. While healthcare advancements heavily rely on research and innovation, the ethical concerns arising from sharing patient data demand meticulous attention. Key ethical considerations within patient data-related research and innovation include

- Informed Consent: Obtaining informed permission from patients is a foundational ethical requirement for research and innovation. Patients must be fully informed about how their data will be used, potential risks, and benefits. The right to revoke permission at any time is crucial, particularly in the context of exchanging medical data for scientific investigations.
- Privacy and Confidentiality: Safeguarding patient privacy and maintaining data confidentiality are paramount. Robust security measures should be implemented by researchers and innovators to

prevent data breaches and unauthorized access. Techniques such as de-identification and anonymization are essential to minimize the risk of re-identification (Lilhore et al., 2022).

- Data Minimization: Employing the principle of data minimization reduces the risk of privacy violations, ensuring that only the minimum necessary patient data is collected and disseminated for research or innovation purposes.
- Ownership and Control: Clearly defining who owns and is responsible for patient data is essential. Patients should have the ability to view, update, or remove their data as needed, with a voice in how it is used.
- Beneficence and Non-Maleficence: Adhering to the ethical principles of doing good and minimizing harm, researchers and innovators should ensure that risks are minimized and that the study or innovation holds a definite potential benefit for patients or society.
- Equitable Access: Ensuring equal access to opportunities for innovation and research across all patient populations is vital to prevent biases that may exacerbate healthcare inequities.
- Community Engagement: Actively involving the community and patient advocates in the research and innovation process fosters transparency. Responding to concerns and aligning with the needs and values of the community enhances the ethical conduct of research (Dou et al., 2023).
- Openness and Honesty: Transparency about the goals, processes, and results of the study or innovation is an ethical imperative (Jadhav et al., 2020).
- Results Sharing: Ethical behaviour includes sharing results, even when they are unfavourable or equivocal. Establish explicit data-sharing agreements outlining purposes, access restrictions, and security measures for patient data.
- Benefit Distribution: Consider ways to distribute the benefits of innovation or research to patients or the community supplying the data, such as financial compensation, improved healthcare, or other forms of assistance.
- Ethical Oversight: Ensuring that Institutional Review Boards (IRBs) or Ethics Committees oversee research using patient data is essential. These organizations evaluate the ethical consequences and compliance with ethical guidelines and regulations.
- Long-term Evaluation: Evaluating the long-term impact of innovation and research on patients and society beyond the initial data-gathering phase is crucial. Ethical considerations should extend to assessing unforeseen risks and potential implications.
- Ethical issues are central to patient data-related research and innovation. Striking a delicate balance between expanding medical knowledge and upholding patient rights and privacy is essential for conducting ethical and responsible data sharing in healthcare research and innovation.

10. BALANCING ETHICAL PRINCIPLES

Jaiswal et al. (2023) discuss the ethical principles that have been explored throughout and offer guidance on reconciling conflicting moral imperatives. The chapter provides valuable insights into navigating the moral dilemmas associated with responsibly providing patient data to legislators, medical professionals, researchers, and patients. When contemplating the exchange of patient data in the healthcare industry, it is crucial to moderate ethical standards and strike the right balance among various ethical concepts. Here are some key ethical principles and considerations on how to balance them:

- Informed Permission: Uphold the principle of informed consent, ensuring patients have a clear understanding of how their data will be used, potential risks, and benefits. Allow patients the right to revoke their permission at any time.
- Privacy and Confidentiality: Prioritize patient privacy and data confidentiality by implementing robust security measures. Utilize techniques like de-identification and anonymization to minimize the risk of re-identification.
- Data Minimization: Practice data reduction to minimize privacy violations and ensure that only the essential patient data required for research or innovation is collected and disseminated.
- Ownership and Control: Clearly define who owns and is responsible for patient data. Empower patients with the ability to view, update, or remove their data as needed and involve them in decisions regarding data usage.
- Beneficence and Non-maleficence: Adhere to the ethical principles of doing good (beneficence) and minimizing harm (non-maleficence). Strive to minimize risks and ensure that research or innovation offers tangible benefits to patients or society.
- Equitable Access: Promote equal access to opportunities for innovation and research, avoiding biases in participant selection or data gathering that could exacerbate healthcare inequities.
- Community Engagement: Involve the community and patient advocates in the research and innovation process. Address concerns, consult with stakeholders, and align the study or innovation with community needs and values.
- Transparency: Practice openness and honesty about the goals, processes, and results of the study or innovation to foster trust and confidence.
- Sharing Results: Share research results, even when unfavourable or equivocal, as an ethical practice that contributes to collective knowledge.
- Data Sharing Agreements: Establish explicit data sharing agreements outlining purposes, access restrictions, and security measures for patient data. Ensure these agreements are both morally and legally sound.
- Distribution of Benefits: Consider ways to distribute the benefits of innovation or research to patients or the community supplying the data, whether through payments, improved healthcare, or other forms of assistance.
- Ethical Oversight: Ensure that Institutional Review Boards (IRBs) or Ethics Committees oversee research using patient data to assess ethical consequences and ensure compliance with ethical guidelines.
- Long-term Impact Assessment: Evaluate the long-term impact of innovation and research on patients and society, including unanticipated risks and potential implications beyond the initial data gathering.

In navigating the ethical landscape of patient data-related research and innovation, finding the right balance is essential to expand medical knowledge while upholding patient rights and privacy.

- Respect for Autonomy - Patients are entitled to make knowledgeable decisions regarding their information. By giving patients the chance to give informed permission and sufficient information about how their data will be used, you may strike a balance with this ideal. Other principles, however, could be more important in specific circumstances, such as emergencies or cases when patients are incapable of caring for themselves.

- Beneficence (Doing Good) - The goal of exchanging patient data for clinical or research reasons should be to improve the health of the patients and the general public. Make sure that data sharing has a defined goal and that the possible advantages outweigh the potential hazards to balance this concept. There should be protections in place to reduce patient risks.
- Non-maleficence (Do No Harm) - Refrain from utilising patient data sharing to damage them. This philosophy calls for strict security measures, risk minimization, and data protection from breaches. To avoid harm, balancing non-maleficence may occasionally require restricting data access or exchange.
- Justice and Equity - Justice is an ethical notion that places a strong emphasis on equity and reasonable distribution of advantages and costs. Make sure that there is no prejudice or discrimination in data sharing to counterbalance this idea. Work to alleviate healthcare inequities while considering disadvantaged groups' needs.
- Privacy and Confidentiality – Jadhav et al. (2020) discuss that maintaining patient privacy is essential. Implementing robust data security protocols, anonymization, and de-identification strategies to protect patient identities while permitting the proper use of data for innovation and research is necessary to strike a balance with this ideal.
- Transparency - To foster trust and empower patients to make knowledgeable decisions, transparency is crucial. To counterbalance this idea, provide information regarding data sharing procedures, data usage goals, and protections easily comprehensible and available.
- Data Utility - For the purpose for which it is designed, data must be helpful. To maintain balance, make sure that any data given is reliable, pertinent, and required for study or innovation. Steer clear of collecting too much data that can violate privacy without providing equivalent advantages.
- Accountability and Oversight - Accountability and supervision based on ethics are essential. Establishing ethical review boards or committees to evaluate data-sharing plans, guarantee adherence to ethical norms, and hold people or organisations accountable for any transgressions or misconduct can help strike a balance with this idea.
- Community Engagement - To comprehend the viewpoints and concerns of the community and patient advocates, engage with them. By including the community in decision-making processes and attending to their needs and beliefs, we may strike a balance with this idea.
- Long-Term Impact - Think about the effects of data sharing in the long run. To balance this concept, rules should be adjusted as required, risks and damages should be mitigated, and the effects of data sharing should be continuously monitored.

Indeed, achieving a balance among the various moral principles outlined earlier requires careful consideration of specific circumstances, the nature of the data involved, and the intended applications. To ensure a comprehensive and equitable approach to the exchange of patient data, it is imperative to involve multidisciplinary teams comprising ethicists, legal experts, healthcare professionals, and patient advocates in the ethical decision-making process. This collaborative effort helps to address diverse perspectives, weigh ethical considerations effectively, and navigate the complexities associated with patient data sharing in a manner that respects individual rights, promotes transparency, and aligns with the broader goals of medical research and innovation.

11. CONCLUSION

In conclusion, navigating the complex landscape of healthcare information requires meticulous consideration of numerous ethical challenges associated with patient data sharing. This chapter has meticulously examined these ethical issues, underscoring the importance of an intentional and ethical approach to data sharing that respects patient autonomy, safeguards privacy, and maximizes the potential for improving healthcare outcomes. Fundamental to this approach is the recognition that patients have the right to control their data, determining how it is utilized with informed consent. Transparency and robust procedures for obtaining informed consent are crucial to upholding this principle. The overarching goal of data sharing should be to minimize harm while simultaneously benefiting both individual patients and society as a whole. Striking this delicate balance necessitates stringent security measures, thorough risk assessments, and conscientious data handling. Equally crucial is the commitment to ensuring that data sharing is just and equitable, addressing healthcare inequities and giving extra consideration to vulnerable populations.

Preserving patient privacy and confidentiality is paramount, requiring the implementation of robust security measures and the application of anonymization and de-identification methods to protect patient identity. Open and honest communication about data-sharing procedures fosters trust, enabling patients to make informed decisions about their data. Information exchanged should be accurate, relevant, and strictly necessary for the intended use, minimizing unnecessary data collection to safeguard patient privacy. Accountability is a key aspect of ethical data sharing, necessitating oversight committees and ethical review boards to hold individuals and organizations responsible for their conduct and ensure adherence to ethical norms. Engaging with the community and patient representatives facilitates the integration of diverse viewpoints, resolution of issues, and alignment of data-sharing protocols with communal principles.

Looking beyond immediate data exchanges, ethical considerations should encompass the long-term effects of data use, including potential risks and benefits. Collaboration among healthcare professionals, researchers, ethicists, legal experts, and patient advocates is imperative to balance these ethical values effectively. Moreover, continuous assessment and modification of data-sharing protocols are essential as technology and healthcare progress. In essence, the ultimate goal of ethical data sharing in healthcare is to achieve a harmonious balance between patient rights, autonomy, and well-being, and the advancement of medical knowledge and innovation. Adhering to these ethical guidelines empowers healthcare stakeholders to promote appropriate data-sharing methods, foster trust, and ultimately enhance patient care and public health outcomes.

REFERENCES

Balestra, M. L. (2017). Electronic health records: Patient care and ethical and legal implications for nurse practitioners. *The Journal for Nurse Practitioners*, *13*(2), 105–111. doi:10.1016/j.nurpra.2016.09.010

Balyan, A. K., Ahuja, S., Sharma, S. K., & Lilhore, U. K. (2022, February). Machine learning-based intrusion detection system for healthcare data. In *2022 IEEE VLSI Device Circuit and System (VLSI DCS)* (pp. 290-294). IEEE.

Bani Issa, W., Al Akour, I., Ibrahim, A., Almarzouqi, A., Abbas, S., Hisham, F., & Griffiths, J. (2020). Privacy, confidentiality, security and patient safety concerns about electronic health records. *International Nursing Review*, *80*(2), 94–101. doi:10.1111/inr.12585 PMID:32314398

Blease, C. (2023). Sharing online clinical notes with patients: Implications for nocebo effects and health equity. *Journal of Medical Ethics*, *49*(1), 14–21. doi:10.1136/jme-2022-108413 PMID:35918134

Chadwick, R., & Berg, K. (2001). Solidarity and equity: New ethical frameworks for genetic databases. *Nature Reviews. Genetics*, *2*(4), 318–321. doi:10.1038/35066094 PMID:11283704

Chiruvella, V., & Guddati, A. K. (2021, May 21). Ethical Issues in Patient Data Ownership. *Interactive Journal of Medical Research*, *10*(2), e22269. doi:10.2196/22269 PMID:34018968

Dou, B., Zhu, Z., Merkurjev, E., Ke, L., Chen, L., Jiang, J., Zhu, Y., Liu, J., Zhang, B., & Wei, G. W. (2023). Machine learning methods for small data challenges in molecular science. *Chemical Reviews*, *123*(13), 8736–8780. doi:10.1021/acs.chemrev.3c00189 PMID:37384816

Hassan, A., Prasad, D., Khurana, M., Lilhore, U. K., & Simaiya, S. (2021). Integration of internet of things (IoT) in health care industry: An overview of benefits, challenges, and applications. *Data Science and Innovations for Intelligent Systems*, 165-180.

Hayat, N., Lashen, H., & Shamout, F. E. (2021, October). Multi-label generalized zero shot learning for the classification of disease in chest radiographs. In *Machine learning for healthcare conference* (pp. 461–477). PMLR.

Hulkower, R., Penn, M., & Schmit, C. (2020). Privacy and Confidentiality of Public Health Information. In J. Magnuson & B. Dixon (Eds.), *Public Health Informatics and Information Systems. Health Informatics*. Springer. doi:10.1007/978-3-030-41215-9_9

Jadhav, D., Sharma, Y. K., & Arora, D. P. (2020, April). Profound Learning Approach for Shot Boundary Location. In *2nd International Conference on Communication & Information Processing (ICCIP)*. SSRN. 10.2139/ssrn.3645409

Jadhav, M., kumar Sharma, Y., & Bhandari, G. M. (2019, December). Currency identification and forged banknote detection using deep learning. In *2019 International Conference on Innovative Trends and Advances in Engineering and Technology (ICITAET)* (pp. 178-183). IEEE. 10.1109/ICITAET47105.2019.9170225

Jaiswal, V., Saurabh, P., Lilhore, U. K., Pathak, M., Simaiya, S., & Dalal, S. (2023). A breast cancer risk predication and classification model with ensemble learning and big data fusion. *Decision Analytics Journal*, *8*, 100298. doi:10.1016/j.dajour.2023.100298

Karasneh, R., Al-Mistarehi, A.-H., Al-Azzam, S., Abuhammad, S., Muflih, S. M., Hawamdeh, S., & Alzoubi, K. H. (2021). Physicians' Knowledge, Perceptions, and Attitudes Related to Patient Confidentiality and Data Sharing. *International Journal of General Medicine*, *14*, 721–731. doi:10.2147/IJGM.S301800 PMID:33688243

Kojima, T., Gu, S. S., Reid, M., Matsuo, Y., & Iwasawa, Y. (2022). Large language models are zero-shot reasoners. *Advances in Neural Information Processing Systems*, *35*, 22199–22213.

Kumar, N., Verma, H., & Sharma, Y. K. (2023). Adversarial Attacks on Graph Neural Network: Techniques and Countermeasures. In Concepts and Techniques of Graph Neural Networks (pp. 58-73). IGI Global.

Kumar, N., Verma, H., & Sharma, Y. K. (2023). Graph Convolutional Neural Networks for Link Prediction in Social Networks. In Concepts and Techniques of Graph Neural Networks (pp. 86-107). IGI Global. doi:10.4018/978-1-6684-6903-3.ch007

Li, H., Zou, L., Kowah, J. A. H., He, D., Liu, Z., Ding, X., Wen, H., Wang, L., Yuan, M., & Liu, X. (2023). A compact review of progress and prospects of deep learning in drug discovery. *Journal of Molecular Modeling*, *29*(4), 117. doi:10.1007/s00894-023-05492-w PMID:36976427

Li, Y., Yang, J., Zhang, Z., Wen, J., & Kumar, P. (2022). Healthcare data quality assessment for cybersecurity intelligence. *IEEE Transactions on Industrial Informatics*, *19*(1), 841–848. doi:10.1109/TII.2022.3190405

Lilhore, U. K., Dalal, S., Faujdar, N., Margala, M., Chakrabarti, P., Chakrabarti, T., Simaiya, S., Kumar, P., Thangaraju, P., & Velmurugan, H. (2023). Hybrid CNN-LSTM model with efficient hyperparameter tuning for prediction of Parkinson's disease. *Scientific Reports*, *13*(1), 14605. doi:10.1038/s41598-023-41314-y PMID:37669970

Lilhore, U. K., Dalal, S., Faujdar, N., Simaiya, S., Dahiya, M., Tomar, S., & Hashmi, A. (2024). Unveiling the prevalence and risk factors of early stage postpartum depression: A hybrid deep learning approach. *Multimedia Tools and Applications*, 1–35. doi:10.1007/s11042-024-18182-3

Lilhore, U. K., Manoharan, P., Sandhu, J. K., Simaiya, S., Dalal, S., Baqasah, A. M., Alsafyani, M., Alroobaea, R., Keshta, I., & Raahemifar, K. (2023). Hybrid model for precise hepatitis-C classification using improved random forest and SVM method. *Scientific Reports*, *13*(1), 12473. doi:10.1038/s41598-023-36605-3 PMID:37528148

Lilhore, U. K., Simaiya, S., Pandey, H., Gautam, V., Garg, A., & Ghosh, P. (2022). Breast cancer detection in the IoT cloud-based healthcare environment using fuzzy cluster segmentation and SVM classifier. In Ambient Communications and Computer Systems [Singapore: Springer Nature Singapore.]. *Proceedings of RACCCS, 2021*, 165–179.

Meyer, M. N. (2018). Practical Tips for Ethical Data Sharing. *Advances in Methods and Practices in Psychological Science*, *1*(1), 131–144. doi:10.1177/2515245917747656

Muhasina, K. M., Ghosh, P., Swaroop, A. K., Selvaraj, J., Prajapati, B. G., Duraiswamy, B., & Palaniswamy, D. S. (2023). 7 Machine Data Mining Learning and. *Artificial Intelligence in Bioinformatics and Chemoinformatics, 117*.

Nayak, S., & Sharma, Y. K. (2023). A modified Bayesian boosting algorithm with weight-guided optimal feature selection for sentiment analysis. *Decision Analytics Journal*, *8*, 100289. doi:10.1016/j.dajour.2023.100289

Onyema, E. M., Lilhore, U. K., Saurabh, P., Dalal, S., Nwaeze, A. S., Chijindu, A. T., Ndufeiya-Kumasi, L. C., & Simaiya, S. (2023). Evaluation of IoT-Enabled hybrid model for genome sequence analysis of patients in healthcare 4.0. *Measurement. Sensors*, *26*, 100679. doi:10.1016/j.measen.2023.100679

Pandey, M., Fernandez, M., Gentile, F., Isayev, O., Tropsha, A., Stern, A. C., & Cherkasov, A. (2022). The transformational role of GPU computing and deep learning in drug discovery. *Nature Machine Intelligence*, *4*(3), 211–221. doi:10.1038/s42256-022-00463-x

Perera, G., Holbrook, A., Thabane, L., Foster, G., & Willison, D. J. (2011). Views on health information sharing and privacy from primary care practices using electronic medical records. *International Journal of Medical Informatics*, *80*(2), 94–101. doi:10.1016/j.ijmedinf.2010.11.005 PMID:21167771

Prabhakar, C., Sekuboyina, A., Paetzold, J. C., Li, H., Amiranashvili, T., & Kleesiek, J. (2021). *Improving Generalized Zero-Shot Learning for Multi-Label Chest X-ray Classification Using Knowledge Graphs.*

Pradeep, S., & Sharma, Y. K. (2019, February). A pragmatic evaluation of stress and performance testing technologies for web-based applications. In *2019 Amity International Conference on Artificial Intelligence (AICAI)* (pp. 399-403). IEEE. 10.1109/AICAI.2019.8701327

Qiu, J., Li, L., Sun, J., Peng, J., Shi, P., Zhang, R., Dong, Y., Lam, K., Lo, F. P.-W., Xiao, B., Yuan, W., Wang, N., Xu, D., & Lo, B. (2023). Large AI models in health informatics: Applications, challenges, and the future. *IEEE Journal of Biomedical and Health Informatics*, *27*(12), 6074–6087. doi:10.1109/JBHI.2023.3316750 PMID:37738186

Qiu, Y., & Wei, G. W. (2023). Artificial intelligence-aided protein engineering: From topological data analysis to deep protein language models. *Briefings in Bioinformatics*, *24*(5), bbad289. doi:10.1093/bib/bbad289 PMID:37580175

Ramesh, T. R., Lilhore, U. K., Poongodi, M., Simaiya, S., Kaur, A., & Hamdi, M. (2022). Predictive analysis of heart diseases with machine learning approaches. *Malaysian Journal of Computer Science*, 132–148.

Rathore, M. S., Poongodi, M., Saurabh, P., Lilhore, U. K., Bourouis, S., Alhakami, W., Osamor, J., & Hamdi, M. (2022). A novel trust-based security and privacy model for the internet of vehicles using encryption and steganography. *Computers & Electrical Engineering*, *102*, 108205. doi:10.1016/j.compeleceng.2022.108205

Ross, M. W., Iguchi, M. Y., & Panicker, S. (2018). Ethical aspects of data sharing and research participant protections. *The American Psychologist*, *73*(2), 138–145. doi:10.1037/amp0000240 PMID:29481107

Sanchez-Fernandez, A., Rumetshofer, E., Hochreiter, S., & Klambauer, G. (2022, April). Contrastive learning of image-and structure-based representations in drug discovery. In ICLR2022 Machine Learning for Drug Discovery.

Schmit, C., Sunshine, G., Pepin, D., Ramanathan, T., Menon, A., & Penn, M. (2017). Transitioning from paper to digital: State statutory and regulatory frameworks for health information technology. *Public Health Reports*, *132*(5), 585–592. doi:10.1177/0033354917722994 PMID:28813620

Sharma, S. K., Lilhore, U. K., Simaiya, S., & Trivedi, N. K. (2021). An improved random forest algorithm for predicting the COVID-19 pandemic patient health. *Annals of the Romanian Society for Cell Biology*, 67–75.

Trinidad, M. G., Platt, J., & Kardia, S. L. R. (2020). The public's comfort with sharing health data with third-party commercial companies. *Humanities & Social Sciences Communications*, *7*(1), 149. doi:10.1057/s41599-020-00641-5 PMID:34337435

Wang, S., Jiang, X., Singh, S., Marmor, R., Bonomi, L., Fox, D., Dow, M., & Ohno-Machado, L. (2016). Genome privacy: Challenges, technical approaches to mitigate risk, and ethical considerations in the United States. *Annals of the New York Academy of Sciences*. doi:10.1111/nyas.13259 PMID:27681358

Zhang, R., Sun, J., Peng, J., Qiu, Y., & Lo, B. (2023). Big data analytics for wearable sensor-based health applications. *IEEE Transactions on Biomedical Engineering*.

Chapter 12
Exploration of Deep Learning and Transfer Learning in Bioinformatics

Yash Mahajan
https://orcid.org/0009-0000-2530-0366
Chandigarh University, India

Muskan Sharma
https://orcid.org/0009-0005-0397-0736
Chandigarh University, India

ABSTRACT

In the dynamic field of bioinformatics, the fusion of deep learning and transfer learning techniques has ushered in a transformative era of discovery. Deep learning, powered by neural networks, has emerged as a formidable tool for deciphering intricate biological data, spanning genomics, proteomics, and metabolomics. Its capacity to automatically discern complex patterns and features within vast datasets has revolutionized our comprehension of biological processes. Transfer learning, a complementary approach, introduces a novel dimension to bioinformatics. By harnessing pre-trained models from domains like computer vision and natural language processing, researchers can expedite their analyses, reducing the demand for extensive labeled data. This cross-domain knowledge transfer expedites the progress of bioinformatics applications, from predicting protein structures to identifying disease biomarkers.

1. INTRODUCTION TO BIOINFORMATICS

A subject that straddles the line between different fields of study, bioinformatics has also transformed our understanding of biology. It sits at the intersection of biology, computing and mathematics as well as data analysis-the goal is to apply IT technology to life's mysteries. This chapter takes you on a journey into the mysterious realm of bioinformatics, exploring its meaning and scope along with-most importantly--its huge importance for modern life sciences.

DOI: 10.4018/979-8-3693-1822-5.ch012

1.1 Definition and Scope of Bioinformatics

Bioinformatics itself focuses on the organization, analysis and interpretation of biological information. Just a few examples of these many biological subfields are genetics, structural biology and proteomics. Its ability to translate this raw biological information into useful knowledge is really its true nature as a tool for joining traditional biological science with the digital world. Most people explain bioinformatics in terms of a translating device that bridges the gap between two languages, computer language and molecular (DNA, RNA) protein language(Löchel & Heider, 2021). What it provides is the approaches and tools needed to understand those vast compendiums of biological data generated by modern technologies. This makes it easier to identify previously unseen connections, structures and theories that could never have been distinguished through mere manual research(Bornes et al., 2023).

Genetics is one of the cornerstones on which bioinformatics relies. Genomics refers to the study of an organism's entire genome, or set of genes. As high-throughput DNA sequencing techniques have come on the scene, so too has genomic data multiplied by leaps and bounds. By applying computer tools and bioinformatics, scientists can crack genetic code, identify genes related to diseases or those that contributed in the past to evolutionary processes(Hua et al., 2023). None of these vocations would be possible without bioinformatics. In addition to genomics and proteomics (the study of a living thing's full complement of protein genes), bioinformatics also embraces structural biology-research into the three dimensions of molecules. Investigating interactions between likely drugs and target proteins can be crucial to the success of new treatments(Lai et al., 2019). Global food security is further enhanced with the assistance provided by bioinformatics in developing genetically altered crops which are more productive and resistant to pests.

Let ' s take a look at the rest of the book. As we do so, you will gain an understanding for how bioinformatics integrates with breakthrough technologies such as deep learning and transfer learning to fully exploit biological data resources. These technologies, combined with a degree of precision and efficiency previously unimaginable to most people, are helping us grasp complex biological processes; predict the structures of proteins; or expedite the process for new treatments(N. Wang & He, 2023).

1.2 Importance of Bioinformatics in Life Sciences

The binding factor in this spider's web of biological sciences is bioinformatics, which allows disparate disciplines to come together and helps academics tackle many of the most serious problems facing today's world. In fact, owing to its importance for so many areas of the biological sciences and spurring innovation on all fronts no one can stress this point enough(Banimfreg, 2023).

1.2.1 Advancing Biomedical Research

One of bioinformatics 's greatest advantages, biomedical study shows. It is hoped that scientists can use computers and databanks to discover genes linked with illness, understand their functions and provide new areas for treatment(Nguyen et al., 2022). Such information provides the foundation for personalized medicine, using a patient's genetic profile to tailor drugs and improve efficacy of treatments while reducing unwanted side-effects. Under this approach, bioinformatics is no longer just a tool in the battle against disease.

1.2.2 Accelerating Drug Discovery

Medical research and development is notoriously a very time-consuming, money Wasting endeavor. It alters this environment, making it easier to quickly screen tens of millions of chemical compounds against target proteins and make predictions about their interactions before defining candidates for experimental confirmation. This is bioinformatics in action(Mousser et al., 2022). This speeds the delivery of life-saving pharmaceuticals to patients and makes drug discovery pipelines shorter, reducing cost.

1.2.3 Enhancing Agriculture and Food Security

In a world dealing with population growth and climate change, bioinformatics is essential to agriculture. By analyzing plant genomes and discovering the genetic components responsible for disease tolerance or resistance to drought, even hail(Iqbal & Kumar, 2023). bioinformaticians are involved in designing crops that may be able one day to withstand harsh conditions. This is to ensure sustainable agriculture and food security.

1.2.4 Unraveling Evolutionary Mysteries

The field of bioinformatics is one that can be very useful both in evolution and biodiversity research. Comparative genomics can reconstruct the evolutionary history of a species, identify genes that remained unchanged over time and better understand how speciation occurs(Bouzid et al., 2023). This expands our understanding of nature and ourselves.

1.2.5 Managing Public Health Crises

The entire world has recently faced the unprecedented public health challenges of a range of pandemics, chief among which is presumably related to novel Coronavirus (COVID-19). Bioinformatics not only made rapid sequencing of the SARS-CoV-2 virus possible, it also enabled the development of diagnostic tools and treatments(H. Wang & Hu, 2023). Why is viral genomic research essential to tracking infectious disease transmission and monitoring mutations?

Thus, bioinformatics can be seen as a discipline that not only brings life to the computer screen but is an awakening for researchers and finds solutions to some of mankind's worst ills. As we enter the following chapters, let us see whether advanced machine learning methods and innovative bioinformatics can combine forces to help open up new vistas in medicine. Let's go on a tour together that will emphasize the vast influence of deep learning and transfer learning in bioinformatics, so as to more solidify our search for answers about enigmas surrounding life itself(Sidorova & Lozano, 2023).

2. INTRODUCTION TO DEEP LEARNING AND TRANSFER LEARNING

The same goes for expertise in biological data, which requires deep learning and transfer learning as the two pillars of machine-based interpretation. These are the tools that have enabled bioinformatics to get a second wind, allowing researchers to better understand biological phenomena at an even more detailed level, predic tthe effects of changes with previously unthinkable accuracy and bring together scattered biological data(Armoogum & Li, 2019). In this chapter, we'll take a journey to understand the core concepts of deep learning; learn about neural net structures; and come to terms with some seasons--including transfer leaning that is necessary for research into bioinformatics.

2.1 Basics of Neural Networks

2.1.1 The Neuron: The Building Block of Neural Networks

The basic computing units of neural networks are called 'neurons'. In Figure [1], how the artificial network modeled on the biological neuron receives a number of inputs, weighs them up and adds them together before sending an output based on some function is explained in English text to describe what happens after various inputs have been sent into it by way of many different pathways? This output becomes an input to further neurons, and thus layers that are connected can perform simultaneous hierarchical processing of information(Hasija, 2023).

Figure 1. Basics of neural networks

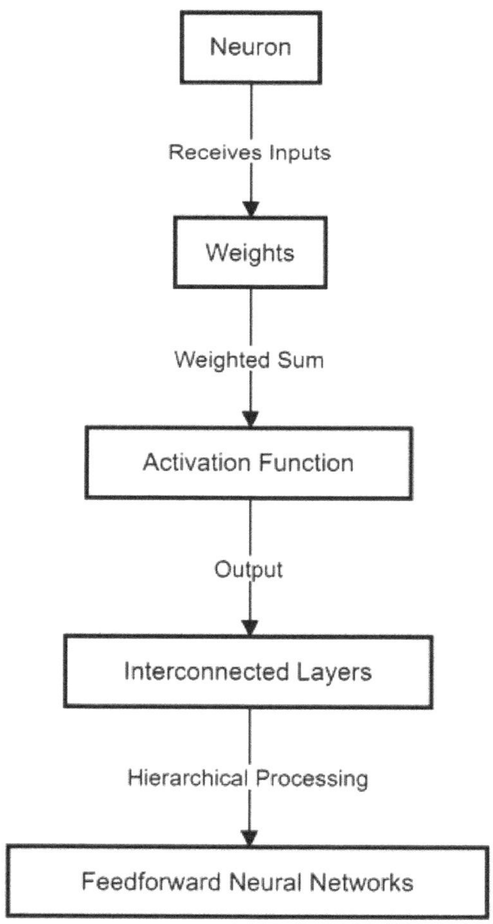

2.1.2 Feedforward Neural Networks (FNNs)

Feedforward networks are the most simple kind of neural network. They consist of an output layer, an input layer and a hidden one or two layers. There are multiple neurons in each layer, and the weights describe connections between them. Such networks are called " feedforward " because there is no feedback, and information moves exclusively in one direction from the input layer to the output layer.

2.1.3 Activation Functions

Artificial neurons must have activation functions to function. They give the network non-linearity, allowing it to represent intricate data interactions. The sigmoid, hyperbolic tangent (tanh), and rectified linear unit (ReLU) are often used as activation functions(Daphal & Koli, 2023). Each has distinctive qualities and is appropriate for certain issues.

2.1.4 Training Neural Networks

During the training process, differences between expected and actual outputs are minimized by adjustment of connection weights. The usual way of doing this is with optimization techniques such as gradient descent, which slowly adjusts weights to find the best values. But in order to learn from these cases, a labeled dataset (in which the correct answers are already known) is required for network training.

2.2 Deep Learning Architectures

2.2.1 Convolutional Neural Networks (CNNs)

Convolutional neural networks, or CNNs for short, have changed image and sequence data analysis. They can be made to detect and identify patterns in data automatically. They are adaptive. The building block of a convolutional neural network is a convolution layer containing filters, which learn how to detect features in the data(Bunyat-zada & Ross, 2023). These filters, which are learned during the training of a network, allow it to identify hierarchical elements in images such as edges or textures and objects.

2.2.2 Recurrent Neural Networks (RNNs)

Since RNNs are designed for sequence data, bioinformatics applications requiring temporal modelling (i.e., chronological hierarchy) find them very useful indeed. Unlike feedforward networks which store no information about their inputs, RNNs have linearly organized connections that trace back to themselves so they can remember frequency of past inputs. With their memory capability, RNNs are ideally suited for such tasks as gene expression analysis or natural language processing(Zafar et al., 2023).

2.2.3 Long Short-Term Memory (LSTM) Networks

The LSTM networks, one of the specialized varieties discussed below that can alleviate this problem and thus benefit deep networks. By adding gating mechanisms, LSTMs can also choose what to remember and forget from previous time steps(Tan et al., 2021). This makes them specially suited to tasks such as protein structure prediction that require the capture of long-range relationships.

2.3 Transfer Learning: Concepts and Significance

2.3.1 What is Transfer Learning?

Transfer learning is a machine learning technique whereby knowledge from one task can be applied to improve results in another related area of endeavor. In the realm of neural networks, it involves beginning new work from continuously trained models. Behind this idea is our experience of studying and constantly drawing on what we have studied to solve the problems in another field.

2.3.2 Transfer Learning in Bioinformatics

In bioinformatics, where data is sparse and labeled examples few to none as mentioned in Figure. [2], transfer learning offers a lifeline of sorts for the practitioner. By starting from pre-trained models resulting from the training of huge and varied sets, bioinformaticians can tailor those tools to biologically specific needs (e.g., protein functionality prediction or drug target interaction prediction)(Nematzadeh et al., 2022). This also reduces the quantity of data required for training, and speeds up model convergence.

Another problem to tackle is the " no free lunch "theory that holds there ' s not one machine learning model which will work for all tasks, and transfer theory relies on tackling this. Of course, bioinformatics researchers can also choose to build on the pre-trained models that already have built in expertise while adapting them for biological data.

Because their use in bioinformatics will be introduced later, this chapter only gives a preliminary introduction to the concepts of deep learning and transfer learning. For example, transfer learning fortifies the capabilities of neural networks by building on cumulative experience; deep learn-ing can provide a powerful toolacro for modeling complex biological data(Nematzadeh et al., 2022). Through the rest

Figure 2. Transfer learning in bioinformatics

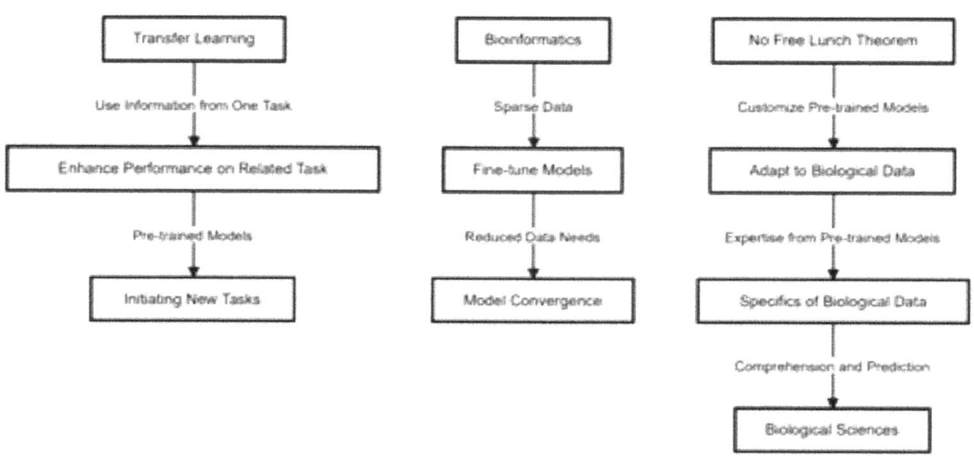

of this book we will explore how these concepts have revolutionized bioinformatics and created new opportunities for understanding and prediction in biology.

3. DEEP LEARNING TECHNIQUES IN BIOINFORMATICS

Bioinformatics involves vast amounts of complex biological data and requires highly innovative calculations to uncover its true treasures. This chapter also explains in detail how convolutional neural networks are used to interpret genetic data and predict protein structure.

3.1 Convolutional Neural Networks (CNNs) in Genomic Data

Life, as told by genes Living creatures 'DNA sequences are the stuff from which that tapestry is woven. The unraveling of such sequences is crucial to the understanding of genetics, evolution and diseases-not an easy task at all. An important ally in this endeavor is the emergence of CNNs which can learn features simply from data(Yu et al., 2023).

3.1.1 CNNs for DNA Sequence Analysis

3.1.1.1 Motif Detection: Uncovering the Genetic Patterns

The motifs are repeated patterns found in portions of the DNA, and these serve as a foundation for gene regulation, protein binding events and evolutionary preservation. But these motifs are hidden away within the nucleotide sequences of random appearance. Through trained filters which recognize these key patterns whether they occur at the beginning, middle or end of a sequence, CNNs are particularly adept at motif detection. It has the ability to identify potential regulatory elements, transcription factor binding sites and other biologically significant DNA patterns. One of its functions is to enable better understanding about genetic mechanisms(Y. Wang et al., 2023).

3.1.1.2 Predicting Functional Regions: The Genomic Landscape

The genomic environment cannot be homogeneous; there are not only codes, but also non-coding regions. These then can be trained to separate these regions, identify splice sites and predict how mutations will affect functionality. This knowledge is important for annotating the genome, discovering disease-related variations and determining the hereditary nature of diseases.

3.1.1.3 Classification and Prediction: Beyond Sequences

The powers of the CNN, however are not limited to single-pass analysis. They excel at regression and multi-class classification applications. For example, these networks can classify DNA sequences into categories such as promoters, enhancers and a section that codes for proteins (Ma et al., 2023). They are also capable of forecasting significant biological properties such as affinity for binding to the gene, RNA secondary structure and protein interactions. Predictions like these are transforming the very way we think about biological problems (Li et al., 2019). They have an enormous influence upon functional genomics research, drug discovery and personalized medical interventions.

Figure 3. CNNs for protein structure prediction

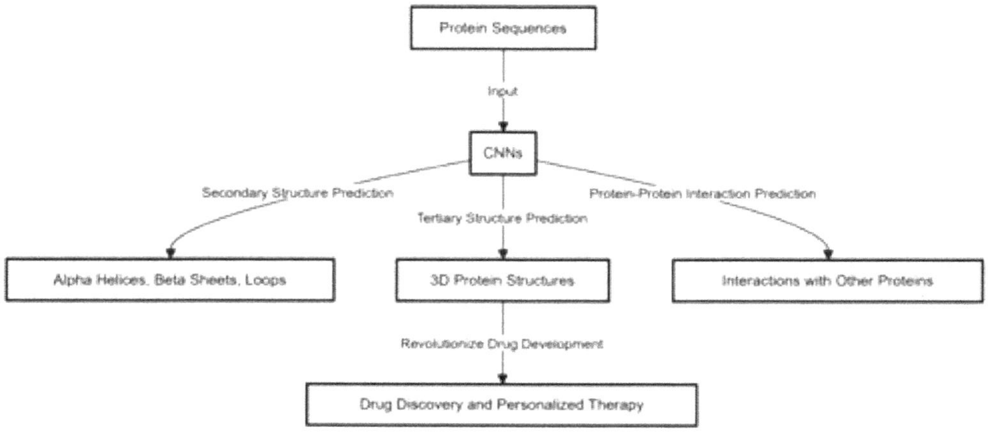

3.1.2 CNNs for Protein Structure Prediction

The biological labor force; per form a variety of functions vital to cellular operation. Discovering their proper roles in diseases and functions will be directly related to an understanding of these 3D structures. How these new CNNs work with the convolving trees that they are linked to, has for predicted protein structure) tricky problem--has changed beyond recognition. Backbone " tree convolutions " (Du et al., 2023.as process mentioned in the Figure. [3].

3.1.2.1 Secondary Structure Prediction: Unveiling the Building Blocks

One example of early progress has been the use of CNNs to predict components (including alpha helices, beta sheets and loops) in secondary protein structure. You et al., 2023). Moreover, these constituents make up the structure of proteins. When trained on large protein databases and structures, CNNs have proven capable of predicting secondary structure with a very high degree of accuracy This provides us with a means of understanding protein structure and function, thereby moving toward the resolution of the riddle that is proteins.

3.1.2.2 Tertiary Structure Prediction: The Holy Grail

Predicting a protein's three-dimensional structure from its amino acid sequence is the ultimate goal of protein bioinformatics. Thanks to the arrival of CNNs a new age in this field has begun. Deep learning methods, such as AlphaFold's tertiary structure prediction algorithm have shown heretofore unthought-of levels of accuracy. With these breakthroughs, one can envision a future in which the structures of proteins could be accurately predicted from their gene sequences. This will have wide-ranging implications for drug design and personalized medicine, as well as fundamental knowledge about diseases at the molecular level.

3.1.2.3 Protein-Protein Interaction Prediction: Collaborative Networks

Independent functioning of proteins is extremely rare, and most require other protein interactions. Researchers have used sequence and structural data to train CNNs, which show great potential as predictors of protein-protein interactions(Barbero-Aparicio et al., 2024). From understanding signaling pathways to determining how disease causes dysfunctions, these types of predictions have numerous different applications. Most notably they can be used to develop individually targeted treatments with incredible precision that will prove revolutionary for medicine at large.

To conclude, convolutional neural networks have opened a new era of data-driven discovery for bioinformatics. As a result, the area has witnessed a revolution through its ability to automatically find features in sequences of DNA and proteins that allow researchers to learn their esoteric knowledge at computer speed. As we proceed with other chapters of this book, which concern applications in bioinformatics and describe a path through to truly breakthrough achievements for the life sciences, deep learning techniques must not be ignored.

3.2 Recurrent Neural Networks (RNNs) for Temporal Data

3.2.1 RNNs for Gene Expression Analysis

Yet since genes themselves are not fixed, gene expression levels may change with time and circumstance. That's because these temporal relationships are crucial to the understanding of regula tory mechanisms driving cellular function. Because the RNN can detect associations that are sequential in nature, it is now an indispensable gene expression analysis tool.

3.2.1.1 Modeling Temporal Dynamics

The temporal relationship between genes is another aspect in which RNNs excel. So they can find fine patterns in time-series gene expression, recording how the genes respond to different environments and at different stages or under certain disease conditions. This skill is also needed to identify key regulators, explain gene regulatory networks and describe the molecular mechanisms underlying biological processes.

3.2.1.2 Time Series Forecasting

Thus RNNs are perfectly suited to time series prediction, which is crucial for gene expression studies. This past data can be employed to predict the expression of genes when experiments are conducted under different conditions, thus improving research techniques. Such types of predictive power are important in the planning and construction of experiments, or for creating effective targeted treatments (for example drug development), personalized medicine.

3.2.1.3 Identifying Biomarkers and Disease Signatures

This means that problems of illness and finding appropriate biomarkers should be approached by analyzing gene expression. This means finding patterns and properties behind gene expression at specific times, and RNNs are essential for this. With this, research will be able to identify disease signatures specific enough for early diagnosis and prognosis and the design of precision medicine treatments.

Figure 4. LSTM networks in bioinformatics

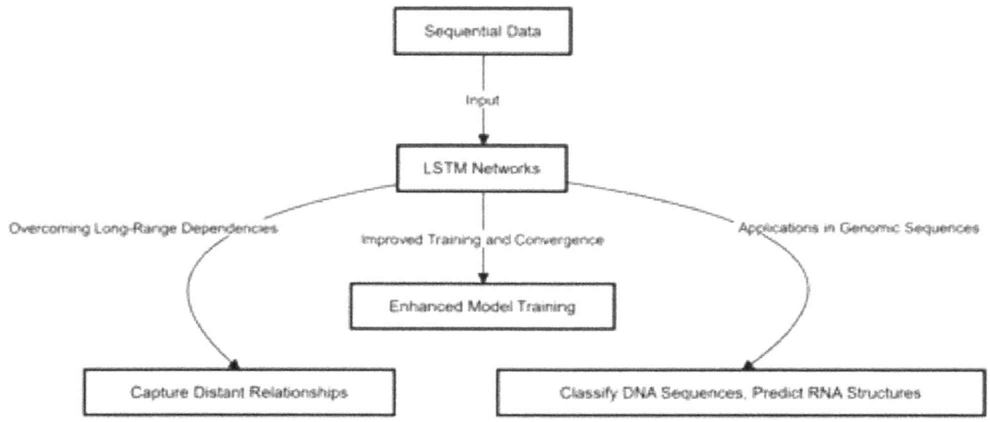

3.2.2 Long Short-Term Memory (LSTM) Networks in Bioinformatics

They excel at sequential learning but can have issues resolving long-range dependencies and gradients during training, as in the case of RNNs. LSTM networks, a specialized kind of RNN that works to overcome these problems, have found widespread application in bioinformatics. As described in Figure. [4], LSTM networks have been used in the field of bioinformatics to achieve heart-like precision (or accuracy) when identifying potentially dangerous people.

3.2.2.1 Overcoming Long-Range Dependencies

LSTM networks can capture and store a sequence of data over very long periods thanks to memory cells. For example, with such tasks as protein folding prediction where the interactions between distant amino acids can affect distinct portions of a given protein's shape, this ability to deal with long-range dependencies becomes extremely valuable.

3.2.2.2 Improved Training and Convergence

LSTMs add gating mechanisms to control information flow in the network. These methods help to make LSTMs even more suitable for use with deep networks by resolving the vanishing gradient problem. This enhances bioinformatics training and convergence, enabling the creation of accurate prediction models.

3.2.2.3 Applications in Genomic Sequences and Beyond

Besides genomic sequence analysis, LSTM networks also can be of assistance in a number of other bioinformatics applications. They excel at distinguishing variations in DNA sequences, estimating RNA secondary structure and performing research on functional genomics. There adaptability and aptitude for describing precise relationships make them an indispensable tool in the bioinformatics arsenal.

The deeper we venture into bioinformatics and the application of deep learning, the more apparent becomes RNNs 'flexibility and LSTM networks' precision at grasping biological time. These tools will help scientists understand the dynamic principles behind gene expression and enable them to make even more accurate predictions, advancing understanding of molecular systems regulating our life. In the chapters that follow, we will focus on current cutting-edge deep learning techniques and applications in bioinformatics to sketch out where new breakthroughs in the life sciences are headed.

3.3 Deep Learning for Functional Annotation

Amidst the vast landscape of biological data, functional annotation serves as a kind of beacon to guide researchers through the labyrinthine network that constitutes life. Deep learning techniques have now become worthy allies, which help us reveal hidden functions and discover new correlations in bioinformatics.

3.3.1 Predicting Gene Function With Deep Learning

But genes, the essence of life that encompasses its genetic code, are not stand-alone items but rather basic elements in biological networks. Bioinformatics tries to understand how these genes operate, and so understanding their functions in various biological processes and the properties of cellular machinery with which they interact.

3.3.1.1 Leveraging Multi-Omics Data

Bite out of gene function prediction Deep learning's strong suit is its ability to ingest and process multi-omics data. Genomics, transcriptome, proteomes and so on are also very easily integrable into deep learning models. From this diverse collection of information, these models can predict gene functions more accurately than previously. A strategy of such breadth goes beyond single-molecule research and allows one to observe gene activities within an overall biological environment. For instance, integrated physiological data can be illuminated by deep learning to reveal regulator relationships between genes and their functions in the cellular processors--illuminating dark corners like diseases or how people may react to stimuli from an environment.

3.3.1.2 Functional Inference From Sequence and Structure

Their gene sequences and protein structures are both rife with functional information. Turns out that deep learning models excel at acquiring these insights. Depending upon their sequences and predicted structures, they can discover functional motifs, identify conserved domains or even predict protein activities. In practice, by understanding the DNA or RNA sequences of the genes researchers can use deep learning to determine whether a given function belongs with them and annotate their genome accordingly. In addition to this functional annotation, deep learning can assemble information about a protein's activities directly from the expected structure and sequence of the molecule. This knowledge is needed to find targets for treatment and clarify the molecular mechanisms of biological processes.

3.3.2 Deep Learning in Drug Discovery

Finding the right drug candidates and then successfully developing them is a difficult job, but it's one that every pharmaceutical company must tackle. As such, deep learning has become a potent catalyst in the search for new drugs and precise treatment of old ones.

3.3.2.1 Drug Target Prediction

Choosing the correct target proteins to act upon is an extremely essential part of drugs research. It has been discovered that deep learning algorithms are extremely effective in targeting potential treatments. He explains that these models use biological data including protein sequences, structures and interactions to suggest for researchers which leads they should follow. This targeted approach not only makes it easier to test potential therapeutic candidates, but the speed of development is accelerated. Deep learning is woven into Drug Discovery as shown in Figure. [5]. This help researchers concentrate on effectiveness, increasing the chances of therapeutic success while cutting down enormously on time and expense to produce a new medication.

3.3.2.2 Drug Compound Design and Optimization

Deep learning is transforming how drugs are designed and optimized. Generative models can greatly accelerate the drug-discovery process by developing new chemical compounds with desired characteristics. In addition, deep learning models can also be used to predict candidates ' therapeutic responses for the relevant target proteins. This makes it more efficient and convenient to select promising substances for further study. With this predictive capability, researchers will be able to design or revise pharmacological molecules with a precision and speed that was not previously feasible. This may well have an epochal impact on drug discovery research processes everywhere around the world15). It also has the possibility of accelerating development of new drugs and providing more effective treatments for various diseases.

3.3.2.3 Drug Repurposing and Personalized Medicine

Outside of traditional drug discovery, deep learning is also being applied to medication repurposing and personalized medicine. Drug repurposing means finding drugs already on the market for new uses. These potentials for drug repurposing are exactly what deep learning models excel at digging up by reviewing many different types of data. Personalized medicine, meanwhile is putting deep learning to work on patient-specific data such as genetic profiles and clinical records. This personalized information will help researchers and healthcare professionals better prescribe local, specific treatments (instead of blanket medication), reduce side effects from treatment while also allowing for more effective responses to patient outcomes.

To summarize, deep learning has revolutionized drug discovery and brought a new age of functional annotation to bioinformatics. Its ability to synthesize information from multiple sources and its prognostic capabilities has enabled it to move into the vanguard of cutting-edge research in both biological sciences and pharmaceuticals, thereby helping accelerate drug development.

4. TRANSFER LEARNING IN BIOINFORMATICS

Getting in gear In bioinformatics, which handles the massive data crunching technologies needed to interpret vast amounts of biological information, there is no end to finding reliable and accurate solutions. As we continue our investigation into deep learning and its applications, another powerful cross-applications techniquethat is catching on recently is called transfer Learning. This heading 4 explores the applications of and revolutionary capability for Transfer Learning in Bioinformatics.

Figure 5. Deep learning in drug discovery

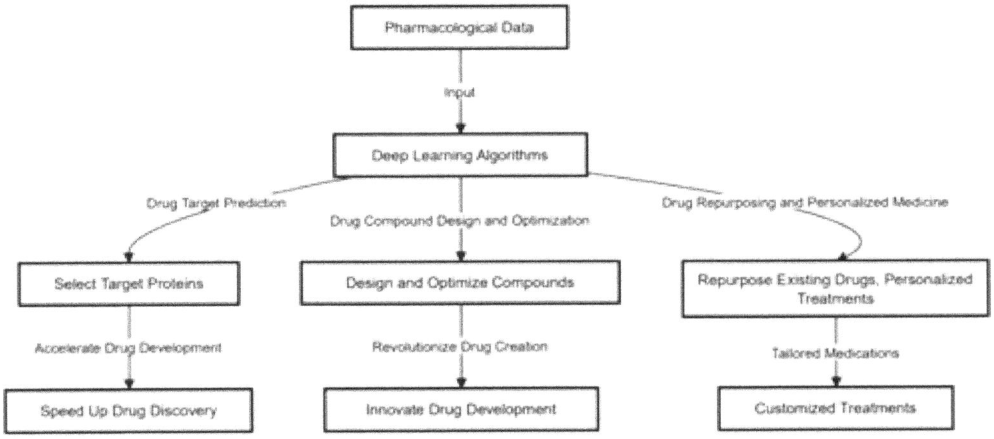

4.1 Transfer Learning Applications in Bioinformatics

Extraction of knowledge from biological data is the crux of bioinformatics. Model Portability, a technique borrowed from the machine learning community, allows us to transfer knowledge gained in one task or data set over into another closely related labor requiring similar skills. All of this raises many possibilities for bioinformatics work.

4.1.1 Transferring Knowledge for Protein Function Prediction

Each protein is a biological workhorse fulfilling some particular job necessary to life's running. The other major shortcoming in bioinformatics is predicting the functions of proteins, which can be applied to many different fields from investigating biological systems themselves and drug development. As illustrate in Figure. [6], How does Transfer Learning guide the protein sequencing and structural mapping project? In this endeavor, transfer learning is like a compass.

4.1.1.1 Leveraging Pretrained Models

Another characteristic of Transfer Learning is the use of pretrained models. When pretrained models are used to predict protein function, they often contain all kinds of knowledge about protein sequences and structures. Trained on massive, varied data sets these models have often learned more complicated patterns and interactions that transcend any particular task.

These pre-trained models can be used only as generalized knowledge. Researchers then just need to optimize them on specific tasks of protein function prediction. This method not only improves the accuracy of predictions, it also speeds up model convergence. This in turn enables one to predict protein function with the accuracy and effectiveness that is otherwise impracticable.

Figure 6. Transfer learning for protein function prediction

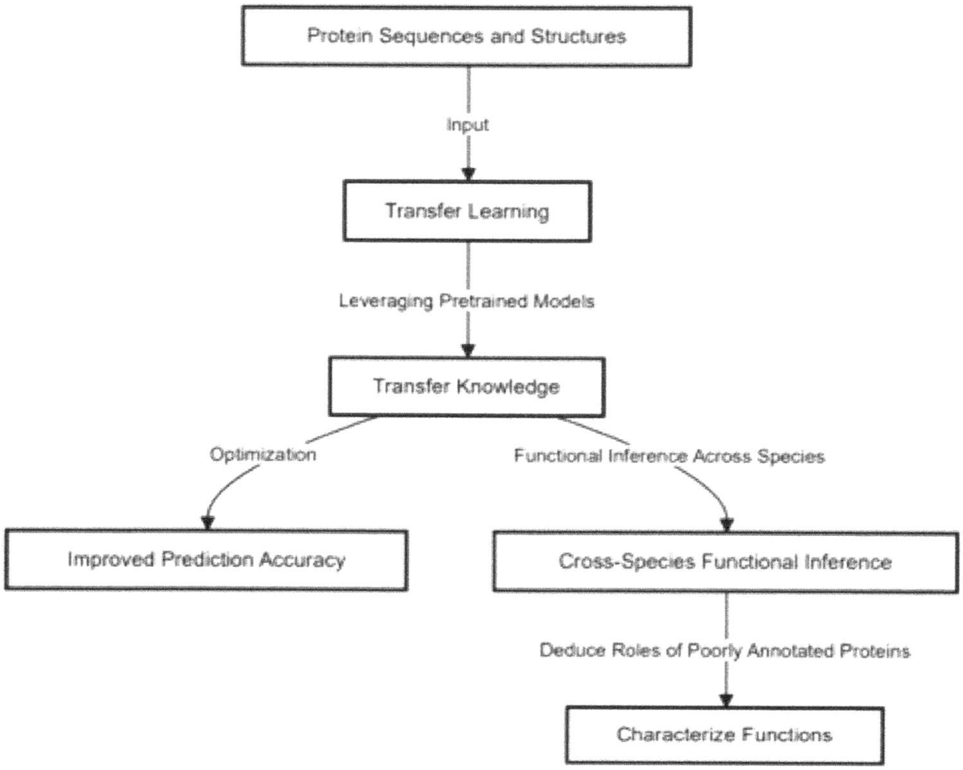

4.1.1.2 Functional Inference Across Species

This huge diversity of protein sequences and functions is representative of life on earth. Cross-species knowledge transfer Transfer learning helps us to negotiate this diversity. For example, if the second species has few annotated data sets, models trained on protein sequences and functions of the first can be adapted to make predictions for it.

This cross-species sharing of knowledge has far-reaching implications. This helps us to infer the functions of poorly annotated proteins in biological processes and gain insights into their biology. Furthermore, this enables us to engage in comparative genomics, mapping out evolutionary relationships and pinning down functions which are shared across species.

4.1.2 Drug Discovery With Transfer Learning

Indeed, the task of locating potential drug targets and bringing them to market is a huge endeavor, relying on accurate estimates and selective screening. In this field, Transfer Learning seems like an oasis of hope that offers high-end techniques with the potential to revolutionize drug development.

4.1.2.1 Transferring Drug-Target Knowledge

The most important stage in drug development is selecting the right target proteins to be pharmacologically targeted. The idea is to find the key that opens a certain biological lock or pathway and grants access to treatment. But target selection is hard work. It can be time and manpower consuming, exhaustive. In terms of this, Transfer Learning really can be a game-changer.

4.1.2.1.1 Knowledge Amplification

By applying transfer learning, that wealth of collective wisdom can be basis for drug-target interactions in the future. For example, if in the past a particular protein that caused some rare disease has been treated successfully with this medication. Understanding this interplay is priceless. One such application of transfer learning could be the identification of potential therapeutic targets for a protein similar to one already mentioned, but biologically unrelated. In short, transfer learning takes advantage of all the insights gained from many hours of study and trial. Providing a road map of drug-target interactions helps us to find our way through this bizarre world. This speeds up the target identification process and increases the possibility of successfully developing effective treatments.

4.1.2.1.2 Precision in Target Selection

The drug discovery process must be extremely precise. Choosing the right target could be key to discovering a successful drug candidate or an innovative therapy. With the help of Transfer Learning, scientists can select proteins to be targeted from data21. By looking for similarities between proteins and examining patterns in drug-protein interactions, the Transfer Learning models can discover drugs that possibly target these proteins. Besides accelerating the drug discovery process, this knowledge-based approach also increases the probabilities of identifying targets that are strongly relevant to biological pathology and likely to be responsive.

4.1.2.2 Rapid Compound Screening

Another case in which Transfer Learning is applied to drug discovery involves compound screening, a crucial step invvetigating the possibility of new drugs. Traditional compound screening involves doing labor-intensive and costly experimental work, often on a large set of compounds. Through Transfer Learning, this process is redesigned to make it faster, simpler and cheaper-to say nothing of the fact that data sources are available.

4.1.2.2.1 Learning from Chemical-Biological Relationships

Pretrained by Transfer Learning, pre-existing associations between chemical structure and biological function can be identified within it. This helps them learn how to identify patterns in the chemical makeup of substances that correspond with certain biological results. That is to say, a model can now understand what makes molecules effective against particular targets. Refining these models on new compound databases allows researchers to speed the screening of compounds for therapeutic potential. Instead of randomly testing thousands of chemicals in the laboratory, researchers can use this model to first choose those with a high probability of success.

4.1.2.2.2 Accelerating Drug Development

But traditional drug development requires time and capital to complete. Clinical trials, toxicity testing and compound improvement are all time-consuming, expensive endeavors. To speed this journey, Transfer Learning shortcuts the initial screening process.

This way, the promising molecules desired for medication can be found early in the development process. Reducing the time and costs associated with experimental screening helps scientists focus on those drugs most likely to succeed. In fact, Transfer Learning in the end is a shortcut for drug development that will hasten new treatments to patients. To conclude, Transfer Learning has a variety of far-reaching uses in drug development. Interestingly, transfer learning brings with it a number of creative extrapolations that can transform the drug development landscape by taking advantage of what has been learned from past interactions between drugs and targets while remodeling compound screening. We will be exploring Transfer Learning in chapters to come, including how this technology can break new ground for bioinformatics and solve some of the most pressing problems in life sciences.

5. FUTURE DIRECTIONS IN DEEP LEARNING AND TRANSFER LEARNING IN BIOINFORMATICS

Now we are entering a phase of revolutionary advances in deep learning and transfer leaning as the bioinformatics landscape continues to evolve. This chapter explores the trend possibilities that are looking toward influencing and inspiring innovation in the future.

5.1 Future Directions in Deep Learning and Transfer Learning in Bioinformatics

The technological development in bioinformatics is accelerating, driven especially by the recent developments of deep learning and transfer learning constructs. Looking into the future, a number of key fields can be identified where research and development is taking place.

5.1.1 Advancements in Deep Learning Models

It is thus widely believed that the rapid emergence of deep-learning models will be critical to future advances in bioinformatics. Backed by increasing computer power and access to data, these models offer us new vistas in our understanding of biological systems.

5.1.1.1 Neural Architecture Search (NAS)

The creation of automated methods for neural architecture search (NAS) is one of the most promising areas for progress. NAS methods independently investigate and identify the best neural network designs for a given bioinformatics problem. This not only hastens model generation but also produces highly effective, specialized models that are superior in fields like structural biology, genomics, and proteomics.

5.1.1.2 Explainable and Interpretable Models

A long-standing issue with deep learning models is their "black-box" character. It is anticipated that future developments will concentrate on improving model interpretability. Explainable AI techniques will make deep learning models' predictions clearer and more usable, especially in clinical and healthcare settings by shedding light on their decision-making procedures.

5.1.2 Integration with Other Bioinformatics Techniques

Although deep learning and transfer learning have shown to be valuable, the future of bioinformatics rests in the interaction of these two methods with more conventional bioinformatics techniques. Integrating deep learning with established techniques creates new opportunities for comprehensive analysis.

5.1.2.1 Multi-Modal Data Integration

In the age of big data, bioinformatics frequently entails the integration of data from several sources, such as clinical records, genomes, proteomics, and metabolomics. Gaining a thorough understanding of intricate biological processes and diseases would require the use of deep learning models that are capable of combining and analyzing data from various sources.

5.1.2.2 Personalized Medicine and Healthcare

One possible research direction is the tailored health care system, with different patient treatments based on an individual's genetic makeup along with his or her lifestyle habits and medical history. Through deep learning and transfer learning, personalization of treatment plans will become possible. These methods aim to operationalize the knowledge acquired from population research in order to clarify treatment strategy and patient care by making predictions about patients' individual conditions. Finally, further developments in deep learning and transfer learning promise to be of tremendous use to bioinformatics research. They hope that in the future other more sophisticated model structures can be developed with even better interpretability and integration capability. Each of these advances may fundamentally change our view of biological systems, speed up the discovery process for new drugs and personalize health care. Let us go on through these new territories--to see just how many revolutionary turning points the biological sciences still have in store for our coming.

6. CONCLUSION

In this long journey into the land of deep learning, transfer learning and their implications for bioinformatics we've seen a tremendous transformation in the life sciences. This level of accuracy and speed means that biological data can finally be organized by scientists and researchers. Deep learning has transformed not only gene function predictions (to name a task that once was deemed unfeasible), but also drug discovery research. Basic treasure troves of multi-omics data have been opened up, notoriously difficult sequences cracked and precious information on the annotation (function) of genes and proteins found.

At the same time, transfer learning has turned into an innovator in bioinformatics. That is to say, by transmitting this precious information that can be used for anticipating the function of a protein and changing everything about how drugs are researched. In addition to speeding up compound screening and simplifying the process of taking new drugs from concept through approval, transfer learning has expanded our ability to identify potential therapeutic targets. Looking into the future, there are even greater prospects for advances in deep learning models. Neural architecture search and enhanced model interpretability are likely to open up new frontiers for the field of bioinformatics. These will be further examined using deep learning and traditional bioinformatics analysis methods, to spur the enticing prospect of personalized medical treatment.

To conclude, deep learning with transfer leaning has utterly transformed the bioinformatics field, not only providing light leading to transformational breakthroughs. Such techniques continue to unravel the fine-grained secrets of life, offering hopes for breakthroughs in both treatment and understanding of our bioaesthetical universe. The biological sciences are brimming with exciting possibilities as we begin the following chapters of our journey, ushering in an era of boundless promise.

REFERENCES

Armoogum, S., & Li, X. (2019). Big Data Analytics and Deep Learning in Bioinformatics With Hadoop. In *Deep Learning and Parallel Computing Environment for Bioengineering Systems* (pp. 17–36). Elsevier., doi:10.1016/B978-0-12-816718-2.00009-9

Banimfreg, B. H. (2023). A comprehensive review and conceptual framework for cloud computing adoption in bioinformatics. *Healthcare Analytics*, *3*, 100190. doi:10.1016/j.health.2023.100190

Barbero-Aparicio, J. A., Olivares-Gil, A., Rodríguez, J. J., García-Osorio, C., & Díez-Pastor, J. F. (2024). Addressing data scarcity in protein fitness landscape analysis: A study on semi-supervised and deep transfer learning techniques. *Information Fusion*, *102*, 102035. doi:10.1016/j.inffus.2023.102035

Bornes, R. S., Montero, J., Correia, A. R. M., & Rosa, N. R. (2023). Use of bioinformatic strategies as a predictive tool in implant-supported oral rehabilitation: A scoping review. *The Journal of Prosthetic Dentistry*, *129*(2), 322.e1–322.e8. doi:10.1016/j.prosdent.2022.12.011 PMID:36710172

Bouzid, A., Almidani, A., Zubrikhina, M., Kamzanova, A., Ilce, B. Y., Zholdassova, M., Yusuf, A. M., Bhamidimarri, P. M., AlHaj, H. A., Kustubayeva, A., Bernstein, A., Burnaev, E., Sharaev, M., & Hamoudi, R. (2023). Integrative bioinformatics and artificial intelligence analyses of transcriptomics data identified genes associated with major depressive disorders including NRG1. *Neurobiology of Stress*, *26*, 100555. doi:10.1016/j.ynstr.2023.100555 PMID:37583471

Bunyat-zada, A. R., & Ross, A. C. (2023). Highlights of bioinformatic tools and methods for validating bioinformatics derived hypotheses for microbial natural products research. *Current Opinion in Chemical Biology*, *76*, 102367. doi:10.1016/j.cbpa.2023.102367 PMID:37453164

Daphal, S. D., & Koli, S. M. (2023). Enhancing sugarcane disease classification with ensemble deep learning: A comparative study with transfer learning techniques. *Heliyon*, *9*(8), e18261. doi:10.1016/j.heliyon.2023.e18261 PMID:37520940

Du, S., Wang, M., Yang, J., Zhao, Y., Wang, J., Yue, M., Xie, C., & Song, H. (2023). An enhanced prediction framework for coalbed methane production incorporating deep learning and transfer learning. *Energy*, *282*, 128877. doi:10.1016/j.energy.2023.128877

Hasija, Y. (2023). A machine learning approach to bioinformatics. In *All About Bioinformatics* (pp. 203–224). Elsevier., doi:10.1016/B978-0-443-15250-4.00010-1

Hua, Y., Xie, D., Zhang, Y., Wang, M., Wen, W., & Sun, J. (2023). Identification and analysis of key genes in adipose tissue for human obesity based on bioinformatics. *Gene*, *888*, 147755. doi:10.1016/j.gene.2023.147755 PMID:37659596

Iqbal, N., & Kumar, P. (2023). From Data Science to Bioscience: Emerging era of bioinformatics applications, tools and challenges. *Procedia Computer Science*, *218*, 1516–1528. doi:10.1016/j.procs.2023.01.130

Lai, K., Twine, N., O'Brien, A., Guo, Y., & Bauer, D. (2019). Artificial Intelligence and Machine Learning in Bioinformatics. In *Encyclopedia of Bioinformatics and Computational Biology* (pp. 272–286). Elsevier. doi:10.1016/B978-0-12-809633-8.20325-7

Li, Y., Huang, C., Ding, L., Li, Z., Pan, Y., & Gao, X. (2019). Deep learning in bioinformatics: Introduction, application, and perspective in the big data era. *Methods (San Diego, Calif.)*, *166*, 4–21. doi:10.1016/j.ymeth.2019.04.008 PMID:31022451

Lilhore, U. K., Manoharan, P., Sandhu, J. K., Simaiya, S., Dalal, S., Baqasah, A. M., Alsafyani, M., Alroobaea, R., Keshta, I., & Raahemifar, K. (2023). Hybrid model for precise hepatitis-C classification using improved random forest and SVM method. *Scientific Reports*, *13*(1), 12473. doi:10.1038/s41598-023-36605-3 PMID:37528148

Lilhore, U. K., Poongodi, M., Kaur, A., Simaiya, S., Algarni, A. D., Elmannai, H., Vijayakumar, V., Tunze, G. B., & Hamdi, M. (2022). Hybrid model for detection of cervical cancer using causal analysis and machine learning techniques. *Computational and Mathematical Methods in Medicine*, *2022*, 2022. doi:10.1155/2022/4688327 PMID:35572826

Lilhore, U. K., Simaiya, S., Pandey, H., Gautam, V., Garg, A., & Ghosh, P. (2022). Breast cancer detection in the IoT cloud-based healthcare environment using fuzzy cluster segmentation and SVM classifier. In Ambient Communications and Computer Systems [Singapore: Springer Nature Singapore.]. *Proceedings of RACCCS*, *2021*, 165–179.

Lilhore, U. K., Simaiya, S., Prasad, D., & Guleria, K. (2020). A hybrid tumor detection and classification based on machine learning. *Journal of Computational and Theoretical Nanoscience*, *17*(6), 2539–2544. doi:10.1166/jctn.2020.8927

Liu, X., Li, J., Ma, J., Sun, H., Xu, Z., Zhang, T., & Yu, H. (2023). Deep Transfer Learning for Intelligent Vehicle Perception: A Survey. *Green Energy and Intelligent Transportation*, *100125*(5), 100125. doi:10.1016/j.geits.2023.100125

Löchel, H. F., & Heider, D. (2021). Chaos game representation and its applications in bioinformatics. *Computational and Structural Biotechnology Journal*, *19*, 6263–6271. doi:10.1016/j.csbj.2021.11.008 PMID:34900136

Ma, Y. J. X., Zschocke, J., Glos, M., Kluge, M., Penzel, T., Kantelhardt, J. W., & Bartsch, R. P. (2023). Automatic sleep-stage classification of heart rate and actigraphy data using deep and transfer learning approaches. *Computers in Biology and Medicine*, *163*, 107193. doi:10.1016/j.compbiomed.2023.107193 PMID:37421734

Mousser, W., Ouadfel, S., & Taleb-Ahmed, A. (2022). Incremental learning of convolutional neural networks in bioinformatics. In *Artificial Neural Networks for Renewable Energy Systems and Real-World Applications* (pp. 223–238). Elsevier. doi:10.1016/B978-0-12-820793-2.00001-X

Nematzadeh, S., Kiani, F., Torkamanian-Afshar, M., & Aydin, N. (2022). Tuning hyperparameters of machine learning algorithms and deep neural networks using metaheuristics: A bioinformatics study on biomedical and biological cases. *Computational Biology and Chemistry*, *97*, 107619. doi:10.1016/j.compbiolchem.2021.107619 PMID:35033837

Nguyen, T. B., Do, D. N., Nguyen-Thi, M.-L., Hoang-The, H., Tran, T.-T., & Nguyen-Thanh, T. (2022). Identification of potential crucial genes and key pathways shared in Inflammatory Bowel Disease and cervical cancer by machine learning and integrated bioinformatics. *Computers in Biology and Medicine*, *149*, 105996. doi:10.1016/j.compbiomed.2022.105996 PMID:36049413

Onyema, E. M., Lilhore, U. K., Saurabh, P., Dalal, S., Nwaeze, A. S., Chijindu, A. T., Ndufeiya-Kumasi, L. C., & Simaiya, S. (2023). Evaluation of IoT-Enabled hybrid model for genome sequence analysis of patients in healthcare 4.0. *Measurement. Sensors*, *26*, 100679. doi:10.1016/j.measen.2023.100679

Ramesh, T. R., Lilhore, U. K., Poongodi, M., Simaiya, S., Kaur, A., & Hamdi, M. (2022). Predictive analysis of heart diseases with machine learning approaches. *Malaysian Journal of Computer Science*, 132–148.

Rathore, M. S., Poongodi, M., Saurabh, P., Lilhore, U. K., Bourouis, S., Alhakami, W., Osamor, J., & Hamdi, M. (2022). A novel trust-based security and privacy model for internet of vehicles using encryption and steganography. *Computers & Electrical Engineering*, *102*, 108205. doi:10.1016/j.compeleceng.2022.108205

Sharma, S. K., Lilhore, U. K., Simaiya, S., & Trivedi, N. K. (2021). An improved random forest algorithm for predicting the COVID-19 pandemic patient health. *Annals of the Romanian Society for Cell Biology*, 67–75.

Sidorova, J., & Lozano, J. J. (2023). Commentary on "A systematic review on machine learning and deep learning techniques in cancer survival prediction": Validation of survival methods. *Progress in Biophysics and Molecular Biology*, *183*, 17–18. doi:10.1016/j.pbiomolbio.2023.08.001 PMID:37579795

Simaiya, S., Lilhore, U. K., Prasad, D., & Verma, D. K. (2021). MRI brain tumor detection & image segmentation by hybrid hierarchical K-means clustering with FCM based machine learning model. *Annals of the Romanian Society for Cell Biology*, 88–94.

Tan, M. S., Cheah, P.-L., Chin, A.-V., Looi, L.-M., & Chang, S.-W. (2021). A review on omics-based biomarkers discovery for Alzheimer's disease from the bioinformatics perspectives: Statistical approach vs machine learning approach. *Computers in Biology and Medicine*, *139*, 104947. doi:10.1016/j.compbiomed.2021.104947 PMID:34678481

Wang, H., & Hu, X. (2023). Deep Learning in Bioinformatics and Biomedicine. *Methods (San Diego, Calif.)*, *209*, 38–39. doi:10.1016/j.ymeth.2022.12.001 PMID:36503039

Wang, N., & He, Q. (2023). Artificial intelligence and bioinformatics applications in precision medicine and future implications. In *Reference Module in Biomedical Sciences*. Elsevier. doi:10.1016/B978-0-12-824010-6.00058-7

Wang, Y., Xu, L., & Zou, Q. (2023). Deep learning methods for bioinformatics and biomedicine. *Methods (San Diego, Calif.)*, *216*, 1–2. doi:10.1016/j.ymeth.2023.06.003 PMID:37295580

Yu, X., Zhou, S., Zou, H., Wang, Q., Liu, C., Zang, M., & Liu, T. (2023). Survey of deep learning techniques for disease prediction based on omics data. *Human Gene*, *35*, 201140. doi:10.1016/j.humgen.2022.201140

Zafar, I., Anwar, S., kanwal, F., Yousaf, W., Un Nisa, F., Kausar, T., ul Ain, Q., Unar, A., Kamal, M. A., Rashid, S., Khan, K. A., & Sharma, R. (2023). Reviewing methods of deep learning for intelligent healthcare systems in genomics and biomedicine. *Biomedical Signal Processing and Control*, *86*, 105263. doi:10.1016/j.bspc.2023.105263

Chapter 13
Exploration of Deep Learning and Transfer Learning Techniques in Bioinformatics

Sumit Bansal
Gurukula Kangri, Haridwar, India

Vandana Sindhi
https://orcid.org/0009-0002-4537-3671
Punjabi University, Patiala, India

Bhim Sain Singla
Punjabi University, Patiala, India

ABSTRACT

The blend of profound learning and more learning techniques has brought about a worldview change in the creating subject of bioinformatics. The revolutionary environment that emerges when the complexity of biological data and artificial intelligence meet is the subject of this book chapter. The section explores profound learning models and moves learning closer, as well as the imaginative applications, hardships, and achievements that have developed at the crossing point of these two powerful fields. The section opens with a point-by-point assessment of profound learning standards and how they are applied to the unique difficulties of bioinformatics datasets. The section likewise digs into the idea of move learning—a strong worldview that utilizes information learned in one space to further develop execution in another. It has been shown that movement learning is powerful. The section likewise dives into the idea of move learning—a strong worldview that involves mastery in one region to further develop execution in another.

1. INTRODUCTION

Leading the way in a technological revolution is bioinformatics, where the incorporation of novel computer techniques holds the potential to simplify the intricate patterns present in biological data (Liang *et*

DOI: 10.4018/979-8-3693-1822-5.ch013

al., 2024). The volume and diversity of biological information is growing quickly, making it increasingly difficult for traditional analytical approaches to extract meaningful insights (Philip Chen and Zhang, 2014). In this context, the combination of bioinformatics and deep learning emerges as a revolutionary force, offering hitherto unheard-of opportunities to analyze the complex patterns concealed inside large datasets.

Deep learning, which is characterized by neural network topologies modelled after the human brain, has shown remarkable promise across a variety of fields, including natural language processing and picture identification (Liu *et al.*, 2017). The accuracy with which the complexities of proteomic, metabolomic, and genomic data can be interpreted is made possible by its application in bioinformatics (Amiri *et al.*, 2024) . By employing hierarchical representations and learning subtle features, deep learning models can uncover hidden connections, forecast biological events, and enhance our understanding of intricate biological systems (Ching *et al.*, 2018).

Another tactical approach to overcoming the limitations of limited labelled biological data is transfer learning (Lu *et al.*, 2015). Transferring data between domains can expedite model training, enhance prediction accuracy, and mitigate the challenges associated with data scarcity that many bioinformatics applications face (Cesario, Comito and Zumpano, 2024). Research into deep learning and transfer learning approaches becomes a crucial path for advancing our understanding of the biological sciences as the boundaries between them become more hazy (Ma *et al.*, 2019).

Beyond just academic interest, this study is important because of its implications for important applications like drug development, customized therapy, and the comprehension of complex biological circuits. Transfer learning paradigms-based deep learning models have the power to revolutionize basic biological science as well as diagnosis and treatment methods. The goal of this book chapter is to clarify the intricacies of this revolutionary convergence, providing academics, practitioners, and students with an in-depth understanding of cutting-edge methods and their implications for bioinformatics. The integration of deep learning and transfer learning holds promise for revolutionizing bioinformatics research and propelling important breakthroughs in the life sciences as we navigate this new territory.

2. FOUNDATIONS OF DEEP LEARNING IN BIOINFORMATICS

An extensive examination of deep learning techniques is covered in this section, along with an overview of CNN, RNN, LSTM, GAN, unsupervised learning, and the various uses of neural networks in bioinformatics. It also discusses the opportunities and difficulties specific to the deep learning discipline. The relationships between the fundamental components of deep learning in bioinformatics are shown in Figure 1.

2.1 Overview of Deep Learning Techniques

Deep learning is a type of machine learning that has transformed difficult data processing and interpretation. It is based on the structure and activities of the human brain and has emerged as a transformative force in a variety of industries. Deep learning algorithms have shown potential in bioinformatics for identifying significant patterns from huge biological datasets (Li et al., 2019). This section comprehensively introduces the core deep learning algorithms, as well as their structures, ideas, and bioinformatics applications.

2.2 Neural Networks: The Foundation of Deep Learning

Deep learning is based on a neural network, a computational model inspired by the neural architecture of the human brain. Neural networks handle information in a hierarchical manner by learning complex patterns and representations from input data via interconnected layers of nodes, or neurons (Thakur, 2021). Deep learning layers are the several levels that allow these networks to extract intricate connections and characteristics from input.

Figure 1. Illustration depicting the interconnected foundations of deep learning in bioinformatics

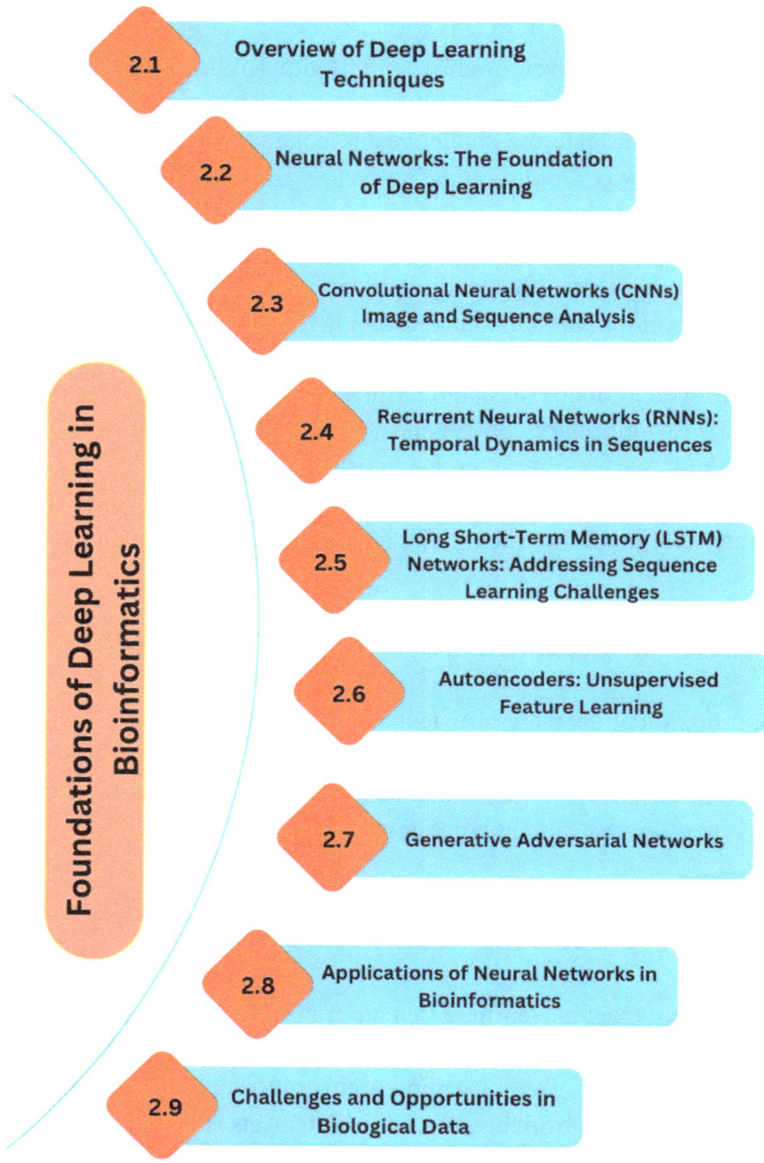

2.3 Convolutional Neural Networks (CNNs): Image and Sequence Analysis

CNNs are a type of neural network that is specifically built to process spatial and temporal input; as such, they are particularly helpful for sequence and image analysis. CNNs excel at tasks such as genetic sequence analysis and image-based diagnostics in bioinformatics. These networks automatically accumulate hierarchical features by utilising convolutional layers, capturing local patterns that are critical for comprehending biological data.

2.4 Recurrent Neural Networks (RNNs): Temporal Dynamics in Sequences

RNNs are designed for sequential data and can mimic temporal dependencies. RNNs are used in bioinformatics for tasks such as protein interaction networks and gene expression profiles when working with time-series data (Monti *et al.*, 2022). These networks are good at capturing dynamic links within biological sequences and can preserve information over time due to their recurrent connections.

2.5 Long Short-Term Memory (LSTM) Networks: Addressing Sequence Learning Challenges

The vanishing gradient problem, which was addressed by the introduction of LSTMs, a form of RNN, is a prevalent challenge when training deep networks on sequential data (Shewalkar, nyavanandi and Ludwig, 2019). LSTMs improve the bioinformatics field's ability to find long-distance correlations between biological sequences, making them excellent for jobs requiring a full understanding of temporal dynamics.

2.6 Autoencoders: Unsupervised Feature Learning

Unsupervised learning methods such as autoencoders aim to learn effective representations of the original data by compressing incoming input into a lower-dimensional space and then retrieving the original data. Autoencoders in bioinformatics assist feature learning and dimensionality reduction, making it easier to derive meaningful representations from high-dimensional biological data (Li *et al.*, 2019).

2.7 Generative Adversarial Networks (GANs): Synthetic Data Generation

A GAN is made up of an adversarially trained discriminator and a generator. GANs are used in bioinformatics to recreate realistic variations, augment datasets, and offer synthetic biological data. This method can be useful when working with limited or imbalanced biological datasets.

2.8 Applications of Neural Networks in Bioinformatics

Because of its ability to automatically find complicated patterns in data, neural networks have become indispensable tools in bioinformatics. The ability of neural networks to extract detailed correlations from biological data has resulted in breakthrough applications in a variety of sectors. This section investigates the various applications of neural networks in bioinformatics, emphasising their critical role in unravelling the secrets of proteomics, genetics, and other sciences.

2.8.1 Genomic Sequence Analysis

Neural networks are required for decoding the complicated code found in DNA sequences. Recurrent neural networks (RNNs) and convolutional neural networks (CNNs) have shown exceptional effectiveness in a wide range of applications, including detecting regulatory elements and predicting gene activity (Barshai, Tripto and Orenstein, 2020). While RNNs are good at modelling temporal characteristics, CNNs are great at capturing spatial correlations in DNA sequences, which helps us understand genomic structures and functions in general.

2.8.2 Protein Structure Prediction

Predicting the structure of proteins is a difficult endeavour that is necessary for understanding their interactions and functions. Deep neural network models, for example, have shown promise in reliably predicting protein structures from amino acid sequences (Ofer, Brandes and Linial, 2021). Drug discovery can be sped up and the molecular mechanisms behind a wide range of illnesses can be better understood by using these models.

2.8.3 Drug Discovery and Design

Neural networks are transforming drug research. These models aid in the identification of potential pharmaceutical candidates by interpreting the complicated relationships between chemical structures and biological action. Generated models, such as generative adversarial networks (GANs) and variational autoencoders, are used to generate novel chemical structures, which speeds up the drug discovery process and paves the door for novel therapeutic treatments.

2.8.4 Disease Diagnosis and Biomarker Discovery

According to (Theodoridis and Kraemer, 2014), neural networks are critical for disease detection and the development of biomarkers that identify specific ailments. Neural networks improve precision medicine procedures by boosting diagnostic accuracy and facilitating the generation of novel biomarkers, from assessing medical imaging data for cancer diagnosis to mining genetic data for disease symptoms.

2.8.5 Personalized Medicine and Treatment Prediction

Neural networks can aid in personalised medicine, which tries to tailor medical treatments to individual characteristics. Neural networks can anticipate the optimal course of treatment by analysing a variety of patient data, including genetics, health records, and lifestyle details (Pai and Bader, 2018). The method reduces side effects while improving patient results.

2.8.6 Metagenomics and Microbiome Analysis

Analysing complex microbial populations is a component of microbiome and metagenomics research. By managing high-dimensional sequencing data, neural networks enhance the discovery of the variety and functional potential of microbiomes. These models provide insights into therapeutic techniques that target the microbiome and help us understand the functions that microbial populations play in both health and disease.

2.9 Challenges and Opportunities in Applying Deep Learning to Biological Data

Because biological datasets are unique and biological processes are complex, applying deep learning to biological data presents several challenges. Some of the main problems in applying deep learning to bioinformatics are as follows:

2.9.1 Data Quality and Quantity

Biological datasets are frequently difficult to get due to factors such as experiment cost, moral quandaries, or the uniqueness of particular circumstances (Auer and Lettre, 2015). This scarcity can make training deep learning models problematic, as they frequently demand massive amounts of different data. Biassed models can originate from class distribution imbalances, such as sick vs healthy samples. When given unbalanced datasets, deep learning algorithms may struggle to generalise well.

2.9.2 Interpretability and Explainability

Deep learning models are sometimes viewed as "black boxes," especially when it comes to complex designs like deep neural networks. Gaining credibility and acceptability in scientific and medical organisations requires an understanding of the biological reasoning underlying a model's predictions (Carlone and Johnson, 2007).

2.9.3 Biological Complexity

The inherent nonlinearity and heterogeneity within biological systems present a formidable obstacle for conventional linear models. The intricate web of connections and dynamic interactions within complex biological systems poses a challenge that deep learning algorithms may struggle to effectively capture. Moreover, the integration of information from diverse sources, such as proteomics, clinical records, and genomes, adds another layer of complexity. Achieving comprehensive insights through deep learning necessitates the adept handling of multimodal data, requiring sophisticated techniques to harmonize and analyze diverse data types. Successfully addressing these challenges is pivotal for unlocking the full potential of deep learning in deciphering the intricacies of biological phenomena.

2.9.4 Computational Resources

High processing Demands: Training deep learning models, especially those with large-scale structures, requires a substantial amount of processing power. For researchers, access to high-performance computer equipment is often a barrier, particularly in places with limited resources.

2.9.5 Transferability and Generalization

Domain Shift: It might be challenging to transfer models between biological domains. Reliable performance can only be ensured by using domain adaptation procedures, as a model trained on one dataset may not generalize well to another.

Deep learning models may overfit (i.e., detect noise in the training set) or underfit (i.e., miss the underlying patterns). The regularization processes and model complexity need to be carefully adjusted to reach the right equilibrium.

3. TRANSFER LEARNING: A BRIDGE TO BIOINFORMATICS

This section, delves into the basics of Fundamentals and Transfer Learning within the context of Bioinformatics, covering motivations, types, applications, challenges, and more.

3.1 Understanding Transfer Learning in the Context of Bioinformatics

In bioinformatics, transfer learning—a machine learning paradigm—has proven to be an effective method for overcoming the limitations posed by sparse biological datasets (Kashyap *et al.*, 2015). By applying knowledge from one task or area to another, transfer learning enhances model performance on a different but related task or domain. The concepts of transfer learning are covered in this section, along with its nuanced application in bioinformatics, with an emphasis on how it could expand a range of biological applications.

3.1.1 Fundamentals of Transfer Learning

To increase performance on a target task or domain, transfer learning involves first training a model on a source task or domain. While classical machine learning usually involves fine-tuning a pre-trained model, transfer learning in deep learning sometimes uses pre-trained neural network weights as initializations for a related task.

3.1.2 Motivation for Transfer Learning in Bioinformatics

In the realm of bioinformatics, the scarcity of data poses a formidable challenge in training deep learning models from the ground up. The limited availability of biological datasets can hinder the development of robust models. To overcome this hurdle, the technique of transfer learning emerges as a crucial solution. By leveraging pre-existing knowledge from larger datasets in closely related domains, transfer learning enhances the capacity of models to generalize effectively to bioinformatics applications. This approach facilitates the adaptation of models across different datasets or domains, accommodating variations in experimental conditions and measurement equipment. The concept of domain adaptation becomes paramount, allowing the seamless transfer of insights gained from one set of data modalities to another, thus enriching the versatility and applicability of bioinformatics models.

3.1.3 Types of Transfer Learning Approaches

Pre-trained models, particularly deep neural networks, offer the promise of serving as efficient feature extractors in bioinformatics. Following training on a source task, the lower layers of the network can encapsulate general features that prove transferable to the target job within the realm of bioinformatics. Fine-tuning represents a pivotal step in this process, where a pre-existing model undergoes further training specifically on the intended bioinformatics task. This iterative refinement enables the model to adapt its acquired representations to the distinctive characteristics of bioinformatics data, enhancing its ability to generate more tailored and accurate outcomes for the intricacies of the target application.

3.1.4 Applications of Transfer Learning in Bioinformatics

Genomic Sequence Analysis harnesses the power of pre-trained models trained on vast datasets to discern generic sequence patterns applicable across diverse biological contexts. Transfer learning plays a pivotal role in tasks like DNA sequence categorization, where knowledge gained from one domain is applied to enhance performance in another. In the realm of Drug Discovery, these models can be fine-tuned using extensive databases of chemical compounds, expediting drug development processes and augmenting the efficiency of virtual screening by predicting molecular features. Additionally, Transfer learning proves invaluable in Protein Function Prediction, allowing the seamless transfer of knowledge acquired from protein-related tasks to models geared towards forecasting intricate aspects such as protein-protein interactions, thereby advancing our understanding of complex biological systems.

3.1.5 Challenges and Considerations

Domain shift poses a significant challenge to model performance, arising from differences between the source and destination domains. This shift can adversely impact the model's adaptability and effectiveness. It becomes imperative to proactively address and analyze domain-specific issues in detail to ensure optimal performance. In the realm of bioinformatics, task relevance becomes paramount, emphasizing the need to ensure that information derived from the source aligns closely with the intended job. This necessitates a profound understanding of the biological underpinnings to facilitate effective learning and accurate application of the model in the targeted bioinformatics tasks.

Lastly, transfer learning is a useful ally in bioinformatics since it offers a fix for the enduring issue of scarce biological data. Transfer learning encourages creativity and speeds up discoveries in the sensitive field of biological data analysis by enabling models to extract significant patterns from bioinformatics datasets by cleverly using knowledge from related fields. As bioinformatics evolves, the careful application of transfer learning has the potential to open up new doors in our understanding of life sciences and personalized medicine.

3.2 Benefits of Transfer Learning for Biological Data

Transfer learning, with its ability to use knowledge gained from one task or domain to improve performance in another, has significant implications for overcoming the issues inherent in biological data processing. This section examines the benefits of transfer learning in the field of bioinformatics, emphasizing its role in overcoming data scarcity, enhancing model generalization, and accelerating progress in a variety of biological applications.

3.2.1 Mitigating Data Scarcity Challenges

Effective utilization of pre-existing knowledge is crucial in overcoming limitations such as experimental constraints, ethical concerns, and the high cost associated with data collection in the realm of bioinformatics. Transfer learning emerges as a valuable strategy to address these challenges by empowering models to harness insights from extensive and diverse datasets in adjacent fields. This approach proves particularly beneficial when biological datasets are inherently restricted in size. Knowledge transfer across domains becomes instrumental in situations where assembling vast datasets is impractical. Transfer learning facilitates the seamless integration of insights from domains abundant in data to those with limited resources, enabling the development of more robust and reliable models. By leveraging pre-existing knowledge, bioinformatics applications can enhance their predictive capabilities and contribute to advancements in understanding complex biological phenomena.

3.2.2 Improving Generalization to Bioinformatics Tasks

Deep neural networks excel in capturing general attributes and representations from source activities, particularly in the realm of bioinformatics. Leveraging these learned traits, models can encode substantial patterns and correlations discerned within intricate biological data sets. This enhanced feature representation proves invaluable in deciphering the complexities of biological processes. Additionally, the adaptability of models to navigate diverse biological contexts is bolstered by transfer learning. Given the inherent variability in biological data arising from factors such as genetic diversity and environmental influences, transfer learning provides a robust framework. It endows models with the flexibility to tackle a spectrum of biological scenarios, furnishing them with a foundational knowledge base that can be fine-tuned for specific tasks, thus enhancing their efficacy in diverse biological applications.

3.2.3 Accelerating Model Training and Convergence

Reduced Training Time is a pivotal advantage in employing transfer learning for bioinformatics tasks. Initializing a model with pre-trained weights from a related task substantially diminishes the time and resources required for training. In the realm of bioinformatics, where investigations are notorious for their costliness and time-intensive nature, this efficiency gain is invaluable. Additionally, the approach offers Faster Convergence, as the model rapidly adapts to the intricacies of bioinformatics work owing to its initiation with pre-existing weights that encapsulate crucial information. This acceleration in convergence is particularly advantageous in expediting the development and optimization of models tailored to intricate biological data, fostering advancements in the field at a swifter pace.

3.2.4 Domain Adaptation for Multi-Modal Data Integration

In the realm of bioinformatics, the integration of diverse data modalities is a pivotal challenge, requiring the harmonization of information derived from various sources such as genomes, proteomics, and clinical records. Transfer learning emerges as a valuable tool in overcoming this complexity, streamlining the adaptation of models to different domains and facilitating the seamless amalgamation of data from multiple origins. This capability enhances the depth of analysis, enabling a more comprehensive exploration of biological phenomena. Additionally, transfer learning plays a crucial role in addressing the inherent heterogeneity in data, accommodating variations in sources and experimental conditions. By leveraging models trained on specific modalities or situations, bioinformatics researchers can attain a more holistic understanding of intricate biological systems, fostering advancements in knowledge and contributing to the overall progress in the field.

In the end, transfer learning is essential to the development of bioinformatics because it offers a calculated solution to the innate challenges brought about by a lack of data and a variety of biological situations. Through the use of pre-existing knowledge and model convergence, transfer learning addresses practical limitations while also advancing bioinformatics into an era of increased efficiency, interpretability, and innovation. With the expansion of transdisciplinary collaboration between machine learning and biology, the advantages of transfer learning are expected to be critical in deciphering the complexities of biological data and stimulating creative discoveries in the life sciences.

4. DEEP LEARNING ARCHITECTURES IN GENOMIC ANALYSIS

This section investigates the representation of genomic data, delving into deep learning approaches and classification methodologies are represented in the Figure 2 below.

4.1 Genomic Data Representation

At the forefront of bioinformatics is the representation of genomic data, which bridges the gap between the computer techniques required for analysis and the intricacies of biological sequences (Potla, Ali and Kapoor, 2021). The representations are still changing, moving from basic nucleotide sequences to multi-dimensional epigenomic information, driven by developments in computer techniques like deep learning. Effective genomic data representation will continue to be essential as research develops because it will help unlock the mysteries of DNA and open the door to important new understandings of genetics, personalized treatment, and life itself.

4.2 Deep Learning Approaches for DNA Sequence Analysis

Understanding genome functions, architecture, and variations is made possible by DNA sequence analysis, which is essential to interpreting the genetic code (Ji *et al.*, 2021). DNA sequence analysis has been transformed by deep learning, which can automatically learn hierarchical features and spot minute patterns that conventional techniques miss. In order to illustrate the revolutionary influence of these technologies on genomics and bioinformatics, this section explores the numerous applications of deep learning in DNA sequence decoding.

4.3 Sequence Classification: Deciphering Function and Variations

Functional annotation and variant calling are pivotal tasks in genomics, both adeptly addressed by deep learning methodologies. In the realm of DNA sequence classification, Convolutional Neural Networks

Figure 2. Deep learning for massive data analysis in plant breeding

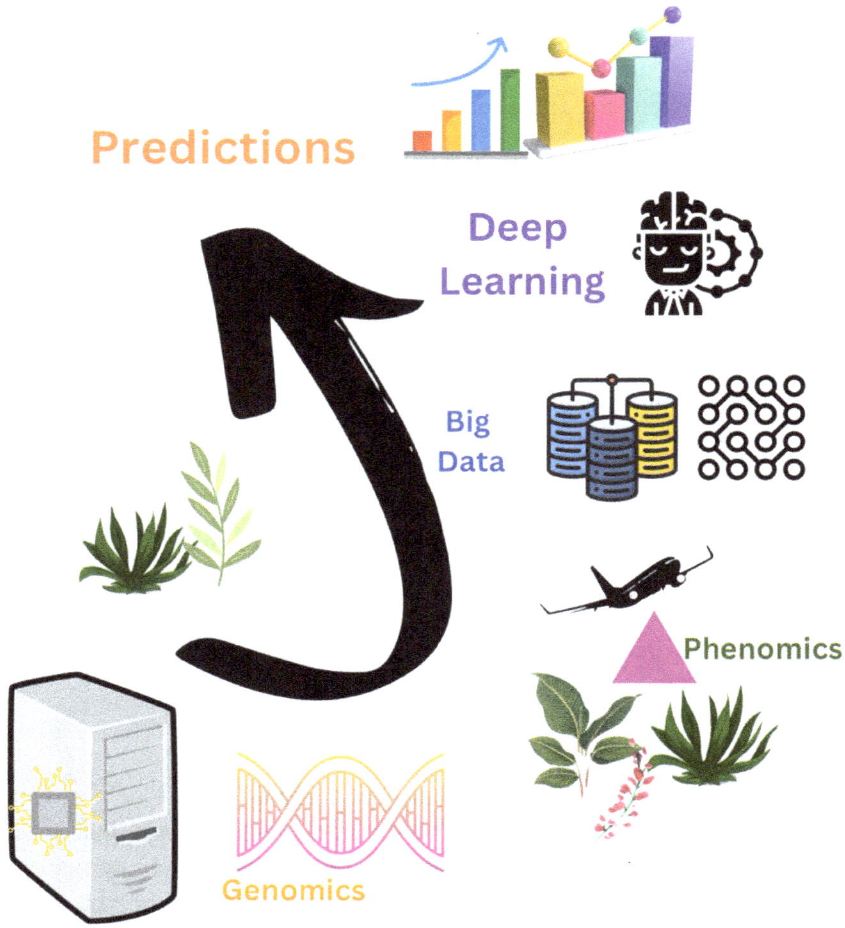

(CNNs) demonstrate exceptional prowess. Their ability to discern regulating regions, coding sequences, and non-coding components stems from their adeptness at learning intricate local patterns and motifs dispersed throughout the genetic sequences. On the other front, the identification of genetic variations is paramount for unraveling the genetic underpinnings of diseases. Deep learning models, including CNNs and Recurrent Neural Networks (RNNs), prove instrumental in this arena. By employing these models, scientists can effectively classify and predict single nucleotide polymorphisms (SNPs) and other variants, facilitating genotyping and advancing personalized medicine initiatives. The integration of deep learning techniques in functional annotation and variant calling not only enhances accuracy but also propels genomics research towards a more nuanced and insightful understanding of genetic landscapes.

4.4 Motif Discovery: Unraveling Regulatory Elements

De Novo Motif Discovery relies on advanced computational techniques, notably unsupervised learning models like Variational Autoencoders (VAEs) and deep learning methods such as Convolutional Neural Networks (CNNs). These algorithms play a pivotal role in unraveling intricate biological patterns by autonomously discerning and extracting sequence features. As part of enhancing interpretability, deep learning models adopt strategies like saliency maps and attention processes. These techniques illuminate the functional significance of discovered regulatory elements and provide valuable insights into the complex orchestration of gene expression, offering a deeper understanding of the underlying mechanisms governing cellular processes. Through the synergy of these computational approaches, researchers gain a powerful toolset for uncovering hidden regulatory motifs and comprehending their biological implications.

4.5 Genomic Sequence Generation: Augmenting Datasets

Generative models such as Long Short-Term Memory (LSTM) networks and Generative Adversarial Networks (GANs) play a pivotal role in the generation of synthetic DNA sequences. These models contribute significantly to enhancing the robustness and generalization of algorithms trained on limited biological data by creating additional synthetic sequences. In the realm of data augmentation, the incorporation of artificially generated sequences into datasets proves invaluable for deep learning models. This augmentation enables the models to capture a broader spectrum of sequence variations, fostering their ability to generalize effectively to novel data instances. Consequently, the synergy between generative models and data augmentation techniques propels advancements in understanding and analyzing complex biological information encoded within DNA sequences.

4.6 DNA Sequence Embeddings: Continuous Representations

Embedding techniques, such as Word2Vec and Seq2Vec, simplify the representation of DNA sequences in continuous vector spaces. These embeddings play a crucial role in facilitating quantitative comparisons and similarity assessments by capturing the semantic relationships inherent in sequences. The ability to translate biological information into numerical vectors enhances the analysis and interpretation of complex genetic data. Additionally, transfer learning becomes a valuable strategy in the realm of bioinformatics, allowing knowledge gained from pre-training models on extensive datasets, which include sequence embeddings, to be transferred to subsequent tasks. This proves particularly beneficial in the face of limited labeled datasets in bioinformatics, providing a powerful approach to leverage existing knowledge and improve the efficiency of sequence analysis and interpretation.

4.7 Predictive Modeling in Genomics Using Deep Neural Networks

This section investigates the use of deep learning in predictive modelling in the genomics domain, demonstrating how these models help to decode the intricacies of genetic information.

4.7.1 Genomic Data Representation: Transforming Sequences Into Insights

In the realm of genomic data analysis, deep neural networks employ transformative techniques such as one-hot encoding to convert nucleotide sequences into numerical matrices. This process is essential for effectively interpreting DNA sequences, allowing for comprehensive analysis. However, the exploration of genomic information extends beyond mere encoding, with embedding layers enabling models to discern and learn intricate semantic relationships among nucleotides. To enhance the richness of genomic representations, the integration of epigenomic features, such as histone modifications and DNA methylation patterns, proves crucial. Deep neural networks excel in unveiling complex correlations between gene regulation and these epigenomic traits, facilitating a more profound understanding of genomic function and contributing to advancements in biological research.

4.7.2 Gene Expression Prediction: Unraveling Transcriptional Dynamics

In the realm of transcriptomic data analysis, the prediction of gene expression levels has become a targeted application of advanced deep neural networks, particularly employing specialized architectures like Recurrent Neural Networks (RNNs) and Long Short-Term Memory (LSTM) networks, as elucidated by (Zhou *et al.*, 2022). These sophisticated models leverage temporal correlations within gene expression profiles, enabling the extraction of dynamic regulatory patterns. Beyond predicting expression levels, deep learning algorithms play a pivotal role in discerning differentially expressed genes across varied conditions. This capability is instrumental in unraveling the intricate molecular mechanisms underpinning diseases, developmental processes, and other biological phenomena, thereby offering valuable insights into the intricate interplay of genetic factors in diverse biological contexts.

4.7.3 Variant Effect Prediction: Decoding Genetic Variations

In the realm of genetics, Single Nucleotide Polymorphisms (SNPs) and variant calling are pivotal areas where deep neural networks play a crucial role. These networks are adept at forecasting the functional consequences of genetic variations, including SNPs and indels. By employing sophisticated models, such as Convolutional Neural Networks (CNNs), these tools unravel the intricate relationship between genetic differences and phenotypic characteristics or disorders. Specifically, in pathogenicity prediction, deep learning architectures are employed to determine the likelihood of a genetic mutation leading to disease. The use of CNNs in this context has significant implications for tailoring pharmaceutical regimens and informing clinical judgments, as it enhances our ability to predict and understand the potential impact of genetic variations on human health.

4.7.4 Protein Structure and Function Prediction: Bridging Genotype to Phenotype

In the realm of structural biology, the application of deep learning techniques has revolutionized our ability to predict intricate protein structures based on genetic data. Leading the way is AlphaFold and similar models that leverage deep neural networks, such as those described by (Gao *et al.*, 2020). These advanced architectures enable the precise 3D modeling of proteins, providing invaluable insights into their relationships and functions. Concurrently, in the field of functional annotation, predictive modeling plays a pivotal role in identifying various genomic elements, including protein-coding regions, regulatory elements, and non-coding RNA. This annotation process equips researchers with crucial functional information, enhancing our understanding of the distinct genetic constituents involved in cellular functions. Together, deep learning in structural biology and functional annotation contribute significantly to unraveling the complexities of genetic information and its impact on biological processes.

4.7.5 Clinical Outcome Prediction: Towards Personalized Medicine

Integration with clinical data is crucial for predicting patient outcomes, treatment responses, and disease risks using deep learning models, as highlighted by (Yousefi *et al.*, 2017). By incorporating both genetic and clinical information, these models contribute to the development of personalized medical plans tailored to individual genetic profiles. Furthermore, the application of deep learning extends to survival analysis, where algorithms assess the likelihood of survival based on comprehensive genetic and clinical data. This approach not only revolutionizes the diagnosis of various diseases, such as cancer, but also influences the planning and customization of treatments, marking a significant advancement in the field of predictive genomics and its impact on healthcare.

In the end, deep neural network-based predictive modelling in genomics is a revolutionary force that bridges the genotype-phenotype gap. By forecasting the patterns of gene expression and deciphering protein structures, these models aid in our understanding of genetics. Additionally, they open the door for individualized medical therapies. As the discipline develops, addressing issues and fusing insights from deep learning models with conventional biological knowledge will propel genomics into an era of remarkable discoveries and applications in medicine and beyond.

5. CASE STUDIES AND APPLICATIONS

This section delves into real-world instances of successful implementations, examining the application of deep learning in bioinformatics. It also considers future directions and emerging trends in this dynamic field.

5.1 Real-World Examples of Successful Implementations

Certainly, several real-world examples showcase the successful implementation of deep learning in various domains. Here are a few notable instances:

5.1.1 AlphaFold: Protein Structure Prediction

In the field of structural biology, AlphaFold, a deep learning model developed by DeepMind, achieved significant advancements in accurately predicting protein structures (AlQuraishi, 2021). It was a major breakthrough in structural biology since the ability to anticipate the structures of proteins is essential for comprehending their functions and developing therapeutics. AlphaFold's performance was demonstrated in the Critical Assessment of Structure Prediction (CASP) competition.

5.1.2 DeepMind Health: Diabetic Retinopathy Detection

An Alphabet startup called DeepMind developed a deep learning technique for identifying diabetic macular edoema and diabetic retinopathy in retinal images. The programme outperformed skilled ophthalmologists after being trained on a big dataset of retinal scans (Ruamviboonsuk *et al.*, 2020). The potential of deep learning in medical picture processing was illustrated by this application.

5.1.3 IBM Watson for Oncology: Cancer Treatment Recommendations

IBM Watson for Oncology is a cognitive computing system used in the healthcare industry that assists medical professionals in treating cancer patients. The system analyses clinical trial data, patient information, and medical literature using deep learning and natural language processing to produce evidence-based treatment recommendations.

5.1.4 Tesla Autopilot: Autonomous Driving

Deep neural networks are used by Tesla's Autopilot system in the automotive industry to perform a number of tasks like autonomous navigation, lane keeping, and object identification. To enable semi-autonomous and autonomous driving, deep learning models process data from sensors such as cameras, radar, and ultrasonic sensors.

5.1.5 Google Photos: Image and Object Recognition

Users can search for photographs based on the content within them using Google photographs, which uses deep learning for object and image recognition in the domain of image processing (Yu *et al.*, 2015). The technology can identify objects, locations, and even individual faces in photos. This illustrates how deep learning can be applied to content indexing and large-scale image processing.

5.1.6 Amazon Alexa and Google Assistant: Natural Language Processing

Voice-activated virtual assistants, like Google Assistant and Amazon Alexa, use deep learning in the domain of natural language processing to comprehend speech and understand natural language. These technologies provide hands-free interaction with smart devices and services by comprehending and reacting to spoken commands.

5.1.7 Netflix and YouTube: Content Recommendation

Within the Entertainment Sector Streaming services like Netflix and YouTube utilise deep learning algorithms to make content recommendations. By providing personalised content based on viewing history, preferences, and behaviour, these systems enhance the overall user experience.

5.1.8 Facebook: Facial Recognition

Facebook is one of the social media platforms that uses deep learning to recognise faces in photos, enabling automatic person tagging. The ability of deep neural networks to learn intricate facial patterns and features makes it possible to quickly and accurately identify persons in images.

These illustrations show how deep learning may be applied and has an impact in a variety of industries, including social media, healthcare, and the automobile industry. We should anticipate more creative uses and effective deep learning implementations in practical settings as technology develops.

5.2 Lessons Learned From Applying Deep Learning in Bioinformatics

Deep learning applications in bioinformatics have yielded useful insights and lessons that contribute to the continued progress of this interdisciplinary subject. Here are some major takeaways from the use of deep learning in bioinformatics:

5.2.1 Data Quality is Crucial: The performance of deep learning models is significantly influenced by the quality and diversity of data. Robust model training in the bioinformatics domain requires data quality assurance through efficient preprocessing, curation, and handling of batch effects, as datasets may be constrained and noisy.

5.2.2 Interdisciplinary Collaboration is Essential: Collaboration between computer scientists, data scientists, biologists, and domain experts are necessary for deep learning in bioinformatics. Creating accurate and physiologically meaningful models requires closing the knowledge gap between computational biology and biology.

5.2.3 Model Interpretability is a Challenge: Interpreting decisions made by deep learning models in bioinformatics remains challenging. Even though these models may predict outcomes with a high degree of accuracy, research is still needed to fully understand the qualities these algorithms pick up and how their predictions affect biology. In order to build confidence in the scientific and medical fields, model interpretability initiatives are essential.

5.2.4 Transfer Learning Mitigates Data Scarcity: One effective way to address the issue of limited biological datasets is by transfer learning, which leverages expertise from related domains. By fine-tuning models on particular bioinformatics tasks and pre-training them on large datasets from many domains, model performance can be increased.

5.2.5 Ethical Considerations are Paramount: Upholding strict ethical standards is necessary while working with genetic and medical data. It's important to properly address issues related to consent, ownership of data, and privacy. The public, researchers, and patients all need to feel confident, and this requires ethical considerations.

5.3 Future Directions and Emerging Trends

Bioinformatics is a dynamic discipline, with continual advances and developing trends influencing the future of research and applications (Mondal, Dey and Sc, 2022). Here are some potential bioinformatics directions and rising trends:

5.3.1 Multi-Modal Data Integration: Integrating several biological data types, including transcriptomics, proteomics, genomes, and clinical data, is a major area of focus. Multi-modal input-handling deep learning algorithms will enable a more thorough understanding of intricate biological processes.

5.3.2 Explainable AI in Bioinformatics: Enhancing the interpretability and explainability of deep learning models is a rapidly expanding subject of research. Techniques that make clear how models generate predictions can increase confidence in their results, particularly in critical domains such as health care.

5.3.3 Graph Neural Networks for Biological Networks: Graph neural networks, or GNNs, are gaining popularity in the analysis of biological networks, including gene regulatory networks and protein-protein interaction networks. Because GNNs are able to capture the relational relationships seen in these networks, they are an excellent choice for jobs requiring graph-structured data.

5.3.4 Precision Medicine and Personalized Genomics: The use of genomes and clinical data in personalized medicine is becoming more and more common. Deep learning models will play a key role in identifying biomarkers, guiding treatment strategies based on genetic makeup, and anticipating individual responses to medicines.

5.3.5 Single-Cell Genomics: Single-cell genomics advances produce high-dimensional data that encapsulates the characteristics of individual cells. Deep learning methods intended for single-cell data analysis will offer fresh perspectives on the mechanisms behind disease, developmental processes, and cellular heterogeneity.

6. CONCLUSION

In conclusion, a new age in bioinformatics has begun with the confluence of deep learning and transfer learning, which has fundamentally changed how we handle the complexity of biological data. This chapter provides a thorough examination of the revolutionary effects of artificial intelligence at the nexus of these dynamic realms. In addition to addressing the difficulties presented by bioinformatics datasets, the investigation of deep learning architectures and transfer learning techniques has advanced a variety of biological applications, most notably genomics, drug discovery, protein structure prediction, and personalized medicine.

The case studies that are provided demonstrate the effective application of these strategies while pointing out some of their possible advantages and disadvantages. The discussion of potential future directions highlights the potential revolution these methods could bring about in the field of bioinformatics study. This chapter is a useful tool for researchers, practitioners, and students who want to use deep learning and transfer learning to understand the complex subtleties of biological data as the fields of computer science and biology collaborate more and more. By this investigation, we add to the continuing conversation that advances both domains towards inventive resolutions and discoveries, promoting a better comprehension of the mutually beneficial interaction between bioinformatics and artificial intelligence.

The studies illustrations that are provided highlight the effective application of these strategies and highlight both their inherent advantages and disadvantages. The field of bioinformatics could undergo a revolution recognition to these techniques, as we explore possible routes for future research. Researchers, practitioners, and students looking to use deep learning and transfer learning for a more sophisticated understanding of biological data will find this chapter to be a useful resource.

REFERENCES

AlQuraishi, M. (2021). Machine learning in protein structure prediction. *Current Opinion in Chemical Biology*, *65*, 1–8. doi:10.1016/j.cbpa.2021.04.005 PMID:34015749

Amiri, Z., Heidari, A., Navimipour, N. J., Esmaeilpour, M., & Yazdani, Y. (2024). The deep learning applications in IoT-based bio- and medical informatics: A systematic literature review. *Neural Computing & Applications*, *3*. doi:10.1007/s00521-023-09366-3

Auer, P. L., & Lettre, G. (2015). Rare variant association studies: Considerations, challenges and opportunities. *Genome Medicine*, *7*(1), 1–11. doi:10.1186/s13073-015-0138-2 PMID:25709717

Barshai, M., Tripto, E., & Orenstein, Y. (2020). Identifying Regulatory Elements via Deep Learning. *Annual Review of Biomedical Data Science*, *3*(1), 315–338. doi:10.1146/annurev-biodatasci-022020-021940

Carlone, H. B., & Johnson, A. (2007). Understanding the science experiences of successful women of color: Science identity as an analytic lens. *Journal of Research in Science Teaching*, *44*(8), 1187–1218. doi:10.1002/tea.20237

Cesario, E., Comito, C. & Zumpano, E. (2024). A survey of the recent trends in deep learning for literature based discovery in the biomedical domain. *Neurocomputing*, *568*(November 2023), p. 127079. . doi:10.1016/j.neucom.2023.127079

Ching, T., Himmelstein, D. S., Beaulieu-Jones, B. K., Kalinin, A. A., Do, B. T., Way, G. P., Ferrero, E., Agapow, P.-M., Zietz, M., Hoffman, M. M., Xie, W., Rosen, G. L., Lengerich, B. J., Israeli, J., Lanchantin, J., Woloszynek, S., Carpenter, A. E., Shrikumar, A., Xu, J., & Greene, C. S. (2018). Opportunities and obstacles for deep learning in biology and medicine. *Journal of the Royal Society, Interface*, *15*(141), 20170387. doi:10.1098/rsif.2017.0387 PMID:29618526

Gao, W., Mahajan, S. P., Sulam, J., & Gray, J. J. (2020). Deep Learning in Protein Structural Modeling and Design. *Patterns (New York, N.Y.)*, *1*(9), 100142. doi:10.1016/j.patter.2020.100142 PMID:33336200

Ji, F., Ma, Q., Zhang, W., Liu, J., Feng, Y., Zhao, P., Song, X., Chen, J., Zhang, J., Wei, X., Zhou, Y., Chang, Y., Zhang, P., Huang, X., Qiu, J., & Pei, D. (2021). A genome variation map provides insights into the genetics of walnut adaptation and agronomic traits. *Genome Biology*, *22*(1), 1–22. doi:10.1186/s13059-021-02517-6 PMID:34706738

KashyapH. (2015). Big Data Analytics in Bioinformatics: A Machine Learning Perspective. arXiv. http://arxiv.org/abs/1506.05101

Li, Y., Huang, C., Ding, L., Li, Z., Pan, Y., & Gao, X. (2019, April). Deep learning in bioinformatics: Introduction, application, and perspective in the big data era. *Methods (San Diego, Calif.)*, *166*, 4–21. doi:10.1016/j.ymeth.2019.04.008 PMID:31022451

Liang, A. (2024). Advancements and applications of single-cell multi-omics techniques in cancer research : Unveiling heterogeneity and paving the way for precision therapeutics. *Biochemistry and Biophysics Reports*, *37*(November 2023). . doi:10.1016/j.bbrep.2023.101589

Lilhore, U. K., Dalal, S., Faujdar, N., Margala, M., Chakrabarti, P., Chakrabarti, T., Simaiya, S., Kumar, P., Thangaraju, P., & Velmurugan, H. (2023). Hybrid CNN-LSTM model with efficient hyperparameter tuning for prediction of Parkinson's disease. *Scientific Reports*, *13*(1), 14605. doi:10.1038/s41598-023-41314-y PMID:37669970

Lilhore, U. K., Dalal, S., Faujdar, N., Simaiya, S., Dahiya, M., Tomar, S., & Hashmi, A. (2024). Unveiling the prevalence and risk factors of early stage postpartum depression: A hybrid deep learning approach. *Multimedia Tools and Applications*, 1–35. doi:10.1007/s11042-024-18182-3

Lilhore, U. K., Manoharan, P., Sandhu, J. K., Simaiya, S., Dalal, S., Baqasah, A. M., Alsafyani, M., Alroobaea, R., Keshta, I., & Raahemifar, K. (2023). Hybrid model for precise hepatitis-C classification using improved random forest and SVM method. *Scientific Reports*, *13*(1), 12473. doi:10.1038/s41598-023-36605-3 PMID:37528148

Lilhore, U. K., Poongodi, M., Kaur, A., Simaiya, S., Algarni, A. D., Elmannai, H., Vijayakumar, V., Tunze, G. B., & Hamdi, M. (2022). Hybrid model for detection of cervical cancer using causal analysis and machine learning techniques. *Computational and Mathematical Methods in Medicine*, *2022*, 2022. doi:10.1155/2022/4688327 PMID:35572826

Lilhore, U. K., Simaiya, S., Pandey, H., Gautam, V., Garg, A., & Ghosh, P. (2022). Breast cancer detection in the IoT cloud-based healthcare environment using fuzzy cluster segmentation and SVM classifier. In Ambient Communications and Computer Systems [Singapore: Springer Nature Singapore.]. *Proceedings of RACCCS*, *2021*, 165–179.

Lilhore, U. K., Simaiya, S., Prasad, D., & Guleria, K. (2020). A hybrid tumor detection and classification based on machine learning. *Journal of Computational and Theoretical Nanoscience*, *17*(6), 2539–2544. doi:10.1166/jctn.2020.8927

Liu, W. (2017). A survey of deep neural network architectures and their applications. Neurocomputing, 234(December 2016), 11–26. doi:10.1016/j.neucom.2016.12.038

Lu, J., Behbood, V., Hao, P., Zuo, H., Xue, S., & Zhang, G. (2015). Transfer learning using computational intelligence: A survey. *Knowledge-Based Systems*, *80*, 14–23. doi:10.1016/j.knosys.2015.01.010

Ma, J., Cheng, J. C. P., Lin, C., Tan, Y., & Zhang, J. (2019). Improving air quality prediction accuracy at larger temporal resolutions using deep learning and transfer learning techniques. *Atmospheric Environment*, *214*(April), 116885. doi:10.1016/j.atmosenv.2019.116885

Mondal, R. K., Dey, A., & Sc, B. (2022). BIOINFORMATICS : PAST, PRESENT AND FUTURE. *International Research Journal of Modernization in Engineering Technology and Science*, *04*(07), 442–454.

Monti, M., Fiorentino, J., Milanetti, E., Gosti, G., & Tartaglia, G. G. (2022). Prediction of Time Series Gene Expression and Structural Analysis of Gene Regulatory Networks Using Recurrent Neural Networks. *Entropy (Basel, Switzerland)*, *24*(2), 141. doi:10.3390/e24020141 PMID:35205437

Ofer, D., Brandes, N., & Linial, M. (2021). The language of proteins: NLP, machine learning & protein sequences. *Computational and Structural Biotechnology Journal*, *19*, 1750–1758. doi:10.1016/j.csbj.2021.03.022 PMID:33897979

Pai, S., & Bader, G. D. (2018). Patient Similarity Networks for Precision Medicine. *Journal of Molecular Biology*, *430*(18), 2924–2938. doi:10.1016/j.jmb.2018.05.037 PMID:29860027

Philip Chen, C. L., & Zhang, C. Y. (2014). Data-intensive applications, challenges, techniques and technologies: A survey on Big Data. *Information Sciences*, *275*, 314–347. doi:10.1016/j.ins.2014.01.015

Potla, P., Ali, S. A., & Kapoor, M. (2021). A bioinformatics approach to microRNA-sequencing analysis. *Osteoarthritis and Cartilage Open*, *3*(1), 100131. doi:10.1016/j.ocarto.2020.100131 PMID:36475076

Ruamviboonsuk, P., Cheung, C. Y., Zhang, X., Raman, R., Park, S. J., & Ting, D. S. W. (2020). Artificial intelligence in ophthalmology: Evolutions in Asia. *Asia-Pacific Journal of Ophthalmology*, *9*(2), 78–84. doi:10.1097/01.APO.0000656980.41190.bf PMID:32349114

Shewalkar, A., Nyavanandi, D., & Ludwig, S. A. (2019). Performance Evaluation of Deep neural networks Applied to Speech Recognition: Rnn, LSTM and GRU. *Journal of Artificial Intelligence and Soft Computing Research*, *9*(4), 235–245. doi:10.2478/jaiscr-2019-0006

Thakur, A. K. A. (2021). Fundamentals of Neural Networks. *International Journal for Research in Applied Science and Engineering Technology*, *9*(VIII), 407–426. doi:10.22214/ijraset.2021.37362

Theodoridis, T., & Kraemer, J. (2014). Early Diagnosis of Complex Diseases by Molecular Biomarkers, Network Biomarkers, and Dynamical Network Biomarkers. *Medicinal Research Reviews*, *34*(3), 455–478. doi:10.1002/med.21293 PMID:23775602

Yousefi, S., Amrollahi, F., Amgad, M., Dong, C., Lewis, J. E., Song, C., Gutman, D. A., Halani, S. H., Velazquez Vega, J. E., Brat, D. J., & Cooper, L. A. D. (2017). Predicting clinical outcomes from large scale cancer genomic profiles with deep survival models. *Scientific Reports*, *7*(1), 1–11. doi:10.1038/s41598-017-11817-6 PMID:28916782

Yu, F. (2015). *LSUN: Construction of a Large-scale Image Dataset using Deep Learning with Humans in the Loop*. doi: https://doi.org//arXiv.1506.03365. doi:10.48550

Zhou, Y., Jia, E., Shi, H., Liu, Z., Sheng, Y., Pan, M., Tu, J., Ge, Q., & Lu, Z. (2022). Prediction of Time-Series Transcriptomic Gene Expression Based on Long Short-Term Memory with Empirical Mode Decomposition. *International Journal of Molecular Sciences*, *23*(14), 7532. doi:10.3390/ijms23147532 PMID:35886880

Chapter 14
Unlocking the Future of Healthcare:
Biomarkers and Personalized Medicine

Baiju B. V.
Vellore Institute of Technology, India

P. Suresh
https://orcid.org/0000-0001-9815-2982
Vellore Institute of Technology, India

G. Subathra
Sathyabama Institute of Science and Technology, India

P. Keerthika
https://orcid.org/0000-0002-9420-6389
Vellore Institute of Technology, India

Kishor Kumar Sadasivuni
Qatar University, Qatar

K. Logeswaran
https://orcid.org/0000-0001-8161-7376
Kongu Engineering College, India

ABSTRACT

Personalized medicine leverages patient-specific biological data to tailor prevention, diagnosis, and treatment. Biomarkers are critical for enabling this precision approach. However, biomarker development faces challenges in discovery, validation, and robust modelling, often requiring extensive labelled data. Machine learning (ML) methods like few-shot and zero-shot learning offer potential solutions by enabling model generalization from limited examples. This chapter provides comprehensive exploration of biomarker types and applications and how few-shot and zero-shot techniques could enhance biomarker prediction tasks. Few-shot learning shows promise for biomarker discovery and validation by transferring knowledge from established biomarkers. Zero-shot learning provides opportunities to detect novel biomarker candidates unconstrained by predefined labels. While nascent, few-shot and zero-shot learning present intriguing paradigms for more efficient biomarker modelling, which could accelerate progress towards personalized medicine.

DOI: 10.4018/979-8-3693-1822-5.ch014

1. OVERVIEW OF CURRENT HEALTHCARE LANDSCAPE

Nowadays, healthcare industries are undergoing continual transformation influenced by factors such as advances in medical technology, changing demographics, and healthcare policy reforms. The ability to efficiently access, manage, and analyse data has assumed paramount significance in current world. In recent times, scientists and researchers have made significant efforts to utilize Artificial Intelligence (AI) in diagnosis, treatment, and medical research. AI completely helps to transform the landscape of healthcare industry and significantly improves overall healthcare. Advancement of AI technology helps in various domains of medical field such as remote health monitoring, disease prediction, and surgical assistance have revolutionized the delivery of healthcare and also surpassing the traditional face-to-face visits (Vatandsoost & Litkouhi, 2019).

AI is transforming healthcare by expediting processes for patients and providers. It enables more customized engagement between patients, clinicians, and health systems. As patient demands and expectations rise, healthcare organizations are utilizing AI and ML to deliver improved patient experiences. With advanced algorithms capable of analysing complex data, AI systems can generate insights to inform clinical decision-making (Yu et al., 2018). This allows for more individualized and targeted care. Additionally, AI can optimize workflows, predict outcomes, and free up clinicians to focus on patients. To continue advancing care, providers need to thoughtfully implement AI that augments human judgment and holistically serves patient needs. AI holds immense potential to transform healthcare, but thoughtfully human-centric design is key to ensure it provides equitable benefit. Table 1 provides a list of acronyms used in this chapter and its expansions.

2. ROLE OF BIOMARKERS

Biomarkers are gaining importance in pharmaceutical research and drug development. They help identify mechanisms of action, predict efficacy and safety signals earlier, and find responsive patients. Specifically, biomarkers assess a drug's biological activity and interactions. They can flag potential toxicity risks preclinical before costly late-stage failures. Biomarkers also enable selecting likely responder subgroups

Table 1. List of acronyms and its expansion

Acronym	Expansion
AI	Artificial Intelligence
ML	Machine Learning
BMI	Body Mass Index
BNP	Brain Natriuretic Peptide
ACE	Angiotensin Converting Enzyme
SNP	Single Nucleotide Polymorphisms
ARB	Angiotensin Receptor Blockers
HIPAA	Health Insurance Portability and Accountability Act
PHI	Personal Health Information

for trials and approval. Though critical for more precise, accelerated drug discovery, biomarkers require extensive verification to validate utility before clinical use (Sarma et al., 2020).

Biomarker concepts and research have evolved over the past five years in response to scientific and clinical advances. Biomarker was coined in the year 1973, to denote the detection and measurement of biological molecules in order to make inferences about health status. Biomarkers serve as markers of biology, disease, or response to treatment. Biomarkers can be treated as quantifiable traits revealing insights into healthy functioning, pathogenesis, and pharmacological reactions (Atkinson et al., 2001). A biomarker, often called a biological biomarker, is a measurement that provides information about the current state of a cell or disease over a specific period of time. These markers play a crucial role as they can act as early indicators, providing a proactive approach to monitoring one's health. The success of personalized medicine and enhanced clinical outcomes hinges significantly on the discovery and utilization of well-defined biomarkers.

In medicine, biomarker refers to proteins analyzed in the blood, whose levels indicate the body's normal or pathological response and response to treatment. In a regular health check, important features such as blood pressure, heart rate, cholesterol, triglycerides and rapid blood sugar are constantly checked. Additionally, physical indicators such as weight, body mass index (BMI) and waist-hip ratio are often used to identify conditions such as obesity and metabolic syndrome. This comprehensive evaluation enables healthcare professionals to diagnose and address a spectrum of health issues, providing a holistic understanding of an individual's well-being. As depicted in Figure 1, a biomarker has specific features that enable it to be used for diagnosing a particular disease condition (Bodaghi et al., 2023).

Figure 1. Principal features of biomarkers

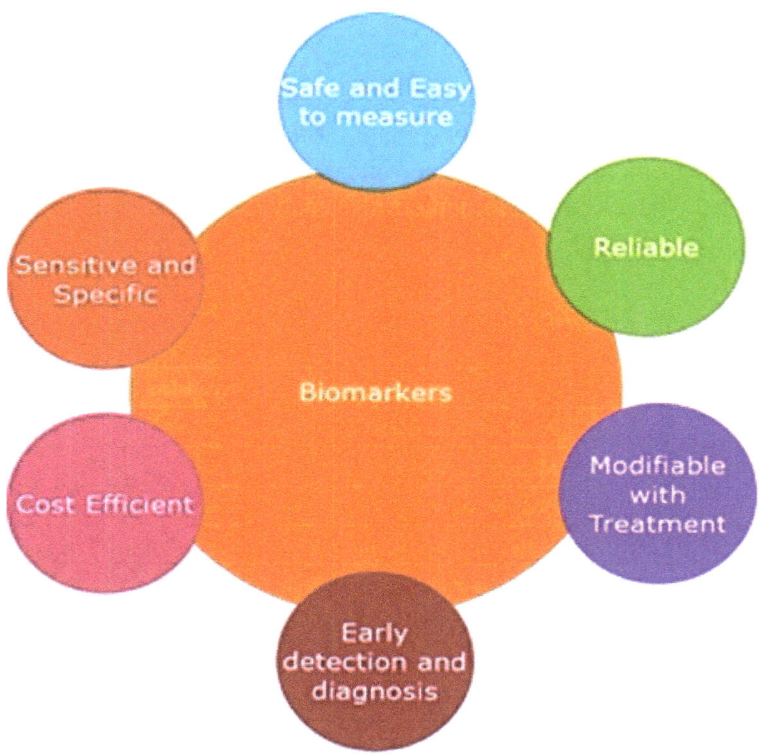

2.1. Categorization of Biomarkers

Biomarkers are categorized based on their primary clinical applications, as shown in Figure 2.

Figure 2. Types of biomarkers based on clinical application

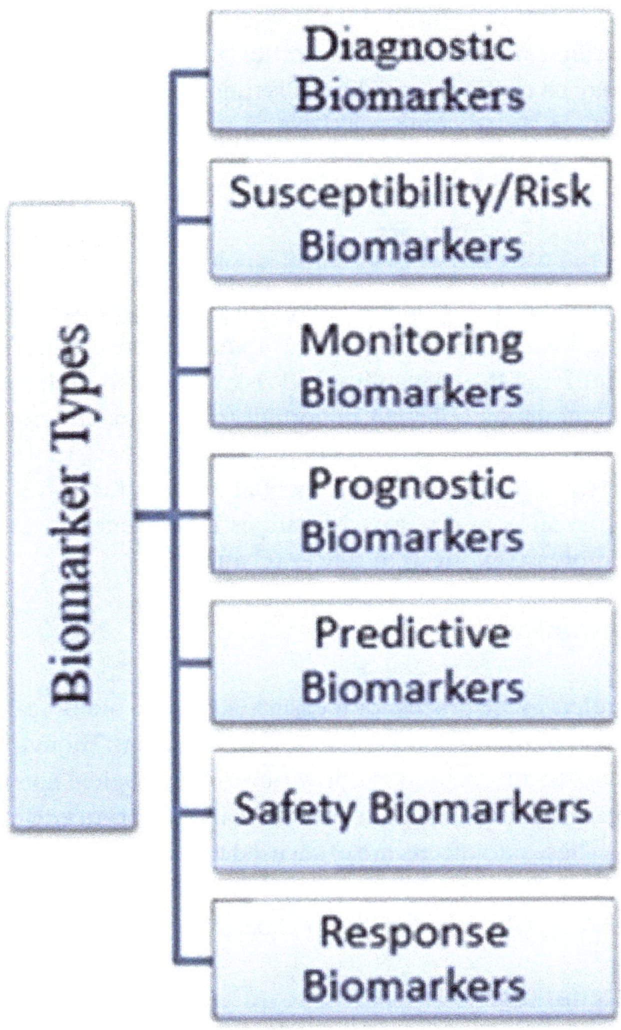

2.1.1. Diagnostic Biomarker

Diagnostic biomarkers are used in clinical and medical research to diagnose diseases by confirming the presence or absence of a particular disease or treatment. These biomarkers can reveal detailed information about the characteristics of the disease. Even when dealing with diagnostic biomarkers that can be measured accurately and reliably in the context of a useful application, assessing biomarker utility remains difficult. An objective is to establish a validation method that ensures the biomarker's consistent, precise, and cost-effective measurement. If a diagnostic biomarker moves from broad application to a specific role in future research or treatment, the specifics of its use should be carefully considered. While diagnostic biomarkers may be useful in one clinical setting, they may provide inaccurate information in a completely different setting. Hemoglobin A_1c (HbA_1c) serves as a diagnostic biomarker for individuals with diabetes (Olsson et al., 2016).

2.1.2. Susceptibility/Risk Biomarkers

Susceptibility or risk biomarkers play a significant role in predicting disease risk and also help to develop prevention approaches in clinical practice. Predictive biomarkers are distinct from prognostic biomarkers in that they focus on identifying the probability and risk associated with the development of a disease prior to its onset. These biomarkers play an important role in epidemiological research because they help identify early markers of disease and environmental factors associated with future diseases. Understanding these predictive elements opens up essential avenues for preventive actions and sheds light on the stages leading to the onset of diseases. Mutations in the genes have been associated to a higher probability of developing breast and ovarian cancer (Califf, 2018).

2.1.3. Monitoring Biomarkers

Monitoring biomarkers refers to the process of measuring specific biological markers in a person's body to assess their health, disease progression, or response to treatment. Biomarkers are measurable indicators that can provide valuable information about various physiological and pathological processes. An example of monitoring biomarkers is to monitor the increases in serum creatinine and/or potassium after medication or treatment. These parameters are often used to estimate the probability of an event. (Califf, 2018). The other biomarkers are brain natriuretic peptide (BNP), which is used to monitor heart failure, and hemoglobin A_1c (HbA_1c), which is used to diagnose and track diabetes.

2.1.4. Prognostic Biomarkers

Prognostic biomarkers are used to predict the likelihood of symptoms occurring in individuals diagnosed with a disease or health problem. These measures include infection or relapse, death, or the development of a new medical condition. Prognostic biomarkers are often used to establish inclusion and exclusion criteria to identify high-risk patients in clinical trials. The statistical power of a study depends on the number of events that occur rather than the overall sample size. The incidence of clinical events increases when patients are selected based on biomarker characteristics to support research. If the treatment has been tested successfully, the difference between the treatment results and the control group is exaggerated, but the significance of the relationship between the treatment and the results is still the same. The

Breast Cancer gene 1 and 2 are employed to assess the probability of a second cancer occurrence in women diagnosed with breast cancer (Möncke-Buchner et al., 2002).

2.1.5. Predictive Biomarkers

A biomarker is classified as predictive when its presence or variation enables the prediction of which patient or group is more likely to experience an effect resulting from exposure to a medical product or environmental agent. Treatment choices are correlated with predictive biomarkers (Fang et al., 2015). The HER2 protein, which suggests that some breast cancer patients might benefit from a particular targeted therapy is an example of a predictive biomarker. Predictive biomarkers in the Chinese Han schizophrenia population, represented by the presence of 12 single nucleotide polymorphisms (SNPs) associated with increased efficacy of olanzapine (Zhou et al., 2019).

2.1.6. Safety Biomarkers

Safety biomarkers are essential in the early phases of drug development to identify potential safety issues before they become clinical problems. These biomarkers are crucial for monitoring the effects of drugs and treatments to ensure their safety and to detect any adverse reactions early in the development process. Serum potassium levels may be used as a safety measure when evaluating patients taking diuretics (which tend to lower blood pressure) or angiotensin-converting enzyme (ACE) inhibitors, angiotensin receptor blockers (ARBs), and aldosterone antagonists (Roush & Sica, 2016).

2.1.7. Response Biomarkers

Response biomarkers also known as Pharmacodynamic biomarker, which is used to identify the occurrence of a biological response in a person following exposure to a medication or environmental solidity. The primary use of response biomarker is to provide critical information to guide clinical management and decide whether to continue treatment. Response biomarkers used to predict clinical trial outcomes that can be classified as candidate surrogate endpoints, potential surrogate endpoints, or validated surrogate endpoints, depending on the strength of the data (Christensen et al., 2021).

2.2. Challenges in Developing Biomarkers

Biomarker research and validation requires extensive time and resources before integration into clinical practice. More rigorous evidence of strong benefits versus risks is typically required for formal biomarker qualification outside of an individual drug approval compared to biomarkers evaluated for specific treatment decisions. This higher bar aims to ensure validated biomarkers provide reliable clinical value and utility across diverse contexts, not just benefit for a singular drug or trial. To qualify novel biomarkers, substantial longitudinal studies are often needed to demonstrate durable prognostic ability and clinical impact across populations. Conducting extensive risk-benefit analysis and collecting comprehensive real-world evidence of performance is key. This level of validation requires greater investment. However, achieving formal qualification enables reliable ongoing use of biomarkers to guide screening, diagnosis, treatment selection and disease monitoring across the healthcare ecosystem. The more strenuous process helps qualify only the most robust and generalizable biomarkers for inclusion into clinical practice guidelines.

Validating novel biomarkers to the point of clinical qualification can substantially increase development costs due to more extensive validation requirements. Lengthier clinical trials may be necessary to demonstrate utility across diverse populations and settings. Expanded analytical testing is also essential to characterize assay performance across laboratories. While these demands increase trial length, complexity, and budgets, the investments are critical to ensure real-world viability and prevent inaccuracies or failures from inadequate validation. Conducting rigorous clinical utility studies reduces downstream costs and patient risks from improper clinical implementation. Although more burdensome initially, comprehensive analytical and clinical validation is key to convincing payers, providers, and oversight bodies of a biomarker's merits and value proposition. The higher bar for qualification aims to validate only the most robust and generalizable biomarkers ready for broad inclusion in clinical practice (Wermuth et al., 2018). Keeping the focus on evidence quality is paramount, even if it necessitates elevated upfront validation expenditures. The reliability of certain biomarkers may pose challenges when it comes to their scientific validation and qualification for future use. Additionally, it's crucial to exercise caution in not misinterpreting biomarker measurements or prematurely attributing a biomarker to a specific disease.

3. INTRODUCTION TO PERSONALIZED MEDICINE

Personalized medicine, also known as individualized medicine or precision medicine, is an evolving medical approach where decisions related to disease prevention, diagnosis, and treatment are tailored to individual patients. This customization is informed by data derived from genetic and genomic information. Personalized medicines effectiveness resides not only in treatment, but also in prevention. Utilizing the genetic profiles of their patients, doctors can make informed decisions in selecting the appropriate medication or therapy, and subsequently administer it with the correct dosage or regimen. Genomics is significantly contributing to the advancement of personalized medicine by providing a detailed molecular insight into the unique variations among individuals (Ielapi et al., 2020). This invaluable information opens up avenues for making precise predictions about an individual's susceptibility to specific diseases, enabling them to tailor a prevention strategy that aligns perfectly with their needs. Depending on the person, the drug could help with the disease, not have any effect, or cause a serious side effect. Knowing how the drug will affect each person could really change healthcare by allowing for personalized treatments for each person (Hua et al., 2022). By considering the diversity of illnesses and external elements like environmental factors, patient requirements, and lifestyle, this approach enables healthcare providers to create treatments and interventions that are more customized and effective.

The rapid development of modern technology and the nature of big data continue to create challenges to the achievement of the short-term and long-term goals of precision medicine (Vaithinathan & Asokan, 2017). Personalized medicine has revolutionised the pharmaceutical industry, giving them the chance to make change not just by creating drugs that target specific molecules, but also by using and reusing existing drugs and combining them. Personalized medicine will change the way we diagnose and treat diseases, and patients will be more involved throughout and after treatment (Mathur & Sutton, 2017).

3.1. Revealing the Core of Personalized Medicine

Revealing the core of personalized medicine involves understanding its key principles and components.

3.1.1. Medical Practitioners

Healthcare can be personalized to include strategies for disease prevention, detection, diagnosis, treatment, and management by utilizing cutting-edge technologies and genetic profile insights. Healthcare is more effective and precise when decisions are made based on an individual's genetic makeup thanks to this personalized approach.

3.1.2. Patients and Consumers

Proactively managing disease risk and treatment strategies through genetic testing, clinical trials, and collaborative relationships with healthcare professionals are proactive and patient-centric approaches. This is actively participating in the process of learning about one's genetic predispositions, investigating novel therapies, and customizing medical decisions to meet one's unique needs. Participation of this kind results in more efficient and individualized healthcare.

3.1.3. Pharmaceutical and Biotechnology Firms

Identifying variation in genes and how it affects the safety and effectiveness of proposed medications is essential to developing personalized therapies and carrying out ground-breaking research. This approach aims at enhancing the safety as well as effectiveness of potential medicines by creating treatments that are specific to each patient's variation in genes.

3.1.4. Enterprises Specialized in Diagnostics

Advanced techniques for analysing and interpreting genetic data can provide a deeper understanding of the genetic structure of diseases and differences in response to treatment between individuals. Elucidating these genetic factors enables more personalized medicine approaches, from molecular subtyping of diseases to optimizing treatment plans based on a patient's genomic profile. However, realizing the full potential of genetic information in healthcare requires cross-disciplinary collaboration, thoughtful integration into clinical practice, and ensuring equitable access. While precision tools for genetic analysis hold promise for unravelling disease mechanisms and guiding targeted therapies, this potential should be harnessed responsibly and equitably to benefit all patients.

3.1.5. Scholars and Academics

Basic and clinical research uncovers new insights into genetics and disease mechanisms, enabling more precise diagnostics and targeted therapeutics. However, translating these findings from bench to bedside requires collaboration and consideration of equitable access. Realizing the full potential of genetics research necessitates responsible harnessing for broad benefit.

3.2. Integrating Personalized Medicine into Healthcare

3.2.1. Principles

Realizing the potential of personalized medicine requires engagement across the healthcare ecosystem to address critical needs. Healthcare providers, payers, employers, policymakers and most importantly patients and families must achieve greater literacy regarding emerging genomic technologies and their appropriate clinical implementation. Thoughtful policies and practices are imperative to uphold patient autonomy through informed consent, while protecting privacy and ethically employing molecular information in care. Open dialogue and evidence gathering should delineate best practices for demonstrating the clinical utility of personalized testing and interventions, so their value is recognized across settings. Investment is required to develop the health IT infrastructure, data systems and management tools that make genomic and clinical information comprehensively available to guide care while remaining user-friendly and patient-centric. Multidisciplinary collaboration is key to establish best practices for effectively and equitably delivering personalized medicine, from streamlining processes to optimizing operations. By working collectively to address these priorities, we can foster appropriate adoption of personalized approaches that improves outcomes across populations (Pritchard et al., 2017).

3.2.2. Personalised Medicine and its Common Challenges

3.2.2.1. Awareness and Education

Several key challenges currently hinder widespread adoption of personalized medicine. Inconsistent terminology and language leads to confusion among stakeholders. Consumer awareness and demand remains low, as the underlying science is often difficult to grasp and correlate to treatment options, while health literacy needs vary. Healthcare providers also lack sufficient knowledge, with limited up-to-date educational resources beyond oncology and genetics. Information on policies, best practices and community support is not readily accessible or utilized. Critical needs exist regarding workforce training for new technologies and modernizing medical school curricula. By addressing these gaps through collaborative efforts to improve genetics literacy, implement updated provider education, streamline policies and processes, boost consumer engagement, and adapt the workforce, we can work to successfully integrate personalized medicine for broad benefit.

3.2.2.2. Patient-Centric Care

The responsible integration of personalized medicine necessitates the resolution of several ethical and legal concerns. Policies around patient consent for using genetic data should be clarified and strengthened to uphold autonomy. Robust data privacy and security measures are needed, as molecular information could be subject to hacking risks. Data sharing policies should balance open science benefits while protecting proprietary interests and individual privacy. Providers must improve engagement with patients in decision-making, accounting for desired level of involvement and preferences in care plans. Genetic and genomic counselling should be made accessible to all families requiring it. Disparities in access and outcomes across patient populations must be proactively addressed, not exacerbated. By collaborating to implement thoughtful policies, improve counselling, engage patients, and promote health equity, we can integrate personalized approaches ethically and equitably.

3.2.2.3. Appreciating Value

Realizing the potential of personalized medicine requires addressing key knowledge gaps regarding clinical utility and value. Payers need clarity on evidence thresholds for determining coverage of emerging diagnostics, with payment aligned to their value in improving care. While economic data is still emerging, further research into clinical and cost-effectiveness can elucidate the value proposition across diverse populations. Practice guidelines and provider adoption also depends on accumulating evidence demonstrating clinical utility and actionability of genetic findings. Healthcare organizations must recognize both the direct return-on-investment and indirect programmatic value of integrating personalized approaches, including preventive care. Multi-stakeholder efforts to incentivize and facilitate responsible data sharing, with appropriate privacy protections, can build the evidence base on outcomes, variability, and value. Open dialogue and collaboration across payers, providers, health systems, and policymakers is critical to establish frameworks and gather key evidence that will support appropriate adoption and reimbursement.

3.2.2.4. Securing Medical Access

Health insurance coverage and reimbursement policies often do not align payment with the value of emerging diagnostics, while fee-for-service models incentivize volume over value-based care. Health IT systems like EHRs need optimization to incorporate genetic data at the point of care. Some providers may be reluctant to adopt personalized approaches due to perceived lack of adequate compensation for the time involved, or challenges in coordinating specialty expertise like genetic counsellors. Clinical guidelines and decision tools need to be updated to reflect new biomarkers and targeted therapies, while also improving understanding of proper context for use to avoid adverse events. Care coordination across medical groups and community organizations around personalized medicine programs and policies needs strengthening. Innovative business models are required to develop sustainable implementation frameworks. Improving access to specialists and emerging diagnostics/treatments in rural areas remains an obstacle. By working collaboratively to address these systemic barriers through improvements in policy, payment models, clinical infrastructure, care coordination, and access, integrated personalized medicine can become a reality.

4. MACHINE LEARNING IN HEALTHCARE

Machine learning is a branch of AI that play a pivotal role in driving positive advancements within the healthcare sector. The market for artificial intelligence in healthcare is expected to grow from $14.6 billion in 2023 to $102.7 billion in 2028, reflecting the evolution of artificial intelligence. This surge is attributed to AI's remarkable ability to analyse and make sense of intricate medical data, showcasing its potential to revolutionize the healthcare industry.

Machine learning excels in handling huge sets of clinical documentation, parsing through raw data to uncover patterns and enhance the accuracy of predicting future medical outcomes. Its applications extend to disease diagnosis, drug development, and therapeutic discoveries, contributing significantly to improved healthcare diagnostics and treatments. At its core, machine learning's fundamental role in healthcare lies in precision medicine, leveraging patient-specific physiological data and treatment parameters to predict the success rates of medical procedures with greater precision.

Figure 3. Machine learning in healthcare sector

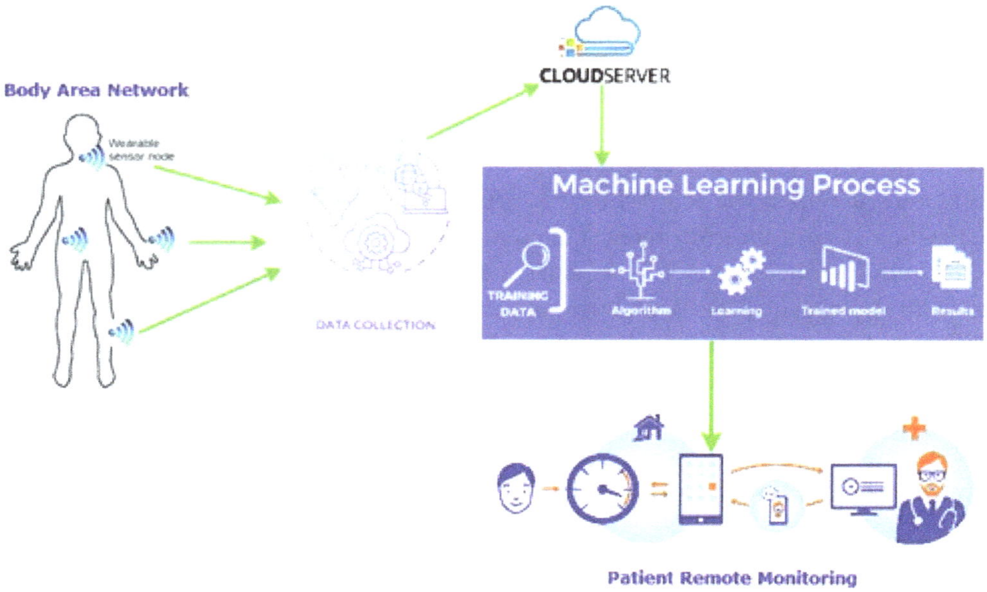

Machine learning uses algorithms to find patterns in data. The inputs are patient traits like demographics, medical history, test results, symptoms, etc. Outcomes like disease progression are also used. By analysing these diverse healthcare data inputs, machine learning models can predict outcomes to guide treatment and gain new insights. But these models need careful design and interpretation to be applied accurately (Jiang et al., 2017). Machine learning consists of two main categories: supervised learning and unsupervised learning. In supervised machine learning, known input and output data are used to train methods to predict future outcomes, while unsupervised machine learning identifies side patterns or hidden patterns in the input data.

4.1. Role of ML in Healthcare

Machine learning techniques can be used to solve many tasks. In the healthcare sector, the transformative impact of machine learning is evident in Figure 3, depicting its potential to reshape the landscape of disease diagnosis, treatment, and prevention (An et al., 2023).

Machine learning in healthcare revolutionizes diagnostics by expertly classifying diseases and offering predictive insights.

- **Classification**: Machine learning algorithms can help to classify and diagnose types of diseases or medical data as needed.
- **Recommendations**: Machine learning algorithms can easily provide medical information without the need to manually retrieve it.
- **Clustering**: Machine learning can help group similar medical cases to analyse the patterns and conduct future research.

- **Prediction:** Machine learning can make predictions about how future occurrences are likely to happen using present data and common patterns.
- **Anomaly detection**: Machine learning can be used to identify anomalies from typical patterns and determine whether additional treatment is necessary.
- **Automation:** Machine learning can efficiently manage routine, time-consuming tasks, such as data entry, appointment scheduling, and inventory management, thereby relieving doctors of these heavy responsibilities.
- **Ranking:** Machine learning can prioritize pertinent information, simplifying the process of locating it.

4.2. Factors for Integrating ML in Healthcare

Machine learning in the healthcare industry relies on personal health information (PHI) and emphasizes the importance of data security and privacy. In addition to the possible harm to one's reputation that could arise from a data breach, strict laws such as the Health Insurance Portability and Accountability Act (HIPAA) place limitations on how patient data may be used and shared. While high-quality, accurate data is imperative, unique challenges exist in keeping healthcare data current for machine learning applications. Medical practices, regulations, clinical codes, and health system rules continuously evolve over time. Machine learning models are only as up-to-date as their training data. If models are not re-trained on the latest data, their performance and reliability quickly deteriorate. This necessitates frequent updating of healthcare data and model re-validation to maintain optimal accuracy. For instance, a machine learning-based medical coding assistance system must be re-trained on contemporary procedure codes. Failure to update could result in coding errors leading to inaccurate billing and flawed reporting. To deploy robust machine learning solutions in the dynamic healthcare domain, maintaining representative, timely training data through continuous data pipeline and model refreshing is essential. Healthcare organizations must invest in this maintenance to avoid degraded model performance when inputs change.

Machine learning models require large volumes of data in order to produce useful insights and accurate predictions. But to guarantee that the model can perform as intended, there must be data specifically, good, trustworthy, and easily available data. Correct, comprehensive, consistent, and use case-relevant data is what defines quality data. To fully utilize a machine learning model, it is essential to have a representative, diverse data set that has been safely and securely extracted from several systems.

4.3. Enhancing the Healthcare Industry With Machine Learning

Machine learning has immense potential to improve patient care, improve health outcomes, as well as increase the efficiency of healthcare delivery. Some of the key benefits machine learning offers for transforming healthcare includes:

- **Personalized Treatment Recommendation:** Machine learning has the capability to examine extensive datasets encompassing patient characteristics, medical histories, genetic information, and treatment outcomes. It can discern patterns and associations within this wealth of information. Customized treatment plans can be developed for every patient using machine learning.

- **Enhanced precision:** Machine learning algorithms can analyse huge sizes of medical data, uncovering significant trends that might avoid human observers. This capability enhances the ability to predict and recognize events more effectively.
- **Drug Discovery & Development:** Drug development is complex and labour-intensive. Machine learning can accelerate it in key ways. By screening chemical databases, predicting compound interactions, and discerning optimal candidates, machine learning enables more efficient drug discovery. Integrated thoughtfully, machine learning can unlock impactful new therapies faster.

4.4. Challenges Faced in Integrating ML in the Healthcare

Machine learning implementation in healthcare faces many challenges and also provide solutions that are needed for effective implementation. Thoughtful integration necessitates addressing concerns around data, ethics, regulation, and clinical adoption. By overcoming these obstacles, machine learning will be able to reach its full potential.

- **Lack of Data Quality and Availability:** The quality of the data fed into machine learning algorithms determines the output. For machine learning models to learn effectively, they rely on processing significant, diverse, and high-quality datasets. Health data often presents challenges as it is distributed across different systems, necessitating enhanced standardization and potentially containing gaps or incomplete information.
- **Heterogeneity of data:** High-quality structured data needs to be easily available to allow for machine learning to be implemented successfully. In the field of health information, deficiencies might jeopardize the accuracy of predictions, causing serious consequences on clinical decision-making. The heterogeneity data in the healthcare sector is one of the challenges that prevent the broad implementation of machine learning. Thousands of handwritten document scans as well as fragmented, duplicate, and missing data can produce insufficient information and impact bad decisions.
- **Provider resistance:** The biggest challenge to machine learning implementation in healthcare is provider resistance. Medical organizations often have to update or replace legacy systems in order to implement machine learning.
- **Regulatory and Ethical Considerations:** In the healthcare sector, strict rules and regulations play an important role such as Health Insurance Portability and Accountability Act (HIPAA) in the US, play a pivotal role. The integration of machine learning into healthcare necessitates meticulous adherence to these regulations to guarantee patient consent and privacy.

5. FEW-SHOT LEARNING

Few-shot learning is a machine learning technique that focuses on learning new concepts from a few examples. Unlike the supervised learning model, this trains on a sample of a large list, learning several aims to expand new categories using a small sample from each category. The goal is to train models to learn how to learn rapidly from limited data. This is achieved through meta-learning, where models first train on large benchmark datasets to acquire generalizable knowledge. They then leverage this knowledge to adapt to novel classes from a scarce number of labelled examples. Early disease detection is a

Figure 4. Few-Shot learning framework

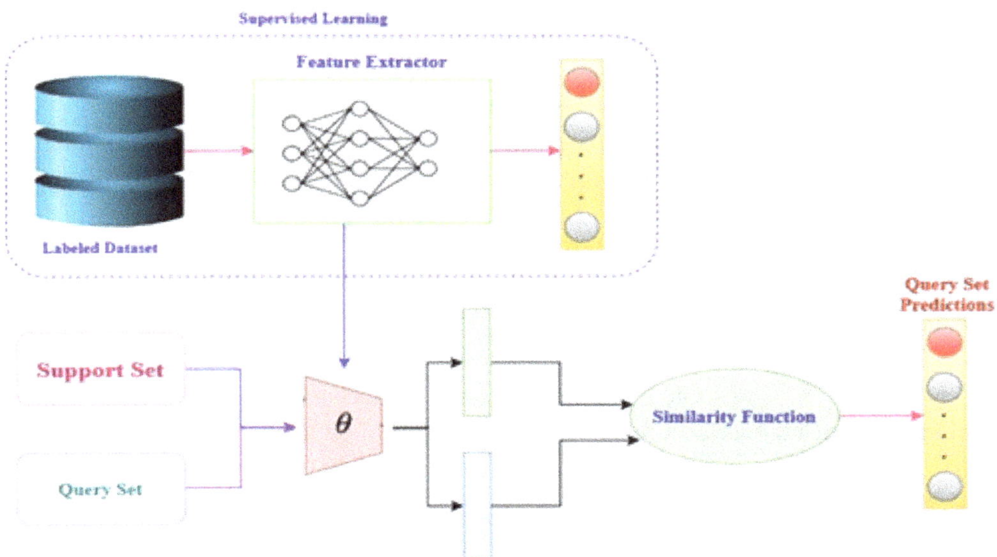

significant challenge in the medical sector today, and as a result, the utilization of few-shot learning is steadily growing in the domain of medical image classification (Munkhdalai & Yu, 2017).

Few-shot learning seeks to imitate human-like versatility in acquiring new concepts from limited experience. The overview of few shot learning framework is shown in the Figure 4. Support Set and Query Set are concepts commonly used in few-shot learning, a type of machine learning where a model is trained to recognize new classes with very limited examples.

The support set is a critical component in few-shot learning, consists of a small number of labelled samples for each novel category or class of data. These labelled examples serve as the foundation upon which a pre-trained model builds its understanding of these new classes. By using this limited but informative set of data, the model can learn to generalize and make predictions for these novel categories, even though it has not seen them during its initial training. The support set essentially provides the model with the necessary cues and information to adapt and extend its knowledge to new, previously unseen classes, making it a valuable tool in tasks like one-shot learning and few-shot classification.

The query set is a crucial component in the context of few-shot learning and meta-learning. It comprises samples drawn from both new and old categories of data, enabling the model to assess its ability to generalize knowledge acquired from a support set. By presenting these novel and familiar instances to the model, we can evaluate its capacity to make predictions, classifications, or inferences based on its prior understanding. This process helps us to evaluate the model's adaptability and transferability of learned information to new tasks or data points.

5.1. Key Enablers for Deploying Few-Shot Learning Systems

Few-Shot Learning models are facilitated by the concept of reliable algorithms that can be created from minimal datasets. The following are some of the factors that are supporting its growing popularity.

- **Sparse Data**: Machine learning algorithms face a notable challenge when confronted with data scarcity, whether it's in the form of supervised data or not, as they strive to generate precise factors and make more refined inferences
- **Anomaly detection**: Few-shot learning enables machines to recognize and interpret rare or uncommon scenarios during their training process.
- **Mitigating data acquisition and computational costs:** Few-shot learning models require less data for training, leading to significant cost savings in data collection and labelling. In addition, the reduced training data leads to lower dimensionality in the training data set, which further reduces the computational cost.

5.2. Approaches for Effective Few-Shot Learning

The three of the most promising strategies for addressing few-shot learning that provide a various ways to tackle the challenge of making accurate predictions with limited labelled data are mentioned below.

- **Meta-Learning:** Meta-learning, often referred to as learning to learn, entails instructing a model to acquire the fundamental structure or meta-knowledge of a task. This approach shows that it is possible to solve the challenge of several studies in which the model is trained for different types of jobs and optimized for new tasks using only a few examples.
- **Transfer Learning:** Transfer learning leverages pre-training models derived from big data and fine-tune them for particular few-shot challenges. Transfer learning leverages the knowledge gained during pre-training.
- **Data Augmentation:** Data augmentation is a method that generates new training data samples by applying diverse transformations to the existing training dataset. One of the main advantages of this method is that it can improve the general capabilities of machine learning models in many computing operations, especially learning two or three times.

5.3. Applications of Few Shot Learning in Healthcare

Few-shot learning has found applications in a variety of fields due to the small datasets required and the cheap cost involved. Here are some notable applications of few-shot learning:

- **Image Classification and Object Recognition**: Models are able to identify and categorize objects in images with few labelled examples in a timely manner. When identifying uncommon or novel objects that are absent from the training dataset, it is extremely helpful.
- **Medical Diagnosis and Healthcare:** Few-shot learning has potential applications in diagnosis and analysis of medical imaging. With little medical data, it can help with rare disease identification, anomaly detection, and patient outcome prediction.
- **Recommendation Systems:** Provide users with individualized product or content recommendations based on a limited number of user interactions or preferences.
- **Drug Discovery:** In pharmaceutical research, small-scale studies play an important role in predicting the effectiveness and safety of drugs with limited experimental data. This enables more efficient screening of compounds with the potential for medical applications.

5.4. Machine Learning and Few-Shot Learning: Unveiling the Contrasts

Machine learning has become a transformative force across industries, including healthcare. But an emerging technique called few-shot learning is gaining traction as a potential alternative paradigm. At a high-level, machine learning uses statistical models trained on large labelled datasets to uncover patterns and make predictions. In contrast, the goal of few-shot learning is to acquire novel ideas from a limited quantity of examples, replicating human-like flexibility.

While machine learning requires massive data to fuel its pattern recognition capabilities, few-shot methods employ specialized meta-learning techniques and architectures to enable rapid adaptation from scarce data. For instance, few-shot image classifiers can recognize new object categories from only 1-5 example images. Machine learning models also struggle to extend beyond the distributions of their training data. Few-shot learning attempts to address this limitation by learning more generalizable feature representations to transfer to novel distributions. In application, machine learning excels at well-defined tasks with abundant data, while few-shot methods currently target problems with limited data scale. Combining them can utilize their complementary strengths.

However, few-shot techniques remain early in maturity with challenges in accuracy and uncertainty quantification. But active research offers promise to address limitations and unlock new use cases benefitting from efficient learning. Further advances in few-shot algorithms are needed to improve adaptability, particularly when new data differs substantially from original training distributions. Overall, few-shot learning offers an intriguing paradigm for more flexible machine learning.

6. ZERO SHOT LEARNING

Zero-shot learning involves building recognition models for previously unseen target classes that weren't labelled during training. This approach transfers information from source classes through labelled samples, utilizing class attributes as auxiliary data. Zero-shot learning comprises a two-stage process such as training and inference.

Zero-shot learning methods aim to acquire knowledge of the intermediate semantic layer and its features, then utilize it during inference to predict a novel data class (Bucher et al., 2016). For Zero-shot Learning to occur, a labelled training set containing both seen and unseen classes is essential. These classes, whether seen or unseen, are interconnected within a high-dimensional vector space referred to as the semantic space. In this space, data from seen classes can be transferred to unseen classes (Zhang et al., 2017).

Zero-shot learning for databases represents an innovative approach to learning database components, capable of accommodating a wide range of tasks without the need to gather training data specifically for supporting a new database. This approach provides the advantages of informational learning and activity-based learning. The fundamental concept of zero-shot learning from data downloads inspired by recent successes in transferring knowledge from models (Hilprecht & Binnig, 2021).

The overview of zero-shot learning is shown in Figure 5. Seen classes refer to the specific categories or labels that have been part of the training dataset used to teach a machine learning model. These classes are familiar to the model, and it has learned to recognize patterns, features, and characteristics associated with these known categories. Training a model with seen classes is the foundational step in the machine learning process, as it enables the model to acquire the knowledge and skills needed to

Figure 5. Overview of zero-shot learning

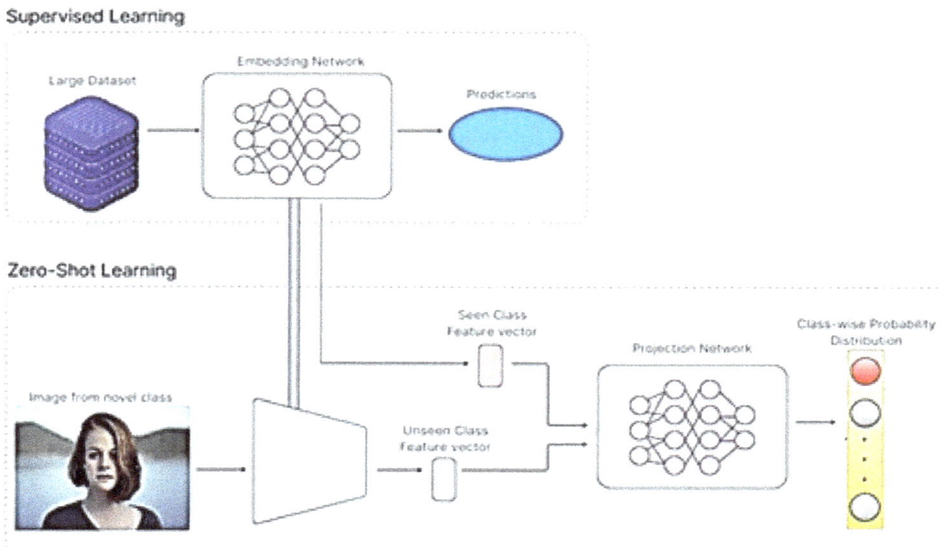

make predictions, classifications, or generate responses within the scope of these known categories. The ability to specify information from visible to invisible categories is a key to zero-shot learning, allowing models to adapt to new devices and operate without direct examples. . Unseen classes represent a fundamental challenge in zero-shot learning, as they encompass categories or labels that the model has never encountered during training. In this context, the model is expected to classify data belonging to these unseen classes without any prior exposure or specific training examples. The model's capability to generalize its knowledge from seen classes to accurately classify instances from these unseen categories is a testament to its ability to understand underlying attributes, relationships, or semantic embedding's that allow for extrapolation beyond the training data.

Zero-shot learning relies on auxiliary information about unseen classes since no labelled examples are available. This includes text descriptions, semantic properties, and word vector embedding's that encapsulate key attributes. Auxiliary data provides abstract knowledge about the unseen classes. The model combines this with its experience on seen classes to infer connections and recognize new classes. Without auxiliary guidance on unseen class properties, zero-shot generalization would not be feasible. Descriptive auxiliary data must be carefully developed to allow for models to learn about classes that have never been identified, even in the absence of any examples.

6.1. Key Challenges Faced in Zero-Shot Learning in Healthcare

The most key challenges faced when applying Zero-Shot Learning in practice are mentioned as follows.

- **Stereotyping:** Few-shot models are biased since they are trained on "seen" classes. When tested on unseen classes, models default to predicting seen classes. This bias worsens if test data has both seen and unseen classes. Reducing bias towards seen classes is key for few-shot model adaptability to diverse new data.

- **Data Privacy:** As healthcare data is highly sensitive, employing zero-shot learning in this domain necessitates rigorous privacy preservation measures to uphold patient confidentiality. Responsible development of secure and ethical zero-shot systems remains imperative for healthcare applications.
- **Clinical Variability:** Considerable variability exists across patient health conditions, presenting symptoms, and treatment effects. To account for such heterogeneity, robust zero-shot models with strong generalization capabilities are vital in healthcare. Developing flexible systems that maintain performance across diverse medical scenarios remains a key priority.
- **Domain Adaptation:** Deploying zero-shot models across different healthcare systems and geographical regions with divergent data characteristics presents significant adaptation challenges. To enable generalizable insights, developing zero-shot systems with the flexibility to align with varied institutional contexts and local distributions is critical. Context-aware, adaptive zero-shot learning is the key for healthcare.

6.2 Zero-Shot Learning Approaches

The two most common approaches used to solve the zero-shot recognition problems are:

6.2.1 Classifier Based Methods

Classifier-based methods are often used for zero-shot learning. A classifier is first trained to map input features to class labels, using only seen classes with examples. Unseen classes have no labelled data. Class embedding's created that encode classes as vectors in a continuous semantic space. This field captures the relationship between visible and invisible groups. During testing, the correlation classifier will use unseen class vectors as input instead of the original labels. It leverages connections learned in the embedding space to classify unseen classes without retraining. However, this approach relies on embedding's adequately representing salient class relationships (Tan et al., 2021).

6.2.2. Instance Based Methods

The aim of the instance based method is to find information that resembles hidden information and then to predict the invisible data using the labels of similar information.. These methods heavily depend on a similarity function, which essentially quantifies how alike different data points are. By comparing an unseen instance with its most similar training examples, instance-based methods offer a way to determine the class or outcome for the new data point. This approach proves especially valuable when the local characteristics or neighbourhood relationships are critical in making accurate predictions, allowing the model to adapt effectively based on its closest matches in the training data.

6.3 Zero-Shot Learning: Contrasts With Traditional Machine Learning

Traditional machine learning typically relies on the availability of extensive labelled datasets to train models effectively, allowing them to learn and make predictions. These models often memorize the data they are exposed to during training and can accurately identify objects in the test dataset that resemble those seen during the training phase. In contrast, zero-shot learning takes a different approach, depending on conditional probabilities and the model's capacity to establish relationships between known classes and entirely new, unseen ones. Instead of memorization, zero-shot learning focuses on the model's ability to recognize and classify new objects based on their inherent attributes and features. This approach reduces the dependency on large training datasets and allows the model to generalize to novel classes it has never encountered before. Zero-shot learning leverages semantic embedding's and representations, where objects are associated with attributes, allowing models to reason about the attributes of new objects they have not seen during training. This approach encourages adaptability, as models can make informed predictions for entirely new classes based on their characteristics.

In zero-shot learning, the model's success hinges on its capacity to understand the underlying semantic relationships between known and unknown classes, enabling it to make accurate inferences. This paradigm is particularly valuable in scenarios where obtaining extensive labelled data is challenging or in applications that require adaptability to emerging or evolving concepts. It stands as a contrast to traditional machine learning methods, which tend to rely heavily on memorization and may struggle when confronted with novel classes.

7. INTEGRATING MACHINE LEARNING FOR BIOMARKER PREDICTIONS IN PERSONALIZED MEDICINE

Few-shot learning and Zero-shot learning, a machine learning paradigm, offer innovative solutions for biomarker predictions, especially in the context of personalized medicine.

7.1. Biomarker Discovery

Biomarker discovery involves the exploration of high-dimensional omics data to identify potential measurable biomarkers that can effectively differentiate between different disease states. Few-shot models trained on established biomarkers can recommend potential candidates for emerging conditions, even when data is scarce. These models can benefit from architectures integrating biological knowledge graphs and pathway databases to inform candidate selection.

Zero-shot approaches introduce a novel perspective by scanning for unusual patterns within the data, enabling the detection of reputed biomarkers without relying on predefined labels. These models have the capability to transfer knowledge acquired from databases containing known biomarkers to nominate candidates for previously unknown diseases or conditions. This approach streamlines the search process by prioritizing exploration based on learned relationships between biomarkers, pathways, and diseases, thereby avoiding exhaustive blind searches.

7.2. Biomarker Validation

Validating newly discovered biomarkers necessitates stringent statistical scrutiny on extensive population samples prior to clinical implementation. Few-shot techniques help create validation plans by learning from proven biomarkers. This method lowers the number of samples needed for validation by using information from reliable biomarkers. However, it's important to think carefully about making sure these validation procedures work well for different groups of people.

Zero-shot techniques offer the potential to diminish the requisite sample sizes for validation by leveraging insights from previously validated biomarkers. Models trained on well-established biomarkers may extend their applicability by suggesting validation criteria and methods for related novel biomarkers with reduced sample requirements. Nevertheless, it is essential to exercise caution when applying these models to distinct patient groups or demographics, as biases may emerge due to limited diversity. Augmenting zero-shot models with iterative human analysis and guidance can augment the generalizability of validation processes.

7.3. Biomarker-Based Disease Diagnosis

Numerous diagnostic tests depend on biomarkers, like risk-associated proteins, for early disease detection. Few-shot techniques help customize models for predicting a patient's diagnosis using their limited medical records. This involves using models that combine information from a patient's medical history, data from a larger population, and insights about biomarkers. Zero-shot techniques can aid in early diagnosis prediction using limited historical patient data. Models trained on established diagnostic biomarkers may identify related biomarkers and risk factors, enabling diagnosis with sparse data.

7.4. Biomarker-Guided Treatments

Therapies are often prescribed according to biomarker profiles indicating treatment response. Few-shot approaches match patients to suitable therapies with minimal examples. These methods employ techniques like memory networks to quickly integrate new biomarker-treatment connections.

Zero-shot methods can link patients to suitable treatments even with limited biomarker-treatment examples. Models leverage learned associations between biomarkers, drug mechanisms, and responses to suggest therapies based on a patient's biomarkers, although challenges involve capturing intricate interactions, addressing heterogeneity, and avoiding overly simplistic reasoning, and using ensembles with causal models can enhance reliability.

8. CHALLENGES AND FUTURE DIRECTIONS FOR RESEARCH

The identification and validation of reliable biomarkers for various diseases is a complex and resource-intensive process, requiring extensive research and clinical trials. Additionally, ensuring the privacy and security of patient data in the era of personalized medicine is a critical concern, as is the need for regulatory frameworks to govern the development and use of these technologies. Furthermore, the cost-effectiveness of personalized treatments and the equitable distribution of healthcare innovations pose significant challenges, with disparities in access to cutting-edge treatments being a potential issue. Nevertheless, overcoming these obstacles promises to revolutionize healthcare by tailoring treatments to individual patients' unique genetic, environmental, and lifestyle factors, offering the potential for more effective and efficient healthcare interventions.

Few-shot learning is explored by applying for human cell classification. The experiments benchmarked various few-shot model architectures and training methodologies using unseen data. The models demonstrated strong performance when evaluated on held-out examples from the original unseen classes. However, significant accuracy drops occurred when transitioning the models to real-world human cell datasets. This highlights a key challenge in few-shot learning degradation when extending to new distributions beyond the original training data. While showing promise on unseen classes, the few-shot models struggled to generalize to the distinct features and complexity of actual biological cell image datasets. Further research into improving few-shot model robustness and generalization is critical before clinical deployment. Real-world evaluation on relevant target data must be part of advancing few-shot techniques for biomedical applications (Walsh et al., 2022).

Zero short learning method and few shot learning method needs to handle the biomarker data that can be highly variable due to factors such as patient demographics, variations in genetics and environmental factors. To ensure the success of machine learning models in healthcare, privacy and ethical issues surrounding the collection, use or sharing of medical information must be addressed as a matter of common intelligence.

9. CONCLUSION

Biomarkers are important for the advancement of research findings, the development of medical products, and the treatment of individuals and populations. Precision medicine is expected to reduce the risk of adverse effects for patients at the same time it will reduce costs and improve the sustainability and efficiency of healthcare systems. Innovative approaches to identifying biomarkers in the area of personalised medicine have been presented by zero shot learning and few short learning, which address challenges that arise from traditional supervised learning methods. Zero shot learning and few short learning provides an efficient framework for biomarker discovery which enables a more effective and individualized techniques in personalized medicine possibly improving the quality of life of patients and advancing the healthcare industry. The future of healthcare resides in our hand that has to be unlocked by continuous commitment and collaborative efforts.

REFERENCES

An, Q., Rahman, S., Zhou, J., & Kang, J. J. (2023). A Comprehensive Review on Machine Learning in Healthcare Industry: Classification, Restrictions, Opportunities and Challenges. In Sensors, 23(9). Sensors (Basel). doi:10.3390/s23094178

Atkinson, A. J., Colburn, W. A., DeGruttola, V. G., DeMets, D. L., Downing, G. J., Hoth, D. F., Oates, J. A., Peck, C. C., Schooley, R. T., Spilker, B. A., Woodcock, J., & Zeger, S. L. (2001). Biomarkers and surrogate endpoints: Preferred definitions and conceptual framework. In Clinical Pharmacology and Therapeutics, 69(3), 89–95. doi:10.1067/mcp.2001.113989

Bodaghi, A., Fattahi, N., & Ramazani, A. (2023). Biomarkers: Promising and valuable tools towards diagnosis, prognosis and treatment of Covid-19 and other diseases. In Heliyon, 9(2). Elsevier Ltd. doi:10.1016/j.heliyon.2023.e13323

Bucher, M., Herbin, S., & Jurie, F. (2016). Improving semantic embedding consistency by metric learning for zero-shot classiffication. Lecture Notes in Computer Science (Including Subseries Lecture Notes in Artificial Intelligence and Lecture Notes in Bioinformatics), 9909 LNCS, (pp. 730–746). Springer. doi:10.1007/978-3-319-46454-1_44

Califf, R. M. (2018). Biomarker definitions and their applications. *Experimental Biology and Medicine, 243*(3), 213–221. doi:10.1177/1535370217750088 PMID:29405771

Christensen, C. H., Chang, J. T., Rostron, B. L., Hammad, H. T., van Bemmel, D. M., Del Valle-Pinero, A. Y., Wang, B., Mishina, E. V., Faulcon, L. M., DePina, A., Brown-Baker, L. N., Kimmel, H. L., Lambert, E., Blount, B. C., Vesper, H. W., Wang, L., Goniewicz, M. L., Hyland, A., Travers, M. J., & Chang, C. M. (2021). Biomarkers of inflammation and oxidative stress among adult former smoker, current e-cigarette users—Results from wave 1 PATH study. *Cancer Epidemiology, Biomarkers & Prevention, 30*(10), 1947–1955. doi:10.1158/1055-9965.EPI-21-0140 PMID:34289969

Fang, B., Mehran, R. J., Heymach, J. V., & Swisher, S. G. (2015). Predictive biomarkers in precision medicine and drug development against lung cancer. In Chinese Journal of Cancer, 34(7), 1–15. BioMed Central. doi:10.1186/s40880-015-0028-4

HilprechtB.BinnigC. (2021). One Model to Rule them All: Towards Zero-Shot Learning for Databases. *CoRR, abs/2105.0.* https://arxiv.org/abs/2105.00642

Hua, J., Li, Y., Liu, C., & Wang, L. (2022). A zero-shot prediction method based on causal inference under non-stationary manufacturing environments for complex manufacturing systems. *Robotics and Computer-integrated Manufacturing, 77*, 102356. doi:10.1016/j.rcim.2022.102356

Ielapi, N., Andreucci, M., Licastro, N., Faga, T., Grande, R., Buffone, G., Mellace, S., Sapienza, P., & Serra, R. (2020). Precision medicine and precision nursing: The era of biomarkers and precision health. In *International Journal of General Medicine* (Vol. 13, pp. 1705–1711). Dove Medical Press Ltd., doi:10.2147/IJGM.S285262

Jiang, F., Jiang, Y., Zhi, H., Dong, Y., Li, H., Ma, S., Wang, Y., Dong, Q., Shen, H., & Wang, Y. (2017). Artificial intelligence in healthcare: Past, present and future. In Stroke and Vascular Neurology (Vol. 2, Issue 4, pp. 230–243). BMJ Specialist Journals. doi:10.1136/svn-2017-000101

Mathur, S., & Sutton, J. (2017). Personalized medicine could transform healthcare (Review). In Biomedical Reports (Vol. 7, Issue 1, pp. 3–5). Spandidos Publications. doi:10.3892/br.2017.922

Möncke-Buchner, E., Reich, S., Mücke, M., Reuter, M., Messer, W., Wanker, E. E., & Krüger, D. H. (2002). Counting CAG repeats in the Huntington's disease gene by restriction endonuclease EcoP15I cleavage. *Nucleic Acids Research, 30*(16), e83. doi:10.1093/nar/gnf082 PMID:12177311

Munkhdalai, T., & Yu, H. (2017). Meta networks. *34th International Conference on Machine Learning, ICML 2017, 5*, 3933–3943. https://proceedings.mlr.press/v70/munkhdalai17a.html

Olsson, B., Lautner, R., Andreasson, U., Öhrfelt, A., Portelius, E., Bjerke, M., Hölttä, M., Rosén, C., Olsson, C., Strobel, G., Wu, E., Dakin, K., Petzold, M., Blennow, K., & Zetterberg, H. (2016). CSF and blood biomarkers for the diagnosis of Alzheimer's disease: A systematic review and meta-analysis. *Lancet Neurology, 15*(7), 673–684. doi:10.1016/S1474-4422(16)00070-3 PMID:27068280

Pritchard, D. E., Moeckel, F., Villa, M. S., Housman, L. T., McCarty, C. A., & McLeod, H. L. (2017). Strategies for integrating personalized medicine into healthcare practice. In Personalized Medicine (Vol. 14, Issue 2, pp. 141–152). Future Medicine Ltd. doi:10.2217/pme-2016-0064

Roush, G. C., & Sica, D. A. (2016). Diuretics for hypertension: A review and update. In American Journal of Hypertension (Vol. 29, Issue 10, pp. 1130–1137). Am J Hypertens. doi:10.1093/ajh/hpw030

Sarma, A., Calfee, C. S., & Ware, L. B. (2020). Biomarkers and Precision Medicine: State of the Art. In Critical Care Clinics (Vol. 36, Issue 1, pp. 155–165). Crit Care Clin. doi:10.1016/j.ccc.2019.08.012

Tan, C., Xu, X., & Shen, F. (2021). A Survey Of zero shot detection: Methods and applications. *Cognitive Robotics*, *1*, 159–167. doi:10.1016/j.cogr.2021.08.001

Vaithinathan, A. G., & Asokan, V. (2017). Public health and precision medicine share a goal. In Journal of Evidence-Based Medicine (Vol. 10, Issue 2, pp. 76–80). J Evid Based Med. doi:10.1111/jebm.12239

Vatandsoost, M., & Litkouhi, S. (2019). The Future of Healthcare Facilities: How Technology and Medical Advances May Shape Hospitals of the Future. *Hospital Practices and Research*, *4*(1), 1–11. doi:10.15171/hpr.2019.01

Walsh, R., Abdelpakey, M. H., Shehata, M. S., & Mohamed, M. M. (2022). Automated human cell classification in sparse datasets using few-shot learning. *Scientific Reports*, *12*(1), 1–11. doi:10.1038/s41598-022-06718-2 PMID:35190567

Wermuth, P. J., Piera-Velazquez, S., Rosenbloom, J., & Jimenez, S. A. (2018). Existing and novel biomarkers for precision medicine in systemic sclerosis. In Nature Reviews Rheumatology (Vol. 14, Issue 7, pp. 421–432). Nature Publishing Group. doi:10.1038/s41584-018-0021-9

Yu, K. H., Beam, A. L., & Kohane, I. S. (2018). Artificial intelligence in healthcare. In Nature Biomedical Engineering (Vol. 2, Issue 10, pp. 719–731). Nature Publishing Group. doi:10.1038/s41551-018-0305-z

Zhang, L., Xiang, T., & Gong, S. (2017). Learning a deep embedding model for zero-shot learning. *Proceedings - 30th IEEE Conference on Computer Vision and Pattern Recognition, CVPR 2017, 2017-Janua*, (pp. 3010–3019). IEEE. 10.1109/CVPR.2017.321

Zhou, W., Xu, Y., Lv, Q., Sheng, Y. H., Chen, L., Li, M., Shen, L., Huai, C., Yi, Z., Cui, D., & Qin, S. (2019). Genetic Association of Olanzapine Treatment Response in Han Chinese Schizophrenia Patients. *Frontiers in Pharmacology*, *10*, 422832. doi:10.3389/fphar.2019.00177 PMID:30886581

Chapter 15
Machine Learning's Potential in Shaping the Future of Bioinformatics Research

V. Dankan Gowda
https://orcid.org/0000-0003-0724-0333
Department of Electronics and Communication Engineering, BMS Institute of Technology and Management, India

Saptarshi Mukherjee
https://orcid.org/0009-0006-1280-4964
Department of Bioinformatics, Maulana Abul Kalam Azad University of Technology, India

Sajja Suneel
https://orcid.org/0009-0007-7759-7924
Department of CSE (Data Science), Institute of Aeronautical Engineering, Dundigal, India

Dinesh Arora
Department of Electronics and Communication, Chandigarh Engineering College, Landran, India

Ujjwal Kumar Kamila
Department of CSE, Asansol Engineering College, India

ABSTRACT

In recent years, the application of machine learning techniques has brought about a profound transformation in the field of bioinformatics. This chapter is dedicated to examining the latest developments in both bioinformatics and machine learning while also exploring potential future directions. Within these pages, the authors delve into the potential advantages of employing machine learning to enhance critical bioinformatics tasks, such as the analysis of genomic sequences and the prediction of protein structures through modeling. The chapter also addresses the challenges faced by researchers when integrating these diverse fields. Nevertheless, optimism prevails in the realm of bioinformatics research, driven by the ever-expanding wealth of biological data and the potential for the development of more sophisticated machine learning models.

DOI: 10.4018/979-8-3693-1822-5.ch015

1. INTRODUCTION

Bioinformatics can be described as an amalgam of many fields like computer science, information engineering, mathematics, and statistics with slight refinement in order to make it easier to handle and interpret the biological data. Deep dive into biology using computer-driven information management technologies. This field is not isolated; it interweaves genomic, proteomics and metabolomics, enhancing it. Think about bioinformatics as a device that assists researchers in organizing complicated and deciphering the enigma behind proteins, genes, and genomes even pointing out variations (R. Kishore Kumar and M. S. C. Senthil Kamalesh.,2023)

In another reality, machine learning, as part of AI, develops super intelligent models capable of analyzing and interpreting information, making decisions. There's a simple rule here: the more data these algorithms swallow, the wiser they get. Machine learning was not just a tool in 2023 as recognized by B. Ram Vishal and K. A. M. Machine learning acts as a powerhouse in dealing with data management, pattern recognition, and predicting something which sometimes outpaces the effort of humans. In many areas including healthcare, banking and autonomous vehicles, this technology is solving complex data problems.

Now, let us mix-up both of them. The presence of machine learning in bioinformatics has taken bioinformatics by storm. Fortunately, advances in sequencing technology and emergence of large biological databanks (D. Palanikkumar, A. Y. Begum, 2023) turned the landscapes of bio information processes around. Traditional data analysis methods? As much as they fight the loadness of contemporary biological information, they lose. The disruptor to this has been entered in form of machine learning.

Bioinformatics has come to rely on these algorithms that can effectively learn from available data and make predictions based on such knowledge as their secret weapon. They are important in analyzing genetic sequences, drug development, disease understanding, among other fields including customized medicine. However, the blend of bioinformatics and machine learning goes beyond innovation it is a revolution.

Enhanced Accuracy and Efficiency: Machine learning models are like high-powered microscopes for data: they observe to a greater extent and do so with more depth. These models are better equipped at handling huge amounts of data and extracting useful information with greater reliability and finesse compared to the more traditional statistical approaches. This is not just about counting; it is unlocking secrets of complex biologic systems for sharper and more accurate predictions.

Predictive Modeling: Predictive modeling in bioinformatics? It's crucial. Consider this as the science's magic wand, especially in identifying the disease or gene expression. This is underscored by a work done by A. Nareshkumar and B. Gururaj (2022). Machine learning is not only about making predictions, but identifying the invisible strands within the huge data shawls and producing better foreseeable designs. It's not a simple move; it's a big stride.

Innovative Solutions: Machinery learning provided new approaches to old bioinformatics issues. However, examples exist where new machine learning approaches are disrupting traditional fields of applications. In this regard, AlphaFold is a case in protein structure prediction that has predicted protein structures at unparalleled precision.

Named Entity Recognition (NER) involves the process of pinpointing and categorizing specific terms from unstructured text into predefined classes. This concept is illustrated in Figure 1, which includes a BioNER framework and an example sentence. As outlined in source [3], NER typically employs three strategies: i) Rule-based, ii) Dictionary-based, and iii) Machine-learning based methods. Rule-

Figure 1. Recognition of named entities in biomedicine

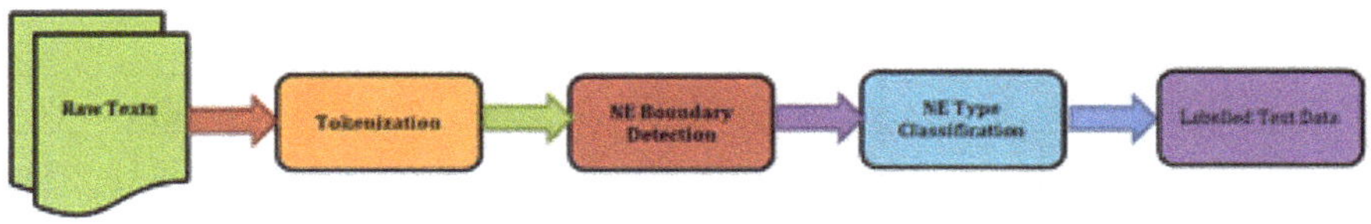

based systems are efficient when entity types can be defined through Orthographic or Morphological patterns(R. Kavitha, and V. Hariram, 2023). However, the process of creating and refining these patterns, along with resolving ambiguities, can be laborious. Additionally, rule-based methods lack robustness and adaptability, requiring constant updates for accurate extraction and differing features across various domains. The dictionary-based method proves effective when the lexicon is comprehensive and current, necessitating preprocessing steps like normalization to align text with the lexicon. However, both rule-based and dictionary-based approaches fall short in recognizing novel entities, failing to extract entities with new patterns or those absent in the existing vocabulary(S. Archana Shreee, G. Jeevitha Sai, 2023).

The machine learning-based method addresses entity recognition by learning unique characteristics that define an entity. This approach predominantly utilizes supervised machine learning algorithms for extracting named entities, as depicted in Figure 2. It requires a substantial amount of annotated data to discern entity features(M. Kranthi, and R. C. Tanguturi, 2023). In this method, entities are extracted using golden standard annotations, and domain resources are employed for disambiguation and accurate labeling. Alternatively, unsupervised algorithms can explore and dissect the text, identifying different entity segments based on shared traits. However, these algorithms require distinct features for effective classification. Techniques such as topic modeling, distributional information, or semantic similarity are employed to group similar entities together and categorize them accordingly(R. C. Tanguturi, and V. C. H, 2023).

Figure 2. Machine learning-based classification framework

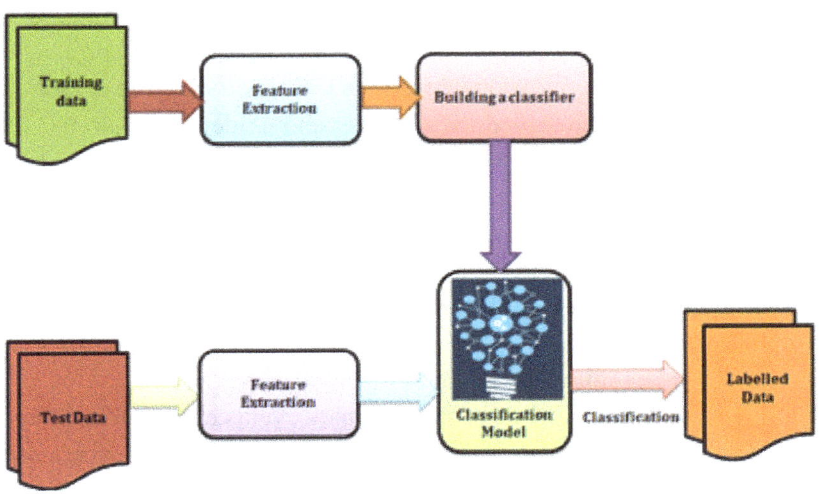

Semi-supervised techniques become valuable when there's an abundance of un-annotated text, but they require the careful selection of initial seed data with specific features for effective learning(R. V. Mailapur and M. K, 2023). A universal method for extracting entities across various domains is not feasible due to the variation in features and classes specific to each domain. To accurately identify a range of entity types, the set of features used must be equally varied. This necessitates the determination of different machine learning algorithms that are suitable for handling such diversity in features and entity types(S. Venkatakiran and N. S. Reddy, 2023).

The task of extracting Named Entities from biomedical texts, known as Biomedical Named Entity Recognition (BM-NER), plays a significant role in fields like information extraction and knowledge discovery. The vast amount of unstructured text in the biomedical field presents a rich opportunity for uncovering new knowledge through Natural Language Processing (NLP) and Machine Learning techniques. The biomedical sector encompasses a variety of entities such as proteins, genes (including DNA and RNA sequences), drugs, and diseases. BM-NER faces unique challenges compared to NER in other domains. These include i) variation in entity size, composition, and frequency, ii) lack of standardized nomenclature for representing chemical structures, iii) the need for precise detection of entity boundaries, iv) dealing with ambiguous abbreviations, and v) the arduous task of creating and updating domain-specific vocabularies due to the continual emergence of new terms(Sadashiva V. Chakrasali, A. Azhagu Jaisudhan Pazhani, 2023). NLP proves instrumental in extracting relevant features for NER, aiding in the extraction and disambiguation of domain-specific entities.

The integration of machine learning into bioinformatics is not just an addition of new tools, but a paradigm shift that is redefining how biological research is conducted(Hombalimath, and N. Patwari, 2023). It is enabling researchers to uncover insights that were previously hidden and is accelerating discoveries in various biological fields. The future of bioinformatics, intertwined with machine learning, promises even greater advancements as both fields continue to evolve.

2. HISTORICAL CONTEXT AND CURRENT STATE

In most cases, the year that closed the twentieth century and marks the opening entry of machine intelligence journey in the bioinformatics sector is usually considered the starting point of this process. The first set included a preliminary study on the analyses of the sequences and protein prediction genes.

Sequence Analysis: The first application of machine learning took place in DNA/protein sequencing. Shortly after this, algorithms that revealed these motifs and tendencies were invented; the discoveries of mysterious interactions among genes followed. These tools include BLAST born in the 90s which changed all sequence alignments completely and heuristics are bioinformatics cornerstone.

Decision trees, neural networks, among other early machine learning models, started to reveal genes concealed in DNA sequences, according to Namitha A R, and Veera Sivakumar Chinamuttevi (YEAR). The models acted as the torches leading towards identification of coding areas and were instrumental in genomics mapping.

As both fields matured, key milestones marked their journey:

Human Genome Project (HGP): One of the landmark events of bioinformatics was the completion of the Human Genome Project (HGP) in 2003. Bioinformetics has grown exponentially in this because of the computational method that was adopted which helped manage a large bulk of data in this project.

Introduction of High-throughput Technologies: These high throughput technologies such as a genome next generation sequencing generated large amount of data that cannot be handled using the former analysis methods. However, machine learning was there in order to process and analyze this flood which led to major advances in proteomics and genomics.

Development of Sophisticated Algorithms: Machine learning evolved with more sophisticated algorithms. Complex tasks like predicting of protein structures and even gene expression patterns could be addressed by CNNs and RNs.

AlphaFold: One notable recent stride is the ALPHAFOLD of DeepMind which was acknowledged in 2021 by GA B Naik, and N J G HG as quoted by Benson. It's a deep learning-based approach for protein structure prediction that has been unravel.

One famous computer method used in processing biological information is an artificial neural networks (ANN), which recreates the network of nerve cells present in our brains. The simple ANN can identify features by itself and it's renowned for being easier compared to probabilistic models. Unlike linear methodologies it is a non-linear approach and fails to show any direct relations between inputs and outcomes. This is what makes it different from other networks such as the ANN with capability to modify features using weight and biases where the initial values for these parameters are given (Jeevans & Sathish, 2020).

Figure 3. Architecture of an artificial neural network

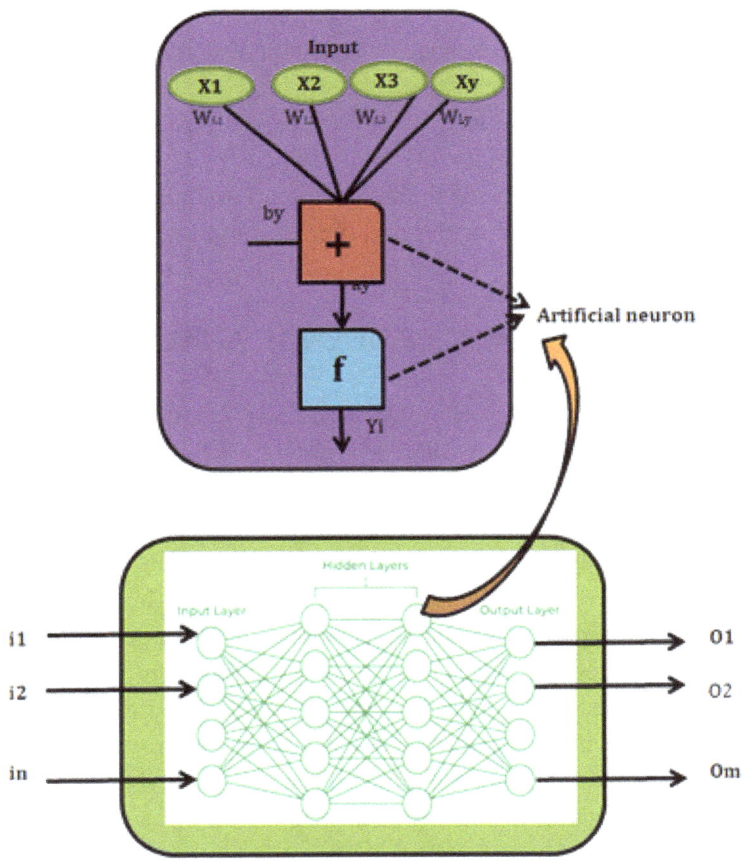

Figure 3 illustrates the ANN structure—its layers and neurons. A neural network is a network that contains several layers, including many hidden layers made up of groups of nodes called neurons. The output layer generates a predictive score from the input layer, and each layer's neurons are governed by mathematical principles. The value of each neuron is calculated based on the values from preceding neurons. Two primary types of ANNs are Convolutional Neural Networks (CNNs) and Recurrent Neural Networks (RNNs). While RNNs are predominantly used in text analysis, CNNs are utilized for analyzing gene sequences. Recently, researchers have increasingly turned to Convolutional Neural Network (CNN) technology for analyzing biological datasets. This approach distinguishes itself from traditional neural networks by the directionality of neuron flow. Standard neural networks typically allow unidirectional flow, facilitating only forward feeding(U. K N and R. V M, 2022). While effective for text and image recognition, some scenarios require full connectivity, leading to more intricate network architectures. Handling large datasets often necessitates these complex structures, which can be a limitation for conventional neural networks. CNNs consist of input, hidden, and output layers, incorporating key stages such as convolution, pooling, fully connected, and normalization layers (M. R. G. and H. Anandaram, 2022). The convolution stage often represents a type of feature extraction process, involving a mathematical cross-correlation executed by a weight matrix within the convolution layer. At the heart of the neural network's design lies a key element, alongside the crucial fully connected layer. Prior research, as outlined by Revanna C and Parismita Sarma in 2022, often turned to the Position Weight Matrix (PWM) for feeding data into these neural networks. But now, in the latest studies, there's a twist: the neural network now feeds on a new kind of nutriment a probability matrix that has been born out two inventive algorithms including CNN. The matrix here, is very useful in locating transcription factor binding sites. Today, machine learning and bioinformatics are bound to each other and this entanglement has helped scholars solve the intricate puzzles presented by huge biological data in the current researches. It is a true fusion taking place with modern technology helping us better manage biomolecular science research. The future is full of promise for breakthroughs as machine learning methods have become more advanced toward the peculiarities of bioinformatics. The future holds the promise for a new age of groundbreaking advancements, discoveries and breakthroughs in the field.

3. ADVANCEMENTS IN GENOMIC SEQUENCE ANALYSIS

Machine learning has played a crucial role in revolutionizing one of the major constituents of bioinformatics–genomic sequence analysis. For scalable, accurate, and efficient methods, one must consider that there are a vast quantity and high level of genomic data involved in this issue. Machine-learning addresses these demands and offers complex algorithms applicable to different genome analysis cases.

Pattern Recognition and Classification: Machine learning systems are very good at finding patterns in data. # It implies that, in terms of genomic sequence analysis this skill is particularly significant since it helps to detect some motives, regulatory elements and even gene structure among the intricate DNA sequences (S. Reddy, et al., 2022). Classification techniques are very important when it comes to identifying different categories or segments like coding and non-coding sectors.

Genome Annotation and Gene Prediction: Machine-learning algorithms are used in genomic sequence annotation and gene function prediction. Such models can be applied to newly sequenced genomes for direct interpretation of functional information on the sequential data.

Figure 4. Enhanced system design for biomedical entity recognition

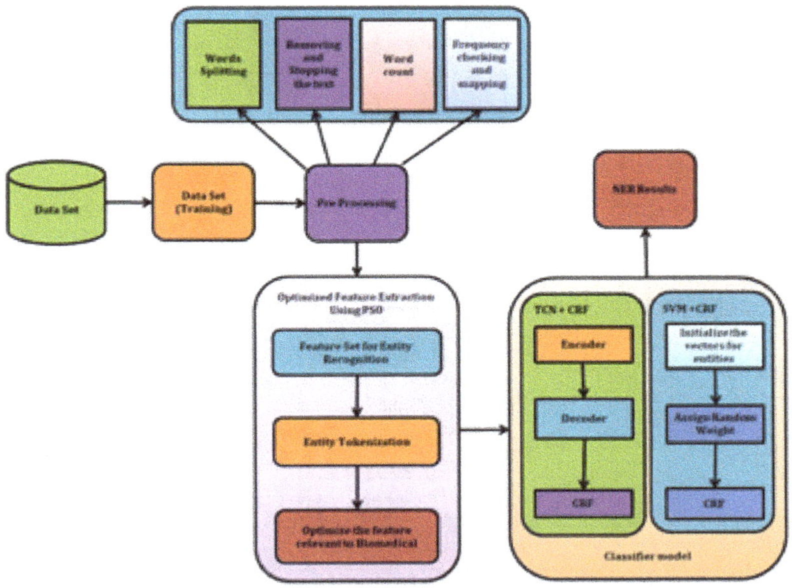

Variant Analysis and Association Studies: Machine learning is important for analyzing genetic differences, such as SNPs, in disease association studies (S. R. Kawale, S. P. Diwan, 2022). The huge database that stores information about some specific features or ailments can be searched using these algorithms, and therefore, genetic markers could be discovered.

This section reveals a novel Biological EBNER system. It integrates a revised particles swarm optimization (IPSO), which is complemented by bio image processing concepts aimed at detecting the biological entities. Its architecture? A complex tapestry with nine crucial strands: As shown vividly in figure 4, these methods include preprocessing, feature extraction, entity encoding, BIO tagging, SVM, CRF, character vector, word vector and TCN. Then, a transformation of the raw dataset is performed in the initial stage of the journey. In the beginning phase of the study, B. Kameswara Rao and Naziya Hussain noted that the process included the word segmentation, eliminating stop words, counting the word frequencies, reducing the repetition by merging similar words (Rao & Hussain, Then is one of the most sophisticated processes in medical imagery; the feature extraction involves identifying the significant attributes essential for entity recognition. This is analogous of sieving gold, distinguishing important tokenized and designed attributes from huge amount. Then we move to PSO algorithm world which is known as one of the most complex optimizers in a multidimensional space working in different fields. According to M. Swathi Pai and B. Naveen K (2020), improving PSO makes it more accurate towards the search targets' pinpointing, balancing between exploitation and exploration. Particles appear in these algorithms dances mostly outside their parameter space and give updated coordinates for particle locations.

It is a novel approach that tokens elements depending on how often they aid in recognizing biomedical information. For instance, this concerns details such as the structure of genes, names of proteins, amino acid sequel, etc., as explained by K. B. Naveen and G. N. Pai in a work published in 2020. Finally, these optimized properties are fed into classifiers. Two types of classification approach are introduced in the resaerch. The first combines an SVM with a CRF. It is good at both linear as well as nonlinear

Figure 5. System configuration

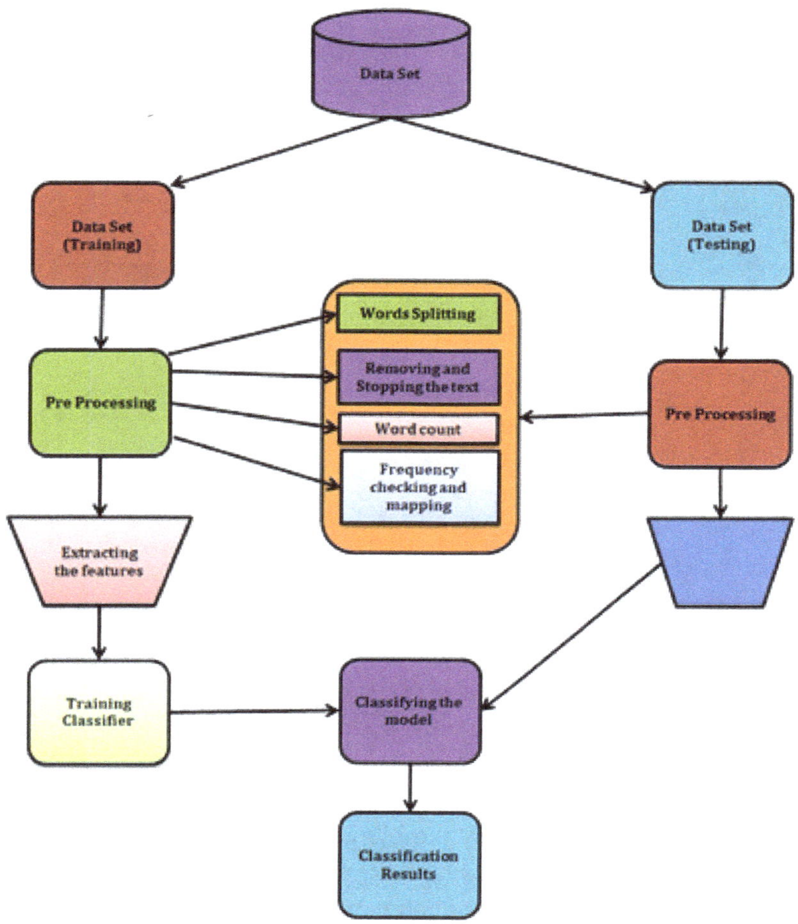

classification, setting up of vectors for the item. However, CRF does not just stop at discriminative classification as CRF. Graphical modeling taking into account the input context for sequential designation. This is usually omitted by traditional approaches. Ramesh Naidu & et al. (2020). As K. Prasad and G. Poornima stated in 2022, SVM is a marginal classifier that learns without contextual awareness unlike CRF. This method combines these algorithms in order to promote entity recognition and proper annotation of biomedical entities.

The methodology also includes temporal convolutional network and conditional random field parallel to SVM-Conditional Random Field. TCN is praised for its low computational costs and ability to handle contextual information with precision when delineating entity borders to enhance localization and better labels. As observed by H. Anandaram and B. K. Verma (2023), both SVM-CRF and TCN-CRF models effectively identify entities and categories in the biomedical domain. The additional computational requirements needed by the CRF are a well-established limitation compared to the space-time efficiency of the TCN. In dealing with these challenges, it uses NLP based on natural language processing specifically for feature extraction and constructs discriminant functions using a supervised learning technique proposed by Prasad and Ashreetha published in 2023. Its performance is assessed via the use of CRFs for markup of the GENIA and CHEMDNR corpus. The model has all-specific features that are linguistic, orthographic, morphological domain specific, local contextual.

The novel approach shown in Figure 5 highlights the advanced design of the system for detecting and recovering biological entities. In this regard, the identification of entities is based on a combination of CRF model and BiO tagging methodology. Thereafter, SVM is used to discriminate these entities' intricate nuances. It is particularly skilful in binary classification which converts the multi-class problem into a sequence of much simpler binary classifications. This smart conversion improves on the learnability of the system while making it more accurate. A multi-class SVM is usually built by combining many small binary SVMs when dealing with a multi-class problem. These assembly can be performed using one versus rest approach or the one-one pairwise binary classifier that is capable of simplifying as well as improving the classification process.

The bias towards one-one pair wise classifier for subtype identification normally stems from class imbalance issues arising from uneven samples size in each class in one-vs-rest technique. The system builds (N) *(n-1)/2 classifiers to discern for 'N' different classes and finally makes the final judgement by major vote. This approach enhances performance but also augments the time complexity which is needed for training the model. A number of distinctive features may not exist within the entities. However, post the initial extraction of the entity term, SVM is one critical part in the intricate dance called entity recognition. It has proved to be essential in identifying entity subtypes among other things.

Diving into case studies, we see machine learning's profound impact on genomic sequence analysis:

The 1000 Genomes Project: It was an ambitious attempt to sequence genomes of a large number of people in order to create a map for human genetic diversity. In this background was the supporting character machine learning that enabled the disentangling of human diversity and risk for diseases.

Cancer Genome Atlas (TCGA): While machine learning was the TCGA's trustable ally delving into numerous tumor DNA sequences. These have helped advance understanding on how to detect various types of cancers, their causes and treatments.

Predicting Gene Expression Levels: Machine learning has always represented means for estimation of gene expression quantities in a sequence of DNA. The functionality is crucial to clarify about the myths surrounding it and consequences upon the cells function and evolution of diseases.

CRISPR/Cas9 Off-Target Effects Prediction: When it comes to genome editing, it is a question of accuracy. Machine-learning models have been developed to ensure CRISPR/Cas9's precision and prevent undesirable mutation. They are critical in ensuring safe and efficient genome engineering.

DeepVariant: Although deep leaning is an old technique, here is where it performs exceptionally well and much enhances the accuracy in genomic variant calling. In this vein, deepVariant serves as an example of where modern machine learning relies on and gets better with DNA studies.

Not surprisingly, these images are impressive but only a start of how Machine Learning assists in genome sequencing analysis. There will be more milestones made in unravelling of genetics and its link with health care and diseases.

4. PREDICTIVE MODELING IN PROTEIN STRUCTURES

Protein structure modeling has been likened as one of the largest mountains in data-rich bioinformatics. Proteins, which are made up of amino acid chains, go through a series of complex foldings to assume appropriate three dimensional structures necessary for their function. These structures provide insights into natural processes, pathogenesis, pharmacotherapy, and therapy. However, this task is beset with challenges:

Complex Folding Patterns: Understanding how proteins fold and transform from a long amino acid chain into a unique 3-dimensional construction depends on interactions across several scales.

Computational Intensity: The simulation of the process of protein folding, however, is computationally intensive and laborious. Traditionally, such methods rely on empirical observations and physical models and can be very expensive and slow.

Data Limitation: Structural data of proteins can be generated through methods such as X-ray crystallography (XC), however this would be impossible to achieve for every protein so there are information gaps.

Enter machine learning, particularly deep learning, as a formidable ally in protein modeling:

AlphaFold: This is a machine learning wonder in which an auto-transformer neural network successfully forecasts protein interactions and structures. CASP saw that AlphaFold outperformed every other method. This is changing how we think about proteins for research and disease-oriented intervention.

Use of Existing Data for Training: AlphaFold is an example of such model which is trained using huge databases comprising of known protein structure. Through this model, predictions of structural properties are made and done so with highest accuracy.

Efficiency and Scalability: Modern approaches beat conventional ones and hence allow the study of a greater number of proteins that are important for drug design.

Integrative Approaches: Machinal learning intersects with experimental data to sharpen protein models fusing computational and experimental powers together.

That is one of the highest-order bioinformatic inventions for implementing machine learning into protein structure prediction. It helps in protein modeling, which is complex and it also makes biology based research and discoveries possible. Protein Structure Prediction is optimistic as there are continuous improvements in AI and in machine processes.

5. EMERGENCE OF MACHINE LEARNING IN THE FIELD OF BIOINFORMATICS

Thus, the machine learning revolution in bioinformatics commences through the application of modern computational approaches to crack complicated biological findings. The case was depicted in Figure 6 with regard to this transformational phase, whereby it adopted the use of data oriented knowledge. Significantly, this modification allows us to understand genetic, protein, and cellular functions at the molecular level.

Personalized medicine also uses machine learning as it provides specifically tailored treatments depending on an individual's genetic make up. Genome sequencing for purposes of identifying a disease's marker, and therapeutic target can be influenced by it as well. Through machine learning, there can be fast traveling in the area of drug discovery and development processes with the capability to predict molecules' interactions and side effects. Technology has been revolutionized by machine learning in diagnosis of disease particularly cancer based on imaging data analysis.

In the fields of systems biology, machine learning unknots the tangles in biological networks. But linking up heterogeneity in sources of information such as genome, proteome and metabolism is another matter entirely. ML fully solves this problem by means of techniques like CRISPR and NGS which provide a large amount of data for machine learning to utilize. Bioinformatics and the Ethical Issues Surround-

Figure 6. Bioinformatics influenced by machine learning

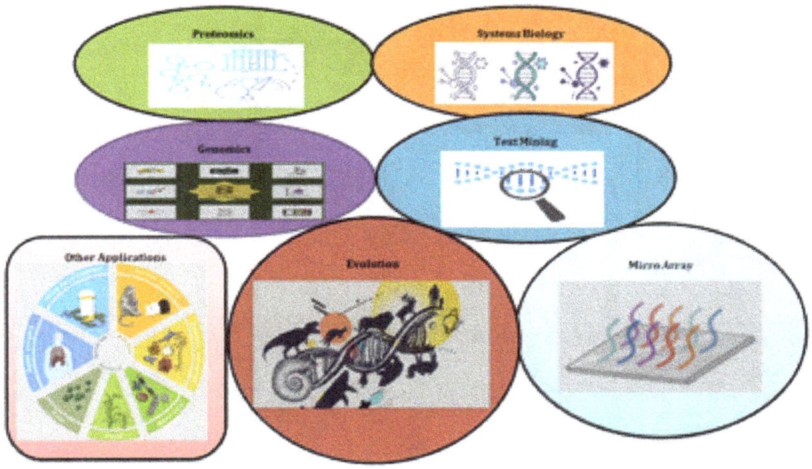

ing biological datai in terms of these technological developments as far as ethical concerns regarding individual biological data are concerned, one has to be very careful when using machine learning in bioinformatics. This, then, is an epoch of more than just technological revolution. It also represents a higher level in terms of morality with regards to the use and applications of data within biology and medicine itself. The explosion of biological data presents twin challenges: high-tech In visbro module, we found powerful databases which provide a million times faster and more accurate than before storing them, organizing them -- then by pointing the touch screen at your target subject will start finding data in accordance with how expertly you ask it questions.

However, the lack of appropriate algorithms for interpreting data from multiple sources as a single cohesive biochemical and biological story about underlying processes has been one major limit in computational biology. Such instruments should not be limited to description of data, and must project models effective for use in system forecasting. Machine learning, one of the fundamental branches of computer science concerning designs for self-learning and data responsive algorithms, is a key component in knowledge discovery about biological systems. Feature learning is part of deep learning which is a subset of machine learning that gets automated. Therefore, algorithms become self-trained for combining heterogeneous data elements within a higher abstract sense and deduce intricate forecasts from vast amounts of inputs such data.

Machine learning has found application in many biological fields like genomics, proteomics, microarrays analysis, systems biology, evolutionary analysis, and text mining showing its potential usefulness and significance in bioriencing.

Bioinformatics, an emerging discipline, holds the key to addressing the current demands in the field of agriculture. This field employs a diverse array of computational techniques, encompassing sequence analysis, data mining, gene discovery, the construction of phylogenetic trees, protein structure and function prediction, and interaction networks, among others. The pervasive application of bioinformatics has made it an indispensable component in virtually every aspect of plant research. Nonetheless, certain unresolved challenges persist within the field. The proliferation of databases and tools that offer specialized datasets and comprehensive annotations for omics technology has proven invaluable for

Figure 7. Integration of deep learning into bioinformatics research

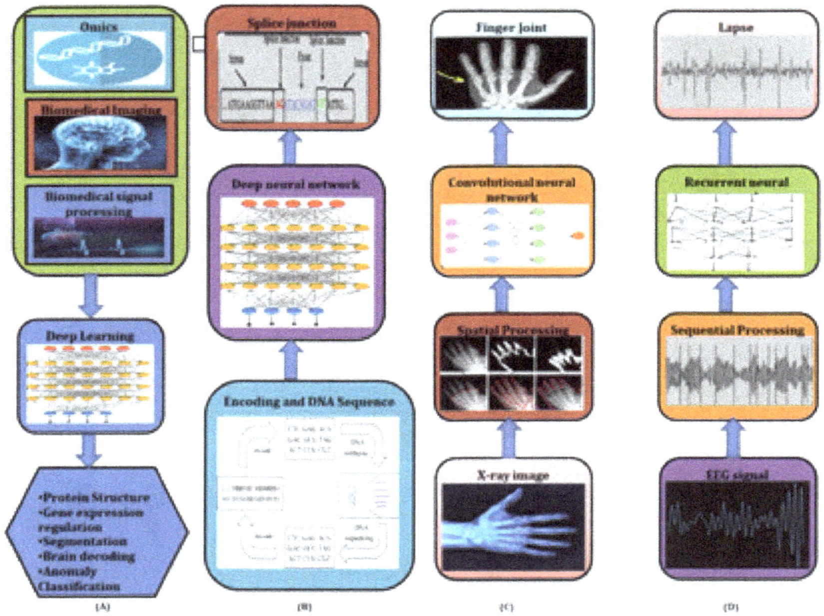

gaining a deeper understanding of functional and expression-related issues, whether related to specific gene families, cellular processes, or plant diseases. As we enter the era of big data in plant science, the advent of powerful technologies like deep learning is streamlining information storage, management, and the extraction of valuable insights from this wealth of data.

Large-scale datasets are the norm in today's world, thus turning big data into insightful information takes on a significance never seen before. This is also true for the discipline of bioinformatics, which is characterized by constant change. Biomedical data of many kinds, including omics data, pictures, and signals, have accumulated considerably. Both business and academics are very interested in them because of their enormous potential in the fields of biological and medical research. For instance, IBM unveiled Watson for Oncology, a platform intended to assess patient medical data and assist physicians in choosing the best course of action. In a similar vein, Google DeepMind, well known for its incredible go player AlphaGo success, has lately entered the healthcare space with the introduction of DeepMind Health. The goal of this program is to create efficient healthcare solutions while maintaining the innovative spirit that drove its earlier successes.

Machine learning is one of the most widely used techniques in the effort to extract information from the massive bioinformatics data warehouses. In order to find hidden patterns, build a model, and then use this model to forecast fresh data, these machine learning algorithms need training data.

The support vector machine, hidden Markov model, Bayesian networks, and Gaussian networks are a few of the well-known methods in this discipline. These algorithms are being used in many different bioinformatics fields, such as proteomics, systems biology, genomics, and many more. Conventional machine learning algorithms are encumbered by limitations when it comes to processing raw data in its unrefined state. Consequently, researchers invest considerable effort in transforming this raw data into high-level abstraction features that are more amenable to analysis, drawing upon their domain expertise.

Figure 8. The potential of deep learning to advance bioinformatics research

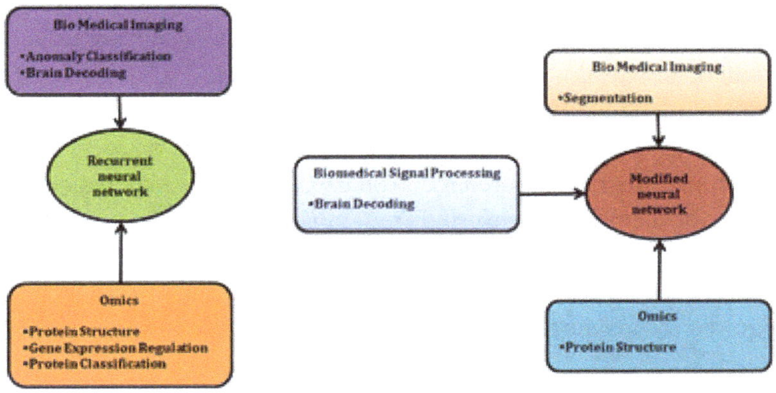

On the other hand, the field of deep learning, which is a more modern version of machine learning, has emerged as a result of many elements coming together, including the growth of huge data, the power of parallel and distributed computing, and the advancement of complex algorithms.

Deep learning algorithms are propelling important developments in a wide range of fields, including as image recognition, audio recognition, and natural language processing. These algorithms have effectively overcome earlier constraints. Certainly, deep learning techniques are widely being used in the field of bioinformatics. Many research efforts have been made to fully use deep learning in the field of bioinformatics, as Figure 7 illustrates. Several prominent examples show that deep learning is making major gains into bioinformatics research. From a broad perspective (A), the versatility of deep learning is evident in its capacity to handle diverse input data and address a wide array of research objectives. For example, in the omics field (B), deep neural networks have been used for the prediction of splice junctions in the DNA sequence data, illustrating their applicability in genomics studies. Deep learning has become one star in this dynamic era of bioinformatics. For example, biomedical imaging (C), where CNNs are useful at tasks such as detection of the finger joints from radiography images. In particular, RNNs have found their way into developing anomaly detectors for EEG signals (D). Such cases underscore the importance of deep learning in bioinformatics giving impetus to unraveling of valuable information from the maze of intricate biological data among the numerous research avenues.

The increasing focus on deep learning for solving intricate biological information further demonstrates that it could uncover key messages. The study categorizes its focus into three distinct domains based on the type of input data: such areas consist of biomedical signal processing, biomedical imaging, and omics. Figure eight presents all areas of bioinformatics with deep learning applications. In addition, the paper uses illustrative examples of sample input data from every category, creating mental images about different applications of deep learning in this area. Bioinformatic puzzles are solved and explored in the colorful world of omics through analyzing and studying transcriptome, proteome, and genomes. The primary inputs here are the raw biological sequences: DNA, RNA, and amino acids. With the advancement of next-generation sequencing technology, these sequences are now accessed and affordable for all. These sequences also serve as a source of position-specific scoring matrices (PSSMs), which are often instrumental in bioinformatics research.

In fact, biomedical imaging is a domain where deep learning application has recently been found to be appropriate. It contains many biological images such as x-rays, PET scans, MRIs and histopathology slides that act as input for deep learning algorithms. These images, which are used on a regular basis in patient care, offer data that algorithmic analysis can easily extract.

Another interesting study area is biomedical signal processing, which involves electrical activity of the human body to address bioinformatics issues. These are various datasets in EMG, EEG, ECOG, ECG, and EOG. Nevertheless, most research focus on EEG data. Since such recordings carry noise and artefacts, it is common for them to be converted from raw signals to wavelet or frequency components before being fed into deep learning engines. Besides, most of these studies involve the addition of human-engineered components such as peak variation and normal decay for better precision and high quality of the input data required for machine learning analysis.

6. DATA AUGMENTATION IN BIOINFORMATICS

One smart machine learning approach is data augmentation that increases the number and variety of training data without requiring new data collection. However, this is done by carefully manipulating old data to create new ones. The importance of data augmentation is multifold:

Boosts Model Robustness: Data augmentation improves the performance of machine learning models by increasing their ability to generalize, leading to less sensitive and more reliable models that are robust towards minor variations.

Thwarts Overfitting: A model that is too skilled in its training data, which includes noise and outliers might fail with the new data. Augmentation brings variation to training data, preventing overfitting.

Solves Limited Data Dilemmas: Data augmentation artificially increases the datasets in bioinformatics, and it provides additional training materials for machine learning models because large datasets can be a rare phenomenon in bioinformatics.

In the bioinformatics landscape, data augmentation morphs to suit different data types and tasks:

Genomic Sequence Augmentation: Augmentation may be needed for DNA/RNA analysis; controlled inserts, deletes, and substitution teach models to look for patterns amidst genetic variances.

Image Data Augmentation in Medical Imaging and Microscopy: Rotation, flip, scale, and crop add to the appeal of biological images. Models can be trained to disregard these differences which are vital in disease diagnosis and cell classification.

Time Series Augmentation for Gene Expression Data: Adding noise, shifting time points, and warping time series may enhance model robustness against expression pattern variation.

Synthetic Data Generation: Artificial intelligence techniques such as generative adversarial networks (GANs) may be used to create brand new, authentic synthetic data points. This works especially well when you do not have a lot of data or it is confidential.

Each of these augmentation strategies equips machine learning models in bioinformatics to perform with greater precision, reliability, and adaptability.

Figure 9 shows how deep learning architectures are categorized into four different groups: modified neural network (MNN), recurrent neural network (RNN), convolutional neural network (CNN), and deep neural network (DNN). Notably, "DNN" is used as a catch-all name for all deep learning architectures in certain research articles. Nonetheless, a narrower definition of "DNN" is used in this study, meaning deep belief network (DBN), stacked auto-encoder (SAE), and multilayer perceptron (MLP). These

Figure 9. Segmentation of bioinformatics research using deep learning into distinct categories

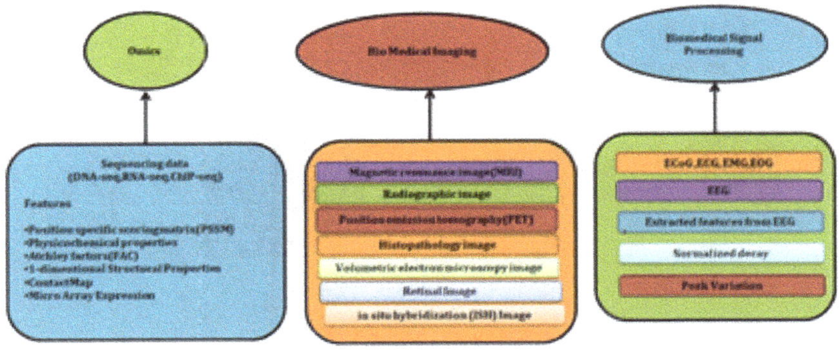

DNN variations use auto-encoders (AEs), restricted Boltzmann machines (RBMs), and perceptrons as the basic building elements of their neural networks.

Conversely, CNN represents a specialized architecture renowned for its exceptional performance in the realm of image recognition. It typically comprises convolution layers, nonlinear layers, and pooling layers, rendering it especially well-suited for processing image data. In contrast, RNN is meticulously engineered to harness sequential information embedded within input data. Achieving this involves integrating cyclic connections among various components, like perceptrons, long short-term memory (LSTM) units, or gated recurrent units (GRUs). The domain for deep learning also includes a spectrum of adapted architectures tailor-made for different data types or tasks. These include the DST-NN, MD-RNN, and CAE. The universe of deep learning includes numerous architectures tailored to match the needs of a given case. CNN is a direct inspiration of human visual cortex. It specializes in processing of two-dimensional images and interpreting diverse arrays of data. To this end, CNNs act as a foundation stone for many image based deep learning applications by taking a similar approach to how the brain responds to visual information.

Complex network of the visual cortex comprises two basic cell types that are simple cells and complex cells. These cells recognize simple figures and shapes in basic cells. Complex cells, on the other hand, use the data from simple cells to identify more complex shapes. Three essential principles have been implemented by CNN to mimic the capability of the visual cortex, which is a natural visual processing system with outstanding capabilities. These concepts include local connection, invariance to location, and invariance to local transitions.

Convolution layers, nonlinear layers, and pooling layers make up the three fundamental parts of CNN's base architecture, as seen in Figure 10. Each convolution layer performs convolutions between local patches and weight vectors called filters to produce sets of local weighted sums, or feature maps, which are used to leverage the inherent correlations within subregions of data. Furthermore, filters are consistently performed throughout the whole dataset to account for the possibility that same patterns will surface regardless of where they exist in the data. This method lowers the amount of parameters that need to be learned, which improves training efficiency while also facilitating strong pattern recognition.

The nonlinear characteristics of the feature maps are then enhanced by nonlinear layers. Max or average subsampling is performed to non-overlapping feature map areas during each pooling layer operation. CNN can handle relatively different but semantically comparable characteristics because to

Figure 10. Fundamental architecture of CNN, comprising the convolution layer, nonlinear layer, and pooling layer

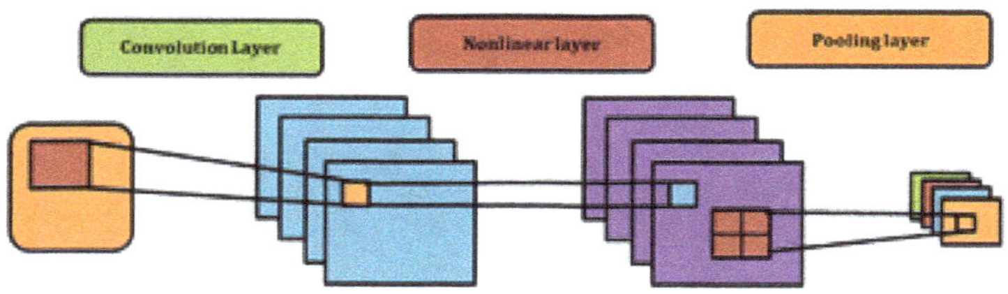

this non-overlapping subsampling technique, which also makes it possible to aggregate local features to find more intricate and higher-level features.

Data augmentation in bioinformatics research as a method to enhance machine learning model performance and reliability, which is particularly advantageous in a field with limited or very complicated data. However, this approach would enhance the quality of biostatistical modeling with greater uses for machine learning by bioinformatics investigators.

7. TRANSFER LEARNING AND ITS APPLICATION

Transfer learning refers to the application of intelligent machine learning techniques, such as using a pre-trained model for one task as a basis for training a new task. The technique involves drawing upon the lessons learned while solving a specific problem and applying them towards resolving another related matter. For example, this usually involves re-training or calibrating an existing pre-trained model based on a massive dataset to address specific issues. This method is specifically crucial when there are limited amounts of data as through the previously obtained features and patterns during the initial training process, the model can still perform with ease.

Key facets of transfer learning include:

Feature Reuse: The model retains some traits acquired from a previous endeavor but refines by removing irrelevant ones.

Fine-Tuning: Such stage consists of fine tuning the model's parameter and internal representation to fit this novel atmosphere through further training on the specifically designed dataset intended for this new problem instance.

Flexibility: Transfer learning is flexible and it can be implemented in various contexts including domains, situations, tasks, or data sets. This involves improving one domain using a pre-existing knowledge and its implementation in another field.

In the realm of bioinformatics, transfer learning has made significant strides:

Genomic Sequence Analysis: Such fine-tuned models could be applied to tasks like detecting disease associated variant and predicting expression level.

Protein Function Prediction: Models initially trained for large-scale protein databases can use transfer learning techniques and specific datasets to predict new protein functionalities through transfer learning.

Medical Image Analysis in Bioinformatics: There are also some clinical applications that can benefit from transfer learning and the use of large image dataset's pre-trained models even though the datasets are usually smaller and specialized.

Drug Discovery and Development: They refine their models using compounds database that contains wider spectrums of compounds so as target specific properties such as dissolution and efficiency.

Metagenomics and Microbiome Analysis: This strategy entails using transfer learning to comprehend intricate microbial assemblages in which metagenomic models instructed to cover a whole plethora of illnesses and ecological research studies.

Transfer learning is one of the most powerful instruments bioinformatics can offer, which allows bioinformaticians to use already available data and models for new, specialized research issues. Such an approach is cost-effective, enhances research capabilities, and provides room for more in-depth analysis. It is a representative example of a cumulative growth concept in bioinformatics, boosting the effectiveness of this branch of science.

8. CHALLENGES AND CONSIDERATIONS

The marriage of machine learning with bioinformatics, while fruitful, ushers in a series of computational conundrums:

Computational Demands: In fact, sophisticated machine learning models particularly deep learning are very demanding on resources. They need powerful processors, plenty of memory and sometimes dedicated hardware such as GPUs or TPUs. Often these are considerable demands for many research entities.

Big Data Management: 'Big data' is often encountered in bioinformatics, huge in size and complexity. There is a wide range of challenges associated with this type of data including data storage, rapid retrieval and processing which may require advanced data management strategies.

Algorithmic Intricacies: In most of the machine learning algorithms training could be a bit complex especially in the deep learning and complex predictive modelling that may take more years. This can bottleneck research progress.

Diverse Data Integration: Nevertheless, bioinformatics data that present as sequences, structures, images, or time series is a special type. Typically, there must be tailored transformation steps that fuse the multiple input data types into homogenous machine learning model.

Beyond computational challenges, ethical and privacy considerations loom large:

Data Privacy: Bioinformatics with respect to private data is particularly focused on ensuring privacy and safety; say sensitive human genetic material. Secure use of data for protecting personal data, follow legal requirements on keeping these data secure and anonymous use.

Consent and Data Use: Consent for using data by the third parties matters a lot. Therefore, it is important to notify study subjects about employing their data, for example, within machine learning scheme with hidden purposes beyond initial assumptions.

Bias and Fairness: However, machine learning algorithms may unintentionally enlarge existing biases used in training.X The bias may arise due to this risk in bioinformatics as they are misleading and give false diagnostics across various cross sections of the country.

Model Interpretability: However, the transparency problem in the context of machine learning especially, deep learning is associated with their "the black box" nature. These models need to be explainable, ensuring trustworthiness and enhancing legitimation in science.P

These issues and concerns have necessitated the appropriate use of machine learning in the context of bioinformatics. More than just technological innovation, it raises moral issues and concerns of legality and socio-economic effect. Nonetheless, finding an equilibrium in this complex issue is as difficult as striking on new discoveries in development of bioinformatics.

9. FUTURE TRENDS AND DIRECTIONS

The machine learning terrain is in a state of constant flux, birthing new trends and methodologies that are reshaping bioinformatics research:

Deep Learning Proliferation: Among other areas in machine learning, particularly deep learning, has made tremendous advancement particularly. Innovative designs like Transformers and GANS are breaking new frontiers of possibilities. Theses models are ideal in handling large amounts of genomics or proteomics data that are complex and multidimensional and thus they perform well in doing so. The advent of an unprecedented turn for bioinformatics research awaits harnessing this power accrued from large-scale biological data.

Rise of Reinforcement Learning (RL): Reinforcement learning (also known as RL) have become quite popular. Drug design in bioinformatics can also be helped with RL which teaches an agent to recognize effective molecular structures through a systematic error-trial mechanism. Such a change in point of view can revolutionize our techniques for addressing complex bioinformatics issues, namely, traveling through immense spaces with respect to particular solutions.

Explainable AI (XAI): The time is now for bioinformatics to move toward making black boxes transparent. Indeed, explainable models constitute one of the prerequisites for developing confidence and understanding regarding the outcomes of critical domains – genetics and personalised medicine. It is important that researchers and physicians should know how AI models arrive at their predictions for AI to be used responsibly and ethically in bioinformatics.

Multi-omic Data Synthesis: The integration of bioinformatics and data fusion from genomic to metabolome. It will give us a better understanding of how biological systems work, and why people get sick. New diagnostics and interventions may allow for further insights into health and diseases when several data formats reveal complex interactions in biological systems.

The ripple effects of these emerging trends in bioinformatics are broad and deep:

Enhanced Predictive Modeling: Deep learning innovations would boost the capabilities of predictive models in bioinformatics, transforming gene expression prediction and modeling of the disease progression areas, among others.

Personalized Medicine Evolution: Machine learning also has a huge potential to revolutionize personalized medicine by improving our ability to characterize different risks of diseases, which lead to developing targeted treatments.

Revolutionizing Drug Discovery: Such as through reinforcement learning, it can facilitate faster chemical space exploration at low cost.

Data Integration and Systems Biology: Studying many biological data sets about complex biological systems will allow us to understand multifactorial diseases that arise out of jointly considered genetics, environment, and life style.

Ethical and Responsible AI: Thus, as machine learning gets in bioinformatics, ethical AI is more required, in order with algorithms being effective, just, understandable and private.

Advancement of machine learning into discovery era of bioinformatics. Modern and vital issues in bio science could be resolved by using high-end machine learning techniques.

10. CONCLUSION

In this chapter, we have surveyed the dynamic terrain in bioinformatics with respect to machine learning, to which machine learning has greatly contributed. It sheds light on how this merging is making an impact to biology science research., Key highlights include:

Historical Foundations: This is what created the basis for the sophisticated machine learning approaches adopted in protein structure prediction and in particular in DNA as well as RNA analyses.

Revolution in Genomic Analysis: Currently, machine learning has revolutionized genomic sequence analysis by enhancing accuracy and efficiency in pattern recognition, gene mapping, and genetic variation scrutiny.

Advancements in Protein Modeling: Recently, major achievements in machine learning include the AlphaFold which makes a huge stride forward predicting of protein structure.

Role of Data Augmentation: Machine learning models in bioinformatics have been made more robust and reliable by using techniques such as data augmentation, particularly where data is scarce or complicated.

Impact of Transfer Learning: Transfer learning in bioinformatics has helped to develop proteins and gene functions in many fields.

Challenges and Ethical Aspects: The use of machine learning raises some computational issues and ethical problems as well, thus requiring more vigilant and innovative ways of addressing them.

As to the future vision for bioinformatics coupled with machine learning development, possibilities abound. Complex neural networks, reinforcement learning, and bioinformatics for explainable and ethical AI will herald new era of bioinformatics. These improvements will widen our knowledge about complicated biological systems and lead to faster breakthroughs for personalized medicine, drug research, and illness understanding.

There is a paradigm shift where machine learning is being applied to bioinformatics, taking research to a higher level. These fields will continue to unravel new revelations and innovations with important ramifications in science and medicine. This collaboration has numerous future opportunities, which will change the way we perceive and utilize biological data within the upcoming years.

REFERENCES

Anandaram, H. N. B. A, N. Gupta & Verma, B. (2023). *IoT Wearable Breast Temperature Assessment System.* 2023 7th International Conference on Computing Methodologies and Communication (ICCMC), Erode, India.

Archana Shreee, S., Maheshwari, B., & Jeevitha Sai, G. (2023). *A Novel Method of Identification of Delirium in Patients from Electronic Health Records Using Machine Learning*. 2023 World Conference on Communication & Computing (WCONF), Raipur, India.

Chakrasali, S. V., Kumar, C., Chaturvedi, A., & Jaisudhan Pazhani, A. A. (2023). Computer vision based healthcare system for identification of diabetes & its types using AI, Measurement. *Sensors (Basel)*, *27*, 10075.

Hombalimath, D. & Patwari, N. (2023). *Symmetrized Feature Selection with Stacked Generalization based Machine Learning Algorithm for the Early Diagnosis of Chronic Diseases*. 2023 5th International Conference on Smart Systems and Inventive Technology (ICSSIT), Tirunelveli, India.

Jeevan, K., & Sathisha, B. M. (2020). Implementation of IoT Based Wireless Electronic Stethoscope. *2020 Third International Conference on Multimedia Processing, Communication & Information Technology (MPCIT)*, (pp. 103-106). IEEE.

Kavitha, R., Kumar, A., Kalpana, V., & Hariram, V. (2023). Artificial Intelligence based Health Monitoring System on IoTH platform. *2023 Second International Conference on Augmented Intelligence and Sustainable Systems (ICAISS)*, Trichy, India.

Kawale, S. R., & Diwan, S. P. (2022). Intelligent Breast Abnormality Framework for Detection and Evaluation of Breast Abnormal Parameters. *2022 International Conference on Edge Computing and Applications (ICECAA),* (pp. 1503-1508). IEEE.

Kishore Kumar, R., Pandidurai, M., & Senthil Kamalesh, M. S. C. (2023). *Design of IoT based Rural Health Helper using Natural Language Processing*. 2023 4th International Conference on Electronics and Sustainable Communication Systems (ICESC), Coimbatore, India.

Kranthi, M., & Tanguturi, R. C. (2023). *Design of Intelligent Medical Integrity Authentication and Secure Information for Public Cloud in Hospital Administration*. 2023 2nd International Conference on Edge Computing and Applications (ICECAA), Namakkal, India.

Lilhore, U. K., Manoharan, P., Sandhu, J. K., Simaiya, S., Dalal, S., Baqasah, A. M., Alsafyani, M., Alroobaea, R., Keshta, I., & Raahemifar, K. (2023). Hybrid model for precise hepatitis-C classification using improved random forest and SVM method. *Scientific Reports*, *13*(1), 12473. doi:10.1038/s41598-023-36605-3 PMID:37528148

Lilhore, U. K., Poongodi, M., Kaur, A., Simaiya, S., Algarni, A. D., Elmannai, H., Vijayakumar, V., Tunze, G. B., & Hamdi, M. (2022). Hybrid model for detection of cervical cancer using causal analysis and machine learning techniques. *Computational and Mathematical Methods in Medicine*, *2022*, 2022. doi:10.1155/2022/4688327 PMID:35572826

Lilhore, U. K., Simaiya, S., Pandey, H., Gautam, V., Garg, A., & Ghosh, P. (2022). Breast cancer detection in the IoT cloud-based healthcare environment using fuzzy cluster segmentation and SVM classifier. In Ambient Communications and Computer Systems [Singapore: Springer Nature Singapore.]. *Proceedings of RACCCS*, *2021*, 165–179.

Lilhore, U. K., Simaiya, S., Prasad, D., & Guleria, K. (2020). A hybrid tumor detection and classification based on machine learning. *Journal of Computational and Theoretical Nanoscience*, *17*(6), 2539–2544. doi:10.1166/jctn.2020.8927

M. R. G. and H. Anandaram. (2022). Extraction of Fetal ECG Using ANFIS and the Undecimated-Wavelet Transform. *2022 IEEE 3rd Global Conference for Advancement in Technology (GCAT),* (pp. 1-5). IEEE.

Mailapur, R. (2023). *Implementation of GUI based Vital Track Ambulance for Patient Health Monitoring.* 2023 8th International Conference on Communication and Electronics Systems (ICCES), Coimbatore, India.

Namitha, A. R., Manu, Y. M., Rashmi, G. R. & Chinamuttevi, V. (2022), IOT Based Smart Health Care System to Monitor Covid-19 Patients. *IJEER, 10*(1), 36-40.

Nareshkumar, A., & Gururaj, B. (2022). An Integrated IoT Technology for Health and Traffic Monitoring System with Smart Ambulance. *2022 IEEE North Karnataka Subsection Flagship International Conference (NKCon)*, Vijaypur, India.

Naveen, K. B., Ramesha, M., & Pai, G. N. (2020). Internet of things: Internet revolution, impact, technology road map and features. *Adv. Math. Sci. J.*, *9*(7), 4405–4414. doi:10.37418/amsj.9.7.11

Onyema, E. M., Lilhore, U. K., Saurabh, P., Dalal, S., Nwaeze, A. S., Chijindu, A. T., Ndufeiya-Kumasi, L. C., & Simaiya, S. (2023). Evaluation of IoT-Enabled hybrid model for genome sequence analysis of patients in healthcare 4.0. *Measurement. Sensors*, *26*, 100679. doi:10.1016/j.measen.2023.100679

Palanikkumar, D., Mary, P. A., & Begum, A. Y. (2023). *A Novel IoT Framework and Device Architecture for Efficient Smart city Implementation*. 2023 7th International Conference on Trends in Electronics and Informatics (ICOEI), Tirunelveli, India.

Prasad, K., Anil Kumar, N., Reddy, N. S., & Ashreetha, B. (2023). Technologies for Comprehensive Information Security in the IoT. *2023 International Conference for Advancement in Technology (ICONAT)*, Goa, India.

Prasad, K., Dekka, S., Tanguturi, R. C., & Poornima, G. (2022). *An Intelligent System for Remote Monitoring of Patients Health and the Early Detection of Coronary Artery Disease*. 2022 International Conference on Smart Generation Computing, Communication and Networking (SMART GENCON), Bangalore, India.

Ram Vishal, B. (2023). *Cardiovascular Disease Prediction Using LSTM Algorithm based On Cytokines.* 2023 4th International Conference for Emerging Technology (INCET), Belgaum, India.

Ramesh, T. R., Lilhore, U. K., Poongodi, M., Simaiya, S., Kaur, A., & Hamdi, M. (2022). Predictive analysis of heart diseases with machine learning approaches. *Malaysian Journal of Computer Science*, 132–148.

Ramesh Naidu, P., Guruprasad, N., & Dankan Gowda, V. (2020). Design and implementation of crypt-cloud system for securing files in cloud. *Adv. Math. Sci. J.*, *9*(7), 4485–4493. doi:10.37418/amsj.9.7.17

Ramesh Naidu, P., Guruprasad, N., & Dankan Gowda, V. (2021). A High-Availability and Integrity Layer for Cloud Storage, Cloud Computing Security: From Single to Multi-Clouds. *Journal of Physics: Conference Series*, *1921*(1), 012072. doi:10.1088/1742-6596/1921/1/012072

Rao, B. K., Chaturvedi, A., & Hussain, N. (2022). Industrial quality healthcare services using Internet of Things and fog computing approach, Measurement. *Sensors (Basel)*, *24*, 100517.

Rathore, M. S., Poongodi, M., Saurabh, P., Lilhore, U. K., Bourouis, S., Alhakami, W., Osamor, J., & Hamdi, M. (2022). A novel trust-based security and privacy model for internet of vehicles using encryption and steganography. *Computers & Electrical Engineering*, *102*, 108205. doi:10.1016/j.compeleceng.2022.108205

Reddy, S., & Patwal, P. P. S. (2022). Data Analytics and Cloud-Based Platform for Internet of Things Applications in Smart Cities. *2022 International Conference on Industry 4.0 Technology (I4Tech)*, (pp. 1-6). IEEE.

Revanna C. (2022). Enhanced Diagnostic Methods for Identifying Anomalies in Imaging of Skin Lesions. *IJEER 10*(4), pp.1077-1085.

Sharma, S. K., Lilhore, U. K., Simaiya, S., & Trivedi, N. K. (2021). An improved random forest algorithm for predicting the COVID-19 pandemic patient health. *Annals of the Romanian Society for Cell Biology*, 67–75.

Simaiya, S., Lilhore, U. K., Prasad, D., & Verma, D. K. (2021). MRI brain tumor detection & image segmentation by hybrid hierarchical K-means clustering with FCM based machine learning model. *Annals of the Romanian Society for Cell Biology*, 88–94.

M. Swathi Pai, M. Shruthi and B. Naveen K, (2020) "Internet of Things: A Survey on Devices, Ecosystem, Components and Communication Protocols," *2020 4th International Conference on Electronics, Communication and Aerospace Technology (ICECA)*, pp. 611-616.

Tanguturi, R. (2023). *Implementation of Machine Learning Approach for Detecting Cardiovascular Diseases*. 2023 3rd International Conference on Intelligent Technologies (CONIT), Hubli, India.

U. K N & R. V M. (2022). *Arduino based COVID-19 Suspect Detection Device*. 2022 6th International Conference on Electronics, Communication and Aerospace Technology, Coimbatore, India.

Venkatakiran, S., Ashreetha, B., & Reddy, N. S. (2023). Implementation of a Machine Learning-based Model for Cardiovascular Disease Post Exposure prophylaxis. *2023 International Conference for Advancement in Technology (ICONAT)*, Goa, India.

Chapter 16
Exploring Few-Shot Learning Approaches for Bioinformatics Advancements

Neha Bhati
https://orcid.org/0009-0008-0171-2786
AVN Innovations, India

Ronak Duggar
AVN Innovations, India

Abdullah Alzahrani
Oakland University, USA

ABSTRACT

This chapter delves deeply into few-shot learning, a rapidly developing area crucial in driving bioinformatics forward. First, the basics of few-shot learning are laid forth, emphasizing the field's applicability to bioinformatics. Case studies showcasing real-world applications in areas as varied as protein structure prediction, drug development, and genomic analysis provide a deep dive into several few-shot learning approaches like meta-learning and transfer learning. The chapter also provides an in-depth analysis of recent developments, highlights current difficulties, and suggests exciting new avenues for exploration. This chapter highlights the rising significance of few-shot learning in bioinformatics and provides insights into its potential to benefit biomedical research.

1. INTRODUCTION

Few-shot and machine learning have established an important new field in bioinformatics. This chapter introduces few-shot learning methods, outlining their theoretical underpinnings and potential applications in bioinformatics. Bioinformatics, which resides at the intersection of biology and data science, uses computer approaches to decipher complex biological data. This chapter lays the platform for an in-depth

DOI: 10.4018/979-8-3693-1822-5.ch016

discussion of crucial methodology, practical applications, and far-reaching ramifications. Still, it also establishes a core understanding of how few-shot learning tactics can be effectively applied within this environment. In the current day, the combination of few-shot learning and machine learning promises to change our understanding of the biological world and propel scientific development to new heights, and this opens the way for future discoveries in bioinformatics (Mendonça et al., 2023).

1.1. What is Few-Shot Learning?

As an exciting and crucial branch of machine learning, few-shot learning is distinguished by its unique ability to equip models with competency in tasks supported by highly minimal labeled data. Unlike the traditional supervised learning approach, this paradigm allows computers to generalize well from a small set of training instances, which often requires large datasets for good training. Bioinformatics is only one field where the growing importance of few-shot learning is making waves. Within the context of bioinformatics research, this chapter sets out to deconstruct the fundamental ideas and sophisticated methodologies that support few-shot learning (Tummala & Suresh, 2023)

1.2. Bioinformatics in Brief

Bioinformatics, a wide-ranging field that bridges biology and computing science, has become central to contemporary life sciences study. It employs a wide range of computational methods, data analysis strategies, and algorithms to decipher the mysterious intricacies of biological processes. Bioinformatics uses data-driven methods to their full potential to unravel the mysteries of the genetic code, reveal the inner workings of complex proteins, and speed up the tortuous process of developing new medications. The subsequent parts of this story provide a condensed review of the essential function played by bioinformatics in expanding our understanding of biological systems. At the same time, it deftly elucidates the dynamic interplay between bioinformatics and few-shot learning, unfolding new insights on how machine learning is positioned to amplify the field's present capabilities and paving the way for forthcoming ground-breaking discoveries (H. Li et al., 2020).

1.3. Chapter Overview

This part plays a conductor to a symphony of structural coherence, providing readers with a clear and illuminating guide across the subsequent chapters' complex terrain. Topics and conversations are presented logically, giving readers a map to follow as they navigate the fascinating maze of few-shot learning in bioinformatics. Within its following chapters, readers may expect an in-depth examination of the fundamental ideas, research methods, and varied uses of few-shot learning. Case studies will be revealed, recent advancements will be clarified, ongoing difficulties will be faced head-on, and most fascinating, a glimpse into the future's possibilities will be offered. This chapter serves a much larger purpose than is usually expected; it acts as an introduction to the symphony of scientific inquiry and discovery that lies ahead, promising to advance bioinformatics studies and add to the fabric of biomedical understanding.

Figure 1. Critical concepts in few-shot learning

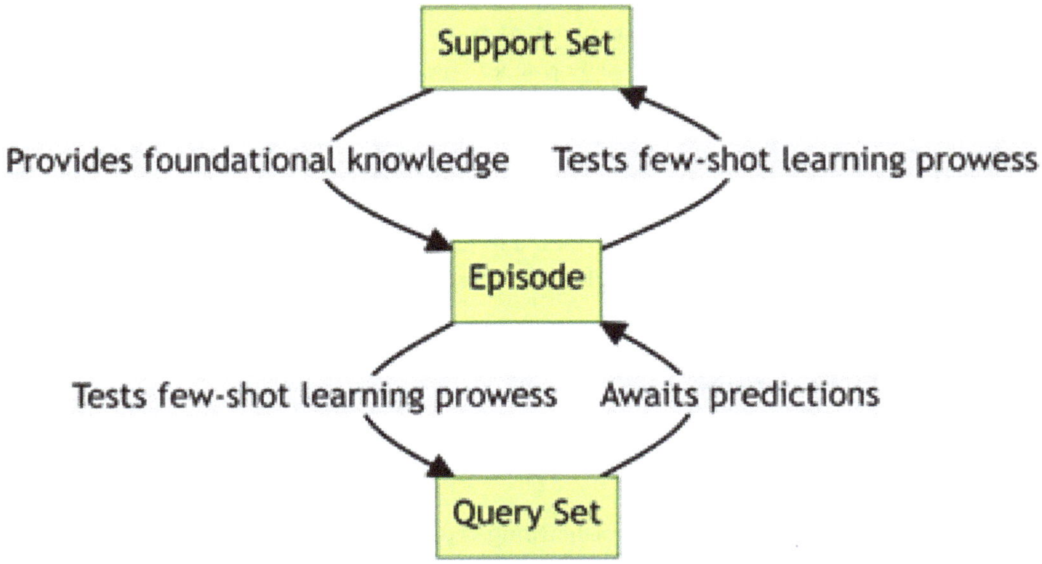

2. BASICS OF FEW-SHOT LEARNING

Within the vast field of machine learning, few-shot learning is a distinct subfield due to its exceptional ability to train models effectively with a small amount of labeled data. Because of this, it possesses features and difficulties distinct from those of traditional supervised learning paradigms. Several concepts are crucial to grasping few-shot learning (Parnami & Lee, 2022).

Figure 1 illustrates the three components of few-shot learning: the Support Set, the Query Set, and the Episode

2.1. Key Concepts

- **Support Set:** The support set is essential in few-shot learning settings. This subset of the training data is carefully curated to include instances labeled with their respective classes or categories. The model's capacity to make predictions is grounded in the information provided by the support set.
- **Query Set:** Unlike the labeled cases in the support set, the query set is waiting for predictions from the model. The model's job is to generalize from the support set to the query set, demonstrating its few-shot learning capabilities, and the knowledge gleaned from the support set is then applied to these unlabeled data.
- **Episode:** Few-shot learning typically follows an episodic structure. A learning task is captured in an episode, typically consisting of a support and query set. The model's few-shot learning prowess is evaluated based on how well it generalizes from the support set to the query set during these episodes.

Table 1. Critical concepts in few-shot learning

Key Concept	Description
Support Set	Samples of training data that have been labeled and contain examples of each class or category.
Query Set	Occurrences without labels that can only be predicted by the model. These samples receive the benefit of the information in the training set.
Episode	In few-shot learning, a support set and query set are combined into a single learning job.

Table 1 provides a summary of key concepts related to few-shot learning, including the "Support Set," "Query Set," and "Episode." It serves as a reference to understand these fundamental concepts within the context of few-shot learning.

2.2. Benefits and Challenges

Few-shot learning has many benefits, making it invaluable in many contexts, including bioinformatics.

- **Reduced Data Dependency:** Few-shot learning reduces the strict data requirements that traditional machine learning algorithms demand, resulting in reduced data dependency. It makes it worthwhile when collecting a large amount of labeled data would require more time, effort, and effort to implement manually.
- **Generalization:** Few-shot learning models are extraordinarily flexible because of their fantastic ability to generalize from minimal examples. Because of this quality, they are more useful in bioinformatics applications where it is crucial to generalize from tiny sample sizes.

However, obstacles remain in the use of few-shot learning:

- **Data Variability:** The inherent complexity and variability of biological data might hinder the efficient training of few-shot learning models. There is still much to learn about how to modify models so that they can handle the complexity of biological data.
- **Algorithm Robustness:** The ongoing goal of research and development is to strengthen the stability and dependability of few-shot learning algorithms. It is still a goal to create models that reliably provide correct predictions in various biological settings and data sets (Song et al., 2023).

2.3. Relevance to Bioinformatics

Few-shot learning's significance in bioinformatics is coming into sharper focus. Since conducting biological experiments is complicated and expensive, researchers using bioinformatics methods often need more labeled data. Few-shot learning is a powerful answer to this problem since it allows models to learn from a small number of labeled samples. Because of this, fundamental bioinformatics tasks like protein classification, gene function prediction, and drug discovery are greatly simplified. Few-shot learning has tremendous relevance in bolstering bioinformatics research and applications, leading to ground-breaking discoveries and breakthroughs in the discipline due to its intrinsic adaptability to varied biological data (Chen et al., 2023).

3. FEW-SHOT LEARNING METHODS

Meta-learning, transfer learning, and a wide variety of algorithms and models are three fundamental techniques that have gained substantial popularity in the field of few-shot learning, each playing a pivotal role in overcoming the issues given by little labeled data.

3.1. Meta-Learning

In the field of few-shot learning, meta-learning is considered a cornerstone notion. The goal of this method is to teach models to learn efficiently. It functions by making models accessible to numerous jobs, each of which includes its unique collection of support and query data. Models utilize this exposure to learnable information and underlying patterns. Models can quickly adjust to novel tasks with only a small amount of labeled data thanks to the gained meta-knowledge. Matching Networks and Prototypical Networks are two notable meta-learning algorithms that learn to generate meaningful similarity metrics among data examples and so perform very well in few-shot classification problems (Sun et al., 2019)

3.2. Transfer Learning

One of the pillars of few-shot learning is transfer learning, which allows for applying skills and knowledge gained from one activity to another. Pre-training tasks include training a model on a large and varied dataset before applying it to a few-shot learning situation. Commonly used pre-trained models, such as BERT and GPT, store a wealth of linguistic knowledge and may be fine-tuned using task-specific labeled data. This method uses the vast knowledge of these pre-trained models, allowing them to adapt to novel tasks with minimal data quickly. When labeled data is poor or unavailable, it is essential to employ transfer learning procedures to improve model performance (Parnami & Lee, 2022)

3.3. Algorithms and Models

Numerous methods and models within the broad context of few-shot learning have been developed to tackle various problems and use cases. These models use various architectures and strategies tailored to the problem, such as convolutional neural networks (CNNs), recurrent neural networks (RNNs), and attention processes. By making effective use of the finite supply of labeled data, few-shot learning models like Siamese Networks, Relation Networks, and Memory-Augmented Neural Networks have proven their superiority on a wide range of tasks (Perez et al., 2021)

In bioinformatics, where a lack of labeled data can be a significant roadblock, strategically implementing these approaches and models is paramount. These methods equip bioinformatics researchers with the keys to the kingdom of machine learning, allowing them to decipher intricate biological phenomena, speed up investigation, and expand the field.

Table 2 provides an overview of few-shot learning algorithms and models, highlighting their key features and typical applications in various domains.

Table 2. Comparison of few-shot learning algorithms and models

Algorithm/Model	Key Features	Applications
Matching Networks	Attention-based, memory-augmented networks	Image recognition, natural language processing
Prototypical Networks	Metric learning, prototypical representations	Image classification, face recognition
Pre-trained Models (e.g., BERT, GPT)	Transfer learning, large-scale pre-training	Text classification, language understanding
Fine-tuning Strategies	Fine-tuning pre-trained models on specific tasks	Various depends on the pre-trained model and task

4. APPLICATIONS IN BIOINFORMATICS

Few-shot learning has emerged as a game-changing strategy in bioinformatics, offering promising solutions to the persistent difficulties caused by a lack of labeled data across essential areas. Here, we examine concrete use cases in which few-shot learning has shown its mettle and the potential for radical improvement.

Figure 2 shows how few-shot learning has had a game-changing impact on bioinformatics, specifically how it has helped fill in data gaps and speed up discoveries in fields including protein analysis, pharmaceutical research, and genomics (H. Li et al., 2020).

Figure 2. Few-shot Learning's impact on bioinformatics

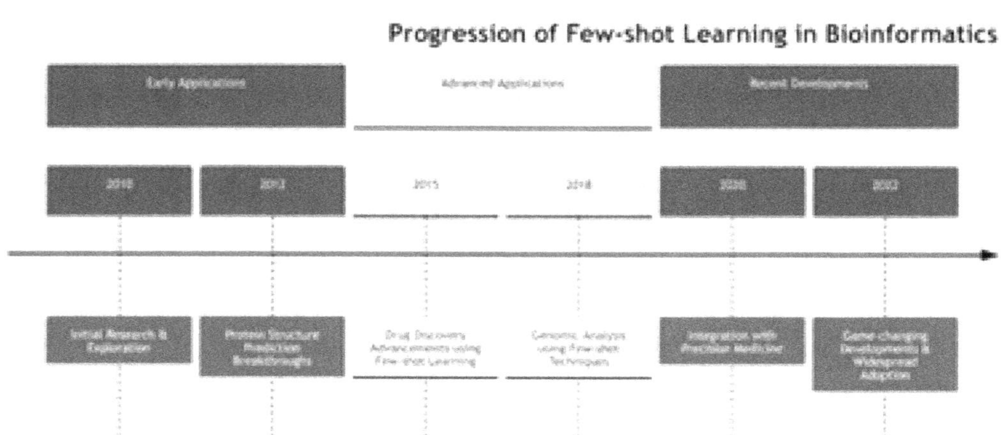

4.1. Protein Structure Prediction

Protein structure prediction is at the forefront of bioinformatics problems because understanding the functions and interactions of proteins in complex biological systems relies on deciphering their three-dimensional configurations. Few-shot learning techniques have left an indelible impact in this field by enabling models to infer protein structures from sparse data. Researchers have made tremendous progress in predicting the precise folding patterns and structural intricacies of proteins by capitalizing on the inherent ability of few-shot models to recognize patterns and similarities among protein sequences and structures. These discoveries have far-reaching implications for medication design, disease understanding, and functional annotation of genes, ultimately leading to game-changing developments in precision medicine (AlQuraishi, 2019).

4.2. Drug Discovery

Few-shot learning has ushered in a new age in the field of drug development, where the rapid analysis of massive chemical datasets is crucial to identifying novel therapeutic molecules. Even when labeled data on specific molecular interactions is lacking, researchers can quickly screen many chemical compounds using few-shot learning algorithms and find prospective candidates for drug development. In light of the urgent healthcare demands of the modern period and the progress in precision medicine, the recent surge in the rate at which new drugs are being discovered is of the utmost importance (K. Li et al., 2020).

4.3. Genomic Analysis

Few-shot learning has also made its impression in the field of genomic analysis, which is notoriously complex and involves a flood of large datasets. Few-shot approaches are potent tools that bioinformaticians use to decipher the baffling complexity of genetic data. These methods have many applications in genomics research, including gene expression profiling, variation detection, and regulatory element ID. To accurately classify genes, discover genomic variants, and anticipate regulatory elements with remarkable reliability, these methods' strength resides in their ability to distill knowledge from small labeled genomic data.

Overall, few-shot learning approaches are accelerating scientific discovery and increasing our understanding of complicated biological phenomena within bioinformatics. These game-changing implementations highlight the great potential of few-shot learning in addressing the difficulties associated with data scarcity, thereby revolutionizing the field of bioinformatics and permanently stamping its mark on biomedical science (Rastogi et al., 2022).

5. CASE STUDY

In this section, we will dive deep into a case study that serves as an example of the use and results of few-shot learning in the complex field of bioinformatics:

Figure 3. Few-shot learning process for protein prediction

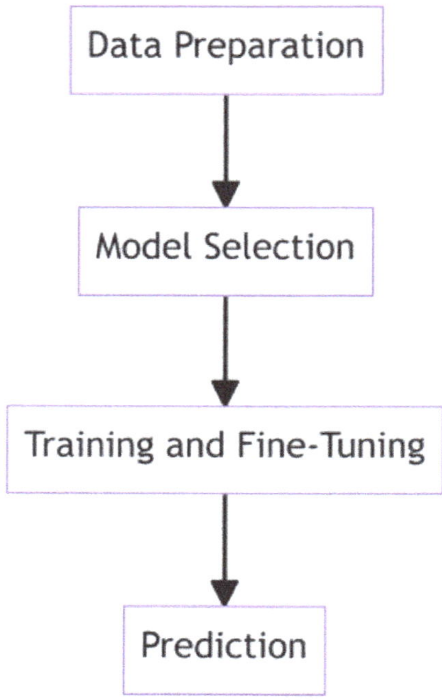

5.1. Predicting Protein Functions

Due to proteins' crucial roles in physiological processes, the prediction of protein activities has emerged as a fundamental challenge in the complex field of bioinformatics. Few-shot learning emerges as a powerful method in this context, allowing protein functions to be predicted with only a few labeled examples. This study sets sail to discover how few-shot learning can be practically applied to predicting protein activities by using sequence and structural data. We create a support set using examples of proteins with known functions culled from annotated sequences and compare it to a query set containing proteins for which we have no functional annotations. Our goal is to extract the functions of the query proteins based on the inherent similarity learned from the support set, and we plan to do this by using the power of few-shot learning methods like Prototypical Networks (Ma et al., 2021).

Figure 3 shows how few-shot learning is applied systematically from data cleaning to predicting protein functions.

5.2. Few-Shot Learning Implementation

A carefully choreographed sequence of linchpin steps must be taken to put few-shot learning for protein function prediction into practice.

- Data Preparation: We start the trip by putting together a carefully curated dataset full of annotated protein sequences and the functions to which they have been assigned. A careful splitting of this dataset into support and query sets accurately recreates the core of the few-shot learning scenario.

- Model Selection Once data preparation is complete, the most crucial choice is which model to use. It requires meticulously selecting a suitable few-shot learning model, such as Prototypical Networks or Matching Networks, equipped with the wisdom to familiarize itself with similarity metrics and to deftly identify proteins based on the sparse ensemble of labeled exemplars.
- Training and Fine-Tuning: The voyage of the selected model begins with extensive training on the support set, followed by fine-tuning. The commonalities between protein sequences and their functions are captured here, and the subtleties that make them such are absorbed. The next step is the fine-tuning phase, where the model's performance is fine-tuned to perfection for its intended use.
- Prediction: The model is now prepared to perform its primary purpose, which is to predict the functionalities intrinsic to proteins within the query set, using the data collected during training and fine-tuning (Yue et al., 2020).

5.3. Results and Insights

The conclusion of this case study leaves us with a wealth of data and a new understanding of how effective few-shot learning may be for predicting protein functions. The few-shot learning approach unveils its remarkable generalization and prediction abilities for the functionalities hidden within unannotated proteins using only a small set of labeled samples. Accuracy, precision, recall, and the F1-score are the measures used to evaluate the model's performance, offering a comprehensive and nuanced look at the model's ability to predict outcomes.

This case study goes beyond traditional measures of performance to highlight the critical role that careful data curation, thoughtful model selection, and expert hyperparameter tuning play in directing the orchestra of few-shot learning when attempting to predict protein functions. It challenges the limits of data availability and sheds light on the path toward knowledge discovery in the life sciences, emphasizing in no uncertain terms the latent potential inherent in few-shot learning as a transformative tool in the arsenal of bioinformatics.

6. ADVANCEMENTS AND CHALLENGES

Figure 4 shows the timeline highlighting the most critical events in few-shot learning, including the debut of game-changing models like CLIP and DALL-E.

Figure 4. Few-shot learning advancements timeline

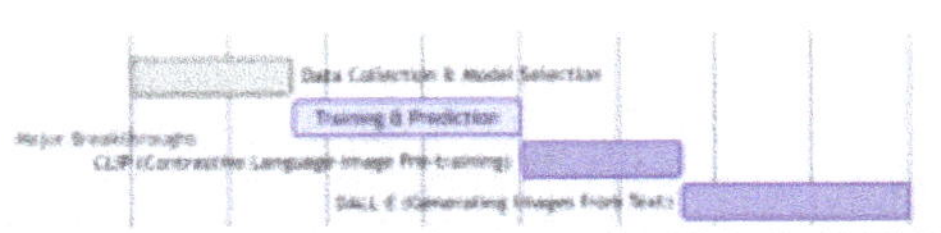

6.1. Recent Developments

There has been a dramatic upswing in few-shot learning in recent years, with several new algorithms and methods emerging. These advancements have made few-shot learning more widely applicable in fields like bioinformatics. Some recent developments that stand out include:

6.1.1. CLIP (Contrastive Language-Image Pre-Training)

Significant interest has been shown in CLIP, a novel few-shot learning model. CLIP is an OpenAI project that aims to unify the understanding of images and text. It is made possible through its training on a large dataset of photos and their associated written descriptions. As a result, CLIP is a noteworthy addition to the few-shot learning landscape, as it can generalize from this training to accomplish a wide variety of tasks with minimum input. CLIP's capacity to link visual and textual information may find use in bioinformatics for things like image-based gene expression analysis and phenotypic prediction.

6.1.2. DALL-E (Generating Images From Text)

Another significant development in few-shot learning is Open AI's DALL-E. It revolutionizes the idea of generative models by creating visuals based on written descriptions. This capacity may cause a dramatic shift in how fields like bioinformatics use picture synthesis. Envision a system that can, from written descriptions, produce visual representations of protein structures or cellular processes, allowing for a more thorough comprehension of these phenomena.

6.2. Limitations and Future Directions

Few-shot learning has come a long way but still has serious obstacles. Resolving these constraints will pave the path for additional development. Some important things to keep in mind for new studies and innovations are as follows:

Table 3 provides an overview of potential areas of study for the future of few-shot learning.

6.2.1. Data Scarcity

The lack of labeled data remains a significant obstacle for few-shot learning. Because of this restriction, few-shot learning is not often used in fields like bioinformatics, where it might be challenging to

Table 3. Future directions in few-shot learning

Research Direction	Description
Data-efficient Algorithms	Develop algorithms that require even fewer labeled examples.
Cross-modal Few-Shot Learning	Extend few-shot learning to handle multiple data modalities.
Transfer Learning Enhancements	Improve transfer learning techniques for better generalization.
Explainable Few-Shot Learning	Develop interpretable models that provide insight into predictions.

get labeled data. Input-efficient algorithms that can learn with less input or novel approaches to data augmentation should be the focus of future studies.

6.2.2. Model Generalization

It is still of the utmost importance to ensure that few-shot learning models can generalize well to various biological data types and domains. In the future, we may look into domain adaption techniques and domain-specific pre-training to enhance model generalization capabilities(Wang et al., 2020).

6.2.3. Interpretability and Explainability

In bioinformatics applications, where decisions can have far-reaching effects, few-shot learning models' interpretability and explainability are paramount. Future studies should aim to create interpretable few-shot learning models and frameworks to comprehend these models' biological consequences further.

6.2.4. Domain Integration

There is great potential in developing few-shot learning models using domain-specific and existing biological knowledge. Such knowledge incorporation can benefit model performance, reliability, and comprehension of intricate biological processes. Possible next steps include creating hybrid models that combine domain expertise with few-shot learning methods.

Models such as CLIP and DALL-E, which represent the state-of-the-art in few-shot learning, have opened up new realms of possibility for the field. Few-shot learning has shown promise in bioinformatics and other fields. However, it can only continue to thrive if problems with data availability, model generalization, interpretability, and domain integration are resolved. Knowledge discovery across the sciences relies heavily on few-shot learning, and its importance will only grow as researchers tirelessly pursue solutions to these problems (Jadon, 2020).

7. CONCLUSION

This chapter ends with a flourish, showering us with a mosaic of understanding that captures the essence of few-shot learning as it applies to bioinformatics. In this final scene, we highlight crucial insights to emphasize the far-reaching significance of this paradigm.

7.1. Key Takeaways

We distill the essence of few-shot learning from the rich tapestry of our investigation by considering:

- The Transformative Power of Limited Opportunities for Learning: Few-shot learning techniques are the vanguard of this revolution, leading the way to model perfection in problems where sparsely labeled data is a limiting factor. Their true worth becomes apparent in bioinformatics when the persistent problem of insufficient data meets its match.

- Meta-Learning and Transfer Learning: The meta-learning and transfer learning approaches, on the other hand, take on the role of instrumental orchestrations within the few-shot learning symphony. They give models the skill of quickly learning new tasks and domains, allowing for a fluid performance on the dynamic stage of bioinformatics.
- Revolutionary Applications: Few-shot learning's revolutionary applications are felt in several key areas of bioinformatics. Few-shot learning scripts are a chronicle of revolution, rewriting the narratives of knowledge discovery in the life sciences, whether it is the elucidation of protein structures, an exploration into the realms of drug development, or the decoding of complicated genetic riddles in genomics analysis.

7.2. Significance for Bioinformatics

Bioinformatics is a diverse field at the intersection of biology and data science, and the significance of few-shot learning has been widely discussed for some time.

- Expediting Critical Tasks: Important Tasks Should Be Done Quickly Through the maze of difficulties associated with protein function prediction, drug development, and genomics analysis, few-shot learning has emerged as a guiding light. It provides a fast track to discoveries and insights even when the source of labeled data is limited.
- A Paragon of Flexibility: Adaptability is sparked by the fire of few-shot learning in bioinformatics. In this setting, models can quickly learn to generalize to new problems, speeding up scientific progress.
- Diverse Insights from Disparate Data: Adding few-shot learning to bioinformaticists' toolkits enables them to use biological data for insight. This cooperation improves our life sciences knowledge by illuminating previously unsuspected connections and relationships.

7.3. Closing Remarks

In conclusion, few-shot learning stands out as a bright star in the constellation of approaches to data scarcity, shedding light on a road paved with breakthroughs in bioinformatics. The more we approach the future, the more it becomes a beacon of hope, illuminating the route towards transformational contributions in biomedical research and beyond and less of a mere instrument.

Cooperation plays a vital role in this adventure. Researchers, data scientists, and domain specialists will need to converge and coordinate their efforts to overcome the difficulties that still lie ahead. When we work together, we can harness the full potential of few-shot learning to shed light on the complex riddles of biological systems. Few-shot learning is shining a bright light on the road to bioinformatics progress, and the new horizons it reveals hold the potential of immeasurable benefits to the fields of medicine and science.

REFERENCES

AlQuraishi, M. (2019). ProteinNet: A standardized data set for machine learning of protein structure. *BMC Bioinformatics*, *20*(1), 1–10. doi:10.1186/s12859-019-2932-0 PMID:31185886

Chen, P., Wang, J., Lin, H., Zhao, D., & Yang, Z. (2023). Few-shot biomedical named entity recognition via knowledge-guided instance generation and prompt contrastive learning. *Bioinformatics (Oxford, England)*, *39*(8), btad496. doi:10.1093/bioinformatics/btad496 PMID:37549065

Jadon, S. (2020). An overview of deep learning architectures in the few-shot learning domain. *arXiv Preprint arXiv:2008.06365*.

Li, H., Tian, S., Li, Y., Fang, Q., Tan, R., Pan, Y., Huang, C., Xu, Y., & Gao, X. (2020). Modern deep learning in bioinformatics. *Journal of Molecular Cell Biology*, *12*(11), 823–827. doi:10.1093/jmcb/mjaa030 PMID:32573721

Li, H., Tian, S., Li, Y., Fang, Q., Tan, R., Pan, Y., Huang, C., Xu, Y., & Gao, X. (2020). Modern deep learning in bioinformatics. *Journal of Molecular Cell Biology*, *12*(11), 823–827. doi:10.1093/jmcb/mjaa030 PMID:32573721

Li, K., Du, Y., Li, L., & Wei, D.-Q. (2020). Bioinformatics approaches for anti-cancer drug discovery. *Current Drug Targets*, *21*(1), 3–17. doi:10.2174/1389450120666190923162203 PMID:31549592

Lilhore, U. K., Manoharan, P., Sandhu, J. K., Simaiya, S., Dalal, S., Baqasah, A. M., Alsafyani, M., Alroobaea, R., Keshta, I., & Raahemifar, K. (2023). Hybrid model for precise hepatitis-C classification using improved random forest and SVM method. *Scientific Reports*, *13*(1), 12473. doi:10.1038/s41598-023-36605-3 PMID:37528148

Lilhore, U. K., Poongodi, M., Kaur, A., Simaiya, S., Algarni, A. D., Elmannai, H., Vijayakumar, V., Tunze, G. B., & Hamdi, M. (2022). Hybrid model for detection of cervical cancer using causal analysis and machine learning techniques. *Computational and Mathematical Methods in Medicine*, *2022*, 2022. doi:10.1155/2022/4688327 PMID:35572826

Lilhore, U. K., Simaiya, S., Pandey, H., Gautam, V., Garg, A., & Ghosh, P. (2022). Breast cancer detection in the IoT cloud-based healthcare environment using fuzzy cluster segmentation and SVM classifier. In Ambient Communications and Computer Systems [Singapore: Springer Nature Singapore.]. *Proceedings of RACCCS*, *2021*, 165–179.

Lilhore, U. K., Simaiya, S., Prasad, D., & Guleria, K. (2020). A hybrid tumor detection and classification based on machine learning. *Journal of Computational and Theoretical Nanoscience*, *17*(6), 2539–2544. doi:10.1166/jctn.2020.8927

Ma, J., Fong, S. H., Luo, Y., Bakkenist, C. J., Shen, J. P., Mourragui, S., Wessels, L. F., Hafner, M., Sharan, R., Peng, J., & Ideker, T. (2021). Few-shot learning creates predictive models of drug response that translate from high-throughput screens to individual patients. *Nature Cancer*, *2*(2), 233–244. doi:10.1038/s43018-020-00169-2 PMID:34223192

Mendonça, S. S., de Sousa Vitória, A. R., de Lima, T. W., Galvão Filho, A. R., & Andrade, C. H. (2023). Exploring New Horizons: Empowering Computer-Assisted Drug Design with Few-shot Learning. *Artificial Intelligence in the Life Sciences*, *4*, 100086. doi:10.1016/j.ailsci.2023.100086

Onyema, E. M., Lilhore, U. K., Saurabh, P., Dalal, S., Nwaeze, A. S., Chijindu, A. T., Ndufeiya-Kumasi, L. C., & Simaiya, S. (2023). Evaluation of IoT-Enabled hybrid model for genome sequence analysis of patients in healthcare 4.0. *Measurement. Sensors*, *26*, 100679. doi:10.1016/j.measen.2023.100679

Parnami, A., & Lee, M. (2022). Learning from few examples: A summary of approaches to few-shot learning. *arXiv Preprint arXiv:2203.04291.*

Perez, E., Kiela, D., & Cho, K. (2021). Authentic few-shot learning with language models. *Advances in Neural Information Processing Systems*, *34*, 11054–11070.

Ramesh, T. R., Lilhore, U. K., Poongodi, M., Simaiya, S., Kaur, A., & Hamdi, M. (2022). Predictive analysis of heart diseases with machine learning approaches. *Malaysian Journal of Computer Science*, 132–148.

Rastogi, S. C., Rastogi, P., & Mendiratta, N. (2022). *Bioinformatics: Methods and Applications-Genomics, Proteomics and Drug Discovery*. PHI Learning Pvt. Ltd.

Rathore, M. S., Poongodi, M., Saurabh, P., Lilhore, U. K., Bourouis, S., Alhakami, W., Osamor, J., & Hamdi, M. (2022). A novel trust-based security and privacy model for internet of vehicles using encryption and steganography. *Computers & Electrical Engineering*, *102*, 108205. doi:10.1016/j.compeleceng.2022.108205

Sharma, S. K., Lilhore, U. K., Simaiya, S., & Trivedi, N. K. (2021). An improved random forest algorithm for predicting the COVID-19 pandemic patient health. *Annals of the Romanian Society for Cell Biology*, 67–75.

Simaiya, S., Lilhore, U. K., Prasad, D., & Verma, D. K. (2021). MRI brain tumor detection & image segmentation by hybrid hierarchical K-means clustering with FCM based machine learning model. *Annals of the Romanian Society for Cell Biology*, 88–94.

Song, Y., Wang, T., Cai, P., Mondal, S. K., & Sahoo, J. P. (2023). A comprehensive survey of few-shot learning: Evolution, applications, challenges, and opportunities. *ACM Computing Surveys*, *55*(13s), 1–40. doi:10.1145/3582688

Sun, Q., Liu, Y., Chua, T.-S., & Schiele, B. (2019). Meta-transfer learning for few-shot learning. *Proceedings of the IEEE/CVF Conference on Computer Vision and Pattern Recognition*, (pp. 403–412). IEEE.

Tummala, S., & Suresh, A. K. (2023). Few-shot learning using explainable Siamese twin network for the automated classification of blood cells. *Medical & Biological Engineering & Computing*, *61*(6), 1–15. doi:10.1007/s11517-023-02804-3 PMID:36800155

Wang, Y., Yao, Q., Kwok, J. T., & Ni, L. M. (2020). Generalizing from a few examples: A survey on few-shot learning. *ACM Computing Surveys*, *53*(3), 1–34. doi:10.1145/3386252

Yue, Z., Zhang, H., Sun, Q., & Hua, X.-S. (2020). Interventional few-shot learning. *Advances in Neural Information Processing Systems*, *33*, 2734–2746.

Chapter 17
Introduction to Bioinformatics and Machine Learning

Rakhi Chauhan
Chitkara University, India

ABSTRACT

ML has revolutionised bioinformatics' difficult biological data analysis. Pattern recognition and biological process categorization help ML systems diagnose diseases, predict protein structures, and investigate gene expression. Few-shot learning and bioinformatics are effective at optimising results with limited resources, overcoming biological dataset access issues. ML and bioinformatics advance precision medicine and drug discovery while improving biological understanding. This chapter examines bioinformatics ML methods like supervised classification, clustering, and probabilistic graphical models to find new insights. Text mining, systems biology, evolution, proteomics, and genomics use deterministic and stochastic heuristics for optimisation. Understanding bioinformatics methods and modern ML technologies while understanding implementation challenges is the study goal. Few-shot learning is highlighted to show its importance. ML and bioinformatics together improve our knowledge and solve real problems, improving research methods in biological and medical sciences.

1. INTRODUCTION

Many applications in bioinformatics lead to a conversion to handle the biological data and that can provide many methods. We can use these methods for extraction ideas that are helpful to glean from big datasets. With the help of bioinformatics, we can add on biology with computer science for consideration of biological events and use machine learning methods. Using machine learning results in correctly diagnosing protein structures. By classifying the models we can identify the patterns that are used to detect and find out the diseases with the help of gene expression data. Machine learning can predict the discovery of drugs through the process of predicting the likely interactions that would occur between medications and targets, as well as assessing the toxicity of substances. The revolution of machine learning and bioinformatics can make us use vast datasets very easily.

DOI: 10.4018/979-8-3693-1822-5.ch017

The use of individualized treatment and finding the hidden biological patterns make us progress to comprehending the difficulties of molecular existence. Bioinformatics and machine learning show many ways to consider the biological system that is used for advancements in the health sector. When the Organisation (HUGO) named the Human Genome Organisation organized global activities in the early 1990s, this marked the beginning of the era of genome research. A complete DNA sequencing of all of the human chromosomes was the goal of the HGP, which was an initiative that was undertaken to sequence the human genome. This model allows us to demonstrate the "Prediction of protein secondary structure by the hidden Markov model," which is a definition of the secondary structure of proteins that were found in the research conducted by Asai and colleagues [1].

Understanding the folding and function of proteins requires several steps, the most important of which is the prediction of the secondary structure of proteins during the process. We are able to estimate the probabilistic framework that is provided by HMM for the description of sequential dependencies using this method. Additionally, we are able to estimate the probabilistic framework by amino acid residues that are associated in the protein sequence. Study [2] by Backofen et al. defines the term "bioinformatics and constraints" as the intersection of bioinformatics and constraint-based techniques. This intersection is referred to as "bioinformatics and constraints." In the discipline of bioinformatics, numerous issues have arisen, and these issues are defined in the same article. Researchers at the aforementioned study identify a wide variety of approaches to technology that is based on limits. The first work focuses on the creation of a predictive model for the secondary structure of proteins, while the second piece emphasises the usefulness of constraint-based methods in the field of bioinformatics. These studies are a reflection of the ongoing efforts that are being made to improve our understanding of biological systems through the application of novel computational methods. These efforts are being made in order to improve biology. We can consider Bioinformatics with the help of following diagram named figure 1:

In the field of Bioinformatics we involve many applications of computer science, biology and statistics. We can consider this with above diagram. In bioinformatics field we may have many complex relationship, structure and process. Computer scientists, physicists, mathematicians, and crystallographers worked together to find solutions to difficult biological problems when they were presented. In order to mimic and analyse biological data, they realized the significance of using computer approaches and acknowledged their capability to do so. These folks were the pioneers of the bioinformatics community and created the first group of people to work together. During that period of time, this community produced computational tools in order to investigate the data by making use of traditional methods from the field of computer science. The standard algorithms, on the other hand, are starting to become increasingly inadequate when it comes to solving practical problems. There is no fundamental explanation at the molecular level for the phenomenon, but it is likely that the genesis of the phenomenon could be linked to the complex character of biological systems. This is in addition to the fact that there is no fundamental explanation. Previously, this was not an option that was taken into consideration. Furthermore, traditional methods of computing are incapable of managing the enormous amount of data that is simultaneously growing at an exponential rate. As a result of their ability to independently acquire information from the data and offer significant insights, machine learning algorithms are rapidly being utilised in the field of bioinformatics. Machine learning techniques have been made possible by the rapid growth of computers, which has made it possible to effectively address biological issues. The reasoning behind this is that the approaches of machine learning entail methods that require a significant amount of computer power. The field of bioinformatics is a multidisciplinary one, meaning that it brings together a number of different fields into a unified whole. This area of study spans a wide variety of fields, including molecular biology, computer science, artificial intelligence, machine learning, dynamic programming, and physical chemistry, among others. It is difficult to precisely define the bounds of bioinformatics study because of the expansive nature of the field and the fact that it is constantly evolving.

The main field of findings focuses on Molecular evolution, molecular modeling, algorithm development, proteomics, and biochemical pathways are some of the topics that are covered in data mining. By finding of Trifonov's publication we can examine the sequence event and progression in field of bioinformatics. The study of Luscombe et. al give us detailed explanation of field of bioinformatics [4]. We can find out and explanations technique of computation and prediction of computer systems tailored for biologists' requirement. This field entails many studies about solving the linked data of molecular biology investigations.

Challenges encountered in the domain of bioinformatics: The preliminary sequence of genome is now accessible. the primary objective of academics working in the subject of bioinformatics is to collect, reveal, and forecast key insights from the databases that hold this information. This is the case because the draft sequence is now available. The International Human Genome Sequence Consortium made an estimate in the year 2001 that the human genome, which is made up of roughly 3 billion base pairs, has between 30,000 and 40,000 genes that code for specific proteins. This estimate was based on the fact that the human genome is constituted of around 3 billion base pairs. The data contains a tremendous quantity of knowledge that is both significant and consequential in terms of human development, physiology, medicine, and evolution. This knowledge is both crucial and consequential. The utilisation of a method that is both clever and effective is an imperative necessity in order to successfully extract information from this data and manage the rapid accumulation of data. Data plays a crucial role in molecular biological research. Molecular biology research utilizes several contemporary experimental methodologies, such as genomic sequence analysis, in order to produce enormous amounts of data. According to the Economist in 1999, a solitary experimental laboratory is capable of producing over 100 terabytes of data every day. The exponential increase of data being uploaded to publicly accessible databases is a direct consequence of the progress made in computing and experimental techniques. The implementation of sophisticated processors with exceptional technological capabilities has considerably accelerated the speed of processing conventional experimental data, while also enabling extensive disk storage. The internet has enabled the exchange and retrieval of experimental data from diverse laboratories. The reasons that help to the widespread growth of biological databases are as follows. As of July 2001, Gen Bank holds over 12 million records of nucleotide sequences, whereas Protein Information Resource (PIR) includes around 233,000 entries of amino acid sequences during the same period. A total of 15,500 three-dimensional structures were included in the PDB as of the month of July in the year before to the current one. There is a table that gives an indication of the sizes of the different databases that are available. Specifically, Tan, A.C., and Gilbert, D. (September 1, 2003) reported their findings. There is a collection of databases that pertain to molecular biology that is both comprehensive and up to date, and scientists have the potential to find this collection through their research and that databases data sizes are defined in table 1.

In accordance to conclusions of this study, this pattern is a defining characteristic of the post genome informatics period, which is distinguished by research that is mostly guided by data. In order to assist scientists in comprehending the fundamental principles of life (biological knowledge) and to facilitate practical applications in the field of biomedicine, he proposes the utilisation of computational methods to gather the data that is stored in the database that is obtained from the study of the genome sequence. This would be done in order to facilitate practical applications in the field of biomedicine. During this particular time period, this is the most critical difficulty that the discipline of bioinformatics is now

Table 1. Databases data size of molecular biology

Database	Data Size	URL
July 2001 edition of Gen Bank	Roughly 12,244,000 sequences	https://www.ncbi.nlm.nih.gov/Genbank/
(July, 2001) Protein Information Resource, often known as PIR	About 233,000 sequences	http://pir.georgetown.edu/
(April, 2001) The PROSITE database, which is a database of protein families and domains	There are 1474 distinct patterns, rules, profiles, and matrices included	http://ca.expasy.org/prosite/
(July, 2001) Protein Data Bank, often known as PDB)	structure code 15531	https://www.rcsb.org/pdb/
Statistical Classification of Proteins, or SCOP, was published in July of 2000.	The 26219th Domain	http://scop.mrc-lmb.cam.ac.uk/scop/
Statistical Classification of Proteins, or SCOP, was published in July of 2000.	The 25320 Domains	http://www.biochem.ucl.ac.uk/cath/
In April of 2001, the Kyoto Encyclopaedia of Genes and Genomes (KEGG) was published.	150 access points	http://www.genome.ad.jp/kegg/

experiencing at the present moment. This evolution has occurred as the era has advanced. The study [5], which illustrates the development of the database notion, draws attention to this expansion. Knowledge base biological databases are developed with the purpose of getting a grasp of the underlying truths that are related with bio molecular data. This is the objective behind their initial design.

It is currently regarded impractical to manually collect information from modern databases, as biologists did in the days before the development of computing. This is due to the fact that scanning the databases does not only take a lot of time but also is inefficient. In addition, the transformation of this information into knowledge is a significant challenge that needs to be successfully conquered.

Moreover, the majority of data is stored in computer systems and will be used through the internet. Luscombe et al. (2001) identified three key responsibilities of bioinformatics in addressing the challenge of data mining:

a) Organizing the data in a manner that is easily available and updatable.
b) Developing novel tools for data analysis.
c) To use the evaluation for understanding the area of expertise.

There are some more challenges in the field of Bioinformatics, after done many analysis we have find out these. Some of them are given below.

- Handling and analyzing large-scale biological data sets, such as those produced by next-generation sequencing technologies, is an example of this.
- The cultivation of computational methods that is dependable and precise for the purpose of forecasting the activity of genes and proteins, as well as for the purpose of enhancing our comprehension of the biological systems that are hidden beneath the surface.

- Something that needs to be addressed is the complexity of biological systems, which hide many levels of organisation, ranging from the molecular to the ecological. This complexity ranges from the molecular to the ecological.
- Data integration is the process of integrating and making meaning of data from a range of sources, such as genomic, transcriptomic, proteomic, and Meta bol omic data, amongst others.
- Using approaches from the field of bioinformatics to find solutions to problems that occur in the real world, such as the development of personalised medicine, the discovery of new medications, and the understanding of the pathos physiology of diseases.
- The administration and analysis of huge and complex data collections, including, amongst other instances, genomic sequences and gene expression data, amongst others.
- The process of designing algorithms that is accurate and efficient for a number of applications in bioinformatics, such as sequence alignment, gene prediction, and phylo genetic reconstruction, as well as other activities that are comparable to these. Integrating and making sense of data that originates from a wide range of sources and levels of information, such as genetic, epigenetic, transcriptomic, and proteomic pieces of information, is the process of data integration. In this particular scenario, "addressing the ethical and legal issues related to the storage, sharing, and use of genomic data".
- One of the challenges that arise in fields such as DNA sequencing and data storage is the difficulty of keeping up with the rapid pace of technological advancements has the goal of simultaneously boosting the clarity of bioinformatics research while concurrently improving its repeatability. When dealing with enormous data sets, it is of utmost importance to handle the scalability challenges that are present in the field of bioinformatics.
- The creation of user-friendly and easily accessible tools for the management and analysis of data in bioinformatics that are always available at all times.
- Keeping up with the fast advancements in technology and approach is a continuous challenge for bioinformatics practitioners.

In order to address these problems, it is crucial to promote cooperation among biologists, computer scientists, statisticians, and other experts, while simultaneously sustaining ongoing endeavors in research and technological progress.

2. MACHINE LEARNING

Machine learning, as per the Oxford English Dictionary (OED), pertains to the ability of a machine to assimilate new information and adapt its functions accordingly. Machine Learning, as defined by Mitchell (Mitchell, 1997), is a methodology that enables systems to enhance their performance by acquiring knowledge from past experiences.

In the field of bioinformatics, computers play a central role as the main research tool, and the term "machine" is commonly used to refer to a computer. However, it is important for the reader to remember that the term "machine" encompasses a wider range of objects; almost any inanimate device capable of carrying out a task could be considered a machine and the study of Hutchinson (1995) states that the fundamental element of the machine learning technique is the utilization of computer algorithms to acquire information from data. Machine learning is a type of computer software that gains knowledge through experience in order to do certain tasks and achieve specific performance goals (Mitchell, 1997). Acquiring knowledge is the fundamental component of intelligence. An ideal solution for handling complex problems, especially in the realm of biological systems, would be a self-learning system that can adapt and improve its performance based on past experiences.

The core premise of learning involves the process of inferring general functionalities based on specific training examples. To facilitate the learning process, the learning agent is provided with a variety of training and test examples that are pertinent to a certain category. The agent is able to acquire knowledge as a result of the training examples, followed by the construction of a hypothesis based on the knowledge that it has acquired. The agent is required to methodically examine all potential hypotheses and identify the most advantageous hypothesis based on the test sets provided. Learning can be classified into three primary categories.

I) In the context of supervised learning, the inputs and outputs of a component are readily observable and evident.
II) In the domain of reinforcement learning, the learning agent has feedback regarding its actions, but it is not explicitly informed about the proper action.
III) In unsupervised learning, the learning agent lacks explicit knowledge of the precise outputs.

The fundamental learning model has four main components of a machine learning agent (Doyle, Figure 2). The constituents are as follows:

a) The learning component is the section of the agent that is accountable for improving its performance.
b) The performance component is accountable for identifying the appropriate course of action.
c) The critic component offers evaluation to the learning components regarding the agent's performance.
d) Problem creator: This aspect of the agent suggests behaviors that may lead to unique or enlightening encounters.

When it comes to the design of a learning system, the process of characterizing the learning problem is of the utmost importance from a design standpoint. Which functions does this educational system serve within the framework of the system as a whole? Can you please provide a detailed explanation of the objective function, as well as the representation that correlates to it? With regard to this learning system, what are the phases that are included in the training procedure exactly? What many kinds of approaches are there to choose from in order to ascertain whether or not the learning system is effective? The basic learning model are described with figure 2.

It is necessary to provide answers to the questions that are presented in the following list before commencing the process of creating the learning system. A typical learning paradigm often includes three distinct categories of learning problems. These challenges can be broken down into three categories. When it comes to the initial challenge, the source of experience, or more precisely, the data sets that are utilised for staff training, is particularly problematic.

The second issue is to the categorization of the activities, the specific purpose of the learning system, and the assessment of the system's capacity to carry out the tasks. During the process of evaluating the learning model, a number of aspects are taken into consideration. These include the accuracy and efficiency of the agent, the quality of the output, the consistency of the learning model, as well as the time and spatial complexity of the tasks. Additional factors include the difficulty of the tasks. Figure 3

Table 2. Six main types of learning

Speed up learning	An instance of deductive learning that does not call for any new input, but which, over time, leads to an improvement in the agent's performance. This can be broken down into two categories: rote learning and generalization. (such as the learning method based on explanations).
Learning by taking advice	Deductive learning is a sort of learning in which the system is able to comprehend new information that is added to its existing knowledge base.
Learning from examples	A method of learning known as inductive learning is one in which concepts are taught through the use of collections of examples that have been labelled. The use of decision trees is one example.
Clustering	A form of learning known as unsupervised inductive learning, in which "natural classes" are discovered for data instances and various classification methods are discovered for them.
Learning by analogy	Through the process of inductive learning, a system is able to transfer knowledge from one database to another database that is based on a different domain.
Discovery	Inductive and deductive learning are two types of learning that refer to the process of learning in which an individual learns without the assistance of a teacher providing instruction. Deduction is a method that is deemed to be correct regardless of whether it proves theorems or reveals concepts about those theorems. On the other hand, when it raises conjectures, it is deemed to be that of an inductive argument.

provides a visual representation of the three fundamental components that are responsible for the machine learning problem we are discussing.

A summary of the six basic types of learning processes is presented in table 2, which can be found here.

Find out more information. E is an abbreviation for experience, T is an abbreviation for tasks, and P is associated with performance evaluation.

A machine learning model is software that can identify patterns or make choices based on a dataset that it has not encountered before. In natural language processing, machine learning models have the ability to analyze and accurately identify the meaning or purpose of sentences or word combinations that have not been encountered before. Supervised learning is a machine learning technique it is training a model by utilizing a labeled dataset, which comprises input-output pairs. This allows the algorithm to have the capacity to construct a correlation between provided inputs and their corresponding outputs.

We can further enhance our understanding of different types of learning by referring to the table 2 provided.

3. MACHINE LEARNING APPROACHES AND THEIR APPLICATIONS IN BIOINFORMATICS RESEARCH

Because of the usefulness of these strategies in solving certain difficulties, the area of bioinformatics makes extensive use of a wide variety of methodology. Humans possess the cognitive ability to comprehend the underlying principles and the regulations that arise from these phenomena.

The clarification of various challenges in biological systems often relies on the use of exemplars, especially empirical evidence. The relationship between the inputs and outputs is still a mystery, despite the fact that humans are able to distinguish the individual inputs and outputs that are associated with each other. The process of protein folding provides evidence for this assertion. The utilization of machine learning techniques is advantageous in this scenario due to their ability to autonomously adapt its internal framework to produce approximation solutions for the specified challenges. The machine learning algorithms used to extract knowledge from data and provide useful insights without relying on explicit programming instructions. Gartner's Hype Cycle for Emerging Technologies reached its highest point in 2017, and Machine Learning attracted substantial interest in the larger field of Artificial Intelligence (AI). In addition, Machine Learning reached its highest peak in the industry. The advanced decision-making capabilities of Google's Alpha Go and Tesla's Autopilot were demonstrated.

The advancements in technology in this field can be attributed to two significant breakthroughs. The computing industry is progressively transitioning to cloud-based systems, which allows for the redistribution of financial resources that were previously allocated to hardware towards computational tasks. Implementing a lease model for access entails the utilization of enhanced and customized equipment. Computers can enhance their learning abilities by consistently assessing and modifying internal evaluation measures through an iterative procedure. When applied in suitable systems, this approach eliminates conventional constraints and enables the advancement of more advanced methodologies. The concept of vertical growth refers to the production of a large number of occurrences, whereas horizontal expansion refers to the acquisition of accurate knowledge regarding the process. They are both working on improving their database skills. Our ability to apply Machine Learning approaches is contingent upon our utilisation of ways for locating interesting datasets. Throughout this session, we will explore the practical uses of sequence annotation, the relationship between disease genes, and the advancement of novel pharmaceuticals.

We can use Machine Learning for analyzation of many task like RNA sequencing (RNAseq) expression data, chromatin accessibility data from DNase I hypersensitive site sequencing (DNase-seq) Libbrecht and Noble (2015) [7]. Many techniques of machine learning entails the capacity described in study Libbrecht and Noble in 2015. One example of this is the work that Ohler and his colleagues have done. Understanding the role of non-coding DNA in the genome (Algama et al., 2017) [9] and locating the precise placements of transcription start sites (TSSs) [8] are two important steps in the biological process. According to study [10] Kummerfeld et. al described the methods of binding defined by transcription factors and transcription factors are the most important controller of gene expression. The study [11] of Hamelryck's work simply analyzed and compared bio macromolecules. This work has capability to get many tasks like synthetic proteins or enzymes and sequencing of molecules.

Machine learning has been employed to research and comprehend the molecular systems that are responsible for hereditary disorders and susceptibilities. This has been accomplished through the move from descriptive applications to interpretative domains. There is a rising knowledge that the intricate interactions that take place between genes and environmental variables have a key role in the development of diseases that are prevalent in humans. This is the reason why this is the case. Typical statistical methods are not the most effective means of discovering interactions of this kind, particularly in instances where the interactions involve more than two genes or when the data is high-dimensional (containing a number of features or independent components). This is especially true in situations when the data is high-dimensional. McKinney and his colleagues conducted research in the year 2006 [12] that gave an explanation for the detection and analysis of a wide variety of illnesses, including RF, MDR, ANNs, and CAs. This research was carried out in order to uncover the causes of these illnesses. This explanation was offered by the data that was acquired from the research to deliver the information. Applications of machine learning have been found in a wide variety of fields, including the study of proteomics, which is also one of the many other fields in which machine learning has been found to have applications.

Proteomics is the primary focus of research in the field of medicine. Proteomics research is utilised in the process of drug discovery for a variety of reasons, including the identification and validation of targets, the discovery of biomarkers for assessing effectiveness and toxicity using readily accessible bodily fluids, and the investigation of the mechanisms that are responsible for therapeutic action or toxicity.

This article provides a comprehensive introduction to both supervised learning and unsupervised learning, which are the two essential approaches in machine learning. Both of these approaches are discussed in detail. A comprehensive discussion is provided for both of these methods. Not only do we investigate predictive models, but we also investigate generative techniques. These techniques were discussed earlier. These models are designed to arrive at a single result for each sample, which is their primary purpose. One of the methods that are included in these approaches is the training of a model to generate unique data that includes features that are comparable to those of the training data. This is one of the techniques that comprise these techniques. Deep Learning, which is sometimes considered to be one of the most significant developments in the field of Machine Learning, is the subject of our final section, which is surely not the least of our sections. To begin, we will discuss the procedures for training and the techniques of quality control that will be implemented immediately. Despite this, we will begin by discussing these procedures. It is something that we would want to propose to the reader that they take a look at the review that was done out by Libbrecht and Noble (2015), which is particularly mentioned as study [13]. The use of this strategy will result in the provision of an explanation that is more detailed. Training a model with data that has been labelled is an example of supervised learning, which is a technique that is utilised in the field of machine learning. The fact that this is the case implies that the data that is included in the input is accompanied by labels that correspond to the output information. Supervised learning is the term used to describe the process of creating a model through training.

A procedure known as "supervised learning" involves training datasets with known labels or targets for each sample, such as disease status or category. This process is referred to as "supervised learning." The process of learning that is being described here is known as "supervised learning." Imagine a situation that is analogous to the one in which a student and a teacher engage, in which the student acquires information by receiving feedback on the accuracy of their responses. In this scenario, the student gathers information. A supervised model makes use of the specific information that is provided in the training set in order to gain knowledge and construct an all-encompassing model that can subsequently be utilised to the task of predicting labels for new data. This is how the model is able to acquire knowledge and develop an all-encompassing model. This is carried out in a manner that is comparable to the example that was presented earlier. You can classify the categories as either categorical, like ethnicity, or continuous, like survival rates. Both of these classifications are possible. Both of these classes are acceptable options. This section contains explanations of both types of labels that are available. Classification is concerned with the circumstance that occurred earlier, whereas regression is concerned with the scenario that occurred later. The significance of both of these scenarios cannot be overstated.

When referring to the inputs that the model is given, the term "features" is typically an appropriate term to use. The term "features" refers to particular kinds of data components that are responsible for providing information about the sample. There are a number of characteristics that can be discovered in a human genome, including expression profiles and genotyping profiles are two examples. When we talk about "supervised learning," we are referring to a technique that involves the process of developing a structured approach to transfer data from a collection of attributes to a set of goals or labels. Supervisory learning is a method that involves the process of transferring data and simply enough, this is what is meant when we talk about "supervised learning." The term "feature selection" refers to one of the subcategories that are included below the overarching category of "supervised learning." Finding the collection of features that have a high link with the labels is the goal in order to get insights into the underlying mechanisms (such as biological insights). This can be accomplished on a number of different levels. Through the process of locating the set of qualities, this can be accomplished. Additionally, in order to get a higher level of accuracy, the goal is to reduce the number of features that are either redundant or noisy.

Supervised machine learning involves developing algorithms that can generate broad patterns and hypotheses by utilizing provided examples to forecast the outcomes of future examples. Supervised machine learning classification methods are designed to categorize data based on existing information. Data science concerns often involve frequent classification tasks. The study [14] described many machine learning methods that are on the basis of for stochastic techniques, instance-based, logic-based and rule based and so many more.

4. UNSUPERVISED LEARNING

The term "unsupervised learning" refers to a type of machine learning that is able to acquire information from data without the requirement for human supervision. In unsupervised learning, we make use of data that has not been tagged, and then we allow the model the freedom to act on the knowledge it has been presented with without providing it with any guidance or direction. This method of learning is frequently referred to as "unsupervised learning." It is possible that we may use the example of Smart Kid to exemplify this concept in order to make this argument more accessible. The provision of any kind of instruction or course of action to him is not required in any given circumstance. For the purpose of this particular type of machine learning, the model does not receive any data that has been tagged in any way.

Clustering and dimensionality reduction are the two primary types of processes that are utilised the most frequently in the context of unsupervised learning applications. It is important to note that both of these procedures are fundamental. Dimensionality reduction and clustering are two other methods that can be utilised. As a consequence of the findings of the study [33], the researchers came to the realization that a great number of individuals conceal patterns of knowledge, such as urban studies. Exploratory data analysis is the most prevalent application of unsupervised learning. This is due to the fact that there are a substantial number of various kinds of datasets that could potentially benefit from this kind of study. Taking into consideration that we are talking about unsupervised learning, the healthcare industry is one of the most important fields. Unsupervised learning allows us to evaluate groups of patients and the responses they have to treatment. This allows us to better understand how patients respond to treatment. We are able to easily carry out analysis as a result of the fact that unsupervised learning does not necessitate the use of any predetermined labels [34]. It is now much simpler to recognise particular drugs thanks to the clustering of the data. The incorporation of unsupervised learning into the research process has become extremely important as a consequence of the genetic information [35]. When it comes to the procedure of assisting gene expression patterns, it becomes useful after receiving assistance. The findings of these investigations will be helpful in the field of medicine for the extraction of a wide variety of unlabeled data, as well as in the comprehension of diseases and the development of novel drugs. These findings will also improve the knowledge of diseases. The basic model of unsupervised learning has given below. In this we can easily illustrate its two type named Clustering and Association.

Clustering: The concept of assessing distance or similarity is central to clustering. Clustering techniques use Euclidean distance, cosine similarity, and correlation coefficients to measure data point similarity. Data variables and research goals often determine the similarity measure.

In general, clustering can be divided into two groups: unsupervised and supervised techniques. Both of these types are described below. Through the use of unsupervised clustering, the data points are clustered together. This method does not require any prior knowledge or labelling of the types or categories in order to function properly. Discovering previously undisclosed data structures and groupings that have been hidden is the objective of this project. Clustering is used to search for new patterns within a tagged collection. This is the most fundamental application of clustering there is. Clustering data is done with the intention of determining the similarities and differences that exist between the clusters that are being studied. In terms of the attributes, the cluster objects that are identical to one another are shared with one another simultaneously. In the context of data analysis, the term "analysis of cluster" refers to the process of grouping with other clusters, which is a component of the analysis process. Clustering is primarily aimed at achieving the statistical approach, and it is just this methodology that enables us to achieve the primary purpose. Data disclosure, which includes the discovery and identification of patterns, is an essential component of clustering. Clustering is a crucial feature of data analysis.

Association: Essentially, association is the process of utilizing a rule-based strategy in order to locate intriguing correlations between data bits and that database, which are stored within vast databases. This is done in order to find relationships between the data bits and the database. In data mining and knowledge discovery, association includes detecting patterns or linkages between variables. Find co-occurrence or correlation patterns between things, events, or traits. This concept helps data analysis, market basket analysis, and recommendation algorithms understand massive datasets. Association involves finding rules or links that explain dataset item relationships. Rules are often "if-then" formulations that presume one object is present or absent. Most relationships are found using association rule mining, which uncovers intriguing data links.

Common association rule metrics include dataset co-occurrence support. Support finds interesting, crucial rules. Another important statistic is confidence, which measures item association strength. High confidence suggests items are tightly connected, making the rule more dependable. Many fields benefit from association rule mining. Identifying frequently purchased items helps merchants organize and market merchandise. Patient data association analysis can discover risk factors and disease conditions. Web mining uses user likes and behaviour to improve recommendation systems. Apriori and FP-growth algorithms mine association rules. User-defined support and confidence thresholds let these algorithms process large datasets and extract links. The majority of the time, it is employed for the aim of assessing retail baskets or transactional records in order to provide an accurate representation of the frequency with which specific items are purchased together. Using these algorithms, client purchasing habits as well as connections between things that were previously hidden are identified. This information is then used to inform recommendation engines or other opportunities for cross-selling. You might be most familiar with these principles from the "Frequently bought together" and "People who bought this item also bought" sections of the online retail site that you choose to shop at.

The integrity of the attributes that were contained in the initial data is not compromised in any way, which allows this to be achieved easily. The main focus is to find pattern and data [15]. In deep learning for learning features without supervision we use machine learning. On the other hand, these learned characteristics are subsequently utilised for actions like as classification and other behaviors that are linked [16].

The rapid progress in deep neural networks, the widespread availability of extensive computing power through cloud computing and distributed computing, and the capacity to store and analyze vast amounts of data have sparked a growing interest in utilizing unsupervised machine learning techniques in the networking domain. The networking industry is very compatible with and receptive to the use of unsupervised machine learning techniques. This is primarily owing to the decentralized decision-making structure of its protocols, the abundance of network data, and the pressing demand for intelligent and cognitive networking. Learning using a small number of training examples:

5. FEW SHOTS LEARNING

The word "FSL" is used in the field of machine learning to refer to a specific group of problems that are differentiated by the variables E, T, and P which are used to describe the problems. (17) [17] In the context of this discussion, the letter "E" refers to a collection of examples that offer a limited amount of supervised information for the variable that is the primary focus of the inquiry. The obstacles that are most commonly encountered in the process of learning a foreign language are typically related with difficulties in supervised learning. To be more explicit, the term "few-shot classification" refers to the method of training classifiers by using a limited number of samples that have been labelled for each category during the training process. A few examples of applications that have been put into practice include the categorization of images [18], the analysis of sentiments expressed in short language [19], and the recognition of objects [20]. Training a classifier h that is capable of accurately predicting the label yi for each input xi is the formal definition of few-shot classification. This training is required in order to get the desired results. It is in accordance with the notations that are provided in Section 1.2 that this is accomplished. In the majority of instances, the N-way-K-shot categorization [21, 18] is taken into consideration. I = KN instances from N classes make up the D train in this classification, with each class having K examples. In other words, the D train is made up of I = KN instances. [21, 22] Few-shot regression is a method that estimates a regression function h by making use of only a limited number of input-output example pairs that are sampled from that function. This technique is referred to as "regression sampling." The value that was observed for the dependent variable y is reflected in the output yi, whereas the value that was seen for the independent variable x is represented by the input xi. In the context of this discussion, the output represents the value of the variable that is dependent. Few-shot reinforcement learning is yet another technique that falls within the category of FSL [23, 24]. In addition to the few-shot supervised learning mode, this sort of learning is also available. This approach focuses on discovering a policy using only a small number of trajectories that contain state-action pairs. The FSL scenario refers to situations where obtaining instances with supervised information is challenging or impossible due to concerns related to privacy, safety, or ethics. An illustrative instance is drug development, a process that aims to unveil the characteristics of novel compounds in order to identify valuable ones as potential medications [25]. Given their potential toxicity, limited efficacy, and low solubility, novel compounds lack substantial biological data as clinical candidates. Therefore, it is crucial to acquire knowledge efficiently from a limited number of samples. Other instances where the desired actions have limited examples are found in FSL translation [65] and cold-start item suggestion [26]. By utilizing FSL, it becomes feasible to acquire appropriate models for these exceptional instances. FSL has made significant contributions not only in the field of bioinformatics but also in many video applications, such as motion prediction, video categorization, action localization, human re-identification, event detection, and object segmentation.

FSL can also alleviate the task of gathering extensive supervised data. For instance, while ResNet surpasses human performance on Image Net, it is necessary to have an adequate number of labeled photos for each class, which can be a time-consuming task to accomplish. FSL has the potential to decrease the amount of data collection required for applications that rely heavily on data. There are a number of tasks that are included in the scope of computer vision and natural language processing. These tasks include image classification [27], image retrieval [28], object tracking [29], gesture recognition [30], image captioning, visual question answering [31], video event detection, language modeling [27], and neural architecture search [32]. This list is not exhaustive.

6. USE OF FEW SHOT LEARNING

The data science domains that are currently experiencing a surge in popularity, such as computer vision, robotics, and many others, have a wide variety of applications for few-shot learning. The methods of character identification, image recognition, and classification can all be accomplished with their assistance. In certain applications of natural language processing (NLP), such as translation, text categorization, sentiment analysis, and so on, they perform admirably. It is possible to employ them in the field of robotics to train robots using a limited number of training sets.

6.1 Example of Zero-Shot Learning

Few-shot learning is a method that involves training a function that predicts similarity between two sets of data. The similarity function is denoted by the notation sim (x, x'). The degree of similarity between the two samples, x and x', is evaluated by this method. The similarity function will return 1 if two samples are identical; this is denoted by the expression sim(x, x')=1. In the event that the samples are distinct, they will return 0. In other words, the equation sim(x, x')=0

Following training, the gained similarity function can be utilised for the purpose of creating predictions for inquiries that have not yet been seen. We are able to obtain the similarity scores by utilizing the similarity function, which allows us to compare the query with each and every sample that is included in the support set. First, locate the sample that has the highest similarity score, and then utilize that sample as the basis for the forecast.

7. CONCLUSION

In the current era of big data, deep learning has become the focal point of global academic and business attention. In the field of bioinformatics, standard machine learning algorithms have achieved significant progress. However, there is also strong anticipation for deep learning to yield good outcomes. This paper provides a comprehensive analysis of bioinformatics research that utilizes deep learning. The analysis focuses on the input data, research aims, and key features of commonly employed deep learning architectures. In order to provide a comprehensive review, we engaged in a detailed discussion regarding potential concerns and areas for further research.

While deep learning may seem highly promising, it is not a universal solution and does not yield exceptional outcomes when directly implemented in the field of bioinformatics. However, there are several issues that need to be taken into account, including the constraints of restricted data quantity and class imbalance, the interpretation of deep learning outcomes, and the selection of the suitable architecture and its hyper parameters. In addition, in order to fully utilize the potential of deep learning, it is crucial to focus on two interesting areas for additional investigation: multimodality and acceleration of deep learning. Therefore, we assert that cautious preparations concerning the matters addressed in the review are crucial for achieving success in such study. This review aims to offer helpful insights and serve as a catalyst for researchers to integrate deep learning into their bioinformatics investigations.

REFERENCES

Al-Shedivat, M., Bansal, T., Burda, Y., Sutskever, I., Mordatch, I., & Abbeel, P. 2018. Continuous adaptation via metalearning in nonstationary and competitive environments. In *International Conference on Le*. IEEE.

Algama, M., Tasker, E., Williams, C., Parslow, A. C., Bryson-Richardson, R. J., & Keith, J. M. (2017). Genome-wide identification of conserved intronic non- coding sequences using a Bayesian segmentation approach. *BMC Genomics*, *18*(1), 1–14. doi:10.1186/s12864-017-3645-2 PMID:28347272

Altae-Tran, H., Ramsundar, B., Pappu, A. S., & Pande, V. (2017). Low data drug discovery with one-shot learning. *ACS Central Science*, *3*(4), 283–293. doi:10.1021/acscentsci.6b00367 PMID:28470045

Asai, K., Hayamizu, S., & Handa, K. (1993). Prediction of protein secondary structure by the hidden Markov model. *Computer Applications in the Biosciences*, *9*(2), 141–146. doi:10.1093/bioinformatics/9.2.141 PMID:8481815

Backofen, R., & Gilbert, D. (2001). Bioinformatics and constraints. *Constraints*, *6*(2/3), 141–156. doi:10.1023/A:1011477420926

Bertinetto, L., Henriques, J. F., Valmadre, J., Torr, P., & Vedaldi, A. (2016). Learning feed-forward one-shot learners. In Advances in Neural Information Processing Systems.

Brock, A., Lim, T., Ritchie, J. M., & Weston, N. (2018). SMASH: One-shot model architecture search through hypernetworks. In *International Conference on Learning Representations*.

Deng, L. (2014, May). A tutorial survey of architectures, algorithms, and applications for deep learning. *APSIPA Transactions on Signal and Information Processing*, *3*(1), 1–29. doi:10.1017/atsip.2013.9

Dong, X., Zhu, L., Zhang, D., Yang, Y., & Wu, F. (2018). Fast parameter adaptation for few-shot image captioning and visual question answering. In *ACM International Conference on Multimedia*, (pp. 54–62). ACM. 10.1145/3240508.3240527

Duan, Y., Andrychowicz, M., Stadie, B., Ho, J., Schneider, J., Sutskever, I., Abbeel, P., & Zaremba, W. 2017. One-shot imitation learning. In Advances in Neural Information Processing Systems.

Eckhardt, C. M., Madjarova, S. J., Williams, R. J., Ollivier, M., Karlsson, J., Pareek, A., & Nwachukwu, B. U. (2023). Unsupervised machine learning methods and emerging applications in healthcare. *Knee Surgery, Sports Traumatology, Arthroscopy : Official Journal of the ESSKA*, *31*(2), 376–381. doi:10.1007/s00167-022-07233-7 PMID:36378293

Fei-Fei, L., Fergus, R., & Perona, P. (2006). One-shot learning of object categories. *IEEE Transactions on Pattern Analysis and Machine Intelligence*, *28*(4), 594–611. doi:10.1109/TPAMI.2006.79 PMID:16566508

Finn, C., Abbeel, P., & Levine, S. (2017). Model-agnostic meta-learning for fast adaptation of deep networks. In *International Conference on Machine Learning*. ACM.

Hamelryck, T. (2009). Probabilistic models and machine learning in structural bioinformatics. *Statistical Methods in Medical Research*, *18*(5), 505–526. doi:10.1177/0962280208099492 PMID:19153168

Kumar, I., Singh, S. P., & Shivam. (2021). Machine learning in bioinformatics. *Bioinformatics (Oxford, England)*, 443–456. doi:10.1016/B978-0-323-89775-4.00020-1

LeCun, Y., Bengio, Y., & Hinton, G. (2015, May). Deep learning. *Nature*, *521*(7553), 436–444. doi:10.1038/nature14539 PMID:26017442

Libbrecht, M., & Noble, W. (2015). Machine learning applications in genetics and genomics. *Nature Reviews. Genetics*, *16*(6), 321–332. doi:10.1038/nrg3920 PMID:25948244

Libbrecht, M. W., & Noble, W. S. (2015). Machine learning applications in genetics and genomics. *Nature Reviews. Genetics*, *16*(6), 321–332. doi:10.1038/nrg3920 PMID:25948244

Lilhore, U. K., Dalal, S., Faujdar, N., Margala, M., Chakrabarti, P., Chakrabarti, T., Simaiya, S., Kumar, P., Thangaraju, P., & Velmurugan, H. (2023). Hybrid CNN-LSTM model with efficient hyperparameter tuning for prediction of Parkinson's disease. *Scientific Reports*, *13*(1), 14605. doi:10.1038/s41598-023-41314-y PMID:37669970

Lilhore, U. K., Dalal, S., Faujdar, N., Simaiya, S., Dahiya, M., Tomar, S., & Hashmi, A. (2024). Unveiling the prevalence and risk factors of early stage postpartum depression: A hybrid deep learning approach. *Multimedia Tools and Applications*, 1–35. doi:10.1007/s11042-024-18182-3

Lilhore, U. K., Manoharan, P., Sandhu, J. K., Simaiya, S., Dalal, S., Baqasah, A. M., Alsafyani, M., Alroobaea, R., Keshta, I., & Raahemifar, K. (2023). Hybrid model for precise hepatitis-C classification using improved random forest and SVM method. *Scientific Reports*, *13*(1), 12473. doi:10.1038/s41598-023-36605-3 PMID:37528148

Lilhore, U. K., Poongodi, M., Kaur, A., Simaiya, S., Algarni, A. D., Elmannai, H., Vijayakumar, V., Tunze, G. B., & Hamdi, M. (2022). Hybrid model for detection of cervical cancer using causal analysis and machine learning techniques. *Computational and Mathematical Methods in Medicine*, *2022*, 2022. doi:10.1155/2022/4688327 PMID:35572826

Lilhore, U. K., Simaiya, S., Pandey, H., Gautam, V., Garg, A., & Ghosh, P. (2022). Breast cancer detection in the IoT cloud-based healthcare environment using fuzzy cluster segmentation and SVM classifier. In Ambient Communications and Computer Systems [Singapore: Springer Nature Singapore.]. *Proceedings of RACCCS, 2021*, 165–179.

Lilhore, U. K., Simaiya, S., Prasad, D., & Guleria, K. (2020). A hybrid tumor detection and classification based on machine learning. *Journal of Computational and Theoretical Nanoscience*, *17*(6), 2539–2544. doi:10.1166/jctn.2020.8927

Luscombe, N., Greenbaum, D., & Gerstein, M. (2001). *What is bioinformatics? An introduction and overview.* IMIA. doi:10.1055/s-0038-1638103

McKinney, B. A., Reif, D. M., Ritchie, M. D., & Moore, J. H. (2006). Machine Learning for Detecting Gene-Gene Interactions. *Applied Bioinformatics*, *5*(2), 77–88. doi:10.2165/00822942-200605020-00002 PMID:16722772

Ohler, U., Liao, G. C., Niemann, H., & Rubin, G. M. (2002). Computational analysis of core promoters in the Drosophila genome. *Genome Biology*, *3*(12), 1–12. doi:10.1186/gb-2002-3-12-research0087 PMID:12537576

Pfister, T., Charles, J., & Zisserman, A. (2014). Domain-adaptive discriminative one-shot learning of gestures. In *European Conference on Computer Vision.*

Singh, A., Thakur, N., & Sharma, A. (2016). *A review of supervised machine learning algorithms.* 2016 3rd International Conference on Computing for Sustainable Global Development (INDIACom), New Delhi, India.

Tang, S., Lomsadze, A., & Borodovsky, M. (2015). Identification of protein coding regions in RNA transcripts. *Nucleic Acids Research*, *43*(12), e78–e78. doi:10.1093/nar/gkv227 PMID:25870408

Triantafillou, E., Zemel, R., & Urtasun, R. 2017. Few-shot learning through an information retrieval lens. In Advances in Neural Information Processing Systems. 2255–2265.

Trifonov, E. N. (2000). Earliest pages of bioinformatics. *Bioinformatics (Oxford, England)*, *16*(1), 5–9. doi:10.1093/bioinformatics/16.1.5 PMID:10812471

Vartak, M., Thiagarajan, A., Miranda, C., Bratman, J., & Larochelle, H. 2017. A meta-learning perspective on cold-start recommendations for items. In Advances in Neural Information Processing Systems.

Vinyals, O., Blundell, C., Lillicrap, T., & Wierstra, D. (2016). Matching networks for one shot learning. In Advances in Neural Information Processing Systems.

Wang, J., & Biljecki, F. (2022). Unsupervised machine learning in urban studies: A systematic review of applications. *Cities (London, England)*, *129*, 103925. doi:10.1016/j.cities.2022.103925

Wang, Y., Yao, Q., Kwok, J., & Ni, L. M. (2019). Generalizing from a Few Examples: A Survey on Few-Shot Learning. *ArXiv.* /abs/1904.05046

Yoon, J., Kim, T., Dia, O., Kim, S., Bengio, Y., & Ahn, S. (2018). Bayesian model-agnostic meta-learning. In Advances in Neural Information Processing Systems, (pp. 7343–7353). IEEE.

Chapter 18
Learning From Scarcity: Unlocking Healthcare Insights With Few-Shot Machine Learning

Pooja Dixit
Manipal University Jaipur, India

Advait Vihan Kommula
Tek-zo Technologies, USA

Pramod Sing Rathore
https://orcid.org/0000-0002-4917-1500
Manipal University Jaipur, India

ABSTRACT

This chapter delves into the transformative synergy of few-shot learning and healthcare, elucidating its impact on medical procedures. Anchored in machine learning fundamentals, it establishes a core framework through a review of algorithms. Addressing challenges of small healthcare datasets, the chapter highlights the pivotal role of few-shot learning. Innovative methods like multimodal integration and federated learning enhance model robustness, offering insights into complex healthcare scenarios. Formal mathematical explanations categorize few-shot learning challenges, opening avenues for a deeper understanding and implementation in medical imaging.

1. INTRODUCTION

The medical field has endured a change because of the deployment of machine learning (ML) techniques, which have altered methods for developing new medications, planning treatments, and diagnosing illnesses. A particular area of machine learning known as few-shot learning (FSL) is applied in situations when there is a shortage of data. It seeks to attain acceptable learning performance despite the training set's small amount of supervised information The input samples in this training set have labels for the

DOI: 10.4018/979-8-3693-1822-5.ch018

matching outputs. As a result, several methods, such as transfer learning, neural networks, and meta-learning, have been developed to improve FSL Wang et. al.(2020).

For supervised learning tasks, a variety of data formats can be used with Few-Shot Learning (FSL). Among them are: (i) Classification: This FSL technique, referred to as N-way-K-shot, counts the number of samples (K) as well as the number of classes (N). On the other hand, one-shot learning (OSL) is employed when there is only one example available, and zero-shot learning (ZSL) is utilised when none are available Qiao et. al. (2017); (ii) Regression: Tasks involving FSL can also be utilised involving regression, in which the objective is to estimate a function using a small sample of input-output pairs this instance, the observed values of the dependent and independent variables are represented in the output and recorded in the input, respectively Zell et. al. (2022). Additionally, FSL is applied to half-supervised or unsupervised learning by weight-orienting prior knowledge and predictions between tasks through reinforcement learning Wang & Chen (2020). The idea of an empirical risk minimizer that lacks reliability is central to the taxonomy of financial literacy. Within this paradigm, the empirical risk—which comes from the model's bad decision-making error decomposition—is used to minimise hypothesis h. The empirical risk can approximate the predicted risk when there are enough supervised samples available, which will result in a good performance and precise predictions Lake et. al. (2015). Numerous techniques have been devised to tackle this issue, and they can be divided into various categories based on how they employ past information. According to Wang and coworkers Wang et. al.(2020). The following categories apply to currently available FSL works: viewpoints depending on which element is increased utilising previous knowledge: (i) Data, by utilising related, labelled, and unlabeled data to use historical information to increase and improve the number of samples can be utilised to leverage the limited amount of data available; (ii) model, which is trained on problems akin to the target problem to gain transferable knowledge and enhance its generalisation abilities; the model, which is constrained by prior knowledge to produce a more condensed hypothesis space between expected and empirical risk; and (iii) When using algorithms particularly created for FSL, predictions can be further refined or It is possible to create new algorithms that can generalise predictions with minimal data. The algorithm, where prior knowledge changes the search method to better reflect empirical and projected risk (Fig. 1) Silva-Mendonça et. al. (2023).

Three essential methods to enhance few-shot prediction accuracy are shown in Fig. 1. First, low data can be supplemented with connected, tagged, and unlabeled data. Secondly, the model can generalise a limited set of data by utilizing insights from similar activities. Finally, an algorithm can be changed for instances with little or no data to improve forecast generality.

A brief introduction to few-shot learning (FSL) theory and practice in the healthcare industry is given, along with an analysis of the challenges and opportunities that machine learning (ML) presents for improving and altering healthcare procedures. A thorough summary of the huge influence of machine learning in healthcare is given in this introduction.

1.1 Significance of Machine Learning in Healthcare

Computer-Assisted Drug Design (CADD) is a paradigm change in drug research that has been spurred by advancements in digitalization, artificial intelligence (AI), and computers. Machine learning (ML) and deep learning (DL) in particular have emerged as excellent AI tools, beating humans in tasks like as drug target identification and the generation of novel drug candidates. Drug discovery has been hastened by a variety of data sources and processing capacity, and machine learning has shown beneficial in solving data-scarce circumstances such as neglected diseases. Few-shot learning (FSL) is one such approach, which allows machine learning (ML) models to draw insights from little amounts of data rather than large datasets. This study looks into the use of FSL in drug development, focusing on its ability to respond swiftly to tasks in the face of limited data, which can lead to faster procedures and higher success rates Silva-Mendonça et. al. (2023).

Machine learning (ML) is essential for precise and resource-efficient disease diagnosis, especially in hypertensive retinopathy (HR). The purpose of this research is to improve computerized HR detection through the use of a unique Few-Shot Learning technique. To combat over fitting on short datasets, the model employs pretrained baseline models for feature embedding, as well as a Convolution Neural Network (CNN) and LSTM-based architecture. To increase generalization, a novel strategy for dynamically creating classification weight vectors for few-shot datasets is proposed. Experiments with publicly available datasets validate the proposed Few-Shot Learning approach in HR detection, revealing its potential Suman et. al. (2023).

1.2 Challenges and Opportunities in Machine Learning for Healthcare

Healthcare has greatly benefited from machine learning (ML), but there are still issues, especially with regard to security and privacy when working with personal data. A key advancement in tackling these issues is Federated Few-shot Learning (FedFSL). In order to protect privacy during cooperative training, Federated Learning (FedL) is employed, producing a globally applicable model that may be used by any client. Modifications to the FedL (CSFedL) and WFedL based on client selection are suggested to help with issues such client performance discrepancies and malicious involvement. A further validation demonstrates the improved performance of Client Selection based WFedFSL (CSWFedFSL). Federated studies highlight the benefits of WFedL, CSFedL, and their combination (CSWFedL). Notably, our approach enhances client privacy protection in one-off situations, significantly contributing to opportunities and resolving issues in the field of healthcare machine learning Xu et. al. (2022). The problem of having enough labelled training data for machine learning and deep learning applications is addressed by zero-shot learning (ZSL), a recommended strategy in requirements engineering (RE). ZSL does this by doing away with the necessity for labeled data. Using contextual word embeddings and transformer-based language models, this novel method correctly categorizes functional terms Alhoshan et. al. (2023).

A recent study recommends an Internet of Things (IoT) solution for real-time electric appliance monitoring in response to the rising demand for precise consumption data in home and building automation. The study highlights the challenges of achieving real-time identification with fewer training datasets, emphasising the importance of cost-effective and efficient data collection in large-scale applications. This is achieved by comparing machine learning algorithms using publicly available datasets with low sampling frequency. The developed IoT system attempts to satisfy these goals by providing affordability and real-time performance for low-powered IoT applications, as well as enabling appliance identification in smart environments Mihailescu et. al. (2020).

A novel AI approach seeks to systematically reduce classification mistakes over time by integrating ongoing enhancements together with measurable assurances. This approach, unlike earlier techniques, allows for the creation of very efficient few-shot AI correction algorithms that are characterised by a linear training complexity and low-cost classification calculations. It can be employed for problem detection and the removal of false positives in industrial settings. Its scalability is demonstrated in embedded contexts with limited computational resources. Its foundations are in the concentration of measure effects and stochastic separation theorems Tyukin et. al. (2021).

The study presents ML-ZSLPGCN, a new technique for multi-label zero-shot categorization. . A Pairnorm-based Graphical Convolution Network is used in the procedure., label features from pictures for training and semantic embedding for labels that are not visible. It makes use of a pairnorm-based normalisation approach to address oversmoothing concerns. In comparison to state-of-the-art methods, experimental findings on the NUSWIDE and MS-COCO datasets show a considerable improvement in performance, with improved memory, accuracy, and F1 score. This method offers a strong remedy for problems with Zero-shot learning in classification tasks with multiple tasks Chauhan & Tiwari (2022).

The problems posed by limited data availability in Deep Learning (DL) applications for Remote Sensing (RS) are the main focus of this work, especially in important socio-environmental domains. Ten cutting-edge deep learning (DL) approaches, including transfer learning and self-supervised learning, are presented in this work. The study stresses that challenges with weak model generalizability and transferability are common. The authors highlight the importance of these strategies in reducing data restrictions and offer a flowchart to make it easier to choose the best approaches in a particular situation. Despite the dearth of reference data, the goal of this thorough assessment is to give engineers and academics the tools they need to use deep learning to efficiently address environmental concerns in the remote sensing sector Safonova et. al. (2023).

This study offers a comprehensive overview of automatically generated machine learning (AutoML) in the healthcare sector, with the goal of enhancing accessibility for medical practitioners with less background knowledge in data science. Examining 101 articles and their accompanying code, the evaluation demonstrates that AutoML techniques can approach or surpass human capabilities in certain machine learning tasks, often more effectively. While acknowledging the potential of machine learning to improve health outcomes and reduce healthcare costs, the research also emphasizes the challenges associated with scaling the technology for use in settings beyond small and medium-sized datasets. In order to automate model composition, selection, and parameterization and promote wider usage in the healthcare industry, autoML is recognized as a critical solution. The study highlights the continued need for additional efforts to enable widespread integration of AutoML in healthcare settings, despite encouraging use cases Waringet et. al. (2020).

This study explores the use of machine learning (ML) and artificial intelligence (AI) in tracking mental health via online social networks. Although the focus of recent developments has been on complex models such as Transformer-based Large Language Models (LLMs), the study tackles issues about the models' opacity in the adoption process within the healthcare industry. The research uses model-agnostic, post hoc surrogate eXplainable AI (XAI) approaches like SHAP and LIME to improve trust and understanding. The effectiveness of XAI approaches in interpreting decisions is demonstrated by extensive experimentation with four datasets and six pretrained LLMs, including domain-adapted models for mental health diagnoses. The results emphasise the significance of include explainability in Transformer-based LLMs for thorough mental health monitoring, and they recommend investigating unsupervised methods such as BERTopic for situations involving annotation difficulties Malhotra & Jindal (2023).

This survey critically looks at the problems in healthcare caused by low-resolution photos produced by Internet of Things devices. Clinical decision-making can be influenced by low-resolution image biases, even with the potential advantages of deep learning and IoT integration. Setting itself apart from prior surveys, it addresses research gaps carefully inspects cutting-edge models, and highlights obstacles to attaining the best possible medical image restoration. The survey gives priority to precise

Table 1. Literature review

Paper	Methodology	Advantages	Disadvantages/Limitations
Suman et. al. (2023)	focuses on NAS while conducting a semantic study of AutoML. investigates data processing for AutoML and assesses NAS using several datasets. talks about NAS techniques and future research avenues..	increases output by automating the model-building process. NAS searches architectures effectively. explores potential avenues for NAS research.	little talk of the difficulties. gives no specifics about how to solve open challenges.
Xu et. al. (2022)	Machine Learning-based examination for Video Capsule Endoscopy (VCE). creation of a framework that combines the clinical relevance graph, taxonomy hierarchy, and raw frames. An analysis of the current VCE techniques.	automated analysis of VCE. Using a taxonomy hierarchy will improve classification. A clinical relevance graph for mapping at the illness level.	little research on the clinical implications. Studies frequently concentrate on finding a single abnormality. issues with bias, class imbalance, and sparsity of the data.
Alhoshan et. al. (2023)	For the three-layer architecture of IoMT, ML-Based IDS.	Enhanced security in addition to conventional safeguards.	Variability, scarce resources, and novel assaults.
Mihailescu et. al. (2020)	Machine learning integration with electrohydrodynamic (EHD) processes.	Accelerate discoveries and streamline the EHD process.	Application of ML for fabrication technologies is lagging.
Tyukin et. al. (2021)	Statistical approaches and DL techniques are utilized in clinical research to apply ML and DL models. Statistical approaches and DL techniques are applied in clinical research to apply ML and DL models.	increased accuracy, dependability, and efficiency.	problems with the unknown data volume.
Chauhan & Tiwari (2021)	Few Shot Learning is used by AFSL-HAR to facilitate effective model adaptation. With a feature extractor and an enhanced discriminator, the FWGAN module is meant to synthesise a variety of samples and increase system performance.	Using a small number of samples for each category on both publicly available and privately created datasets, AFSL-HAR demonstrates high accuracy (98.9% and 94.7%) in identifying novel activities, offering a practical method for model modification.	There was little discussion of AFSL-HAR's scalability and resource needs, and it wasn't investigated whether or not the FWGAN module could be implemented practically in real-world situations.
Safonova et. al. (2023)	AI and ML integration of DT and XR technologies in healthcare.	development of sophisticated 3D patient models, enhancing medical interventions.	The creation of precise 3D models and the smooth integration of DT and XR technologies could present technological difficulties that call for technical know-how and troubleshooting.
Waring et. al. (2020)	DL model pre-training on unlabeled sensor data is conducted via self-supervised learning. Pay attention to strain gauge data's RUL estimation for aerospace fatigue. Artificial datasets evaluate the accuracy of predictions.	Pre-training the model effectively, improving RUL prediction significantly, and using less computing power than with untrained models.	Industrial application challenges, an emphasis on synthetic datasets, and possible restrictions on extrapolating findings to different systems.
Malhotra& Jindal (2023)	thorough examination of privacy, attacks, algorithms, and FL principles. examination of FL security, attack vectors, vulnerability sources, and privacy techniques. covers both fundamentals and cutting-edge research on privacy and threats in Florida.	Florida uses decentralised learning to protect privacy. Survey enhances awareness of security, privacy, and potential assaults by consolidating FL information.	Investigation of FL security implications is necessary; there may be flaws in current algorithms. Dealing with security risks and guaranteeing regular uptake in IoT applications are challenges.
Nayem et. al. (2023)	Methodology for the survey that includes a thorough analysis of FL backdoor attack strategies and defences. thorough explanation of all methods and analyses to improve comprehension.	gives a detailed analysis of Florida's backdoor attack landscape. insights about strategies, obstacles, safeguards, and next steps. increases knowledge and awareness of possible risks in FL.	limited investigation of FL backdoor attacks. Possibilities for weaknesses in handling new attack strategies. More investigation is required in FL to strengthen defences and handle new threats.
Liu et. al. (2022)	Few-shot learning combined with weakly supervised training using a multitask DCNN on 9 NCIP patients. Further fine-tuned with 291 labeled samples.	novel method that provides probability and lesion segmentation through few-shot learning. demonstrates promise as a COVID-19 diagnostic instrument.	Insufficient training data and possible gaps in the way COVID-19 variants are being addressed. Additional verification is required for varied demographics.

super-resolution techniques to enhance the quality of medical photographs and bolster the effectiveness of deep learning models in healthcare applications. It prepares the ground for improving medical image restoration for more useful healthcare applications by addressing its drawbacks and unanswered issues Silva-Mendonça et. al. (2023).

2. LITERATURE REVIEW

2.1 Foundations of Few-Shot Learning

Artificial intelligence relies heavily on few-shot learning and machine learning, which are essential for allowing computers to learn and predict. Below is an outline of their main concepts:

2.2 Few-Shot Learning

The new data is categorised using a few labelled datasets using the model's prior training using few shot learning, a machine learning model's structural representation and an instance of meta-learning. In the past, computer vision was intended to be the exclusive area of focus for FSL research. Even still, the lack of gathered benchmarks and natural language data is slowing down NLP growth. The use of few-shot learning in medical photos is receiving a lot of focus since domain-related technicalities, expressions, and coalition are creating challenges Here, previous knowledge is used to a restricted set of data in chronological order to train the prototype before it is effectually generalised. The data are then processed one step further for fine-tuning. Three steps use the prior knowledge: obtaining parameters, suppressing hypothesis space, and augmenting training data Snell et. al.

2.3 Crucial Few-Shot Learning Terms

In order to get the idea of few-shot learning, let's look at two sets ($C\ train \cap C\ test = \varphi$), where C train has more training datasets than C test (X_train^N,Y_train^N), but C test has relatively less training datasets. Class categories associated with clinical patterns might be studied in order to compute the medical domain's parallels from this information. For the purpose of generalising a model with fewer datasets, few-shot learning is the most efficient approach available. In FSL, k is the number of labeled samples that make up the n number of support sets, and n is the total number of classes that the trained model has generalisation capability. The task becomes more challenging the more 'Values there are; conversely, the task becomes more challenging the less k values there are. Additionally, one-shot learning is demonstrated for $k=1$ When $k=0$, zero-shot learning occurs. The training set of the model consists of n training instances and k classifications, where $n \ll N$ is a subset of the classes $n \times k << |Ctrain|$ that make up the training set. In Fig. 2, they are specified as the Sb is a support set. However there are more examples external of the Sb support set, including query set Qb, which makes use of the support set for data collection and feature extraction in order to make generalizations. It is made up of conventional and recent facts. The validation of the supervised learning algorithm examples depends equally on these two data sources consequently; changing the support set S b necessitates changing the inquiry set Qb. To estimate the query set φ, one needs get familiar with the following model parameters, $f\varphi: RD \rightarrow Rd$ ($D >> d$) Snell et. al. (2017).

2.4 Algorithm Based on Few Shot Learning Classification

Fig. 3 illustrates that the field of few-shot learning for medical image categorizations is a growth-oriented field. Based on their differences, inputs are grouped during the classification process. Accurate findings are obtained by classifying the labeled data using a similarity technique. This method focuses on finding and retrieving feature data with a smaller sample. Fewer data are used in the K-shot N-way method of few-shot categorization. Figure 3 summarizes the few-shot image classification techniques that are currently in use Zhang et. al. (2020).

2.5 Prototypical Networks

One kind of FSL model that has proven useful in many different situations is the pseudonym (PN). Prototypes, which are illustrative samples of each class, form the foundation of PNs. Prototypes are used by PNs to categories fresh data points. PNs were initially described in Finn et. al. (2017). PNs' basic tenet is that a prototype—that is, the mean of all the training in a class, represents that class. PNs determine the distance between a new data point and each prototype in order to classify it. The class of the prototype closest to the data point receives its assignment.

PNs are superior to other FSL models in a number of ways. PNs are first of all reasonably easy to implement. Second, training PNs is effective. Third, PNs have demonstrated efficacy across a range of applications.

Figure 4 shows how, after network training, the class's internal embedding is compared with the query of unseen samples. The features are dimensionally lowered and compared using a comparable method for more insights.

2.6 Matching Networks

Another classification methodology, matching networks Zell et. al. (2022), uses memory augmentation to try and understand how a classifier functions from fewer data. This network combines the benefits of both a normal neural network, it needs thousands of examples in order to understand the structure, and the k-nearest neighbour technique, which requires no training dataset. The formula shows the attention mechanism α, which contrasts ˆx with xi in order to retrieve information from memory.

As a result, the attention mechanism essentially calculates the cosine difference between examples from the support set and fresh examples. As a result, the SoftMax function is used in the process to normalise the output in the 0–1 range. After that, An embedding function g is used to map the support set. which is often a bidirectional LSTM. Based on other set elements and each unique xi, the method g for encoding uses the entire set S, resulting in g (xi, S). Let's say that when there are few training instances, a tiny batch B is taken into account. As illustrated in Fig. 5 via encoding function f, the matching network normally extracts the data within the supplied base set S and minimises mistake Finn et. al. (2017).

2.7 Relational Networks

Relational networks (RNs) are a subset of neural networks that use plug-in components to handle relational reasoning. This network's structure aims to establish a co-relation between two input samples via reasoning. The following is the function that the relational network uses:

$$RN(O) = f_\phi \sum_{i,j} g_\theta(O_i, O_j)$$

O is seen as a collection of items utilised by gθ and fφ functions. The relationship between object pairs is determined by the function gθ. Assume that dimensionality reduction via complexity n2 is utilised to remove every items only once from the input object pair using a conventional Multi-Layer Perceptron model. Its independence from input or output order is another noteworthy feature. This expresses how the items relate to one another..A relational network is not defined by just text or just pictures. This network only looks for input in the form of an embedding. It is possible to make the model dynamic by ignoring the illustrative object semantics. Relation networks are used to conduct classification tasks based on distance measurements. For fixed feature representation, two common metrics are typically utilised: the standard (linear) Mahalanobis metric and the fixed metric (Euclidean or cosine). Nevertheless, using similarity functionsThe relationship network has been trained to comprehend irregular metrics and deep embeddings. This is especially helpful since fixed metrics emphasise selecting fixed metrics such as Euclidean, cosine, or Mahalanobis and operate on the premise of linear separability following the embedding. The Relation network, on the other hand, creates an asymmetric resemblance measure that performs more adept at integrating and categorising matching and incongruity pairings.. Additionally, this network implements a adaptive approximation method with a model's efficacy. For a fair evaluation, the network uses the same architectural configuration since most single-shot learning paradigms For the incorporation module, utilise four convolutional layers. In Fig. 6. Specifically, each convolutional block has three distinct layers: a 3x3 convolution with 64 filters a ReLU nonlinearity layer, and a batch normalisation layerAdditionally, a 2x2 max-pooling layer is present in the initial pair of blocks but absent from the latter two. Consequently, the network may be used to generate Maps of features produced as a result for the further convolutional layers of the relation module. Moreover, ReLU the non-linear and normalisation by batch, 2x2 max-pooling, and 3x3 convolution with 64 filters for each convolutional block make up the relationship modules, which consists of two Two completely linked layers plus convolutional modules. All connected segments, with the exception of the section of information, are ReLU to be able to keep the relationship score within an acceptable range Finn et. al. (2023).

2.8 Model-Agnostic Meta-Learning

Unaffected by Models The Meta-Learning paradigm is highly regarded due to its ability to adapt and solve a wide variety of issues. The approach becomes more nimble in its execution as data availability rises. A tiny data set is trained using the meta-learning technique to get rid of over-fitting. Reducing function loss for newly allocated tasks is the real goal of this approach. Assume that the parameter vector θ is employed for a specific job selected during the training phase from the support set. After that, a small number of swatches are used to train the network to produce feedback specifically for that task. The performance and parameters of the model are then assessed by exercising a test set. Moreover, the parameterized function linked to the parameter vector θ is defined as fθ. When a new task Ti is inserted into the model, its parameter vector θ becomes θ′ i. Therefore, this job is completed by fine-tuning a parameter where α indicates the learning rate, L shows the Ti task's lack method, and fθ stands for the model (refer to Eq. 1). From now on, This approach employs batch-wise processing with a limited amount of samples Zell et. al. (2017).

2.9 Formal Mathematical Statements of Few Shot Learning Problems

The primarily interesting problem is how AI models learn from training instances, but also how they generalise the valuable training data and retain label data. The ongoing process of finding the answers to these centralised questions makes the deep learning model more effective. One of the hottest subjects in the world of deep learning right now is few-shot learning Zell et. al. (2017). Several models and approaches are shown to be helpful for increased generalisation using a small number of training data instances. The mathematical foundations of some generalised few-shot learning problems and their solutions will be covered in this chapter. The generalised few-shot learning issues are divided in to two categories: gaining knowledge from fresh instances and acquiring new skills from fewer datasets, in order to provide a solid mathematical foundation.

2.10 Learning From New Examples (Problem 1)

Problem Statement:The aim is to create a model M that can correctly categorise new samples (Xnew) from a different distribution given a restricted dataset Dtrain of medical images and related labels.

Solution:A meta-education algorithm that can swiftly adjust to a new cases in addition little data must be trained as part of the solution. The goal of few-shot learning methods such as Model-Agnostic Meta-Learning (MAML) is to quickly learn a model initialization from a small sample size.

Theorem: Let D_{train}={(xi,yi)} be a dataset with medical image and label data, and let M(θ) be a model parameterized by θ. The goal of the MAML technique is to find an initialization of θ, such that the model M(θ) generates correct predictions for a new example (Xnew,Ynew) after one or more gradient updates Finn et. al. (2023).

3. LEARNING FROM NEW CLASS (PROBLEM 2)

3.1 Problem Statement

The goal is to create a new class with a limited number of samples and train a model MM that, when given a dataset Dtrain that contains medical images of classes that already exist, can correctly identify instances of this new class.

3.2 Solution

This difficulty can be resolved by applying a few-shot learning technique such as Prototypical Networks, which learns a metric system in which instances of the same grade are close to one another.

3.3 Theorem

Let Dtrain={(xi,yi)} be a dataset with medical image and label data, and let M(θ) be a model parameterized by θ. The Prototypical Networks approach looks for a region where the distance between instances of the same class is minimised in order to reliably identify new classes with few samples Finn et. al. (2023).

4. CHALLENGES AND SOLUTIONS

4.1 Challenges

4.1.2 Limited Labeled Healthcare Data for Few-Shot Learning

Challenge: Obtaining high-quality annotated healthcare data in this scenario is extremely difficult, because few-shot learning relies on a restricted number of labelled occurrences.

Solution: Promote collaboration between healthcare companies to create shared datasets for few-shot learning, look into ways to create synthetic data, and implement active learning techniques to maximize data efficiency.

4.1.3 Interpretability of Few-Shot Learning Models

Challenge: Because few-shot learning models are inherently complicated, it may be challenging to offer unambiguous interpretations, which is crucial when making healthcare decisions.

Solution: To ensure that the models' judgments align with clinical reasoning, develop interpretable few-shot learning models, employ explainability strategies, and collaborate with medical experts.

4.1.4 Integration with Clinical Workflows

Challenge: Introducing few-shot learning models into the way that clinical workflows are now done could upend accepted wisdom and make physician acceptance more difficult.

Solution: By include medical experts in the model-creation process, developing user-friendly interfaces, and conducting pilot studies, it will be easier to integrate few-shot learning models into routine clinical decision-making.

4.1.5 Data Privacy and Ethical Concerns

Challenge: Using ML models, particularly in few-shot learning scenarios, poses ethical and privacy concerns due to the sensitive nature of healthcare data.

Solution: To ensure patient privacy, strictly adhere to data anonymization protocols, abide by moral standards, and ensure that healthcare regulations such as HIPAA are complied with.

4.1.6 Limited Generalization in Diverse Healthcare Settings

Challenge: It could be hard for few-shot learning models to generalise in a range of healthcare settings where patient demographics, equipment, and practices change.

Solution: Make use of transfer learning techniques, test model efficacy in many healthcare settings, and refine few-shot learning models using a range of datasets that represent various healthcare scenarios.

4.2 Suggestions for Overcoming these Challenges

4.2.1 Establish Collaborative Research Initiatives

Promote collaborations between educational institutions, healthcare facilities, and tech firms to create an environment that is conducive to data sharing and collaborative few-shot learning research.

4.2.2 Educate and Involve Healthcare Professionals

To ensure that the technology is successfully incorporated into clinical practice, organise workshops and training sessions to educate medical professionals about few-shot learning, involve them in the creation of models, and address any concerns they may have.

4.2.3 Adopt Standardized Protocols and Guidelines

Develop and promote established practices for few-shot learning in the healthcare sector, concentrating on model interpretability, data protection, and ethical considerations.

4.2.4 Invest in Explainable AI Research

Allocate funds to studies on explainable AI techniques specific to few-shot learning, which permit open decision-making in healthcare-related applications.

4.2.5 Facilitate Cross-Domain Knowledge Transfer

Support initiatives that guarantee the adaptability and durability of few-shot learning in a variety of clinical scenarios by facilitating the transfer of models and knowledge between various healthcare domains.

By resolving these issues and putting the recommended solutions into practice, this chapter can lessen the inherent hurdles connected with the integration of ML technologies into the healthcare industry while offering insightful information about the field.

5. FUTURE DIRECTIONS

Because the healthcare sector requires improved performance and real-world application, few-shot learning has a bright future. The abstract and previous research indicate that ongoing efforts are needed to improve and optimise the current algorithms to guarantee their efficacy and efficiency in real healthcare settings. Enhancing few-shot learning models' interpretability is essential for tackling deep learning systems' intrinsic black box issue, especially in therapeutic settings. Before deploying few-shot learning algorithms in clinical settings, it is imperative to establish accurate baselines and consistent evaluation standards to ensure their seamless integration. Enhancing applications like picture segmentation and classification will be the main focus of future research, with a particular emphasis on medical imaging as a crucial field of study. The lack of data in the healthcare sector presents problems that need for innovative thinking and study into transfer learning strategies, especially with regard to labelled training data. Integrating few-shot learning with currently in use technologies, such electronic health record systems and diagnostic tools, is necessary to improve processes in the healthcare industry as a whole.

Managing particular issues with medical text data, such as intricate semantics and specialised language, is necessary for the proper implementation of few-shot learning in the medical sector. Future research and development programmes should prioritise ethical challenges such as data security and patient privacy. Few-shot learning has a bright future in healthcare, but medical practitioners and machine learning scientists need to keep working together to properly use the technology and transform medical processes.

Implementing few-shot learning in the medical industry effectively requires managing specific problems with medical text data, such as complex semantics and specialized terminology. Data security and patient privacy are two ethical challenges that should be given top priority in future research and development projects. In order to fully utilize the technology and revolutionize medical procedures, machine learning experts and medical practitioners must continue to collaborate if few-shot learning is to have a bright future in the field of healthcare.

6. CONCLUSION

This chapter explores in detail the potential revolutionary power of few-shot learning in addressing data scarcity issues in the healthcare industry. Few-shot learning shows promise as a flexible approach that works well in a variety of healthcare environments for patient classification, diagnostic prediction, and medical picture analysis. Its ability to provide meaningful insights and precise predictions from sparse training samples, so transforming medical procedures, is a significant discovery.

In decentralized healthcare ecosystems, federated learning integration with few-shot learning is emphasized as a possible path forward for patient privacy protection. This cooperative method improves the robustness of machine learning models while also guaranteeing data security. It is well known that the confluence of few-shot learning and multimodal data integration provides a more thorough understanding of complicated healthcare settings by combining genetic data, clinical notes, and medical imaging to enhance patient outcomes. The chapter highlights few-shot learning as a revolutionary force and imagines how it will have a significant influence on medical decision-making, offering physician's never-before-seen assistance with patient care, diagnosis, and treatment planning. The conclusion, which predicts a paradigm change towards timely, accurate, and personalized healthcare treatments, fervently urges for the implementation of few-shot learning in medical operations. This investigation into the nexus of few-shot learning and healthcare represents a frontier where technological innovation and patient-centered care collide, opening the door to a future in which artificial intelligence and medicine work together to transform the delivery of healthcare.

REFERENCES

Alhoshan, W., Ferrari, A., & Zhao, L. (2023). Zero-shot learning for requirements classification: An exploratory study. *Information and Software Technology*, *159*, 107202. doi:10.1016/j.infsof.2023.107202

Balyan, A. K., Ahuja, S., Sharma, S. K., & Lilhore, U. K. (2022, February). Machine learning-based intrusion detection system for healthcare data. In *2022 IEEE VLSI Device Circuit and System (VLSI DCS)* (pp. 290-294). IEEE.

Chauhan, V., & Tiwari, A. (2022). Pairnorm based Graphical Convolution Network for zero-shot multi-label classification. *Engineering Applications of Artificial Intelligence*, *114*, 105012. doi:10.1016/j.engappai.2022.105012

Finn, C., Abbeel, P., & Levine, S. (2017). Model-agnostic meta-learning for fast adaptation of deep networks. In *International Conference on Machine Learning*, (pp. 1126–1135). PMLR.

Hassan, A., Prasad, D., Khurana, M., Lilhore, U. K., & Simaiya, S. (2021). Integration of internet of things (IoT) in health care industry: An overview of benefits, challenges, and applications. *Data Science and Innovations for Intelligent Systems,* 165-180.

Jaiswal, V., Saurabh, P., Lilhore, U. K., Pathak, M., Simaiya, S., & Dalal, S. (2023). A breast cancer risk predication and classification model with ensemble learning and big data fusion. *Decision Analytics Journal*, *8*, 100298. doi:10.1016/j.dajour.2023.100298

Lake, B. M., Salakhutdinov, R., & Tnenbaum, J. B. (2015). Human-level concept learning through probabilistic program induction. *Science*, *350*(6266), 1332–1338. doi:10.1126/science.aab3050 PMID:26659050

Lilhore, U. K., Dalal, S., Faujdar, N., Margala, M., Chakrabarti, P., Chakrabarti, T., Simaiya, S., Kumar, P., Thangaraju, P., & Velmurugan, H. (2023). Hybrid CNN-LSTM model with efficient hyperparameter tuning for prediction of Parkinson's disease. *Scientific Reports*, *13*(1), 14605. doi:10.1038/s41598-023-41314-y PMID:37669970

Lilhore, U. K., Dalal, S., Faujdar, N., Simaiya, S., Dahiya, M., Tomar, S., & Hashmi, A. (2024). Unveiling the prevalence and risk factors of early stage postpartum depression: A hybrid deep learning approach. *Multimedia Tools and Applications*, 1–35. doi:10.1007/s11042-024-18182-3

Lilhore, U. K., Manoharan, P., Sandhu, J. K., Simaiya, S., Dalal, S., Baqasah, A. M., Alsafyani, M., Alroobaea, R., Keshta, I., & Raahemifar, K. (2023). Hybrid model for precise hepatitis-C classification using improved random forest and SVM method. *Scientific Reports*, *13*(1), 12473. doi:10.1038/s41598-023-36605-3 PMID:37528148

Lilhore, U. K., Poongodi, M., Kaur, A., Simaiya, S., Algarni, A. D., Elmannai, H., Vijayakumar, V., Tunze, G. B., & Hamdi, M. (2022). Hybrid model for detection of cervical cancer using causal analysis and machine learning techniques. *Computational and Mathematical Methods in Medicine*, *2022*, 2022. doi:10.1155/2022/4688327 PMID:35572826

Lilhore, U. K., Simaiya, S., Pandey, H., Gautam, V., Garg, A., & Ghosh, P. (2022). Breast cancer detection in the IoT cloud-based healthcare environment using fuzzy cluster segmentation and SVM classifier. In Ambient Communications and Computer Systems [Singapore: Springer Nature Singapore.]. *Proceedings of RACCCS, 2021*, 165–179.

Malhotra, A., & Jindal, R. (2023). XAI transformer based approach for interpreting depressed and suicidal user behavior on online social networks. *Cognitive Systems Research*, *101186*. doi:10.1016/j.cogsys.2023.101186

Mihailescu, R. C., Hurtig, D., & Olsson, C. (2020). End-to-end anytime solution for appliance recognition based on high-resolution current sensing with few-shot learning. *Internet of Things : Engineering Cyber Physical Human Systems*, *11*, 100263. doi:10.1016/j.iot.2020.100263

Nayem, J., Hasan, S. S., Amina, N., Das, B., Ali, M. S., Ahsan, M. M., & Raman, S. (2023). Few Shot Learning for Medical Imaging: A Comparative Analysis of Methodologies and Formal Mathematical Framework. arXiv preprint arXiv:2305.04401. doi:10.1007/978-3-031-47772-0_4

Onyema, E. M., Lilhore, U. K., Saurabh, P., Dalal, S., Nwaeze, A. S., Chijindu, A. T., Ndufeiya-Kumasi, L. C., & Simaiya, S. (2023). Evaluation of IoT-Enabled hybrid model for genome sequence analysis of patients in healthcare 4.0. *Measurement. Sensors*, *26*, 100679. doi:10.1016/j.measen.2023.100679

Qiao, S., Liu, C., Shen, W., & Yuille, A. (2017). Few-shot image recognition by predicting parameters from activations. arXiv preprint arXiv:1706.03466. https://doi.org//arXiv.1706.03466 doi:10.48550

Ramesh, T. R., Lilhore, U. K., Poongodi, M., Simaiya, S., Kaur, A., & Hamdi, M. (2022). Predictive analysis of heart diseases with machine learning approaches. *Malaysian Journal of Computer Science*, 132–148.

Rathore, M. S., Poongodi, M., Saurabh, P., Lilhore, U. K., Bourouis, S., Alhakami, W., Osamor, J., & Hamdi, M. (2022). A novel trust-based security and privacy model for internet of vehicles using encryption and steganography. *Computers & Electrical Engineering*, *102*, 108205. doi:10.1016/j.compeleceng.2022.108205

Safonova, A., Ghazaryan, G., Stiller, S., Main-Knorn, M., Nendel, C., & Ryo, M. (2023). Ten deep learning techniques to address small data problems with remote sensing. *International Journal of Applied Earth Observation and Geoinformation*, *125*, 103569. doi:10.1016/j.jag.2023.103569

Santoro, A., Bartunov, S., Botvinick, M., Wierstra, D., & Lillicrap, T. (2016). Meta-learning with memory-augmented neural networks. In *International Conference on Machine Learning*, (pp. 1842–1850). PMLR.

Sharma, S. K., Lilhore, U. K., Simaiya, S., & Trivedi, N. K. (2021). An improved random forest algorithm for predicting the COVID-19 pandemic patient health. *Annals of the Romanian Society for Cell Biology*, 67–75.

Simaiya, S., Lilhore, U. K., Prasad, D., & Verma, D. K. (2021). MRI brain tumor detection & image segmentation by hybrid hierarchical K-means clustering with FCM based machine learning model. *Annals of the Romanian Society for Cell Biology*, 88–94.

Snell, J., Swersky, K., & Zemel, R. S. (2017). Prototypical Networks for Few-Shot Learning. In Advances in neural information processing systems, 30, 4077-4087.

Suman, S., Tiwari, A. K., Ingale, T., & Singh, K. (2023). Automated detection of Hypertensive Retinopathy using few-shot learning. *Biomedical Signal Processing and Control*, *86*(Part C), 105310. doi:10.1016/j.bspc.2023.105310

Tyukin, I. Y., Gorban, A. N., McEwan, A. A., Meshkinfamfard, S., & Tang, L. (2021). Blessing of dimensionality at the edge and geometry of few-shot learning. *Information Sciences*, *564*, 124–143. doi:10.1016/j.ins.2021.01.022

Wang, Q., & Chen, K. (2020). Multi-label zero-shot human action recognition via joint latent ranking embedding. *Neural Networks*, *122*, 1–23. doi:10.1016/j.neunet.2019.09.029 PMID:31675624

Wang, Y., Yao, Q., Kwok, J. T., & Ni, L. M. (2020). Generalizing from a Few Examples: A Survey on Few-shot Learning. *ACM Computing Surveys*, *53*(6), 1–34. doi:10.1145/3214306

Waring, J., Lindvall, C., & Umeton, R. (2020). Automated machine learning: Review of the state-of-the-art and opportunities for healthcare. *Artificial Intelligence in Medicine*, *104*, 101822. doi:10.1016/j.artmed.2020.101822 PMID:32499001

Xu, X., Niu, S., Wang, Z., Li, D., Yang, H., & Du, W. (2022). Client selection based weighted federated few-shot learning. *Applied Soft Computing*, *128*, 109488. doi:10.1016/j.asoc.2022.109488

Zell, A., Sumbul, G., & Demir, B. (2022). Deep Metric Learning-Based Semi-Supervised Regression With Alternate Learning. In *IEEE International Conference on Image Processing (ICIP)*. IEEE. 10.1109/ICIP46576.2022.9897939

Zhang, F., Wang, Q., & Li, X. (2020). Deep meta-relation network for visual few-shot learning. In ICASSP 2020-2020 *IEEE International Conference on Acoustics, Speech and Signal Processing (ICASSP),* (pp. 1509–1513). IEEE. 10.1109/ICASSP40776.2020.9053330

Zhang, F., Wang, Q., & Li, X. (2020). Deep meta-relation network for visual few-shot learning. In *ICASSP 2020-2020 IEEE International Conference on Acoustics, Speech and Signal Processing*. IEEE. 10.1109/ICASSP40776.2020.9053330

APPENDIX

Figures

Figure 1. Components of few short problem

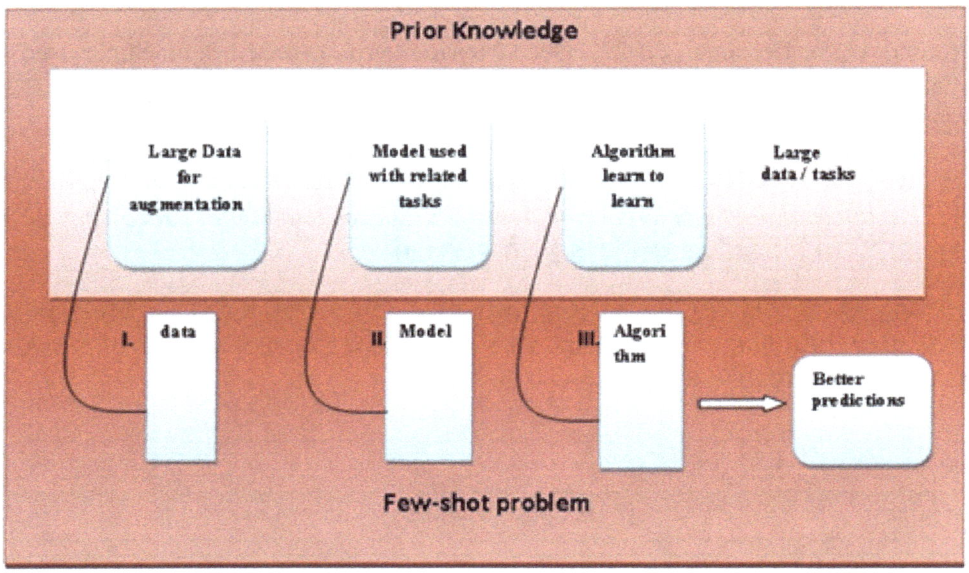

Figure 2. A straightforward depiction of a few-shot learning model with a query set and support set. Decisions are made based on a similarity function and features taken from both datasets.

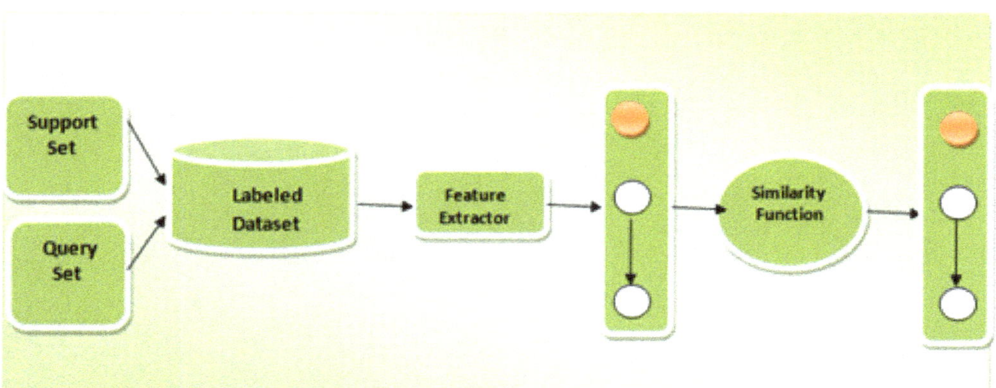

Figure 3. An illustration showing many one-shot classification-based methods for relational networks, model-agnostic meta-learning, matching networks, and prototypical networks

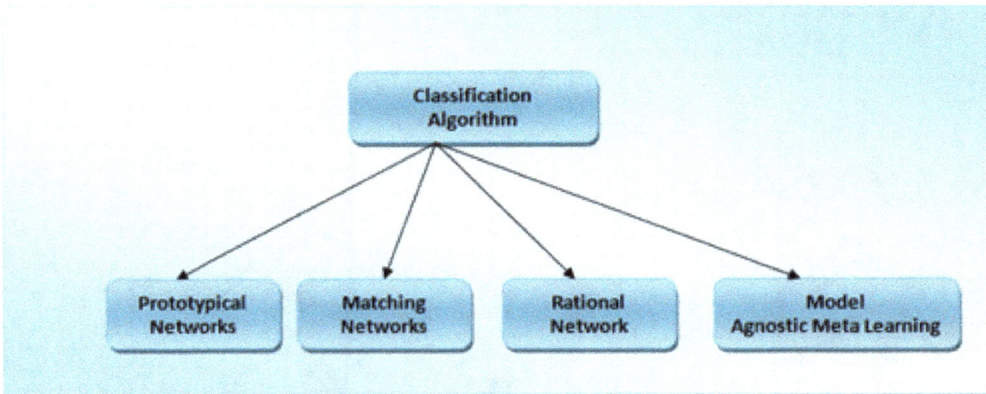

Figure 4. Diagram of a typical network following sample selection, including the class prototype, query-to-prototype distance, loss and propagation calculations, and training commencement

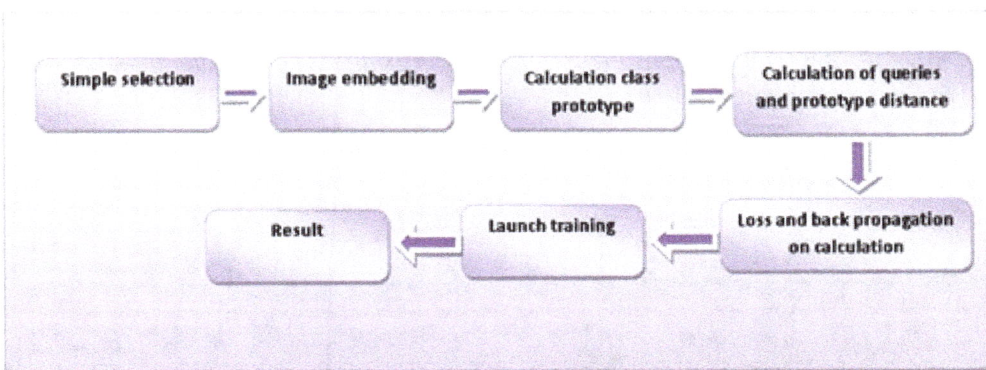

Figure 5. The block diagram of the corresponding network shows how the set of supports is fed into a coding function (G) for mapping and a system of attention (X) to identify examples that have not yet been seen and determine the probability function (F) of the matching rates with the given label

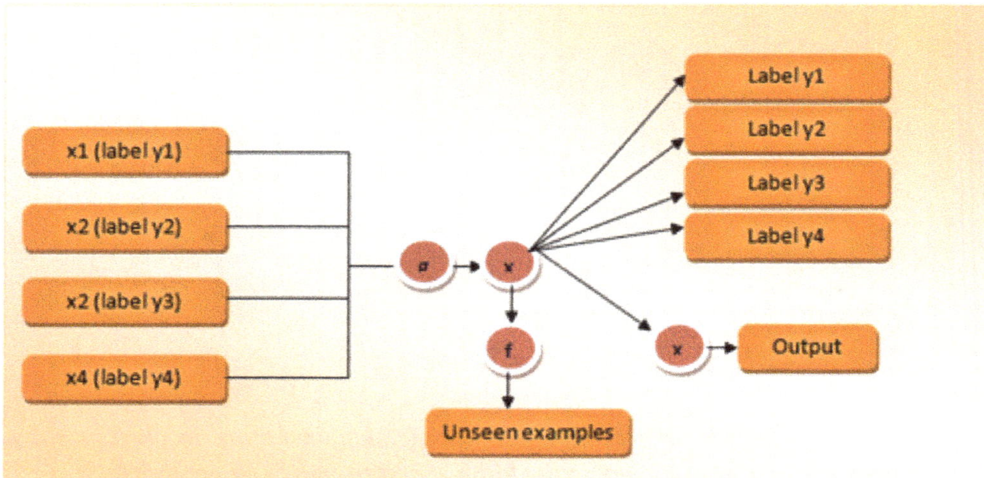

Figure 6. An example of a related a network architecture utilising six convolutional modules in few-shot learning (a) for embedding purposes (b)

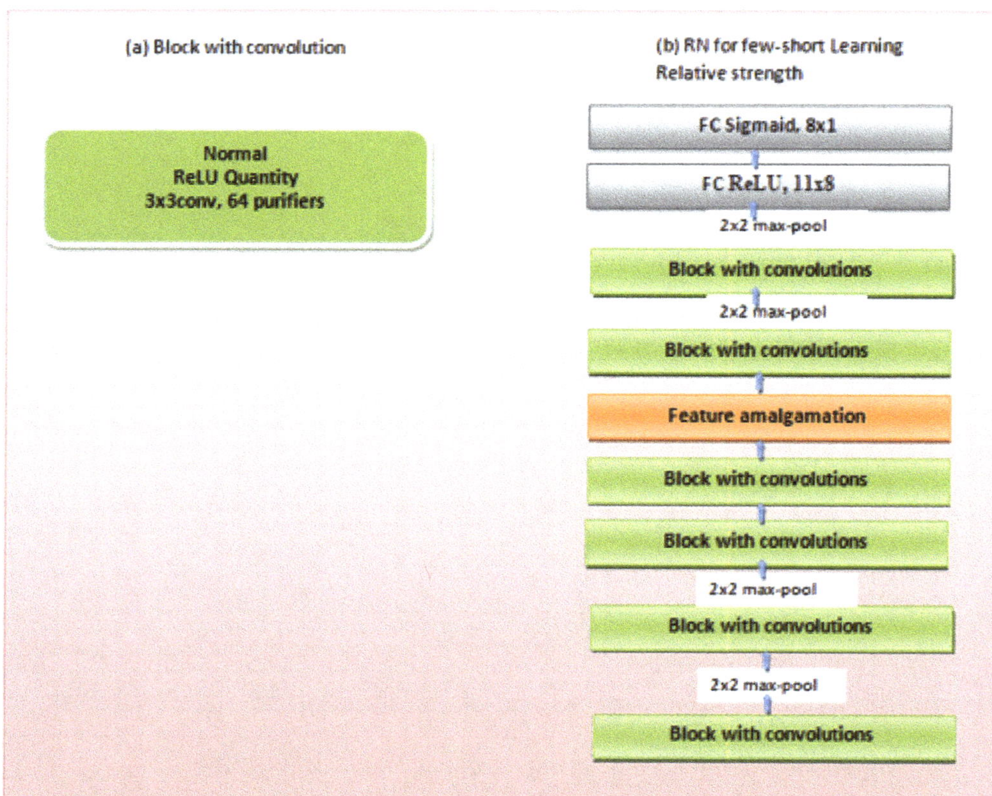

Compilation of References

Abbasniya, M. R., Sheikholeslamzadeh, S. A., Nasiri, H., & Emami, S. (2022). Classification of breast tumors based on histopathology images using deep features and ensemble of gradient boosting methods. *Computers & Electrical Engineering*, *103*, 108382. doi:10.1016/j.compeleceng.2022.108382

Abdar, M., Pourpanah, F., Hussain, S., Rezazadegan, D., Liu, L., Ghavamzadeh, M., Fieguth, P., Cao, X., Khosravi, A., Acharya, U. R., Makarenkov, V., & Nahavandi, S. (2021). A review of uncertainty quantification in deep Learning: Techniques, applications and challenges. *Information Fusion*, *76*, 243–297. doi:10.1016/j.inffus.2021.05.008

Adeli, A., & Neshat, M. (2010, March). *A fuzzy expert system for heart disease diagnosis*. In *Proceedings of international multi conference of engineers and computer scientists,* Hong Kong.

Adeli, E., Li, X., Kwon, D., Zhang, Y., & Pohl, K. M. (2019). Logistic regression confined by cardinality-constrained sample and feature selection. *IEEE Transactions on Pattern Analysis and Machine Intelligence*, *42*(7), 1713–1728. doi:10.1109/TPAMI.2019.2901688 PMID:30835210

Adilin Beatris. (2019). *All About the Basics of Big Data: History*. Types and Applications.

Afzal, S., Hassan, M., Ullah, S., Abbas, H., Tawakkal, F., & Khan, M. A. (2022). Breast cancer; discovery of novel diagnostic biomarkers, drug resistance, and therapeutic implications. *Frontiers in Molecular Biosciences*, *9*, 783450. doi:10.3389/fmolb.2022.783450 PMID:35265667

Agiwal, M., Roy, A., & Saxena, N. (2016). Next generation 5G wireless networks: A comprehensive survey. *IEEE Communications Surveys and Tutorials*, *18*(3), 1617–1655. doi:10.1109/COMST.2016.2532458

Agrahari, A., & Rao, D. T. V. D. (2017). A review paper on Big Data: Technologies, tools and trends. *Int Res J Eng Technol*, *4*(10), 10.

Ahmed, S. (2022). *Classification of Plant Disease from Leaf Images Using Few-Shot Learning* [Doctoral dissertation, Department of Computer Science and Engineering (CSE), Islamic University of Technology (IUT), Board Bazar, Gazipur, Bangladesh]

Ahmed, S., Ilyas, M., & Raja, M. Y. A. (2018, March). Internet of Things: application in smart healthcare. In *Proceedings of the 9th International Conference on Society and Information Technology (IICSIT 2018)* (pp. 19-24). IEEE.

Al Delfi. H. H., Mahdi, Q. S., & Celik, B. (2023, January). The side-effects of 5G between hypothesis and theory. In AIP Conference Proceedings (Vol. 2554, No. 1, p. 080001). AIP Publishing LLC.

Alamdar, F., Mohammadi, F. S., & Amiri, A. (2019). Twin bounded weighted relaxed support vector machines. *IEEE Access : Practical Innovations, Open Solutions*, *7*, 22260–22275. doi:10.1109/ACCESS.2019.2897891

Alex, S., Dhanaraj, K. J., & Deepthi, P. P. (2022). Private and energy-efficient decision tree-based disease detection for resource-constrained medical users in mobile healthcare network. *IEEE Access : Practical Innovations, Open Solutions*, *10*, 17098–17112. doi:10.1109/ACCESS.2022.3149771

Algama, M., Tasker, E., Williams, C., Parslow, A. C., Bryson-Richardson, R. J., & Keith, J. M. (2017). Genome-wide identification of conserved intronic non- coding sequences using a Bayesian segmentation approach. *BMC Genomics*, *18*(1), 1–14. doi:10.1186/s12864-017-3645-2 PMID:28347272

Alhasson, H. F., & Obara, B. (2016). *2D Adaptive Grid-Based Image Analysis Approach for Biological Networks*. IEEE. doi:10.1109/BIBE.2016.17

Alhoshan, W., Ferrari, A., & Zhao, L. (2023). Zero-shot learning for requirements classification: An exploratory study. *Information and Software Technology*, *159*, 107202. doi:10.1016/j.infsof.2023.107202

AlQuraishi, M. (2019). ProteinNet: A standardized data set for machine learning of protein structure. *BMC Bioinformatics*, *20*(1), 1–10. doi:10.1186/s12859-019-2932-0 PMID:31185886

AlQuraishi, M. (2021). Machine learning in protein structure prediction. *Current Opinion in Chemical Biology*, *65*, 1–8. doi:10.1016/j.cbpa.2021.04.005 PMID:34015749

Al-Shedivat, M., Bansal, T., Burda, Y., Sutskever, I., Mordatch, I., & Abbeel, P. 2018. Continuous adaptation via meta-learning in nonstationary and competitive environments. In *International Conference on Le*. IEEE.

Alshehri, F., & Muhammad, G. (2020). A comprehensive survey of the Internet of Things (IoT) and AI-based smart healthcare. *IEEE Access : Practical Innovations, Open Solutions*, *9*, 3660–3678. doi:10.1109/ACCESS.2020.3047960

Altae-Tran, H., Ramsundar, B., Pappu, A. S., & Pande, V. (2017). Low data drug discovery with one-shot learning. *ACS Central Science*, *3*(4), 283–293. doi:10.1021/acscentsci.6b00367 PMID:28470045

Ameur, Y., Aziz, R., Audigier, V., & Bouzefrane, S. (2022, September). Secure and non-interactive k-NN classifier using symmetric fully homomorphic encryption. In *International Conference on Privacy in Statistical Databases* (pp. 142-154). Cham: Springer International Publishing. 10.1007/978-3-031-13945-1_11

Aminizadeh, S., Heidari, A., Toumaj, S., Darbandi, M., Navimipour, N. J., Rezaei, M., Talebi, S., Azad, P., & Unal, M. (2023). The applications of machine learning techniques in medical data processing based on distributed computing and the Internet of Things. *Computer Methods and Programs in Biomedicine*, *241*, 107745. doi:10.1016/j.cmpb.2023.107745 PMID:37579550

Amiri, Z., Heidari, A., Navimipour, N. J., Esmaeilpour, M., & Yazdani, Y. (2024). The deep learning applications in IoT-based bio- and medical informatics: A systematic literature review. *Neural Computing & Applications*, *3*. doi:10.1007/s00521-023-09366-3

An, Q., Rahman, S., Zhou, J., & Kang, J. J. (2023). A Comprehensive Review on Machine Learning in Healthcare Industry: Classification, Restrictions, Opportunities and Challenges. In Sensors, 23(9). Sensors (Basel). doi:10.3390/s23094178

Anandan, J., Dwivedi, D. P., & Govindaraj, V. (2023). Clinical phenotypes of COPD and their impact on quality of life: A cross-sectional study. In *Respiratory Medicine* (Vol. 220, p. 107452). Elsevier BV., doi:10.1016/j.rmed.2023.107452

Anandaram, H. N. B. A, N. Gupta & Verma, B. (2023). *IoT Wearable Breast Temperature Assessment System*. 2023 7th International Conference on Computing Methodologies and Communication (ICCMC), Erode, India.

Anandaram, H. N. B. A, N. Gupta and B. K. Verma, (2023). *IoT Wearable Breast Temperature Assessment System*. 2023 7th International Conference on Computing Methodologies and Communication (ICCMC), Erode, India.

Anandaram, H. N. B. A., Gupta, N., & Verma, B. (2023). *IoT Wearable Breast Temperature Assessment System.* 2023 7th International Conference on Computing Methodologies and Communication (ICCMC), Erode, India.

Anderson, B., Ouyang, D., D'Agostino, A., Bonin, B., Smith, E., Kraushaar, V., Rudman, S. L., & Ho, D. E. (2023). *Leveraging Genomic Sequencing Data to Evaluate Disease Surveillance Strategies. In iScience*. Elsevier BV. doi:10.1016/j.isci.2023.108488

Ansori, A. N., Antonius, Y., Susilo, R. J., Hayaza, S., Kharisma, V. D., Parikesit, A. A., Zainul, R. J. K., Jakhmola, V., Saklani, T., Rebezov, M., Ullah, M. E., Maksimiuk, N., Derkho, M., & Burkov, P. (2023). Application of CRISPR-Cas9 genome editing technology in various fields: A review. *Narra J*, *3*(2), e184–e184. doi:10.52225/narra.v3i2.184

Archana Shreee, S., Maheshwari, B., & Jeevitha Sai, G. (2023). *A Novel Method of Identification of Delirium in Patients from Electronic Health Records Using Machine Learning.* 2023 World Conference on Communication & Computing (WCONF), Raipur, India.

Archana Shreee, S., Maheshwari, B., & Jeevitha Sai, G. (2023). A Novel Method of Identification of Delirium in Patients from Electronic Health Records Using Machine Learning. 2023 World Conference on Communication & Computing (WCONF), RAIPUR, India.

Aridas, C. K., Karlos, S., Kanas, V. G., Fazakis, N., & Kotsiantis, S. B. (2019). Uncertainty based under-sampling for learning naive bayes classifiers under imbalanced data sets. *IEEE Access : Practical Innovations, Open Solutions*, *8*, 2122–2133. doi:10.1109/ACCESS.2019.2961784

Armoogum, S., & Li, X. (2019). Big Data Analytics and Deep Learning in Bioinformatics With Hadoop. In *Deep Learning and Parallel Computing Environment for Bioengineering Systems* (pp. 17–36). Elsevier., doi:10.1016/B978-0-12-816718-2.00009-9

Arunachalam, P., Janakiraman, N., Sivaraman, A. K., Balasundaram, A., Vincent, R., Rani, S., ... Rajesh, M. (2022). Synovial Sarcoma Classification Technique Using Support Vector Machine and Structure Features. *Intelligent Automation & Soft Computing*, *32*(2), 1241–1259. doi:10.32604/iasc.2022.022573

Asai, A., Konno, M., Taniguchi, M., Vecchione, A., & Ishii, H. (2021). Computational healthcare: Present and future perspectives. *Experimental and Therapeutic Medicine*, *22*(6), 1–13. doi:10.3892/etm.2021.10786 PMID:34659497

Asai, K., Hayamizu, S., & Handa, K. (1993). Prediction of protein secondary structure by the hidden Markov model. *Computer Applications in the Biosciences*, *9*(2), 141–146. doi:10.1093/bioinformatics/9.2.141 PMID:8481815

Athota, L., Shukla, V. K., Pandey, N., & Rana, A. (2020, June). Chatbot for healthcare system using artificial intelligence. In *2020 8th International conference on reliability, infocom technologies and optimization (trends and future directions) (ICRITO)* (pp. 619-622). IEEE. 10.1109/ICRITO48877.2020.9197833

Atkinson, A. J., Colburn, W. A., DeGruttola, V. G., DeMets, D. L., Downing, G. J., Hoth, D. F., Oates, J. A., Peck, C. C., Schooley, R. T., Spilker, B. A., Woodcock, J., & Zeger, S. L. (2001). Biomarkers and surrogate endpoints: Preferred definitions and conceptual framework. In Clinical Pharmacology and Therapeutics, 69(3), 89–95. doi:10.1067/mcp.2001.113989

Auer, P. L., & Lettre, G. (2015). Rare variant association studies: Considerations, challenges and opportunities. *Genome Medicine*, *7*(1), 1–11. doi:10.1186/s13073-015-0138-2 PMID:25709717

Avolio, M., & Fuduli, A. (2020). A semiproximal support vector machine approach for binary multiple instance learning. *IEEE Transactions on Neural Networks and Learning Systems*, *32*(8), 3566–3577. doi:10.1109/TNNLS.2020.3015442 PMID:32822307

Awwalu, J., Garba, A. G., Ghazvini, A., & Atuah, R. (2015). Artificial intelligence in personalized medicine application of AI algorithms in solving personalized medicine problems. *International Journal of Computer Theory and Engineering*, *7*(6), 439–443. doi:10.7763/IJCTE.2015.V7.999

Backofen, R., & Gilbert, D. (2001). Bioinformatics and constraints. *Constraints*, *6*(2/3), 141–156. doi:10.1023/A:1011477420926

Bai, Y., Wang, G., Wei, J., Dai, C., Xu, X., Cai, X., Wu, B., Sun, W., Xu, Q., & Jiao, S. (2019). *Using clinical genomic sequencing to guide personalized cancer therapy in China*. PME. doi:10.2217/pme-2018-0056

Baik, S., Choi, M., Choi, J., Kim, H., & Lee, K. M. (2023). Learning to learn task-adaptive hyperparameters for few-shot learning. *IEEE Transactions on Pattern Analysis and Machine Intelligence*. PMID:37030677

Bajorath, J. (2021). State-of-the-art of artificial intelligence in medicinal chemistry. doi:10.2144/fsoa-2021-0030

Balaji, V. R., Suganthi, S. T., Rajadevi, R., Kumar, V. K., Balaji, B. S., & Pandiyan, S. (2020). Skin disease detection and segmentation using dynamic graph cut algorithm and classification through Naive Bayes classifier. *Measurement*, *163*, 107922. doi:10.1016/j.measurement.2020.107922

Baldi, P. (2007). *Machine Learning Challenges in Chemoinformatics and Drug Screening and Design*. Future Design. doi:10.1109/ICMLA.2007.125

Balestra, M. L. (2017). Electronic health records: Patient care and ethical and legal implications for nurse practitioners. *The Journal for Nurse Practitioners*, *13*(2), 105–111. doi:10.1016/j.nurpra.2016.09.010

Balyan, A. K., Ahuja, S., Sharma, S. K., & Lilhore, U. K. (2022, February). Machine learning-based intrusion detection system for healthcare data. In *2022 IEEE VLSI Device Circuit and System (VLSI DCS)* (pp. 290-294). IEEE.

Balyan, A. K., Ahuja, S., Sharma, S. K., & Lilhore, U. K. (2022, February). Machine learning-based intrusion detection system for healthcare data. In 2022 IEEE VLSI Device Circuit and System (VLSI DCS) (pp. 290-294). IEEE.

Bani Issa, W., Al Akour, I., Ibrahim, A., Almarzouqi, A., Abbas, S., Hisham, F., & Griffiths, J. (2020). Privacy, confidentiality, security and patient safety concerns about electronic health records. *International Nursing Review*, *80*(2), 94–101. doi:10.1111/inr.12585 PMID:32314398

Banimfreg, B. H. (2023). A comprehensive review and conceptual framework for cloud computing adoption in bioinformatics. *Healthcare Analytics*, *3*, 100190. doi:10.1016/j.health.2023.100190

Barbero-Aparicio, J. A., Olivares-Gil, A., Rodríguez, J. J., García-Osorio, C., & Díez-Pastor, J. F. (2024). Addressing data scarcity in protein fitness landscape analysis: A study on semi-supervised and deep transfer learning techniques. *Information Fusion*, *102*, 102035. doi:10.1016/j.inffus.2023.102035

Bari Antor, M., Jamil, A. H. M., Mamtaz, M., Monirujjaman Khan, M., Aljahdali, S., Kaur, M., Singh, P., & Masud, M. (2021). A comparative analysis of machine learning algorithms to predict alzheimer's disease. *Journal of Healthcare Engineering*, *2021*, 2021. doi:10.1155/2021/9917919 PMID:34336171

Barshai, M., Tripto, E., & Orenstein, Y. (2020). Identifying Regulatory Elements via Deep Learning. *Annual Review of Biomedical Data Science*, *3*(1), 315–338. doi:10.1146/annurev-biodatasci-022020-021940

Belle, A., Thiagarajan, R., Soroushmehr, S. M., Navidi, F., Beard, D. A., & Najarian, K. (2015). Big data analytics in healthcare. *BioMed Research International*. PMID:26229957

Belpomme, D., Hardell, L., Belyaev, I., Burgio, E., & Carpenter, D. O. (2018). Thermal and non-thermal health effects of low intensity non-ionizing radiation: An international perspective. *Environmental Pollution*, *242*, 643–658. doi:10.1016/j.envpol.2018.07.019 PMID:30025338

Bendre, N., Desai, K., & Najafirad, P. (2021). *Generalized Zero-Shot Learning Using Multimodal Variational Auto-Encoder With Semantic Concepts. 2021 IEEE International Conference on Image Processing (ICIP)*, Anchorage, AK, USA. 10.1109/ICIP42928.2021.9506108

Bertinetto, L., Henriques, J. F., Valmadre, J., Torr, P., & Vedaldi, A. (2016). Learning feed-forward one-shot learners. In Advances in Neural Information Processing Systems.

Bhandari, N., Devra, S., & Singh, K. (2017). Evolution of cellular network: from 1G to 5G. *International journal of engineering and techniques, 3*(5), 98-105.

Blanco-Gonzalez, A., Cabezón, A., Seco-Gonzalez, A., Conde-Torres, D., Antelo-Riveiro, P., Piñeiro, Á., & García-Fandiño, R. (2022). The Role of AI in Drug Discovery: Challenges, OpportunitiesandStrategies. *Pharmaceuticals (Basel, Switzerland)*, *16*(6), 891. doi:10.3390/ph16060891 PMID:37375838

Blease, C. (2023). Sharing online clinical notes with patients: Implications for nocebo effects and health equity. *Journal of Medical Ethics*, *49*(1), 14–21. doi:10.1136/jme-2022-108413 PMID:35918134

Bodaghi, A., Fattahi, N., & Ramazani, A. (2023). Biomarkers: Promising and valuable tools towards diagnosis, prognosis and treatment of Covid-19 and other diseases. In Heliyon, 9(2). Elsevier Ltd. doi:10.1016/j.heliyon.2023.e13323

Bornes, R. S., Montero, J., Correia, A. R. M., & Rosa, N. R. (2023). Use of bioinformatic strategies as a predictive tool in implant-supported oral rehabilitation: A scoping review. *The Journal of Prosthetic Dentistry*, *129*(2), 322.e1–322.e8. doi:10.1016/j.prosdent.2022.12.011 PMID:36710172

Bouzid, A., Almidani, A., Zubrikhina, M., Kamzanova, A., Ilce, B. Y., Zholdassova, M., Yusuf, A. M., Bhamidimarri, P. M., AlHaj, H. A., Kustubayeva, A., Bernstein, A., Burnaev, E., Sharaev, M., & Hamoudi, R. (2023). Integrative bioinformatics and artificial intelligence analyses of transcriptomics data identified genes associated with major depressive disorders including NRG1. *Neurobiology of Stress*, *26*, 100555. doi:10.1016/j.ynstr.2023.100555 PMID:37583471

Brauneck, A., Schmalhorst, L., Kazemi Majdabadi, M. M., Bakhtiari, M., Völker, U., Baumbach, J., Baumbach, L., & Buchholtz, G. (2023). Federated machine learning, privacy-enhancing technologies, and data protection laws in medical research: Scoping review. *Journal of Medical Internet Research*, *25*, e41588. doi:10.2196/41588 PMID:36995759

Brock, A., Lim, T., Ritchie, J. M., & Weston, N. (2018). SMASH: One-shot model architecture search through hypernetworks. In *International Conference on Learning Representations.*

Brown, C., Nazeer, R., Gibbs, A., Le Page, P., Mitchell, A. R., & Mitchell, A. R. (2023). Breaking bias: The role of artificial intelligence in improving clinical decision-making. *Cureus*, *15*(3). doi:10.7759/cureus.36415 PMID:37090406

Bucher, M., Herbin, S., & Jurie, F. (2016). Improving semantic embedding consistency by metric learning for zero-shot classiffication. Lecture Notes in Computer Science (Including Subseries Lecture Notes in Artificial Intelligence and Lecture Notes in Bioinformatics), 9909 LNCS, (pp. 730–746). Springer. doi:10.1007/978-3-319-46454-1_44

Bunyat-zada, A. R., & Ross, A. C. (2023). Highlights of bioinformatic tools and methods for validating bioinformatics derived hypotheses for microbial natural products research. *Current Opinion in Chemical Biology*, *76*, 102367. doi:10.1016/j.cbpa.2023.102367 PMID:37453164

Bushberg, J. T., Chou, C. K., Foster, K. R., Kavet, R., Maxson, D. P., Tell, R. A., & Ziskin, M. C. (2020). IEEE Committee on Man and Radiation—COMAR Technical Information Statement: Health and safety issues concerning exposure of the general public to electromagnetic energy from 5G wireless communications networks. *Health Physics*, *119*(2), 236–246. doi:10.1097/HP.0000000000001301 PMID:32576739

Cacciamani, G. E., Chen, A., Gill, I. S., & Hung, A. J. (2023). Artificial intelligence and urology: Ethical considerations for urologists and patients. *Nature Reviews. Urology*, 1–10. PMID:37524914

Califf, R. M. (2018). Biomarker definitions and their applications. *Experimental Biology and Medicine*, *243*(3), 213–221. doi:10.1177/1535370217750088 PMID:29405771

Cao, H., Bernard, S., Sabourin, R., & Heutte, L. (2019). Random forest dissimilarity based multi-view learning for radiomics application. *Pattern Recognition*, *88*, 185–197. doi:10.1016/j.patcog.2018.11.011

Cao, Y., Geddes, T. A., Yang, J. Y. H., & Yang, P. (2020). Ensemble deep Learning in bioinformatics. *Nature Machine Intelligence*, *2*(9), 500–508. doi:10.1038/s42256-020-0217-y

Carbonell, J. G., Michalski, R. S., & Mitchell, T. M. (1983). An overview of machine learning. *Machine Learning*, 3–23.

Carlberg, M., Hedendahl, L., Koppel, T., & Hardell, L. (2019). High ambient radiofrequency radiation in Stockholm city, Sweden. *Oncology Letters*, *17*(2), 1777–1783. PMID:30675237

Carlone, H. B., & Johnson, A. (2007). Understanding the science experiences of successful women of color: Science identity as an analytic lens. *Journal of Research in Science Teaching*, *44*(8), 1187–1218. doi:10.1002/tea.20237

Cassidy, J. (2020). *Applications of Machine Learning in Drug Discovery I: Target Discovery and Small Molecule Drug Design*. In Tech Open. doi:10.5772/intechopen.93159

Cesario, E., Comito, C. & Zumpano, E. (2024). A survey of the recent trends in deep learning for literature based discovery in the biomedical domain. *Neurocomputing*, *568*(November 2023), p. 127079. . doi:10.1016/j.neucom.2023.127079

Chadwick, R., & Berg, K. (2001). Solidarity and equity: New ethical frameworks for genetic databases. *Nature Reviews. Genetics*, *2*(4), 318–321. doi:10.1038/35066094 PMID:11283704

Chafai, N., Bonizzi, L., Botti, S., & Badaoui, B. (2023). *Emerging applications of machine learning in genomic medicine and healthcare*. Taylor & Francis. doi:10.1080/10408363.2023.2259466

Chai, Z., Yang, X., Liu, Z., Lei, Y., Zheng, W., Ji, M., & Zhao, J. (2019). *Correlation Analysis-Based Neural Network Self-Organizing Genetic Evolutionary Algorithm*. IEEE. doi:10.1109/ACCESS.2019.2942035

Chakrasali, S. V., Kumar, C., Chaturvedi, A., & Jaisudhan Pazhani, A. A. (2023). Computer vision based healthcare system for identification of diabetes & its types using AI, Measurement. *Sensors (Basel)*, *27*, 10075.

Chandan, R. K., Suman, P. N., & Sinha, K. (2021). The Environmental Impact of 5G Technology on Humans and Animals. In Handbook of Research on Knowledge and Organization Systems in Library and Information Science (pp. 48-68). IGI Global.

Chauhan, V., & Tiwari, A. (2022). Pairnorm based Graphical Convolution Network for zero-shot multi-label classification. *Engineering Applications of Artificial Intelligence*, *114*, 105012. doi:10.1016/j.engappai.2022.105012

Chavva, I. R., Crawford, A. L., Mazurek, M. H., Yuen, M. M., Prabhat, A. M., Payabvash, S., Sze, G., Falcone, G. J., Matouk, C. C., de Havenon, A., Kim, J. A., Sharma, R., Schiff, S. J., Rosen, M. S., Kalpathy-Cramer, J., Iglesias Gonzalez, J. E., Kimberly, W. T., & Sheth, K. N. (2022). Deep learning applications for acute stroke management. *Annals of Neurology*, *92*(4), 574–587. doi:10.1002/ana.26435 PMID:35689531

Chen, B., Chen, T., Zeng, X., Zhang, W., Lu, Q., Hou, Z., & Helal, S. (2023). DFML: Dynamic federated meta-learning for rare disease prediction. *IEEE/ACM Transactions on Computational Biology and Bioinformatics*. PMID:37022085

Chen, F., & Gadekallu, T. R. (2023). Multi-dimensional data optimal classification algorithm for quality evaluation of distance teaching in universities. *Mobile Networks and Applications*, 1–11. doi:10.1007/s11036-023-02186-8

Chen, J., Geng, Y., Chen, Z., Pan, J. Z., He, Y., Zhang, W., Horrocks, I., & Chen, H. (2023). Zero-Shot and Few-Shot Learning With Knowledge Graphs: A Comprehensive Survey. *Proceedings of the IEEE*, *111*(6), 653–685. doi:10.1109/JPROC.2023.3279374

Chen, P., Wang, J., Lin, H., Zhao, D., & Yang, Z. (2023). Few-shot biomedical named entity recognition via knowledge-guided instance generation and prompt contrastive learning. *Bioinformatics (Oxford, England)*, *39*(8), btad496. doi:10.1093/bioinformatics/btad496 PMID:37549065

Chen, Y. C., Chang, C. C., Hung, C. C., Lin, J. F., & Hsu, S. Y. (2022). SecDT: Privacy-Preserving Outsourced Decision Tree Classification Without Polynomial Forms in Edge-Cloud Computing. *IEEE Transactions on Signal and Information Processing Over Networks*, *8*, 1037–1048. doi:10.1109/TSIPN.2022.3233193

Chen, Y., Mao, Q., Wang, B., Duan, P., Zhang, B., & Hong, Z. (2022). Privacy-preserving multi-class support vector machine model on medical diagnosis. *IEEE Journal of Biomedical and Health Informatics*, *26*(7), 3342–3353. doi:10.1109/JBHI.2022.3157592 PMID:35259122

Cherney, R., Major, R., & Fitzpatrick, T. (2023). Qualify AI Drug Discovery Tools through FDA ISTAND Program to Model Responsible Drug Discovery AI and Mitigate Dual Use Concerns. *Journal of Science Policy & Governance*, *22*(3). http://www.sciencepolicyjournal.org/uploads/5/4/3/4/5434385/cherney_etal_jspg_22-3.pdf. doi:10.38126/JSPG220302

Ching, T., Himmelstein, D. S., Beaulieu-Jones, B. K., Kalinin, A. A., Do, B. T., Way, G. P., Ferrero, E., Agapow, P.-M., Zietz, M., Hoffman, M. M., Xie, W., Rosen, G. L., Lengerich, B. J., Israeli, J., Lanchantin, J., Woloszynek, S., Carpenter, A. E., Shrikumar, A., Xu, J., & Greene, C. S. (2018). Opportunities and obstacles for deep learning in biology and medicine. *Journal of the Royal Society, Interface*, *15*(141), 20170387. doi:10.1098/rsif.2017.0387 PMID:29618526

Chiruvella, V., & Guddati, A. K. (2021, May 21). Ethical Issues in Patient Data Ownership. *Interactive Journal of Medical Research*, *10*(2), e22269. doi:10.2196/22269 PMID:34018968

Cho, W., & Kim, E. (2022). Improving Augmentation Efficiency for Few-Shot Learning. *IEEE Access : Practical Innovations, Open Solutions*, *10*, 17697–17706. doi:10.1109/ACCESS.2022.3151057

Christensen, C. H., Chang, J. T., Rostron, B. L., Hammad, H. T., van Bemmel, D. M., Del Valle-Pinero, A. Y., Wang, B., Mishina, E. V., Faulcon, L. M., DePina, A., Brown-Baker, L. N., Kimmel, H. L., Lambert, E., Blount, B. C., Vesper, H. W., Wang, L., Goniewicz, M. L., Hyland, A., Travers, M. J., & Chang, C. M. (2021). Biomarkers of inflammation and oxidative stress among adult former smoker, current e-cigarette users—Results from wave 1 PATH study. *Cancer Epidemiology, Biomarkers & Prevention*, *30*(10), 1947–1955. doi:10.1158/1055-9965.EPI-21-0140 PMID:34289969

Christopher, B., Mary, Y. S., Khandaker, M. U., Bradley, D. A., Chew, M. T., & Jojo, P. J. (2020). Effects of mobile phone radiation on certain hematological parameters. *Radiation Physics and Chemistry*, *166*, 108443. doi:10.1016/j.radphyschem.2019.108443

Conneau, A., Ma, M., Khanuja, S., Zhang, Y., Axelrod, V., Dalmia, S, & Bapna, A. (2023, January). Fleurs: Few-shot learning evaluation of universal representations of speech. In *2022 IEEE Spoken Language Technology Workshop (SLT)* (pp. 798-805). IEEE. 10.1109/SLT54892.2023.10023141

Cyranoski, D. (2021). AI drug discovery booms in China. *Nature Biotechnology*, *39*(8), 785–786. doi:10.1038/s41587-021-01016-0 PMID:34345035

Dahiya, N., Sharma, Y. K., Rani, U., Hussain, S., Nabilal, K. V., Mohan, A., & Nuristani, N. (2023). Hyper-parameter tuned deep learning approach for effective human monkeypox disease detection. *Scientific Reports*, *13*(1), 15930. doi:10.1038/s41598-023-43236-1 PMID:37741892

Daphal, S. D., & Koli, S. M. (2023). Enhancing sugarcane disease classification with ensemble deep learning: A comparative study with transfer learning techniques. *Heliyon*, *9*(8), e18261. doi:10.1016/j.heliyon.2023.e18261 PMID:37520940

Davenport, T., & Kalakota, R. (2019). The potential for artificial intelligence in healthcare. *Future Healthcare Journal*, *6*(2), 94–98. doi:10.7861/futurehosp.6-2-94 PMID:31363513

de Vocht, F., & Albers, P. (2022). The population health effects from 5G: Controlling the narrative. *Frontiers in Public Health*, *10*, 1082031. doi:10.3389/fpubh.2022.1082031 PMID:36600933

DeepThink Health. 2020. Deepthinkhealth.com. https://www.deepthinkhealth.com

del Peral-Rosado, J. A., Raulefs, R., López-Salcedo, J. A., & Seco-Granados, G. (2017). Survey of cellular mobile radio localization methods: From 1G to 5G. *IEEE Communications Surveys and Tutorials*, *20*(2), 1124–1148. doi:10.1109/COMST.2017.2785181

Demir, E. (2022, August). Big Biological Data: Need for a Reorientation of the Governance Framework. In *2022 IEEE Conference on Computational Intelligence in Bioinformatics and Computational Biology (CIBCB)* (pp. 1-7). IEEE. 10.1109/CIBCB55180.2022.9863047

Deng, L. (2014, May). A tutorial survey of architectures, algorithms, and applications for deep learning. *APSIPA Transactions on Signal and Information Processing*, *3*(1), 1–29. doi:10.1017/atsip.2013.9

Deshwal, D., Sangwan, P., Dahiya, N., Lilhore, U. K., Dalal, S., & Simaiya, S. (2023). [Detection using Hybrid CNN-RNN Architecture with Transfer Learning from X-Rays. *Current Medical Imaging*.]. *COVID*, 19.

Detopoulou, P., Voulgaridou, G., Moschos, P., Levidi, D., Anastasiou, T., Dedes, V., Diplari, E.-M., Fourfouri, N., Giaginis, C., Panoutsopoulos, G. I., & Papadopoulou, S. K. (2023). Artificial intelligence, nutrition, and ethical issues: A mini-review. *Clinical Nutrition Open Science*, *50*, 46–56. doi:10.1016/j.nutos.2023.07.001

Devyatkin, D. A. (2023). Estimation of vegetation indices with Random Kernel Forests. *IEEE Access : Practical Innovations, Open Solutions*, *11*, 29500–29509. doi:10.1109/ACCESS.2023.3261129

Didi, K., & Zecevic, M. (2022). On How AI Needs to Change to Advance the Science of Drug Discovery. https://dx.doi.org//arXiv.2212.12560 doi:10.48550

Dong, S., Liu, D., Ouyang, R., Zhu, Y., Li, L., Li, T., & Liu, J. (2019). Second-order markov assumption based bayes classifier for networked data with heterophily. *IEEE Access : Practical Innovations, Open Solutions*, *7*, 34153–34161. doi:10.1109/ACCESS.2019.2892757

Dong, X., Zhu, L., Zhang, D., Yang, Y., & Wu, F. (2018). Fast parameter adaptation for few-shot image captioning and visual question answering. In *ACM International Conference on Multimedia*, (pp. 54–62). ACM. 10.1145/3240508.3240527

Dorn, M., Buriol, L., & Lamb, L. (2011). *Combining Machine Learning and Optimization Techniques to Determine 3-D Structures of Polypeptides*. IJCAI. doi:10.5591/978-1-57735-516-8/IJCAI11-469

Dou, B., Zhu, Z., Merkurjev, E., Ke, L., Chen, L., Jiang, J., Zhu, Y., Liu, J., Zhang, B., & Wei, G. W. (2023). Machine learning methods for small data challenges in molecular science. *Chemical Reviews*, *123*(13), 8736–8780. doi:10.1021/acs.chemrev.3c00189 PMID:37384816

Dou, H., Shen, F., Zhao, J., & Mu, X. (2023). Understanding neural network through neuron level visualization. In *Neural Networks* (Vol. 168, pp. 484–495). Elsevier BV. doi:10.1016/j.neunet.2023.09.030

Duan, Y., Andrychowicz, M., Stadie, B., Ho, J., Schneider, J., Sutskever, I., Abbeel, P., & Zaremba, W. 2017. One-shot imitation learning. In Advances in Neural Information Processing Systems.

Du, S., Wang, M., Yang, J., Zhao, Y., Wang, J., Yue, M., Xie, C., & Song, H. (2023). An enhanced prediction framework for coalbed methane production incorporating deep learning and transfer learning. *Energy*, *282*, 128877. doi:10.1016/j.energy.2023.128877

Dwivedi, D. N., Mahanty, G., & Pathak, Y. K. (2023). AI Applications for Financial Risk Management. In M. Irfan, M. Elmogy, M. Shabri Abd. Majid, & S. El-Sappagh (Eds.), The Impact of AI Innovation on Financial Sectors in the Era of Industry 5.0 (pp. 17-31). IGI Global. doi:10.4018/979-8-3693-0082-4.ch002

Dwivedi, D. N., Pandey, A. K., & Dwivedi, A. D. (2023). Examining the emotional tone in politically polarized Speeches in India: An In-Depth analysis of two contrasting perspectives. *SOUTH INDIA JOURNAL OF SOCIAL SCIENCES, 21*(2), 125-136. https://journal.sijss.com/index.php/home/article/view/65

Dwivedi, D. N. (2024). The Use of Artificial Intelligence in Supply Chain Management and Logistics. In D. Sharma, B. Bhardwaj, & M. Dhiman (Eds.), *Leveraging AI and Emotional Intelligence in Contemporary Business Organizations* (pp. 306–313). IGI Global.

Dwivedi, D. N., & Mahanty, G. (2024). AI-Powered Employee Experience: Strategies and Best Practices. In M. Rafiq, M. Farrukh, R. Mushtaq, & O. Dastane (Eds.), *Exploring the Intersection of AI and Human Resources Management* (pp. 166–181). IGI Global. doi:10.4018/979-8-3693-0039-8.ch009

Dwivedi, D. N., Mahanty, G., & Vemareddy, A. (2022). How Responsible Is AI?: Identification of Key Public Concerns Using Sentiment Analysis and Topic Modeling. [IJIRR]. *International Journal of Information Retrieval Research, 12*(1), 1–14. doi:10.4018/IJIRR.298646

Eckhardt, C. M., Madjarova, S. J., Williams, R. J., Ollivier, M., Karlsson, J., Pareek, A., & Nwachukwu, B. U. (2023). Unsupervised machine learning methods and emerging applications in healthcare. *Knee Surgery, Sports Traumatology, Arthroscopy : Official Journal of the ESSKA*, *31*(2), 376–381. doi:10.1007/s00167-022-07233-7 PMID:36378293

El-Hajj, A. M., & Naous, T. (2020, September). Radiation Analysis in a Gradual 5G Network Deployment Strategy. In *2020 IEEE 3rd 5G World Forum (5GWF)* (pp. 448-453). IEEE.

Elswick, C. M., Strong, M. J., Joseph, J. R., Saadeh, Y., Oppenlander, M., & Park, P. (2020). Robotic-assisted spinal surgery: Current generation instrumentation and new applications. *Neurosurgery Clinics*, *31*(1), 103–110. PMID:31739920

Eluwole, O. T., Udoh, N., Ojo, M., Okoro, C., & Akinyoade, A. J. (2018). From 1G to 5G, what next? *IAENG International Journal of Computer Science*, *45*(3).

Ettari, S. V., Roden, E., Ahuja, V., & Samadani, U. (2022). Oculogica: An Eye-Catching Innovation in Health Care and the Privacy Implications of Artificial Intelligence and Machine Learning in Diagnostics for the Human Brain. SMU Sci. & Tech. L. Rev., 25, 23.

Fang, B., Mehran, R. J., Heymach, J. V., & Swisher, S. G. (2015). Predictive biomarkers in precision medicine and drug development against lung cancer. In Chinese Journal of Cancer, 34(7), 1–15. BioMed Central. doi:10.1186/s40880-015-0028-4

Fei-Fei, L., Fergus, R., & Perona, P. (2006). One-shot learning of object categories. *IEEE Transactions on Pattern Analysis and Machine Intelligence*, *28*(4), 594–611. doi:10.1109/TPAMI.2006.79 PMID:16566508

Feng, Y., Huang, X., Yang, P., Yu, J., & Sang, J. (2022). *Non-generative Generalized Zero-shot Learning via Task-correlated Disentanglement and Controllable Samples Synthesis.* 2022 IEEE/CVF Conference on Computer Vision and Pattern Recognition (CVPR), New Orleans, LA, USA. 10.1109/CVPR52688.2022.00913

Feng, W., Dauphin, G., Huang, W., Quan, Y., & Liao, W. (2019). New margin-based subsampling iterative technique in modified random forests for classification. *Knowledge-Based Systems, 182*, 104845. doi:10.1016/j.knosys.2019.07.016

Fernández-Durán, J. J., & Gregorio-Domínguez, M. M. (2024). Regression models for circular data based on nonnegative trigonometric sums. In *Journal of Statistical Planning and Inference* (Vol. 230, p. 106114). Elsevier BV. doi:10.1016/j.jspi.2023.106114

Finn, C., Abbeel, P., & Levine, S. (2017). Model-agnostic meta-learning for fast adaptation of deep networks. In *International Conference on Machine Learning*, (pp. 1126–1135). PMLR.

Finn, C., Abbeel, P., & Levine, S. (2017). Model-agnostic meta-learning for fast adaptation of deep networks. In *International Conference on Machine Learning*. ACM.

Food and Drug Administration & National Institutes of Health. (n.d.). *BEST (biomarkers, endpoints, and other tools) resource.* NCBI. https://www.ncbi.nlm.nih.gov/books/NBK326791

Frank, J. W. (2021). Electromagnetic fields, 5G and health: What about the precautionary principle? *Journal of Epidemiology and Community Health, 75*(6), 562–566. doi:10.1136/jech-2019-213595 PMID:33468601

Fu, Y., Yin, Z., Su, M., Wu, Y., & Liu, G. (2020). Construction and reasoning approach of belief rule-base for classification base on decision tree. *IEEE Access : Practical Innovations, Open Solutions, 8*, 138046–138057. doi:10.1109/ACCESS.2020.3012453

Gao, W., Mahajan, S. P., Sulam, J., & Gray, J. J. (2020). Deep Learning in Protein Structural Modeling and Design. *Patterns (New York, N.Y.), 1*(9), 100142. doi:10.1016/j.patter.2020.100142 PMID:33336200

Gaur, L., Solanki, A., Wamba, S. F., & Jhanjhi, N. Z. (Eds.). (2021). *Advanced AI techniques and applications in bioinformatics.* CRC Press. doi:10.1201/9781003126164

Ge, Y., Guo, Y., Das, S., Al-Garadi, M. A., & Sarker, A. (2023). Few-shot learning for medical text: A review of advances, trends, and opportunities. *Journal of Biomedical Informatics, 144*, 104458. doi:10.1016/j.jbi.2023.104458 PMID:37488023

Ghosh, A., Chakraborty, D., & Law, A. (2018). Artificial intelligence in Internet of things. *CAAI Transactions on Intelligence Technology, 3*(4), 208–218. doi:10.1049/trit.2018.1008

Ghumbre, S., Patil, C., & Ghatol, A. (2011, December). Heart disease diagnosis using support vector machine. In *International conference on computer science and information technology (ICCSIT') Pattaya* (pp. 84-88). IEEE.

Giebel, G. D., & Gissel, C. (2019). Accuracy of mHealth devices for atrial fibrillation screening: Systematic review. *JMIR mHealth and uHealth, 7*(6), e13641. doi:10.2196/13641 PMID:31199337

Gonzalez-Usigli, H. A. (2023, September 29). *Parkinson Disease (PD).* MSD Manual Consumer Version. https://www.msdmanuals.com/en-in/home/brain,-spinal-cord,-and-nerve-disorders/movement-disorders/parkinson-disease-pd

Goyal, J., & Singla, K., Akashdeep, & Singh, S. (2020). A survey of wireless communication technologies from 1G to 5G. In *Second International Conference on Computer Networks and Communication Technologies: ICCNCT 2019* (pp. 613-624). Springer International Publishing. 10.1007/978-3-030-37051-0_69

Goyal, S. (2022). Effective software defect prediction using support vector machines (SVMs). *International Journal of System Assurance Engineering and Management, 13*(2), 681–696. doi:10.1007/s13198-021-01326-1

Grafton, F., Ho, J. J., Ranjbarvaziri, S., Farshidfar, F., Budan, A., Steltzer, S. S., Maddah, M., Loewke, K., Green, K., Patel, S., Hoey, T., & Mandegar, M. (2021). Deep learning detects cardiotoxicity in a high-content screen with induced pluripotent stem cell-derived cardiomyocytes. *eLife*, *10*, e68714. Advance online publication. doi:10.7554/eLife.68714 PMID:34338636

Guan, X., Mural, R., & Uberbacher, E. (1994). *Protein structure prediction using hybrid AI methods*. IEEE. doi:10.1109/CAIA.1994.323633

Guan, J., Lu, Z., Xiang, T., Li, A., Zhao, A., & Wen, J.-R. (2021). Zero and Few Shot Learning With Semantic Feature Synthesis and Competitive Learning. *IEEE Transactions on Pattern Analysis and Machine Intelligence*, *43*(7), 2510–2523. doi:10.1109/TPAMI.2020.2965534 PMID:31940521

Guleria, K., Sharma, A., Lilhore, U. K., & Prasad, D. (2020). Breast cancer prediction and classification using supervised learning techniques. *Journal of Computational and Theoretical Nanoscience*, *17*(6), 2519–2522. doi:10.1166/jctn.2020.8924

Gultekin, D. H., & Siegel, P. H. (2020). Absorption of 5G radiation in brain tissue as a function of frequency, power and time. *IEEE Access : Practical Innovations, Open Solutions*, *8*, 115593–115612. doi:10.1109/ACCESS.2020.3002183

Guo, C., Zhang, J., Liu, Y., Xie, Y., Han, Z., & Yu, J. (2020). Recursion enhanced random forest with an improved linear model (RERF-ILM) for heart disease detection on the internet of medical things platform. *IEEE Access : Practical Innovations, Open Solutions*, *8*, 59247–59256. doi:10.1109/ACCESS.2020.2981159

Guo, Q. (2023). Bioinformatics analysis of the diversity of gut microbiota and different microbiota on insulin resistance in diabetes mellitus patients. In *Heliyon* (p. e22117). Elsevier BV. doi:10.1016/j.heliyon.2023.e22117

Gupta, A., Dwivedi, D. N., & Shah, J. (2023). Applying Machine Learning for Effective Customer Risk Assessment. In: Artificial Intelligence Applications in Banking and Financial Services. Future of Business and Finance. Springer, Singapore. doi:10.1007/978-981-99-2571-1_6

Gupta, A., Dwivedi, D. N., & Shah, J. (2023). Artificial Intelligence-Driven Effective Financial Transaction Monitoring. In: Artificial Intelligence Applications in Banking and Financial Services. Future of Business and Finance. Springer, Singapore. doi:10.1007/978-981-99-2571-1_7

Gupta, A., Dwivedi, D. N., & Shah, J. (2023). Financial Crimes Management and Control in Financial Institutions. In: Artificial Intelligence Applications in Banking and Financial Services. Future of Business and Finance. Springer, Singapore. doi:10.1007/978-981-99-2571-1_2

Gupta, A., Dwivedi, D. N., & Shah, J. (2023). Machine Learning-Driven Alert Optimization. In: Artificial Intelligence Applications in Banking and Financial Services. Future of Business and Finance. Springer, Singapore. doi:10.1007/978-981-99-2571-1_8

Gupta, A., Dwivedi, D. N., & Shah, J. (2023). Overview of Money Laundering. In: Artificial Intelligence Applications in Banking and Financial Services. Springer, Singapore. doi:10.1007/978-981-99-2571-1_1

Gupta, A., Dwivedi, D. N., & Shah, J. (2023). Overview of Technology Solutions. In: Artificial Intelligence Applications in Banking and Financial Services. Future of Business and Finance. Springer, Singapore. doi:10.1007/978-981-99-2571-1_3

Gupta, A., Dwivedi, D. N., & Shah, J. (2023). Setting up a Best-In-Class AI-Driven Financial Crime Control Unit (FCCU). In: Artificial Intelligence Applications in Banking and Financial Services. Future of Business and Finance. Springer, Singapore. doi:10.1007/978-981-99-2571-1_11

Gu, X., Cao, Z., Jolfaei, A., Xu, P., Wu, D., Jung, T. P., & Lin, C. T. (2021). EEG-based brain-computer interfaces (BCIs): A survey of recent studies on signal sensing technologies and computational intelligence approaches and their applications. *IEEE/ACM Transactions on Computational Biology and Bioinformatics*, *18*(5), 1645–1666. doi:10.1109/TCBB.2021.3052811 PMID:33465029

Guxens, M., Vermeulen, R., Steenkamer, I., Beekhuizen, J., Vrijkotte, T. G., Kromhout, H., & Huss, A. (2019). Radiofrequency electromagnetic fields, screen time, and emotional and behavioural problems in 5-year-old children. *International Journal of Hygiene and Environmental Health*, *222*(2), 188–194. doi:10.1016/j.ijheh.2018.09.006 PMID:30314943

Hamelryck, T. (2009). Probabilistic models and machine learning in structural bioinformatics. *Statistical Methods in Medical Research*, *18*(5), 505–526. doi:10.1177/0962280208099492 PMID:19153168

Hamet, P., & Tremblay, J. (2017). Artificial intelligence in medicine. *Metabolism: Clinical and Experimental*, *69*, S36–S40. doi:10.1016/j.metabol.2017.01.011 PMID:28126242

Haq, A. U., Li, J. P., Memon, M. H., Malik, A., Ahmad, T., Ali, A., ... Shahid, M. (2019). Feature selection based on L1-norm support vector machine and effective recognition system for Parkinson's disease using voice recordings. *IEEE Access : Practical Innovations, Open Solutions*, *7*, 37718–37734. doi:10.1109/ACCESS.2019.2906350

Hardell, L., & Carlberg, M. (2020). [Comment] Health risks from radiofrequency radiation, including 5G, should be assessed by experts with no conflicts of interest. *Oncology Letters*, *20*(4), 1–1. doi:10.3892/ol.2020.11876 PMID:32774475

Hardell, L., & Nilsson, M. (2023). Case report: The microwave syndrome after installation of 5g emphasizes the need for protection from radiofrequency radiation. *Annals of Case Reports*, *8*, 1112.

Hardell, L., & Nyberg, R. (2020). [Comment] Appeals that matter or not on a moratorium on the deployment of the fifth generation, 5G, for microwave radiation. *Molecular and Clinical Oncology*, *12*(3), 247–257. doi:10.3892/mco.2020.1984 PMID:32064102

Harrer, S., Shah, P., Antony, B., & Hu, J. (2019). Artificial intelligence for clinical trial design. *Trends in Pharmacological Sciences*, *40*(8), 577–591. doi:10.1016/j.tips.2019.05.005 PMID:31326235

Harry, V. N., Semple, S. I., Parkin, D. E., & Gilbert, F. J. (2010). Use of new imaging techniques to predict tumour response to therapy. *The Lancet. Oncology*, *11*(1), 92–102. doi:10.1016/S1470-2045(09)70190-1 PMID:20129132

Hasan, H. J., Yosef, A. J., & Hachem, H. A. (2021). 5G radiation and potential risks to the environment and human health. [TURCOMAT]. *Turkish Journal of Computer and Mathematics Education*, *12*(6), 1689–1693.

Hashemzadeh, H., Shojaeilangari, S., Allahverdi, A., Rothbauer, M., Ertl, P., & Naderi-manesh, H. (2021). A combined microfluidic deep learning approach for lung cancer cell high throughput screening toward automatic cancer screening applications. *Scientific Reports*, *11*(1), 9804. doi:10.1038/s41598-021-89352-8 PMID:33963232

Hasija, Y. (2023). A machine learning approach to bioinformatics. In *All About Bioinformatics* (pp. 203–224). Elsevier., doi:10.1016/B978-0-443-15250-4.00010-1

HaslumJ. F.LardeauC.KarlssonJ.TurkkiR.LeuchowiusK. J.SmithK.MüllersE. (2023). Cell Painting-based bioactivity prediction boosts high-throughput screening hit-rates and compound diversity. bioRxiv doi:10.1101/2023.04.03.535328

Hassan, A., Prasad, D., Khurana, M., Lilhore, U. K., & Simaiya, S. (2021). Integration of internet of things (IoT) in health care industry: An overview of benefits, challenges, and applications. *Data Science and Innovations for Intelligent Systems*, 165-180.

Havas, M. (2013). Radiation from wireless technology affects the blood, the heart, and the autonomic nervous system1. *Reviews on Environmental Health*, *28*(2-3), 75–84. doi:10.1515/reveh-2013-0004 PMID:24192494

Hayat, N., Lashen, H., & Shamout, F. E. (2021, October). Multi-label generalized zero shot learning for the classification of disease in chest radiographs. In *Machine learning for healthcare conference* (pp. 461–477). PMLR.

Helmick, H., Rodriguez, N., & Kokini, J. L. (2023). Utilization of creep ringing and bioinformatic modelling in study of cold denatured pea protein emulsions. In Innovative Food Science & Emerging Technologies (Vol. 88, p. 103420). Elsevier BV. https://doi.org/ doi:10.1016/j.ifset.2023.103420

Hemalatha, R., Rhesa, M., & Revathi, S. (2020). 5G mobile communication, challenges: A survey. *Bull Sci Res*, *35*(42), 35.

Herington, J., McCradden, M. D., Creel, K., Boellaard, R., Jones, E. C., Jha, A. K., Rahmim, A., Scott, P. J. H., Sunderland, J. J., Wahl, R. L., Zuehlsdorff, S., & Saboury, B. (2023). Ethical considerations for artificial intelligence in medical imaging: Deployment and governance. *Journal of Nuclear Medicine*, *64*(10), 1509–1515. doi:10.2967/jnumed.123.266110 PMID:37620051

Hidayat, T. H. J., Ruldeviyani, Y., Aditama, A. R., Madya, G. R., Nugraha, A. W., & Adisaputra, M. W. (2022). Sentiment analysis of twitter data related to Rinca Island development using Doc2Vec and SVM and logistic regression as classifier. *Procedia Computer Science*, *197*, 660–667. doi:10.1016/j.procs.2021.12.187

HilprechtB.BinnigC. (2021). One Model to Rule them All: Towards Zero-Shot Learning for Databases. *CoRR, abs/2105.0.* https://arxiv.org/abs/2105.00642

Hinrikus, H., Koppel, T., Lass, J., Orru, H., Roosipuu, P., & Bachmann, M. (2022). Possible health effects on the human brain by various generations of mobile telecommunication: A review based estimation of 5G impact. *International Journal of Radiation Biology*, *98*(7), 1210–1221. doi:10.1080/09553002.2022.2026516 PMID:34995145

Hirschberg, J., & Manning, C. D. (2015). Advances in natural language processing. *Science*, *349*(6245), 261–266. doi:10.1126/science.aaa8685 PMID:26185244

Hodara, H., & Skaljo, E. (2021). From 1G to 5G. *Fiber and Integrated Optics*, *40*(2-3), 85–183. doi:10.1080/01468030.2021.1919358

Hoffman, Chris (2019). *What is 5G, and how fast will it be?* How-To Geek LLC.

Hogeweg, P. (2011). The roots of bioinformatics in theoretical biology. *PLoS Computational Biology*, *7*(3), e1002021. doi:10.1371/journal.pcbi.1002021 PMID:21483479

Hollington, J., & Jansen, M. (2023, September 9). How fast is 5G? What you need to know about 5G speeds. *Digital Trends*. https://www.digitaltrends.com/mobile/how-fast-is-5g/

Hombalimath, D. & Patwari, N. (2023). *Symmetrized Feature Selection with Stacked Generalization based Machine Learning Algorithm for the Early Diagnosis of Chronic Diseases.* 2023 5th International Conference on Smart Systems and Inventive Technology (ICSSIT), Tirunelveli, India.

Hombalimath, D. Palanikkumar, & Patwari, N. (2023). *Symmetrized Feature Selection with Stacked Generalization based Machine Learning Algorithm for the Early Diagnosis of Chronic Diseases.* 2023 5th International Conference on Smart Systems and Inventive Technology (ICSSIT), Tirunelveli, India.

Houssein, E. H., Abdelminaam, D. S., Hassan, H. N., Al-Sayed, M. M., & Nabil, E. (2021). A hybrid barnacles mating optimizer algorithm with support vector machines for gene selection of microarray cancer classification. *IEEE Access : Practical Innovations, Open Solutions*, *9*, 64895–64905. doi:10.1109/ACCESS.2021.3075942

Hu, Y., Zhao, T., Zhang, N., Zhang, Y., & Cheng, L. (2000). *A Review of Recent Advances and Research on Drug Target Identification Methods.* Bentham Science. doi:10.2174/1389200219666180925091851

Hua, J., Li, Y., Liu, C., & Wang, L. (2022). A zero-shot prediction method based on causal inference under non-stationary manufacturing environments for complex manufacturing systems. *Robotics and Computer-integrated Manufacturing*, *77*, 102356. doi:10.1016/j.rcim.2022.102356

Huang, A., Xu, R., Chen, Y., & Guo, M. (2023). Research on multi-label user classification of social media based on ML-KNN algorithm. *Technological Forecasting and Social Change*, *188*, 122271. doi:10.1016/j.techfore.2022.122271

Huang, S., Yang, J., Fong, S., & Zhao, Q. (2020). Artificial intelligence in cancer diagnosis and prognosis: Opportunities and challenges. *Cancer Letters*, *471*, 61–71. doi:10.1016/j.canlet.2019.12.007 PMID:31830558

Huang, Z., & Chen, D. (2021). A breast cancer diagnosis method based on VIM feature selection and hierarchical clustering random forest algorithm. *IEEE Access : Practical Innovations, Open Solutions*, *10*, 3284–3293. doi:10.1109/ACCESS.2021.3139595

Hua, Y., Xie, D., Zhang, Y., Wang, M., Wen, W., & Sun, J. (2023). Identification and analysis of key genes in adipose tissue for human obesity based on bioinformatics. *Gene*, *888*, 147755. doi:10.1016/j.gene.2023.147755 PMID:37659596

Hufnagl, B., Steiner, D., Renner, E., Löder, M. G., Laforsch, C., & Lohninger, H. (2019). A methodology for the fast identification and monitoring of microplastics in environmental samples using random decision forest classifiers. *Analytical Methods*, *11*(17), 2277–2285. doi:10.1039/C9AY00252A

Hulkower, R., Penn, M., & Schmit, C. (2020). Privacy and Confidentiality of Public Health Information. In J. Magnuson & B. Dixon (Eds.), *Public Health Informatics and Information Systems. Health Informatics*. Springer. doi:10.1007/978-3-030-41215-9_9

Ielapi, N., Andreucci, M., Licastro, N., Faga, T., Grande, R., Buffone, G., Mellace, S., Sapienza, P., & Serra, R. (2020). Precision medicine and precision nursing: The era of biomarkers and precision health. In *International Journal of General Medicine* (Vol. 13, pp. 1705–1711). Dove Medical Press Ltd., doi:10.2147/IJGM.S285262

Iliashenko, O., Bikkulova, Z., & Dubgorn, A. (2019). Opportunities and challenges of artificial intelligence in healthcare. In *E3S Web of Conferences* (Vol. 110, p. 02028). EDP Sciences. doi:10.1051/e3sconf/201911002028

Imhoff, M., Webb, A., & Goldschmidt, A. (2001). On behalf of the ESICM. *Intensive Care Medicine*, *27*(1), 179–186. doi:10.1007/PL00020869

Iqbal, M. S., Abd-Alrazaq, A., & Househ, M. (2022). Artificial Intelligence Solutions to Detect Fraud in Healthcare Settings: A Scoping Review. *Advances in Informatics, Management and Technology in Healthcare*, 20-23.

Iqbal, M. J., Javed, Z., Sadia, H., Qureshi, I. A., Irshad, A., Ahmed, R., Malik, K., Raza, S., Abbas, A., Pezzani, R., & Sharifi-Rad, J. (2021). Clinical applications of artificial intelligence and machine learning in cancer diagnosis: Looking into the future. *Cancer Cell International*, *21*(1), 1–11. doi:10.1186/s12935-021-01981-1 PMID:34020642

Iqbal, N., & Kumar, P. (2023). From Data Science to Bioscience: Emerging era of bioinformatics applications, tools and challenges. *Procedia Computer Science*, *218*, 1516–1528. doi:10.1016/j.procs.2023.01.130

Iquebal, M. A., Jaiswal, S., Mukhopadhyay, C. S., Sarkar, C., Rai, A., & Kumar, D. (2015). Applications of bioinformatics in plant and agriculture. *PlantOmics: the omics of plant science*, 755-789.

Iroju, O. G., & Olaleke, J. O. (2015). A systematic review of natural language processing in healthcare. *International Journal of Information Technology and Computer Science*, *8*(8), 44–50. doi:10.5815/ijitcs.2015.08.07

Izquierdo-Verdiguier, E., & Zurita-Milla, R. (2020). An evaluation of Guided Regularized Random Forest for classification and regression tasks in remote sensing. *International Journal of Applied Earth Observation and Geoinformation*, *88*, 102051. doi:10.1016/j.jag.2020.102051

Jadhav, D., Sharma, Y. K., & Arora, D. P. (2020, April). Profound Learning Approach for Shot Boundary Location. In *2nd International Conference on Communication & Information Processing (ICCIP)*. SSRN. 10.2139/ssrn.3645409

Jadhav, M., Kumar Sharma, Y., & Bhandari, G. M. (2019, December). Currency identification and forged banknote detection using deep learning. In *2019 International Conference on Innovative Trends and Advances in Engineering and Technology (ICITAET)* (pp. 178-183). IEEE. 10.1109/ICITAET47105.2019.9170225

Jadon, S. (2020). An overview of deep learning architectures in the few-shot learning domain. *arXiv Preprint arXiv:2008.06365.*

Jain, G. (n.d.). *Application of Machine Learning in Drug Discovery and Development Lifecycle. AI Publications.* doi:10.22161/ijmpd.6.6.4

Jain, V., Phophalia, A., & Mitra, S. K. (2022). HML-RF: Hybrid multi-label random forest. *IEEE Access : Practical Innovations, Open Solutions*, *10*, 22902–22914. doi:10.1109/ACCESS.2022.3154420

Jaiswal, V., Saurabh, P., Lilhore, U. K., Pathak, M., Simaiya, S., & Dalal, S. (2023). A breast cancer risk prediction and classification model with ensemble learning and big data fusion. *Decision Analytics Journal*, *8*, 100298. doi:10.1016/j.dajour.2023.100298

Jayawardena, S., Epps, J., & Ambikairajah, E. (2020). Ordinal logistic regression with partial proportional odds for depression prediction. *IEEE Transactions on Affective Computing*.

Jeevan, K., & Sathisha, B. M. (2020). Implementation of IoT Based Wireless Electronic Stethoscope. *2020 Third International Conference on Multimedia Processing, Communication & Information Technology (MPCIT)*, (pp. 103-106). ACM.

Jiang, F., Jiang, Y., Zhi, H., Dong, Y., Li, H., Ma, S., Wang, Y., Dong, Q., Shen, H., & Wang, Y. (2017). Artificial intelligence in healthcare: Past, present and future. In Stroke and Vascular Neurology (Vol. 2, Issue 4, pp. 230–243). BMJ Specialist Journals. doi:10.1136/svn-2017-000101

Jiang, B., Lu, Y., Zhang, B., & Lu, G. (2023). Few-Shot Learning for Image Denoising. *IEEE Transactions on Circuits and Systems for Video Technology*, *33*(9), 4741–4753. doi:10.1109/TCSVT.2023.3248585

Jiang, J., Ma, X., Ouyang, D., & Williams, R. O. III. (2022). Emerging artificial intelligence (ai) technologies used in the development of solid dosage forms. *Pharmaceutics*, *14*(11), 2257. doi:10.3390/pharmaceutics14112257 PMID:36365076

Jiang, T., Gradus, J. L., & Rosellini, A. J. (2020). Supervised machine learning: A brief primer. *Behavior Therapy*, *51*(5), 675–687. doi:10.1016/j.beth.2020.05.002 PMID:32800297

Jiao, L., Geng, X., & Pan, Q. (2019). BP \$ k \$ NN: \$ k \$-nearest neighbor classifier with pairwise distance metrics and belief function theory. *IEEE Access : Practical Innovations, Open Solutions*, *7*, 48935–48947. doi:10.1109/ACCESS.2019.2909752

Ji, F., Ma, Q., Zhang, W., Liu, J., Feng, Y., Zhao, P., Song, X., Chen, J., Zhang, J., Wei, X., Zhou, Y., Chang, Y., Zhang, P., Huang, X., Qiu, J., & Pei, D. (2021). A genome variation map provides insights into the genetics of walnut adaptation and agronomic traits. *Genome Biology*, *22*(1), 1–22. doi:10.1186/s13059-021-02517-6 PMID:34706738

Jiménez-Luna, J., Grisoni, F., Weskamp, N., & Schneider, G. (2021). *Artificial intelligence in drug discovery: recent advances and future perspectives*. Taylor & Francis. doi:10.1080/17460441.2021.1909567

Johnson, K. B., Wei, W.-Q., Weeraratne, D., Frisse, M., Misulis, K., Rhee, K., Zhao, J., & Snowdon, J. (2020). *Precision Medicine, AIandthe Future of Personalized Health Care*. ASCPT. doi:10.1111/cts.12884

Kanimozhi, P., & Aruldoss Albert Victoire, T. (2022). Oppositional tunicate fuzzy C-means algorithm and logistic regression for intrusion detection on cloud. *Concurrency and Computation*, *34*(4), e6624. doi:10.1002/cpe.6624

Karabacak, M., & Margetis, K. (2023). Embracing Large Language Models for Medical Applications: Opportunities and Challenges. *Cureus*, *15*(5). doi:10.7759/cureus.39305 PMID:37378099

Karasneh, R., Al-Mistarehi, A.-H., Al-Azzam, S., Abuhammad, S., Muflih, S. M., Hawamdeh, S., & Alzoubi, K. H. (2021). Physicians' Knowledge, Perceptions, and Attitudes Related to Patient Confidentiality and Data Sharing. *International Journal of General Medicine*, *14*, 721–731. doi:10.2147/IJGM.S301800 PMID:33688243

Karimi, M., Wu, D., Wang, Z., & Shen, Y. (2019). DeepAffinity: Interpretable deep learning of compound–protein affinity through unified recurrent and convolutional neural networks. *Bioinformatics (Oxford, England)*, *35*(18), 3329–3338. doi:10.1093/bioinformatics/btz111 PMID:30768156

Karipidis, K., Mate, R., Urban, D., Tinker, R., & Wood, A. (2021). 5G mobile networks and health—A state-of-the-science review of the research into low-level RF fields above 6 GHz. *Journal of Exposure Science & Environmental Epidemiology*, *31*(4), 585–605. doi:10.1038/s41370-021-00297-6 PMID:33727687

Kasbe, T., & Pippal, R. S. (2017, August). Design of heart disease diagnosis system using fuzzy logic. In *2017 International Conference on Energy, Communication, Data Analytics and Soft Computing (ICECDS)* (pp. 3183-3187). IEEE. 10.1109/ICECDS.2017.8390044

KashyapH. (2015). Big Data Analytics in Bioinformatics: A Machine Learning Perspective. arXiv. http://arxiv.org/abs/1506.05101

Kaushal, A., Kaushal, D., & Pathak, Y. Practicing Medicine and Ethical Issues Applying Artificial Intelligence. In *Ethical Issues in AI for Bioinformatics and Chemoinformatics* (pp. 87–106). CRC Press.

Kavitha, R., Kumar, A., Kalpana, V., & Hariram, V. (2023). Artificial Intelligence based Health Monitoring System on IoTH platform. *2023 Second International Conference on Augmented Intelligence and Sustainable Systems (ICAISS)*, Trichy, India.

Kawale, S. R., & Diwan, S. P. (2022). Intelligent Breast Abnormality Framework for Detection and Evaluation of Breast Abnormal Parameters. *2022 International Conference on Edge Computing and Applications (ICECAA),* (pp. 1503-1508). ACM.

Kawale, S. R., & Diwan, S. P. (2022). Intelligent Breast Abnormality Framework for Detection and Evaluation of Breast Abnormal Parameters. *2022 International Conference on Edge Computing and Applications (ICECAA),* (pp. 1503-1508). IEEE.

Kewsuwun, N., & Kajornkasirat, S. (2022). A sentiment analysis model of agritech startup on Facebook comments using naive Bayes classifier. *International Journal of Electrical & Computer Engineering (2088-8708), 12*(3).

Khalid, M. N., Abdullah, A., Ijaz, Z., Naheed, N., Hamad, A., Sheir, M. A., & Khan, M. D. (2021). Application and Potential Use of Advanced Bioinformatics Techniques in Agriculture and Animal Sciences. *Ind. J. Pure App. Biosci*, *9*(3), 237–246. doi:10.18782/2582-2845.8702

Khan, M. S. I., Islam, N., Uddin, J., Islam, S., & Nasir, M. K. (2022). Water quality prediction and classification based on principal component regression and gradient boosting classifier approach. *Journal of King Saud University. Computer and Information Sciences*, *34*(8), 4773–4781. doi:10.1016/j.jksuci.2021.06.003

Kianoush, S., Savazzi, S., Rampa, V., Costa, L., & Tolochenko, D. (2023). A Random Forest Approach to Body Motion Detection: Multisensory Fusion and Edge Processing. *IEEE Sensors Journal*, *23*(4), 3801–3814. doi:10.1109/JSEN.2022.3232085

Kim, K., Lee, Y. S., Kim, N., Choi, H. D., & Lim, K. M. (2022). 5G Electromagnetic Radiation Attenuates Skin Melanogenesis In Vitro by Suppressing ROS Generation. *Antioxidants*, *11*(8), 1449. doi:10.3390/antiox11081449 PMID:35892650

Kishore Kumar, R., Pandidurai, M., & Senthil Kamalesh, M. S. C. (2023). *Design of IoT based Rural Health Helper using Natural Language Processing*. 2023 4th International Conference on Electronics and Sustainable Communication Systems (ICESC), Coimbatore, India.

Kojima, T., Gu, S. S., Reid, M., Matsuo, Y., & Iwasawa, Y. (2022). Large language models are zero-shot reasoners. *Advances in Neural Information Processing Systems*, *35*, 22199–22213.

Kostoff, R. N., Heroux, P., Aschner, M., & Tsatsakis, A. (2020). Adverse health effects of 5G mobile networking technology under real-life conditions. *Toxicology Letters*, *323*, 35–40. doi:10.1016/j.toxlet.2020.01.020 PMID:31991167

Krämer, K. (2023). Daily briefing: Alzheimer's makes brain cells undergo'cellular suicide'. *Nature*. doi:10.1038/d41586-023-02965-z

Kranthi, M., & Tanguturi, R. C. (2023). *Design of Intelligent Medical Integrity Authentication and Secure Information for Public Cloud in Hospital Administration*. 2023 2nd International Conference on Edge Computing and Applications (ICECAA), Namakkal, India.

Kraus, W. L. (2015). Would you like a hypothesis with those data? Omics and the age of discovery science. *Molecular Endocrinology (Baltimore, Md.)*, *29*(11), 1531–1534. doi:10.1210/me.2015-1253 PMID:26524008

Kuang, Z., Zhao, Y., & Yang, X. (2023). Machine learning approaches for plant miRNA prediction: Challenges, advancements, and future directions. In Agriculture Communications, 1(2). Elsevier BV. doi:10.1016/j.agrcom.2023.100014

Kumar, K., & Gandhi, H. (2023). *Artificial Intelligence for the Management of Breast Cancer: An Overview*. Bentham Science. doi:10.2174/0115701638262066231030052520

Kumar, N., Verma, H., & Sharma, Y. K. (2023). Adversarial Attacks on Graph Neural Network: Techniques and Countermeasures. In Concepts and Techniques of Graph Neural Networks (pp. 58-73). IGI Global.

Kumar, N., Verma, H., & Sharma, Y. K. (2023). Graph Convolutional Neural Networks for Link Prediction in Social Networks. In Concepts and Techniques of Graph Neural Networks (pp. 86-107). IGI Global. doi:10.4018/978-1-6684-6903-3.ch007

Kumar, I., Singh, S. P., & Shivam. (2021). Machine learning in bioinformatics. *Bioinformatics (Oxford, England)*, 443–456. doi:10.1016/B978-0-323-89775-4.00020-1

Kumari, L. V., & Sai, Y. P. (2022). Classification of ECG beats using optimized decision tree and adaptive boosted optimized decision tree. *Signal, Image and Video Processing*, *16*(3), 695–703. doi:10.1007/s11760-021-02009-x

Kumar, R., Rani, S., & Khangura, S. S. (Eds.). (2023). *Machine Learning for Sustainable Manufacturing in Industry 4.0: Concept, Concerns and Applications*. CRC Press. doi:10.1201/9781003453567

Kunduru, A. R. (2023). Machine Learning in Drug Discovery: A Comprehensive Analysis of Applications, Challenges, and Future Directions. *International Journal on Orange Technologies*, *5*(8), 29–37.

Lai, K., Twine, N., O'Brien, A., Guo, Y., & Bauer, D. (2019). Artificial Intelligence and Machine Learning in Bioinformatics. In *Encyclopedia of Bioinformatics and Computational Biology* (pp. 272–286). Elsevier. doi:10.1016/B978-0-12-809633-8.20325-7

Lake, B. M., Salakhutdinov, R., & Tnenbaum, J. B. (2015). Human-level concept learning through probabilistic program induction. *Science*, *350*(6266), 1332–1338. doi:10.1126/science.aab3050 PMID:26659050

Lancellotti, C., Cancian, P., Savevski, V., Kotha, S. R. R., Fraggetta, F., Graziano, P., & Di Tommaso, L. (2021). Artificial intelligence & tissue biomarkers: Advantages, risks and perspectives for pathology. *Cells*, *10*(4), 787. doi:10.3390/cells10040787 PMID:33918173

Landau, A. Y., Ferrarello, S., Blanchard, A., Cato, K., Atkins, N., Salazar, S., Patton, D. U., & Topaz, M. (2022). Developing machine learning-based models to help identify child abuse and neglect: Key ethical challenges and recommended solutions. *Journal of the American Medical Informatics Association : JAMIA*, *29*(3), 576–580. doi:10.1093/jamia/ocab286 PMID:35024859

LeCun, Y., Bengio, Y., & Hinton, G. (2015). Deep learning. *nature, 521*(7553), 436-444.

LeCun, Y., Bengio, Y., & Hinton, G. (2015, May). Deep learning. *Nature*, *521*(7553), 436–444. doi:10.1038/nature14539 PMID:26017442

Lee, M. (2023). Recent Advances in Deep Learning for Protein-Protein Interaction Analysis: A Comprehensive Review. *Molecules (Basel, Switzerland)*, *28*(13), 5169. doi:10.3390/molecules28135169 PMID:37446831

Lei, D., Tang, J., Li, Z., & Wu, Y. (2019). Using low-rank approximations to speed up kernel logistic regression algorithm. *IEEE Access : Practical Innovations, Open Solutions*, *7*, 84242–84252. doi:10.1109/ACCESS.2019.2924542

Li, Z., Wang, X., & Zhang, T. (2020). *From 5G to 5G+*. Singapore: Springer Singapore., doi:10.1007/978-981-15-6819-0_3

Liang, A. (2024). Advancements and applications of single-cell multi-omics techniques in cancer research : Unveiling heterogeneity and paving the way for precision therapeutics. *Biochemistry and Biophysics Reports*, *37*(November 2023). . doi:10.1016/j.bbrep.2023.101589

Libbrecht, M., & Noble, W. (2015). Machine learning applications in genetics and genomics. *Nature Reviews. Genetics*, *16*(6), 321–332. doi:10.1038/nrg3920 PMID:25948244

Li, H., Parikh, N. A., & He, L. (2018). A novel transfer learning approach to enhance deep neural network classification of brain functional connectomes. *Frontiers in Neuroscience*, *12*, 491. doi:10.3389/fnins.2018.00491 PMID:30087587

Li, H., Tian, S., Li, Y., Fang, Q., Tan, R., Pan, Y., Huang, C., Xu, Y., & Gao, X. (2020). Modern deep learning in bioinformatics. *Journal of Molecular Cell Biology*, *12*(11), 823–827. doi:10.1093/jmcb/mjaa030 PMID:32573721

Li, H., Zou, L., Kowah, J. A. H., He, D., Liu, Z., Ding, X., Wen, H., Wang, L., Yuan, M., & Liu, X. (2023). A compact review of progress and prospects of deep learning in drug discovery. *Journal of Molecular Modeling*, *29*(4), 117. doi:10.1007/s00894-023-05492-w PMID:36976427

Li, K., Du, Y., Li, L., & Wei, D.-Q. (2020). Bioinformatics approaches for anti-cancer drug discovery. *Current Drug Targets*, *21*(1), 3–17. doi:10.2174/1389450120666190923162203 PMID:31549592

Li, L., & Liu, Z. P. (2022). A connected network-regularized logistic regression model for feature selection. *Applied Intelligence*, *52*(10), 11672–11702. doi:10.1007/s10489-021-02877-3

Lilhore, U. K., Dalal, S., Faujdar, N., Margala, M., Chakrabarti, P., Chakrabarti, T., Simaiya, S., Kumar, P., Thangaraju, P., & Velmurugan, H. (2023). Hybrid CNN-LSTM model with efficient hyperparameter tuning for prediction of Parkinson's disease. *Scientific Reports*, *13*(1), 14605. doi:10.1038/s41598-023-41314-y PMID:37669970

Lilhore, U. K., Dalal, S., Faujdar, N., Simaiya, S., Dahiya, M., Tomar, S., & Hashmi, A. (2024). Unveiling the prevalence and risk factors of early stage postpartum depression: A hybrid deep learning approach. *Multimedia Tools and Applications*, 1–35. doi:10.1007/s11042-024-18182-3

Lilhore, U. K., Imoize, A. L., Lee, C. C., Simaiya, S., Pani, S. K., Goyal, N., Kumar, A., & Li, C. T. (2022). Enhanced convolutional neural network model for cassava leaf disease identification and classification. *Mathematics*, *10*(4), 580. doi:10.3390/math10040580

Lilhore, U. K., Manoharan, P., Sandhu, J. K., Simaiya, S., Dalal, S., Baqasah, A. M., Alsafyani, M., Alroobaea, R., Keshta, I., & Raahemifar, K. (2023). Hybrid model for precise hepatitis-C classification using improved random forest and SVM method. *Scientific Reports*, *13*(1), 12473. doi:10.1038/s41598-023-36605-3 PMID:37528148

Lilhore, U. K., Poongodi, M., Kaur, A., Simaiya, S., Algarni, A. D., Elmannai, H., Vijayakumar, V., Tunze, G. B., & Hamdi, M. (2022). Hybrid model for detection of cervical cancer using causal analysis and machine learning techniques. *Computational and Mathematical Methods in Medicine*, *2022*, 2022. doi:10.1155/2022/4688327 PMID:35572826

Lilhore, U. K., Simaiya, S., Dalal, S., & Damaševičius, R. (2023). An intelligent waste classification model using hybrid CNN-LSTM with transfer learning for a sustainable environment. *Multimedia Tools and Applications*, 1–25.

Lilhore, U. K., Simaiya, S., Pandey, H., Gautam, V., Garg, A., & Ghosh, P. (2022). Breast cancer detection in the IoT cloud-based healthcare environment using fuzzy cluster segmentation and SVM classifier. In Ambient Communications and Computer Systems [Singapore: Springer Nature Singapore.]. *Proceedings of RACCCS, 2021*, 165–179.

Lilhore, U. K., Simaiya, S., Pandey, H., Gautam, V., Garg, A., & Ghosh, P. (2022). Breast cancer detection in the IoT cloud-based healthcare environment using fuzzy cluster segmentation and SVM classifier. In *Ambient Communications and Computer Systems.* Springer.

Lilhore, U. K., Simaiya, S., Prasad, D., & Guleria, K. (2020). A hybrid tumor detection and classification based on machine learning. *Journal of Computational and Theoretical Nanoscience*, *17*(6), 2539–2544. doi:10.1166/jctn.2020.8927

Li, M., Vanberkel, P., & Zhong, X. (2022). Predicting ambulance offload delay using a hybrid decision tree model. *Socio-Economic Planning Sciences*, *80*, 101146. doi:10.1016/j.seps.2021.101146

Li, M., Wu, Z., Wang, W., Lu, K., Zhang, J., Zhou, Y., Chen, Z., Li, D., Zheng, S., Chen, P., & Wang, B. (2021). Protein-protein interaction sites prediction based on an under-sampling strategy and random forest algorithm. *IEEE/ACM Transactions on Computational Biology and Bioinformatics*, *19*(6), 3646–3654. doi:10.1109/TCBB.2021.3123269 PMID:34705656

Lin, A., Wu, Q., Heidari, A. A., Xu, Y., Chen, H., Geng, W., Li, C., & Li, C. (2019). Predicting intentions of students for master programs using a chaos-induced sine cosine-based fuzzy K-nearest neighbor classifier. *IEEE Access : Practical Innovations, Open Solutions*, *7*, 67235–67248. doi:10.1109/ACCESS.2019.2918026

Lin, C., Lin, C. S., Lee, D. J., Lee, C. C., Chen, S. J., Tsai, S. H., Kuo, F.-C., Chau, T., & Lin, S. H. (2021). Artificial Intelligence–assisted electrocardiography for early diagnosis of thyrotoxic periodic paralysis. *Journal of the Endocrine Society*, *5*(9), bvab120. doi:10.1210/jendso/bvab120 PMID:34308091

Lindorff-Larsen, K., & Kragelund, B. B. (2021). On the potential of machine learning to examine the relationship between sequence, structure, dynamics and function of intrinsically disordered proteins. *Journal of Molecular Biology*, *433*(20), 167196. doi:10.1016/j.jmb.2021.167196 PMID:34390736

Lin, J. C. (2020). 5G communication technology and coronavirus disease [Health Matters]. *IEEE Microwave Magazine*, *21*(9), 16–19. doi:10.1109/MMM.2020.2999236

Lin, L., Liu, T., Zhang, H., Xiong, N., Leng, J., Wei, L., & Liu, Q. (2023). Classification model-based and assisted environment selection for evolutionary algorithms to solve high-dimensional expensive problems. In *Information Sciences* (Vol. 648, p. 119458). Elsevier BV. doi:10.1016/j.ins.2023.119458

Li, P., Liu, F., Jiao, L., Li, S., Li, L., Liu, X., & Huang, X. (2023). Knowledge transduction for cross-domain few-shot learning. *Pattern Recognition*, *141*, 109652. doi:10.1016/j.patcog.2023.109652

Li, S., Yang, Q., Jiang, H., Cortés-Vecino, J. A., & Zhang, Y. (2020). Parasitologist-level classification of apicomplexan parasites and host cell with deep cycle transfer learning (DCTL). *Bioinformatics (Oxford, England)*, *36*(16), 4498–4505. doi:10.1093/bioinformatics/btaa513 PMID:32413103

Li, T. H. S., Chiu, H. J., & Kuo, P. H. (2022). Hepatitis C Virus Detection Model by Using Random Forest, Logistic-Regression and ABC Algorithm. *IEEE Access : Practical Innovations, Open Solutions*, *10*, 91045–91058. doi:10.1109/ACCESS.2022.3202295

Liu, W. (2017). A survey of deep neural network architectures and their applications. Neurocomputing, 234(December 2016), 11–26. doi:10.1016/j.neucom.2016.12.038

Liu, G., & Jiang, D. (2016). 5G: Vision and requirements for mobile communication system towards the year 2020. *Chinese Journal of Engineering*, *2016*, 8. doi:10.1155/2016/5974586

Liu, Q., Chen, J., Wang, Y., Li, S., Jia, C., Song, J., & Li, F. (2021). DeepTorrent: A deep learning-based approach for predicting DNA N4-methylcytosine sites. *Briefings in Bioinformatics*, *22*(3), bbaa124. doi:10.1093/bib/bbaa124 PMID:32608476

Liu, X., Li, J., Ma, J., Sun, H., Xu, Z., Zhang, T., & Yu, H. (2023). Deep Transfer Learning for Intelligent Vehicle Perception: A Survey. *Green Energy and Intelligent Transportation*, *100125*(5), 100125. doi:10.1016/j.geits.2023.100125

Li, W., Wang, Z., Yang, X., Dong, C., Tian, P., Qin, T., Huo, J., Shi, Y., Wang, L., Gao, Y., & Luo, J. (2023). Libfewshot: A comprehensive library for few-shot learning. *IEEE Transactions on Pattern Analysis and Machine Intelligence*, *45*(12), 14938–14955. doi:10.1109/TPAMI.2023.3312125 PMID:37669193

Li, Y., Huang, C., Ding, L., Li, Z., Pan, Y., & Gao, X. (2019). Deep learning in bioinformatics: Introduction, application, and perspective in the big data era. *Methods (San Diego, Calif.)*, *166*, 4–21. doi:10.1016/j.ymeth.2019.04.008 PMID:31022451

Li, Y., Yang, J., Zhang, Z., Wen, J., & Kumar, P. (2022). Healthcare data quality assessment for cybersecurity intelligence. *IEEE Transactions on Industrial Informatics*, *19*(1), 841–848. doi:10.1109/TII.2022.3190405

Li, Z., Xiao, Y., Peng, J., Locke, D., Holmes, D., Li, L., Hamilton, S., Cook, E., Myer, L., Vanderwall, D. E., Cloutier, N., Siddiqui, A. M., Whitehead, P., Bishop, R., Zhao, L., & Cvijic, M. (2020). Quantifying drug tissue biodistribution by integrating high content screening with deep-learning analysis. *Scientific Reports*, *10*(1), 14408. doi:10.1038/s41598-020-71347-6 PMID:32873881

Lobe, B., Morgan, D., & Hoffman, K. A. (2020). Qualitative data collection in an era of social distancing. *International Journal of Qualitative Methods*, *19*, 1609406920937875. doi:10.1177/1609406920937875

Löchel, H. F., & Heider, D. (2021). Chaos game representation and its applications in bioinformatics. *Computational and Structural Biotechnology Journal*, *19*, 6263–6271. doi:10.1016/j.csbj.2021.11.008 PMID:34900136

López, I., Félix, N., Rivera, M., Alonso, A., & Maestú, C. (2021). What is the radiatiotn before 5G? A correlation study between measurements in situ and in real time and epidemiological indicators in Vallecas, Madrid. *Environmental Research*, *194*, 110734. doi:10.1016/j.envres.2021.110734 PMID:33434609

Louk, M. H. L., & Tama, B. A. (2023). Dual-IDS: A bagging-based gradient boosting decision tree model for network anomaly intrusion detection system. *Expert Systems with Applications*, *213*, 119030. doi:10.1016/j.eswa.2022.119030

Lu, D., Yue, Y., Hu, Z., Xu, M., Tong, Y., & Ma, H. (2023). Effective detection of Alzheimer's disease by optimizing fuzzy K-nearest neighbors based on salp swarm algorithm. *Computers in Biology and Medicine*, *159*, 106930. doi:10.1016/j.compbiomed.2023.106930 PMID:37087779

Lu, J., Behbood, V., Hao, P., Zuo, H., Xue, S., & Zhang, G. (2015). Transfer learning using computational intelligence: A survey. *Knowledge-Based Systems*, *80*, 14–23. doi:10.1016/j.knosys.2015.01.010

Luo, F., Wang, M., Liu, Y., Zhao, X. M., & Li, A. (2019). DeepPhos: Prediction of protein phosphorylation sites with deep learning. *Bioinformatics (Oxford, England)*, *35*(16), 2766–2773. doi:10.1093/bioinformatics/bty1051 PMID:30601936

Luscombe, N., Greenbaum, D., & Gerstein, M. (2001). *What is bioinformatics? An introduction and overview.* IMIA. doi:10.1055/s-0038-1638103

M. R. G. & Anandaram. H. (2022). Extraction of Fetal ECG Using ANFIS and the Undecimated-Wavelet Transform. *2022 IEEE 3rd Global Conference for Advancement in Technology (GCAT),* (pp. 1-5). IEEE.

M. R. G. and H. Anandaram. (2022). Extraction of Fetal ECG Using ANFIS and the Undecimated-Wavelet Transform. *2022 IEEE 3rd Global Conference for Advancement in Technology (GCAT),* (pp. 1-5). IEEE.

M. Swathi Pai, M. Shruthi and B. Naveen K, (2020) "Internet of Things: A Survey on Devices, Ecosystem, Components and Communication Protocols," *2020 4th International Conference on Electronics, Communication and Aerospace Technology (ICECA)*, pp. 611-616.

Madapana, N., & Wachs, J. (2021). ZF-SSE: A Unified Sequential Semantic Encoder for Zero-Few-Shot Learning. *2021 16th IEEE International Conference on Automatic Face and Gesture Recognition (FG 2021)*, (pp. 1-8). IEEE. 10.1109/FG52635.2021.9667025

Madona, P., Zafitrah, Y., Sari, J. N., Zain, M. M., & Valzon, M. (2021, November). Classification of ECG Signals Using the Naïve Bayes Classification Method and Its Implementation in Android-Based Smart Health Care. In *2021 International Conference on Computer Science and Engineering (IC2SE)* (Vol. 1, pp. 1-7). IEEE. 10.1109/IC2SE52832.2021.9791475

Mahapatra, S., Gupta, V. R., Sahu, S. S., & Panda, G. (2021). Deep neural network and extreme gradient boosting based Hybrid classifier for improved prediction of Protein-Protein interaction. *IEEE/ACM Transactions on Computational Biology and Bioinformatics*, *19*(1), 155–165. doi:10.1109/TCBB.2021.3061300 PMID:33621179

Mahmood, H., Shaban, M., Rajpoot, N., & Khurram, S. A. (2021). Artificial Intelligence-based methods in head and neck cancer diagnosis: An overview. *British Journal of Cancer*, *124*(12), 1934–1940. doi:10.1038/s41416-021-01386-x PMID:33875821

Mailapur, R. (2023). *Implementation of GUI based Vital Track Ambulance for Patient Health Monitoring.* 2023 8th International Conference on Communication and Electronics Systems (ICCES), Coimbatore, India.

Ma, J., Cheng, J. C. P., Lin, C., Tan, Y., & Zhang, J. (2019). Improving air quality prediction accuracy at larger temporal resolutions using deep learning and transfer learning techniques. *Atmospheric Environment*, *214*(April), 116885. doi:10.1016/j.atmosenv.2019.116885

Ma, J., Fong, S. H., Luo, Y., Bakkenist, C. J., Shen, J. P., Mourragui, S., Wessels, L. F., Hafner, M., Sharan, R., Peng, J., & Ideker, T. (2021). Few-shot learning creates predictive models of drug response that translate from high-throughput screens to individual patients. *Nature Cancer*, *2*(2), 233–244. doi:10.1038/s43018-020-00169-2 PMID:34223192

Majhi, V., Paul, S., & Jain, R. (2019, February). Bioinformatics for healthcare applications. In 2019 Amity international conference on artificial intelligence (AICAI) (pp. 204-207). IEEE. doi:10.1109/AICAI.2019.8701277

Malhotra, A., & Jindal, R. (2023). XAI transformer based approach for interpreting depressed and suicidal user behavior on online social networks. *Cognitive Systems Research, 101186.* doi:10.1016/j.cogsys.2023.101186

ManchadoE.HuangC.-H.TasdemirN.TschaharganehD. F.WilkinsonJ.LoweS. (2017). A Pipeline for Drug Target Identification and Validation. https://dx.doi.org/ doi:10.1101/sqb.2016.81.031096

Marr, B. (2017). *First FDA approval for clinical cloud-based deep learning in healthcare. Forbes*. Forbes Publishing Company.

Marsland S. (2015). *Machine learning: an algorithmic perspective*. CRC press.

Mathur, S., & Sutton, J. (2017). Personalized medicine could transform healthcare (Review). In Biomedical Reports (Vol. 7, Issue 1, pp. 3–5). Spandidos Publications. doi:10.3892/br.2017.922

Matthew, U. O., & Kazaure, J. S. (2021). Chemical polarization effects of electromagnetic field radiation from the novel 5G network deployment at ultra high frequency. *Health and Technology*, *11*(2), 305–317. doi:10.1007/s12553-020-00501-x PMID:33527067

Ma, Y. J. X., Zschocke, J., Glos, M., Kluge, M., Penzel, T., Kantelhardt, J. W., & Bartsch, R. P. (2023). Automatic sleep-stage classification of heart rate and actigraphy data using deep and transfer learning approaches. *Computers in Biology and Medicine*, *163*, 107193. doi:10.1016/j.compbiomed.2023.107193 PMID:37421734

Mayorga-Ruiz, I., Jiménez-Pastor, A., Fos-Guarinos, B., López-González, R., García-Castro, F., & Alberich-Bayarri, Á. (2019). The role of AI in clinical trials. *Artificial Intelligence in Medical Imaging: Opportunities, applications and risks*, 231-243.

McKinney, B. A., Reif, D. M., Ritchie, M. D., & Moore, J. H. (2006). Machine Learning for Detecting Gene-Gene Interactions. *Applied Bioinformatics*, *5*(2), 77–88. doi:10.2165/00822942-200605020-00002 PMID:16722772

McLaughlin, M. F., Donoviel, D. B., & Jones, J. A. (2017). Novel indications for commonly used medications as radiation protectants in spaceflight. *Aerospace Medicine and Human Performance*, *88*(7), 665–676. doi:10.3357/AMHP.4735.2017 PMID:28641684

Memiş, S., Enginoğlu, S., & Erkan, U. (2022). Fuzzy parameterized fuzzy soft k-nearest neighbor classifier. *Neurocomputing*, *500*, 351–378. doi:10.1016/j.neucom.2022.05.041

Mendonça, S. S., de Sousa Vitória, A. R., de Lima, T. W., Galvão Filho, A. R., & Andrade, C. H. (2023). Exploring New Horizons: Empowering Computer-Assisted Drug Design with Few-shot Learning. *Artificial Intelligence in the Life Sciences*, *4*, 100086. doi:10.1016/j.ailsci.2023.100086

Mewada, A., & Dewang, R. K. (2023). SA-ASBA: A hybrid model for aspect-based sentiment analysis using synthetic attention in pre-trained language BERT model with extreme gradient boosting. *The Journal of Supercomputing*, *79*(5), 5516–5551. doi:10.1007/s11227-022-04881-x

Meyer, M. N. (2018). Practical Tips for Ethical Data Sharing. *Advances in Methods and Practices in Psychological Science*, *1*(1), 131–144. doi:10.1177/2515245917747656

Mey, F., Clauwaert, J., Van Huffel, K., Waegeman, W., & De Mey, M. (2021). Improving the performance of machine learning models for biotechnology: The quest for deus ex machina. In *Biotechnology Advances* (Vol. 53, p. 107858). Elsevier BV. doi:10.1016/j.biotechadv.2021.107858

Mihailescu, R. C., Hurtig, D., & Olsson, C. (2020). End-to-end anytime solution for appliance recognition based on high-resolution current sensing with few-shot learning. *Internet of Things : Engineering Cyber Physical Human Systems*, *11*, 100263. doi:10.1016/j.iot.2020.100263

Miller, A. B., Sears, M. E., Morgan, L. L., Davis, D. L., Hardell, L., Oremus, M., & Soskolne, C. L. (2019). Risks to health and well-being from radio-frequency radiation emitted by cell phones and other wireless devices. *Frontiers in Public Health*, *7*, 223. doi:10.3389/fpubh.2019.00223 PMID:31457001

Min, S., Lee, B., & Yoon, S. (2017). Deep learning in bioinformatics. *Briefings in Bioinformatics*, *18*(5), 851–869. PMID:27473064

Min, X., Chen, N., Chen, T., & Jiang, R. (2016, December). DeepEnhancer: Predicting enhancers by convolutional neural networks. *2016 IEEE International Conference on Bioinformatics and Biomedicine (BIBM)* (pp. 637-644). IEEE. 10.1109/BIBM.2016.7822593

Miranda-Vega, J. E., Rivas-López, M., & Fuentes, W. F. (2020). K-nearest neighbor classification for pattern recognition of a reference source light for machine vision system. *IEEE Sensors Journal*, *21*(10), 11514–11521. doi:10.1109/JSEN.2020.3024094

Moein, M. M., Saradar, A., Rahmati, K., Mousavinejad, S. H. G., Bristow, J., Aramali, V., & Karakouzian, M. (2023). Predictive models for concrete properties using machine learning and deep learning approaches: A review. *Journal of Building Engineering*, *63*, 105444. doi:10.1016/j.jobe.2022.105444

Mohammad, R. M. A. (2022). An enhanced multiclass support vector machine model and its application to classifying file systems affected by a digital crime. *Journal of King Saud University. Computer and Information Sciences*, *34*(2), 179–190. doi:10.1016/j.jksuci.2019.10.010

Möncke-Buchner, E., Reich, S., Mücke, M., Reuter, M., Messer, W., Wanker, E. E., & Krüger, D. H. (2002). Counting CAG repeats in the Huntington's disease gene by restriction endonuclease EcoP15I cleavage. *Nucleic Acids Research*, *30*(16), e83. doi:10.1093/nar/gnf082 PMID:12177311

Mondal, R. K., Dey, A., & Sc, B. (2022). BIOINFORMATICS : PAST, PRESENT AND FUTURE. *International Research Journal of Modernization in Engineering Technology and Science*, *04*(07), 442–454.

Monti, M., Fiorentino, J., Milanetti, E., Gosti, G., & Tartaglia, G. G. (2022). Prediction of Time Series Gene Expression and Structural Analysis of Gene Regulatory Networks Using Recurrent Neural Networks. *Entropy (Basel, Switzerland)*, *24*(2), 141. doi:10.3390/e24020141 PMID:35205437

Moon, J. H. (2020). Health effects of electromagnetic fields on children. *Clinical and Experimental Pediatrics*, *63*(11), 422–428. doi:10.3345/cep.2019.01494 PMID:32683815

Moshood, T. D., Shittu, R. A., & Abidin, T. (2020). Covid-19 and 5G radiation are two parallel lines: A systematic review. *Int. J. Innov. Sci. Res. Technol*, *5*(7), 744–751. doi:10.38124/IJISRT20JUL597

Moskowitz, J. M. (2019). We Have No Reason to Believe 5G Is Safe. *Scientific American*.

Mousser, W., Ouadfel, S., & Taleb-Ahmed, A. (2022). Incremental learning of convolutional neural networks in bioinformatics. In *Artificial Neural Networks for Renewable Energy Systems and Real-World Applications* (pp. 223–238). Elsevier. doi:10.1016/B978-0-12-820793-2.00001-X

Muhasina, K. M., Ghosh, P., Swaroop, A. K., Selvaraj, J., Prajapati, B. G., Duraiswamy, B., & Palaniswamy, D. S. (2023). 7 Machine Data Mining Learning and. *Artificial Intelligence in Bioinformatics and Chemoinformatics, 117.*

Munkhdalai, T., & Yu, H. (2017). Meta networks. *34th International Conference on Machine Learning, ICML 2017, 5*, 3933–3943. https://proceedings.mlr.press/v70/munkhdalai17a.html

Murray, N. M., Unberath, M., Hager, G. D., & Hui, F. K. (2020). Artificial intelligence to diagnose ischemic stroke and identify large vessel occlusions: A systematic review. *Journal of Neurointerventional Surgery*, *12*(2), 156–164. doi:10.1136/neurintsurg-2019-015135 PMID:31594798

Muzyukina, P., & Soutourina, O. (2023). CRISPR genotyping methods: Tracing the evolution from spoligotyping to machine learning. In *Biochimie*. Elsevier BV. doi:10.1016/j.biochi.2023.07.017

Mylrea, M., & Robinson, N. (2023). Artificial Intelligence (AI) Trust Framework and Maturity Model: Applying an Entropy Lens to Improve Security, Privacy, and Ethical AI. *Entropy (Basel, Switzerland)*, *25*(10), 1429. doi:10.3390/e25101429 PMID:37895550

Nadarzynski, T., Miles, O., Cowie, A., & Ridge, D. (2019). Acceptability of artificial intelligence (AI)-led chatbot services in healthcare: A mixed-methods study. *Digital Health*, *5*, 2055207619871808. doi:10.1177/2055207619871808 PMID:31467682

Nadkarni, P. M., Ohno-Machado, L., & Chapman, W. W. (2011). Natural language processing: An introduction. *Journal of the American Medical Informatics Association : JAMIA*, *18*(5), 544–551. doi:10.1136/amiajnl-2011-000464 PMID:21846786

Naik, G. (2021). Covid-19 Prevention Kit Based on an Infrared Touchless Thermometer and Distance Detector. *2021 5th International Conference on Electronics, Communication and Aerospace Technology (ICECA)*, (pp. 358-362). IEEE.

Namitha, A. R., Manu, Y. M., & Rashmi, G. R. (2022), IOT Based Smart Health Care System to Monitor Covid-19 Patients. *IJEER, 10*(1), 36-40.

Namitha, A. R., Manu, Y. M., Rashmi, G. R. & Chinamuttevi, V. (2022), IOT Based Smart Health Care System to Monitor Covid-19 Patients. *IJEER, 10*(1), 36-40.

Naqvi, M. R., Jaffar, M. A., Aslam, M., Shahzad, S. K., Iqbal, M. W., & Farooq, A. (2020, June). Importance of big data in precision and personalized medicine. In *2020 International Congress on Human-Computer Interaction, Optimization and Robotic Applications (HORA)* (pp. 1-6). IEEE. 10.1109/HORA49412.2020.9152842

Narayanan, R. R., Durga, N., & Nagalakshmi, S. (2022). Impact of Artificial Intelligence (AI) on Drug Discovery and Product Development. *Indian Journal of Pharmaceutical Education and Research*, *56*(3s), s387–s397. https://ijper.org/sites/default/files/IndJPhaEdRes-56-3s-s387.pdf. doi:10.5530/ijper.56.3s.146

Nareshkumar, A., & Gururaj, B. (2022). An Integrated IoT Technology for Health and Traffic Monitoring System with Smart Ambulance. *2022 IEEE North Karnataka Subsection Flagship International Conference (NKCon)*, Vijaypur, India.

Navathe, S. (2007). *Text Mining and Ontology Applications in Bioinformatics and GIS*. IEEE. doi:10.1109/ICMLA.2007.122

Naveen, K. B., Ramesha, M., & Pai, G. N. (2020). Internet of things: Internet revolution, impact, technology road map and features. *Adv. Math. Sci. J.*, *9*(7), 4405–4414. doi:10.37418/amsj.9.7.11

Nayak, S., & Sharma, Y. K. (2023). A modified Bayesian boosting algorithm with weight-guided optimal feature selection for sentiment analysis. *Decision Analytics Journal*, *8*, 100289. doi:10.1016/j.dajour.2023.100289

Nayem, J., Hasan, S. S., Amina, N., Das, B., Ali, M. S., Ahsan, M. M., & Raman, S. (2023). Few Shot Learning for Medical Imaging: A Comparative Analysis of Methodologies and Formal Mathematical Framework. arXiv preprint arXiv:2305.04401. doi:10.1007/978-3-031-47772-0_4

Nematzadeh, S., Kiani, F., Torkamanian-Afshar, M., & Aydin, N. (2022). Tuning hyperparameters of machine learning algorithms and deep neural networks using metaheuristics: A bioinformatics study on biomedical and biological cases. *Computational Biology and Chemistry*, *97*, 107619. doi:10.1016/j.compbiolchem.2021.107619 PMID:35033837

Nguyen, T. B., Do, D. N., Nguyen-Thi, M.-L., Hoang-The, H., Tran, T.-T., & Nguyen-Thanh, T. (2022). Identification of potential crucial genes and key pathways shared in Inflammatory Bowel Disease and cervical cancer by machine learning and integrated bioinformatics. *Computers in Biology and Medicine*, *149*, 105996. doi:10.1016/j.compbiomed.2022.105996 PMID:36049413

NyamekeF. E.BinS.Ahiaklo-KuzR. K.PeprahR. O. Few-Shot Learning: A Step for Cash Crops Disease Classification. Available at SSRN 4080599. doi:10.2139/ssrn.4080599

O'connor, J. P., Aboagye, E. O., Adams, J. E., Aerts, H. J., Barrington, S. F., Beer, A. J., Boellaard, R., Bohndiek, S. E., Brady, M., Brown, G., Buckley, D. L., Chenevert, T. L., Clarke, L. P., Collette, S., Cook, G. J., deSouza, N. M., Dickson, J. C., Dive, C., Evelhoch, J. L., & Waterton, J. C. (2017, March). Imaging biomarker roadmap for cancer studies. *Nature Reviews. Clinical Oncology*, *14*(3), 169–186. doi:10.1038/nrclinonc.2016.162 PMID:27725679

Ochal, M., Patacchiola, M., Vazquez, J., Storkey, A., & Wang, S. (2023). Few-shot learning with class imbalance. *IEEE Transactions on Artificial Intelligence*, *4*(5), 1348–1358. doi:10.1109/TAI.2023.3298303

Ofer, D., Brandes, N., & Linial, M. (2021). The language of proteins: NLP, machine learning & protein sequences. *Computational and Structural Biotechnology Journal*, *19*, 1750–1758. doi:10.1016/j.csbj.2021.03.022 PMID:33897979

Ohler, U., Liao, G. C., Niemann, H., & Rubin, G. M. (2002). Computational analysis of core promoters in the Drosophila genome. *Genome Biology*, *3*(12), 1–12. doi:10.1186/gb-2002-3-12-research0087 PMID:12537576

Olsson, B., Lautner, R., Andreasson, U., Öhrfelt, A., Portelius, E., Bjerke, M., Hölttä, M., Rosén, C., Olsson, C., Strobel, G., Wu, E., Dakin, K., Petzold, M., Blennow, K., & Zetterberg, H. (2016). CSF and blood biomarkers for the diagnosis of Alzheimer's disease: A systematic review and meta-analysis. *Lancet Neurology*, *15*(7), 673–684. doi:10.1016/S1474-4422(16)00070-3 PMID:27068280

Onyema, E. M., Lilhore, U. K., Saurabh, P., Dalal, S., Nwaeze, A. S., Chijindu, A. T., Ndufeiya-Kumasi, L. C., & Simaiya, S. (2023). Evaluation of IoT-Enabled hybrid Model for genome sequence analysis of patients in healthcare 4.0. *Measurement. Sensors*, *26*, 100679. doi:10.1016/j.measen.2023.100679

Orbai, A. M. (2022). *What Are Common Symptoms of Autoimmune Disease? Johns Hopkins Medicine website.* Accessed June.

Osareh, A., & Shadgar, B. (2010). Machine learning techniques to diagnose breast cancer. *5th international symposium on health informatics and bioinformatics 2010 Apr 20* (pp. 114-120). IEEE. 10.1109/HIBIT.2010.5478895

Ouadah, A., Zemmouchi-Ghomari, L., & Salhi, N. (2022). Selecting an appropriate supervised machine learning algorithm for predictive maintenance. *International Journal of Advanced Manufacturing Technology*, *119*(7-8), 4277–4301. doi:10.1007/s00170-021-08551-9

Padhi, A., Agarwal, A., Saxena, S. K., & Katoch, C. D. S. (2023). Transforming clinical virology with AI, machine learning and deep learning: A comprehensive review and outlook. *Virusdisease*, *34*(3), 1–11. doi:10.1007/s13337-023-00841-y PMID:37780897

Pai, S., & Bader, G. D. (2018). Patient Similarity Networks for Precision Medicine. *Journal of Molecular Biology*, *430*(18), 2924–2938. doi:10.1016/j.jmb.2018.05.037 PMID:29860027

Palanikkumar, D., Mary, P. A., & Begum, A. Y. (2023). A Novel IoT Framework and Device Architecture for Efficient Smart city Implementation. 2023 7th International Conference on Trends in Electronics and Informatics (ICOEI), Tirunelveli, India.

Pallathadka, H., Wenda, A., Ramirez-Asís, E., Asís-López, M., Flores-Albornoz, J., & Phasinam, K. (2023). Classification and prediction of student performance data using various machine learning algorithms. *Materials Today: Proceedings*, *80*, 3782–3785. doi:10.1016/j.matpr.2021.07.382

Panagopoulos, D. J. (2019). Comparing DNA damage induced by mobile telephony and other types of man-made electromagnetic fields. *Mutation Research/Reviews in Mutation Research*, *781*, 53–62. doi:10.1016/j.mrrev.2019.03.003 PMID:31416578

Pandey, M., Fernandez, M., Gentile, F., Isayev, O., Tropsha, A., Stern, A. C., & Cherkasov, A. (2022). The transformational role of GPU computing and deep learning in drug discovery. *Nature Machine Intelligence*, *4*(3), 211–221. doi:10.1038/s42256-022-00463-x

Parashar, A., Parashar, A., Ding, W., Shabaz, M., & Rida, I. (2023). Data Pre-processing and Feature Selection Techniques in Gait Recognition: A Comparative Study of Machine Learning and Deep Learning Approaches. *Pattern Recognition Letters*, *172*, 65–73. doi:10.1016/j.patrec.2023.05.021

Parekh, A. E., Shaikh, O. A., Simran, F., Manan, S., & Hasibuzzaman, M. A. (2023). *Artificial intelligence (AI) in personalized medicine: AI-generated personalized therapy regimens based on genetic and medical history: short communication*. doi:10.1097/MS9.0000000000001320

Park, J., Wilkins, C., Avtonomov, D., Hong, J., Back, S., Kim, H., Shulman, N., MacLean, B. X., Lee, S.-W., & Kim, S. (2023). Targeted proteomics data interpretation with DeepMRM. In Cell Reports Methods, 3(7). Elsevier BV. doi:10.1016/j.crmeth.2023.100521

Parnami, A., & Lee, M. (2022). Learning from few examples: A summary of approaches to few-shot learning. *arXiv Preprint arXiv:2203.04291*.

Patel, M. B., Patel, J., Patel, D., Prajapati, P., & Prajapati, J. Basic Ethical Issues in Bioinformatics and Chemoinformatics. In *Ethical Issues in AI for Bioinformatics and Chemoinformatics* (pp. 12–38). CRC Press. doi:10.1201/9781003353751-2

Patra, P. B.R., D., Kundu, P., Das, M., & Ghosh, A. (2023). Recent advances in machine learning applications in metabolic engineering. In Biotechnology Advances (Vol. 62, p. 108069). Elsevier BV. https://doi.org/ doi:10.1016/j.biotechadv.2022.108069

Paul, D., Sanap, G., Shenoy, S., Kalyane, D., Kalia, K., & Tekade, R. K. (2021). Artificial intelligence in drug discovery and development. *Drug Discovery Today*, *26*(1), 80–93. doi:10.1016/j.drudis.2020.10.010 PMID:33099022

Pawlak, R., Krawiec, P., & Zurek, J. (2019). On measuring electromagnetic fields in 5G technology. *IEEE Access : Practical Innovations, Open Solutions*, *7*, 29826–29835. doi:10.1109/ACCESS.2019.2902481

Pepic, I., Feldt, R., Ljungström, L., Torkar, R., Dalevi, D., Maurin Söderholm, H., Andersson, L.-M., Axelson-Fisk, M., Bohm, K., Sjöqvist, B. A., & Candefjord, S. (2021). Early detection of sepsis using artificial intelligence: A scoping review protocol. *Systematic Reviews*, *10*(1), 1–7. doi:10.1186/s13643-020-01561-w PMID:33453724

Perera, G., Holbrook, A., Thabane, L., Foster, G., & Willison, D. J. (2011). Views on health information sharing and privacy from primary care practices using electronic medical records. *International Journal of Medical Informatics*, *80*(2), 94–101. doi:10.1016/j.ijmedinf.2010.11.005 PMID:21167771

Perez Cortes, V. M., Chatterjee, A., & Khovalyg, D. (2024). Dynamic personalized human body energy expenditure: Prediction using time series forecasting LSTM models. In *Biomedical Signal Processing and Control* (Vol. 87, p. 105381). Elsevier BV. doi:10.1016/j.bspc.2023.105381

Perez, E., Kiela, D., & Cho, K. (2021). Authentic few-shot learning with language models. *Advances in Neural Information Processing Systems*, *34*, 11054–11070.

Pfister, T., Charles, J., & Zisserman, A. (2014). Domain-adaptive discriminative one-shot learning of gestures. In *European Conference on Computer Vision.*

Philip Chen, C. L., & Zhang, C. Y. (2014). Data-intensive applications, challenges, techniques and technologies: A survey on Big Data. *Information Sciences*, *275*, 314–347. doi:10.1016/j.ins.2014.01.015

Philippidis, A. (2021). From Model to Molecule: NVIDIA Doubles Down on AI Drug Discovery. *GEN Edge*, *3*(1), 238–246. doi:10.1089/genedge.3.1.038

Physicians' Health Initiative for Radiation and Environment (PHIRE). (n.d.). *Safety limits and political conflicts of interest*. PHIRE. https://phiremedical.org/safety-limits-and-political-conflicts-of-interest/

Picard, M., Scott-Boyer, M. P., Bodein, A., Périn, O., & Droit, A. (2021). Integration strategies of multi-omics data for machine learning analysis. *Computational and Structural Biotechnology Journal*, *19*, 3735–3746. doi:10.1016/j.csbj.2021.06.030 PMID:34285775

Pidstrigach, J., & Reich, S. (2023). Affine-invariant ensemble transform methods for logistic regression. *Foundations of Computational Mathematics*, *23*(2), 675–708. doi:10.1007/s10208-022-09550-2

Pio, G., Mignone, P., Magazzù, G., Zampieri, G., Ceci, M., & Angione, C. (2022). Integrating genome-scale metabolic modelling and transfer learning for human gene regulatory network reconstruction. *Bioinformatics (Oxford, England)*, *38*(2), 487–493. doi:10.1093/bioinformatics/btab647 PMID:34499112

Pölzl, C. (2011). EMF recommendations specific for children? *Progress in Biophysics and Molecular Biology*, *107*(3). 467–472. doi:10.1016/j.pbiomolbio.2011.09.018 PMID:21967839

Potla, P., Ali, S. A., & Kapoor, M. (2021). A bioinformatics approach to microRNA-sequencing analysis. *Osteoarthritis and Cartilage Open*, *3*(1), 100131. doi:10.1016/j.ocarto.2020.100131 PMID:36475076

Pourpanah, F., Abdar, M., Luo, Y., Zhou, X., Wang, R., Lim, C. P., Wang, X.-Z., & Wu, Q. M. J. (2023). A Review of Generalized Zero-Shot Learning Methods. *IEEE Transactions on Pattern Analysis and Machine Intelligence*, *45*(4), 4051–4070. doi:10.1109/TPAMI.2022.3191696 PMID:35849673

Prabhakar, C., Sekuboyina, A., Paetzold, J. C., Li, H., Amiranashvili, T., & Kleesiek, J. (2021). *Improving Generalized Zero-Shot Learning for Multi-Label Chest X-ray Classification Using Knowledge Graphs.*

Pradeep, S., & Sharma, Y. K. (2019, February). A pragmatic evaluation of stress and performance testing technologies for web-based applications. In *2019 Amity International Conference on Artificial Intelligence (AICAI)* (pp. 399-403). IEEE. 10.1109/AICAI.2019.8701327

Pradeep, S., Sharma, Y. K., Lilhore, U. K., Simaiya, S., Kumar, A., Ahuja, S., Margala, M., Chakrabarti, P., & Chakrabarti, T. (2023). Developing an SDN security model (EnsureS) based on lightweight service path validation with batch hashing and tag verification. *Scientific Reports*, *13*(1), 17381. doi:10.1038/s41598-023-44701-7 PMID:37833379

Prasad Dhore, D. Y. K. S., & Kumbharkar, D. P. (2021). Resourceful Investigation of Malnutrition with Machine Learning Using CNN Procedure. International Journal of Advanced Research in Science. *Tongxin Jishu*, *3*(2), 12–19.

Prasad, K., Dekka, S., Tanguturi, R. c., & Poornima, G. (2022). *An Intelligent System for Remote Monitoring of Patients Health and the Early Detection of Coronary Artery Disease*. 2022 International Conference on Smart Generation Computing, Communication and Networking (SMART GENCON), Bangalore, India.

Prasad, K., Dekka, S., Tanguturi, R. C., & Poornima, G. (2022). *An Intelligent System for Remote Monitoring of Patients Health and the Early Detection of Coronary Artery Disease*. 2022 International Conference on Smart Generation Computing, Communication and Networking (SMART GENCON), Bangalore, India.

Prasad, K., Anil Kumar, N., Reddy, N. S., & Ashreetha, B. (2023). Technologies for Comprehensive Information Security in the IoT. *2023 International Conference for Advancement in Technology (ICONAT)*, Goa, India.

Pritchard, D. E., Moeckel, F., Villa, M. S., Housman, L. T., McCarty, C. A., & McLeod, H. L. (2017). Strategies for integrating personalized medicine into healthcare practice. In Personalized Medicine (Vol. 14, Issue 2, pp. 141–152). Future Medicine Ltd. doi:10.2217/pme-2016-0064

Priya, S., Tripathi, G., Singh, D., Jain, P., & Kumar, A. (2022). Machine learning approaches and their applications in drug discovery and design. https://dx.doi.org/ doi:10.1111/cbdd.14057

Priya, A., Garg, S., & Tigga, N. P. (2020). Predicting anxiety, depression and stress in modern life using machine learning algorithms. *Procedia Computer Science*, *167*, 1258–1267. doi:10.1016/j.procs.2020.03.442

Qian, L., Zheng, H., Zhou, H., Qin, R., & Li, J. (2013). Classification of time series gene expression in clinical studies via integration of biological network. *PLoS One*, *8*(3), e58383. doi:10.1371/journal.pone.0058383 PMID:23516469

Qiu, J., Li, L., Sun, J., Peng, J., Shi, P., Zhang, R., Dong, Y., Lam, K., Lo, F. P.-W., Xiao, B., Yuan, W., Wang, N., Xu, D., & Lo, B. (2023). Large AI models in health informatics: Applications, challenges, and the future. *IEEE Journal of Biomedical and Health Informatics*, *27*(12), 6074–6087. doi:10.1109/JBHI.2023.3316750 PMID:37738186

Qiu, Y., & Wei, G. W. (2023). Artificial intelligence-aided protein engineering: From topological data analysis to deep protein language models. *Briefings in Bioinformatics*, *24*(5), bbad289. doi:10.1093/bib/bbad289 PMID:37580175

Rabcan, J., Levashenko, V., Zaitseva, E., Kvassay, M., & Subbotin, S. (2019). Application of fuzzy decision tree for signal classification. *IEEE Transactions on Industrial Informatics*, *15*(10), 5425–5434. doi:10.1109/TII.2019.2904845

Raha, K., Wollacott, A., Italia, M. J., & Desjarlais, J. (n.d.). *Prediction of amino acid sequence from structure*. Wiley. doi:10.1110/ps.9.6.1106

Rahman, S., Khan, S., & Porikli, F. (2018). A Unified Approach for Conventional Zero-Shot, Generalized Zero-Shot, and Few-Shot Learning. *IEEE Transactions on Image Processing*, *27*(11), 5652–5667. doi:10.1109/TIP.2018.2861573 PMID:30072324

Ram Vishal, B. & Shankar, M. (2023). *Cardiovascular Disease Prediction Using LSTM Algorithm based On Cytokines*. 2023 4th International Conference for Emerging Technology (INCET), Belgaum, India.

Ram Vishal, B. (2023). *Cardiovascular Disease Prediction Using LSTM Algorithm based On Cytokines*. 2023 4th International Conference for Emerging Technology (INCET), Belgaum, India.

Ramakrishnan, M., & Athikary, K. G. (2023). Health effects of fifth-generation technologies. *International Journal of Environmental Health Engineering*, *12*(1), 3. doi:10.4103/ijehe.ijehe_18_22

Ramesh Naidu, P., Guruprasad, N., & Dankan Gowda, V. (2020). Design and implementation of cryptcloud system for securing files in cloud. *Adv. Math. Sci. J.*, *9*(7), 4485–4493. doi:10.37418/amsj.9.7.17

Ramesh Naidu, P., Guruprasad, N., & Dankan Gowda, V. (2021). A High-Availability and Integrity Layer for Cloud Storage, Cloud Computing Security: From Single to Multi-Clouds. *Journal of Physics: Conference Series*, *1921*(1), 012072. doi:10.1088/1742-6596/1921/1/012072

Ramesh, T. R., Lilhore, U. K., Poongodi, M., Simaiya, S., Kaur, A., & Hamdi, M. (2022). Predictive analysis of heart diseases with machine learning approaches. *Malaysian Journal of Computer Science*, •••, 132–148.

Rao, B. K., Chaturvedi, A., & Hussain, N. (2022). Industrial quality healthcare services using Internet of Things and fog computing approach, Measurement. *Sensors (Basel)*, *24*, 100517.

Rasouli, J. J., Shao, J., Neifert, S., Gibbs, W. N., Habboub, G., Steinmetz, M. P., Benzel, E., & Mroz, T. E. (2021). Artificial intelligence and robotics in spine surgery. *Global Spine Journal*, *11*(4), 556–564. doi:10.1177/2192568220915718 PMID:32875928

Rastogi, S. C., Rastogi, P., & Mendiratta, N. (2022). *Bioinformatics: Methods and Applications-Genomics, Proteomics and Drug Discovery*. PHI Learning Pvt. Ltd.

Rathore, M. S., Poongodi, M., Saurabh, P., Lilhore, U. K., Bourouis, S., Alhakami, W., Osamor, J., & Hamdi, M. (2022). A novel trust-based security and privacy model for internet of vehicles using encryption and steganography. *Computers & Electrical Engineering*, *102*, 108205. doi:10.1016/j.compeleceng.2022.108205

Reddy, S., & Patwal, P. P. S. (2022). Data Analytics and Cloud-Based Platform for Internet of Things Applications in Smart Cities. *2022 International Conference on Industry 4.0 Technology (I4Tech)*, (pp. 1-6). IEEE.

Redivo, E., Viroli, C., & Farcomeni, A. (2023). Quantile-distribution functions and their use for classification, with application to naïve Bayes classifiers. *Statistics and Computing*, *33*(2), 55. doi:10.1007/s11222-023-10224-4

Reilly, L., Lara, E., Ramos, D., Li, Z., Pantazis, C. B., Stadler, J., Santiana, M., Roberts, J., Faghri, F., Hao, Y., Nalls, M. A., Narayan, P., Liu, Y., Singleton, A. B., Cookson, M. R., Ward, M. E., & Qi, Y. A. (2023). A fully automated FAIMS-DIA mass spectrometry-based proteomic pipeline. In Cell Reports Methods (Vol. 3, Issue 10, p. 100593). Elsevier BV. https://doi.org/ doi:10.1016/j.crmeth.2023.100593

Restrepo, J. C., Dueñas, D., Corredor, Z., & Liscano, Y. (2023). *Advances in Genomic Data and Biomarkers: Revolutionizing NSCLC Diagnosis and Treatment*. MDPI. doi:10.3390/cancers15133474

Revanna C R, Rao, K., & Sarma, P. (2022), Enhanced Diagnostic Methods for Identifying Anomalies in Imaging of Skin Lesions. IJEER 10(4), pp.1077-1085.

Revanna C. (2022). Enhanced Diagnostic Methods for Identifying Anomalies in Imaging of Skin Lesions. *IJEER 10*(4), pp.1077-1085.

Revanna, C. R., Kameswara Rao, B., & Sarma, P. (2022). Enhanced Diagnostic Methods for Identifying Anomalies in Imaging of Skin Lesions. *IJEER 10*(4), 1077-1085.

Rghioui, A., Lloret, J., Harane, M., & Oumnad, A. (2020). A smart glucose monitoring system for diabetic patient. *Electronics (Basel)*, *9*(4), 678. doi:10.3390/electronics9040678

Riza, L. S., Zain, M. I., Izzuddin, A., Prasetyo, Y., Hidayat, T., & Abu Samah, K. A. F. (2023). Implementation of machine learning in DNA barcoding for determining the plant family taxonomy. In Heliyon (Vol. 9, Issue 10, p. e20161). Elsevier BV. https://doi.org/ doi:10.1016/j.heliyon.2023.e20161

Rokade, M. D., & Sharma, Y. K. (2021, March). MLIDS: A machine learning approach for intrusion detection for real-time network dataset. In *2021 International conference on emerging smart computing and informatics (ESCI)* (pp. 533-536). IEEE. 10.1109/ESCI50559.2021.9396829

Ross, M. W., Iguchi, M. Y., & Panicker, S. (2018). Ethical aspects of data sharing and research participant protections. *The American Psychologist*, *73*(2), 138–145. doi:10.1037/amp0000240 PMID:29481107

Roush, G. C., & Sica, D. A. (2016). Diuretics for hypertension: A review and update. In American Journal of Hypertension (Vol. 29, Issue 10, pp. 1130–1137). Am J Hypertens. doi:10.1093/ajh/hpw030

Ruamviboonsuk, P., Cheung, C. Y., Zhang, X., Raman, R., Park, S. J., & Ting, D. S. W. (2020). Artificial intelligence in ophthalmology: Evolutions in Asia. *Asia-Pacific Journal of Ophthalmology*, *9*(2), 78–84. doi:10.1097/01.APO.0000656980.41190.bf PMID:32349114

Russell, S. J., & Norvig, P. (2010). *Artificial intelligence a modern approach.*

Sachi, S., Singh, A. K., Jain, A., Devi, S., Sharma, Y. K., & Athithan, S. (2023, July). Hate Speech Detection Using the GPT-2 and Natural Language Processing. In *2023 Intelligent Methods, Systems, and Applications (IMSA)* (pp. 276-280). IEEE.

Sadoughi, F., Kazemy, Z., Hamedan, F., Owji, L., Rahmanikatigari, M., & Azadboni, T. T. (2018). Artificial intelligence methods for the diagnosis of breast cancer by image processing: a review. *Breast Cancer: Targets and Therapy*, 219-230.

Safonova, A., Ghazaryan, G., Stiller, S., Main-Knorn, M., Nendel, C., & Ryo, M. (2023). Ten deep learning techniques to address small data problems with remote sensing. *International Journal of Applied Earth Observation and Geoinformation*, *125*, 103569. doi:10.1016/j.jag.2023.103569

Saha, S., Sarker, P. S., Al Saud, A., Shatabda, S., & Newton, M. H. (2022). Cluster-oriented instance selection for classification problems. *Information Sciences*, *602*, 143–158. doi:10.1016/j.ins.2022.04.036

Sakly, H., Said, M., Jennane, R., & Tagina, M. (2023). Epidemiological Study of Cardiopathies and Valvulopathies using Binary Logistic Regression. *Journal of Data. Information & Management*, 1–13.

Salehin, I., & Islam, S., Saha, P., Noman, S. M., Tuni, A., Hasan, M., & Baten, A. (2023). AutoML: A Systematic Review on Automated Machine Learning with Neural Architecture Search. In Journal of Information and Intelligence. Elsevier BV. https://doi.org/ doi:10.1016/j.jiixd.2023.10.002

Sanchez-Fernandez, A., Rumetshofer, E., Hochreiter, S., & Klambauer, G. (2022, April). Contrastive learning of image- and structure-based representations in drug discovery. In ICLR2022 Machine Learning for Drug Discovery.

Sandhya, M. P., Raju, G. P. A., & Iyappan, M. (2023). Bioinformatics: Exploring the Past, Present and Future. *MULTIDISCIPLINARY*, *127*, 149.

Santoro, A., Bartunov, S., Botvinick, M., Wierstra, D., & Lillicrap, T. (2016). Meta-learning with memory-augmented neural networks. In *International Conference on Machine Learning*, (pp. 1842–1850). PMLR.

Sanyal, R., Kar, D., & Sarkar, R. (2021). Carcinoma type classification from high-resolution breast microscopy images using a hybrid ensemble of deep convolutional features and gradient boosting trees classifiers. *IEEE/ACM Transactions on Computational Biology and Bioinformatics*, *19*(4), 2124–2136. doi:10.1109/TCBB.2021.3071022 PMID:33819160

Sarker, I. H. (2021). Machine learning: Algorithms, real-world applications and research directions. *SN Computer Science*, *2*(3), 160. doi:10.1007/s42979-021-00592-x PMID:33778771

Sarma, A., Calfee, C. S., & Ware, L. B. (2020). Biomarkers and Precision Medicine: State of the Art. In Critical Care Clinics (Vol. 36, Issue 1, pp. 155–165). Crit Care Clin. doi:10.1016/j.ccc.2019.08.012

Schinkel, M., Paranjape, K., Panday, R. N., Skyttberg, N., & Nanayakkara, P. W. (2019). Clinical applications of artificial intelligence in sepsis: A narrative review. *Computers in Biology and Medicine*, *115*, 103488. doi:10.1016/j.compbiomed.2019.103488 PMID:31634699

Schmit, C., Sunshine, G., Pepin, D., Ramanathan, T., Menon, A., & Penn, M. (2017). Transitioning from paper to digital: State statutory and regulatory frameworks for health information technology. *Public Health Reports*, *132*(5), 585–592. doi:10.1177/0033354917722994 PMID:28813620

Sciorio, R., Tramontano, L., & Esteves, S. C. (2022). Effects of mobile phone radiofrequency radiation on sperm quality. *Zygote (Cambridge, England)*, *30*(2), 159–168. doi:10.1017/S096719942100037X PMID:34384508

Seeja, G., Selvakumar, A. A., & Henry, V. B. (2023). A Novel Approach for Disaster Victim Detection Under Debris Environments Using Decision Tree Algorithms with Deep Learning Features. *IEEE Access : Practical Innovations, Open Solutions*, *11*, 54760–54772. doi:10.1109/ACCESS.2023.3281461

Selvaratnam, B., Oliynyk, A. O., & Mar, A. (2023). Interpretable Machine Learning in Solid-State Chemistry, with Applications to Perovskites, Spinels, and Rare-Earth Intermetallics: Finding Descriptors Using Decision Trees. *Inorganic Chemistry*, *62*(28), 10865–10875. doi:10.1021/acs.inorgchem.3c01153 PMID:37390482

Senior, A., Evans, R., Jumper, J., Kirkpatrick, J., Sifre, L., Green, T., Qin, C., Zídek, A., Nelson, A. W. R., Bridgland, A., & Hassabis, D. (2020). *Improved protein structure prediction using potentials from deep learning*. doi:10.1038/s41586-019-1923-7

Sethi, J. K., & Mittal, M. (2022). Efficient weighted naive bayes classifiers to predict air quality index. *Earth Science Informatics*, *15*(1), 541–552. doi:10.1007/s12145-021-00755-7

Shah, A. S. (2022, January). A survey from 1G to 5G including the advent of 6G: architectures, multiple access techniques, and emerging technologies. In *2022 IEEE 12th Annual Computing and Communication Workshop and Conference (CCWC)* (pp. 1117-1123). IEEE.

Shah, D., Singh, A., & Prasad, S. S. (2022, November). Sentimental Analysis Using Supervised Learning Algorithms. In *2022 3rd International Conference on Computation, Automation and Knowledge Management (ICCAKM)* (pp. 1-6). IEEE. 10.1109/ICCAKM54721.2022.9990320

Sharma, S., & Sharma, D. (2018). *Intelligently Applying Artificial Intelligence in Chemoinformatics*. doi:10.2174/1568026619666181120150938

Sharma, V., Singh, A., Chauhan, S., Sharma, P. K., Chaudhary, S., Sharma, A., Porwal, O., & Fuloria, N. (2023). *Role of Artificial Intelligence in Drug Discovery and Target Identification in Cancer*. doi:10.2174/1567201821666230905090621

Sharma, Y. K., & Khatal Sunil, S. (2019). Health Care Patient Monitoring using IoT and Machine Learning. In *IOSR Journal of Engineering (IOSR JEN)*.

Sharma, S. K., Lilhore, U. K., Simaiya, S., & Trivedi, N. K. (2021). An improved random forest algorithm for predicting the COVID-19 pandemic patient health. *Annals of the Romanian Society for Cell Biology*, 67–75.

Sharma, Y. K., Athithan, S., Sachi, S., Singh, A. K., Jain, A., & Devi, S. (2023, July). Copy and Move Forged Image Detection by Deep Learning. In *2023 World Conference on Communication & Computing (WCONF)* (pp. 1-6). IEEE. 10.1109/WCONF58270.2023.10235115

Sheng, J., Wu, S., Zhang, Q., Li, Z., & Huang, H. (2022). A Binary Classification Study of Alzheimer's Disease Based on a Novel Subclass Weighted Logistic Regression Method. *IEEE Access : Practical Innovations, Open Solutions*, *10*, 68846–68856. doi:10.1109/ACCESS.2022.3186888

Shewalkar, A., Nyavanandi, D., & Ludwig, S. A. (2019). Performance Evaluation of Deep neural networks Applied to Speech Recognition: Rnn, LSTM and GRU. *Journal of Artificial Intelligence and Soft Computing Research*, *9*(4), 235–245. doi:10.2478/jaiscr-2019-0006

Shivthare, S. K., Sharma, Y. K., & Patil, R. D. (2023). Multi-modal Biometric in Computer Vision. *Multi-modal Biometric and Machine Learning Technologies: Applications for Computer Vision*, 1-29.

Shoaib, M., Shah, B., Sayed, N., Ali, F., Ullah, R., & Hussain, I. (2023). Deep learning for plant bioinformatics: An explainable gradient-based approach for disease detection. *Frontiers in Plant Science*, *14*, 14. doi:10.3389/fpls.2023.1283235 PMID:37900739

Shoaib, T., & Parveen, D. S. (2022). Deep Learning Driven Drug Discovery and Use of Machine Learning Strategies. *RE:view*, *13*(SO3). doi:10.47750/pnr.2022.13.S03.112

Sidorova, J., & Lozano, J. J. (2023). Commentary on "A systematic review on machine learning and deep learning techniques in cancer survival prediction": Validation of survival methods. *Progress in Biophysics and Molecular Biology*, *183*, 17–18. doi:10.1016/j.pbiomolbio.2023.08.001 PMID:37579795

Silva, J. S., & Saraiva, A. (2015). *A methodology for applying social network analysis metrics to biological interaction networks*. ACM., doi:10.1145/2808797.2808824

Simaiya, S., Lilhore, U. K., Prasad, D., & Verma, D. K. (2021). MRI brain tumor detection & image segmentation by hybrid hierarchical K-means clustering with FCM based machine learning model. *Annals of the Romanian Society for Cell Biology*, 88–94.

Simkó, M., & Mattsson, M. O. (2019). 5G wireless communication and health effects—A pragmatic review based on available studies regarding 6 to 100 GHz. *International Journal of Environmental Research and Public Health*, *16*(18), 3406. doi:10.3390/ijerph16183406 PMID:31540320

Singh, A., Thakur, N., & Sharma, A. (2016). *A review of supervised machine learning algorithms*. 2016 3rd International Conference on Computing for Sustainable Global Development (INDIACom), New Delhi, India.

Singh, K., & Kumar, A., Sharma, Y. K., & Rai, A. K. (2023). AIoT-based e-commerce. *AIoT Technologies and Applications for Smart Environments*, 215.

Singh, K., Kumar, A., Sharma, Y. K., & Rai, A. K. (2023). AIoT-based e-commerce. *AIoT Technologies and Applications for Smart Environments*, 215.

Singh, K., & Malhotra, D. (2023). Meta-Health: Learning-to-Learn (Meta-learning) as a Next Generation of Deep Learning Exploring Healthcare Challenges and Solutions for Rare Disorders: A Systematic Analysis. *Archives of Computational Methods in Engineering*, *30*(7), 1–32. doi:10.1007/s11831-023-09927-8

Singh, P., Singh, N., & Deka, G. C. (2021). Prospects of machine learning with Blockchain in healthcare and agriculture. In *Multidisciplinary functions of Blockchain technology in AI and IoT applications* (pp. 178–208). IGI Global. doi:10.4018/978-1-7998-5876-8.ch009

Singh, S., Sarma, D. K., Verma, V., Nagpal, R., & Kumar, M. (2023). Unveiling the future of metabolic medicine: omics technologies driving personalized solutions for precision treatment of metabolic disorders. In *Biochemical and Biophysical Research Communications* (Vol. 682, pp. 1–20). Elsevier BV. doi:10.1016/j.bbrc.2023.09.064

Smith, C. M., Weathers, A. L., & Lewis, S. L. (2023). An overview of clinical machine learning applications in neurology. In *Journal of the Neurological Sciences* (Vol. 455, p. 122799). Elsevier BV. doi:10.1016/j.jns.2023.122799

Snell, J., Swersky, K., & Zemel, R. S. (2017). Prototypical Networks for Few-Shot Learning. In Advances in neural information processing systems, 30, 4077-4087.

Sofri, T., Rahim, H. A., Abdulmalek, M., Abd Rani, K., Omar, M. H., Yasin, M. N. M., & Soh, P. J. (2021). Health Effects of 5G Base Station Exposure: A Systematic Review. *IEEE Access : Practical Innovations, Open Solutions, 10*, 41639–41656. doi:10.1109/ACCESS.2021.3139385

Song, Y., Wang, T., Cai, P., Mondal, S. K., & Sahoo, J. P. (2023). A comprehensive survey of few-shot learning: Evolution, applications, challenges, and opportunities. *ACM Computing Surveys*, *55*(13s), 1–40. doi:10.1145/3582688

Sonmez, M. E., Eczacıoglu, N., Gumuş, N. E., Aslan, M. F., Sabanci, K., & Aşikkutlu, B. (2022). Convolutional neural network-Support vector machine based approach for classification of cyanobacteria and chlorophyta microalgae groups. *Algal Research*, *61*, 102568. doi:10.1016/j.algal.2021.102568

Sood, R., & Garg, A. (2014). Digital society from 1G to 5G: A comparative study. [IJAIEM]. *International Journal of Application or Innovation in Engineering & Management*, *3*(2), 186–193.

Spyns, P. (1996). Natural language processing in medicine: an overview. *Methods of information in medicine, 35*(04/05), 285-301.

Strimbu, K., & Tavel, J. A. (2010). What are biomarkers? *Current Opinion in HIV and AIDS*, *5*(6), 463–466. doi:10.1097/COH.0b013e32833ed177 PMID:20978388

Suman, S., Tiwari, A. K., Ingale, T., & Singh, K. (2023). Automated detection of Hypertensive Retinopathy using few-shot learning. *Biomedical Signal Processing and Control*, *86*(Part C), 105310. doi:10.1016/j.bspc.2023.105310

Sunaryono, D., Sarno, R., & Siswantoro, J. (2022). Gradient boosting machines fusion for automatic epilepsy detection from EEG signals based on wavelet features. *Journal of King Saud University. Computer and Information Sciences*, *34*(10), 9591–9607. doi:10.1016/j.jksuci.2021.11.015

Sun, Q., Liu, Y., Chua, T.-S., & Schiele, B. (2019). Meta-transfer learning for few-shot learning. *Proceedings of the IEEE/CVF Conference on Computer Vision and Pattern Recognition*, (pp. 403–412). IEEE.

Swapna, M., Sharma, Y. K., & Prasad, B. M. G. (2020). A survey on face recognition using convolutional neural network. In *Data Engineering and Communication Technology: Proceedings of 3rd ICDECT-2K19* (pp. 649–661). Springer Singapore. doi:10.1007/978-981-15-1097-7_54

Swapna, M., Sharma, Y. K., & Prasadh, B. M. G. (2020). CNN Architectures: Alex Net, Le Net, VGG, Google Net, Res Net. *Int. J. Recent Technol. Eng*, *8*(6), 953–960. doi:10.35940/ijrte.F9532.038620

Swathi Pai, M. Shruthi, M., & Naveen, K. (2020). Internet of Things: A Survey on Devices, Ecosystem, Components and Communication Protocols. *2020 4th International Conference on Electronics, Communication and Aerospace Technology (ICECA)*, (pp. 611-616). IEEE.

Szarmach, M., & Czarnowski, I. (2023, June). Decision Tree-Based Algorithms for Detection of Damage in AIS Data. In *International Conference on Computational Science* (pp. 17-32). Cham: Springer Nature Switzerland. 10.1007/978-3-031-36024-4_2

Talukder, A., & Haas, R. (2021, June). AIoT: AI meets IoT and web in smart healthcare. In *Companion Publication of the 13th ACM Web Science Conference 2021* (pp. 92-98). ACM.

Tan, C., Xu, X., & Shen, F. (2021). A Survey Of zero shot detection: Methods and applications. *Cognitive Robotics*, *1*, 159–167. doi:10.1016/j.cogr.2021.08.001

Tang, A., Woldemariam, S., Roger, J., & Sirota, M. (2022). Translational Bioinformatics to Enable Precision Medicine for All: Elevating Equity across Molecular, Clinical, and Digital Realms. *Yearbook of Medical Informatics*, *31*(01), 106–115. doi:10.1055/s-0042-1742513 PMID:36463867

Tang, J., Ma, T., & Luo, Q. (2020). Trends prediction of big data: A case study based on fusion data. *Procedia Computer Science*, *174*, 181–190. doi:10.1016/j.procs.2020.06.073

Tang, S., Lomsadze, A., & Borodovsky, M. (2015). Identification of protein coding regions in RNA transcripts. *Nucleic Acids Research*, *43*(12), e78–e78. doi:10.1093/nar/gkv227 PMID:25870408

Tanguturi, R. (2023). *Implementation of Machine Learning Approach for Detecting Cardiovascular Diseases*. 2023 3rd International Conference on Intelligent Technologies (CONIT), Hubli, India.

Tanguturi, R. C. (2023). Implementation of Machine Learning Approach for Detecting Cardiovascular Diseases. *2023 3rd International Conference on Intelligent Technologies (CONIT)*, Hubli, India.

Tanguturi, R. C. L. S. V. S. L, S. R. C. K & V. C. H. (2023). Implementation of Machine Learning Approach for Detecting Cardiovascular Diseases. *2023 3rd International Conference on Intelligent Technologies (CONIT)*, Hubli, India.

Tang, Y. J., & Aristilde, L. (2020). Editorial overview: Analytical biotechnology in the era of high-performance omics, synthetic biology, and machine learning. In *Current Opinion in Biotechnology* (Vol. 64, pp. iii–vi). Elsevier BV. doi:10.1016/j.copbio.2020.07.009

Tan, M. S., Cheah, P.-L., Chin, A.-V., Looi, L.-M., & Chang, S.-W. (2021). A review on omics-based biomarkers discovery for Alzheimer's disease from the bioinformatics perspectives: Statistical approach vs machine learning approach. *Computers in Biology and Medicine*, *139*, 104947. doi:10.1016/j.compbiomed.2021.104947 PMID:34678481

Thakur, A. K. A. (2021). Fundamentals of Neural Networks. *International Journal for Research in Applied Science and Engineering Technology*, *9*(VIII), 407–426. doi:10.22214/ijraset.2021.37362

The World Bank. (2020). *Health Overview: Development News, Research, Data*. World Bank.. https://www. worldbank.org/en/topic/health/overview#1

Theodoridis, T., & Kraemer, J. (2014). Early Diagnosis of Complex Diseases by Molecular Biomarkers, Network Biomarkers, and Dynamical Network Biomarkers. *Medicinal Research Reviews*, *34*(3), 455–478. doi:10.1002/med.21293 PMID:23775602

Theodosiou, A. A., & Read, R. C. (2023). Artificial intelligence, machine learning and deep learning: Potential resources for the infection clinician. *The Journal of Infection*, *87*(4), 287–294. doi:10.1016/j.jinf.2023.07.006 PMID:37468046

Tian, S., Yang, W., Le Grange, J. M., Wang, P., Huang, W., & Ye, Z. (2019). Smart healthcare: Making medical care more intelligent. *Global Health Journal (Amsterdam, Netherlands)*, *3*(3), 62–65. doi:10.1016/j.glohj.2019.07.001

Tracy, J. M., Özkanca, Y., Atkins, D. C., & Ghomi, R. H. (2020). Investigating voice as a biomarker: Deep phenotyping methods for early detection of Parkinson's disease. *Journal of Biomedical Informatics*, *104*, 103362. doi:10.1016/j.jbi.2019.103362 PMID:31866434

Tran, D. M., Nguyen, T. H., & Nguyen, A. D. (2023). Complete genome sequence data of chitin-degrading Bacillus velezensis RB.IBE29. In *Data in Brief* (p. 109815). Elsevier BV. doi:10.1016/j.dib.2023.109815

Triantafillou, E., Zemel, R., & Urtasun, R. 2017. Few-shot learning through an information retrieval lens. In Advances in Neural Information Processing Systems. 2255–2265.

Trifonov, E. N. (2000). Earliest pages of bioinformatics. *Bioinformatics (Oxford, England)*, *16*(1), 5–9. doi:10.1093/bioinformatics/16.1.5 PMID:10812471

Trinidad, M. G., Platt, J., & Kardia, S. L. R. (2020). The public's comfort with sharing health data with third-party commercial companies. *Humanities & Social Sciences Communications*, *7*(1), 149. doi:10.1057/s41599-020-00641-5 PMID:34337435

Tripathi, M., Nath, A., Singh, T., Ethayathulla, A. S., & Kaur, P. (2021). Evolving scenario of big data and Artificial Intelligence (AI) in drug discovery. *Molecular Diversity*, *25*(3), 1827–1844. doi:10.1007/s11030-021-10256-w PMID:34159484

Tripathi, S., & Jha, M. (2022). Advancements in Neuroradiology via Artificial Intelligence and Machine Learning. *International Journal of Bioinformatics and Intelligent Computing*, *1*(2), 130–141. doi:10.61797/ijbic.v1i2.159

Tsai, E., Shakbatyan, R., Evans, J., Rossetti, P., Graham, C., Sharma, H., Lin, C.-F., & Lebo, M. S. (2016). Bioinformatics Workflow for Clinical Whole Genome Sequencing at Partners HealthCare. *Personalized Medicine*, *6*(1), 12. doi:10.3390/jpm6010012 PMID:26927186

Tummala, S., & Suresh, A. K. (2023). Few-shot learning using explainable Siamese twin network for the automated classification of blood cells. *Medical & Biological Engineering & Computing*, *61*(6), 1–15. doi:10.1007/s11517-023-02804-3 PMID:36800155

Turon, G., Hlozek, J., Woodland, J. G., Chibale, K., & Duran-Frigola, M. (2023). First fully-automated AI/ML virtual screening cascade implemented at a drug discovery centre in Africa. *Nature Communications*, *14*(1), 1–10. doi:10.1038/s41467-023-41512-2 PMID:37714843

Tyukin, I. Y., Gorban, A. N., McEwan, A. A., Meshkinfamfard, S., & Tang, L. (2021). Blessing of dimensionality at the edge and geometry of few-shot learning. *Information Sciences*, *564*, 124–143. doi:10.1016/j.ins.2021.01.022

U. K N & R. V M. (2022). *Arduino based COVID-19 Suspect Detection Device.* 2022 6th International Conference on Electronics, Communication and Aerospace Technology, Coimbatore, India.

Uddin, S., Khan, A., Hossain, M. E., & Moni, M. A. (2019). Comparing different supervised machine learning algorithms for disease prediction. *BMC Medical Informatics and Decision Making*, *19*(1), 1–16. doi:10.1186/s12911-019-1004-8 PMID:31864346

Uguz, D. U., Berief, F., Leonhardt, S., & Antink, C. H. (2020, September). Classification of 12-lead ECGs using gradient boosting on features acquired with domain-specific and domain-agnostic methods. In 2020 Computing in Cardiology (pp. 1-4). IEEE.

Uwinama, I., Rusanganwa, J., & Ingabire, E. (2023). The Moral Implications of Big Data and Machine Learning in Healthcare: A Review. *American Journal of Technology*, *2*(1), 45–53.

Vadapalli, S., Abdelhalim, H., Zeeshan, S., & Ahmed, Z. (2022). *Artificial intelligence and machine learning approaches using gene expression and variant data for personalized medicine.* doi:10.1093/bib/bbac191

Vaithinathan, A. G., & Asokan, V. (2017). Public health and precision medicine share a goal. In Journal of Evidence-Based Medicine (Vol. 10, Issue 2, pp. 76–80). J Evid Based Med. doi:10.1111/jebm.12239

Vartak, M., Thiagarajan, A., Miranda, C., Bratman, J., & Larochelle, H. 2017. A meta-learning perspective on cold-start recommendations for items. In Advances in Neural Information Processing Systems.

Vatandsoost, M., & Litkouhi, S. (2019). The Future of Healthcare Facilities: How Technology and Medical Advances May Shape Hospitals of the Future. *Hospital Practices and Research*, *4*(1), 1–11. doi:10.15171/hpr.2019.01

Vemula, D., Jayasurya, P., Sushmitha, V., Kumar, Y. N., & Bhandari, V. (2022). CADD, AI and ML in Drug Discovery: A Comprehensive Review. *European Journal of Pharmaceutical Sciences*, *176*, 106324. doi:10.1016/j.ejps.2022.106324 PMID:36347444

Venkatakiran, S., Ashreetha, B., & Reddy, N. S. (2023). Implementation of a Machine Learning-based Model for Cardiovascular Disease Post Exposure prophylaxis. *2023 International Conference for Advancement in Technology (ICONAT)*, Goa, India.

Vinyals, O., Blundell, C., Lillicrap, T., & Wierstra, D. (2016). Matching networks for one shot learning. In Advances in Neural Information Processing Systems.

Vu, D. H., Vu, T. S., & Luong, T. D. (2022). An efficient and practical approach for privacy-preserving Naive Bayes classification. *Journal of information Security and Applications, 68*, 103215.

Vu, D. H. (2022). Privacy-preserving Naive Bayes classification in semi-fully distributed data model. *Computers & Security*, *115*, 102630. doi:10.1016/j.cose.2022.102630

Walsh, R., Abdelpakey, M. H., Shehata, M. S., & Mohamed, M. M. (2022). Automated human cell classification in sparse datasets using few-shot learning. *Scientific Reports*, *12*(1), 1–11. doi:10.1038/s41598-022-06718-2 PMID:35190567

Walters, W., & Barzilay, R. (2021). Critical assessment of AI in drug discovery. *Expert Opinion on Drug Discovery*, *16*(5), 497–507. doi:10.1080/17460441.2021.1915982 PMID:33870801

Wang, J., Gao, D., Zhu, S., Wang, S., & Liu, H. (2023). Fault diagnosis method of photovoltaic array based on support vector machine. *Energy sources, part a: recovery, utilization, and environmental effects, 45*(2), 5380-5395.

Wang, Y., Yao, Q., Kwok, J., & Ni, L. M. (2019). Generalizing from a Few Examples: A Survey on Few-Shot Learning. *ArXiv*. /abs/1904.05046

Wang, C., Liu, S., Yang, H., Guo, J., Wu, Y., & Liu, J. (2023). Ethical considerations of using ChatGPT in health care. *Journal of Medical Internet Research*, *25*, e48009. doi:10.2196/48009 PMID:37566454

Wang, D., Zeng, S., Xu, C., Qiu, W., Liang, Y., Joshi, T., & Xu, D. (2017). MusiteDeep: A deep-learning framework for general and kinase-specific phosphorylation site prediction. *Bioinformatics (Oxford, England)*, *33*(24), 3909–3916. doi:10.1093/bioinformatics/btx496 PMID:29036382

Wang, H., & Hu, X. (2023). Deep Learning in Bioinformatics and Biomedicine. *Methods (San Diego, Calif.)*, *209*, 38–39. doi:10.1016/j.ymeth.2022.12.001 PMID:36503039

Wang, J., & Biljecki, F. (2022). Unsupervised machine learning in urban studies: A systematic review of applications. *Cities (London, England)*, *129*, 103925. doi:10.1016/j.cities.2022.103925

Wang, N., & He, Q. (2023). Artificial intelligence and bioinformatics applications in precision medicine and future implications. In *Reference Module in Biomedical Sciences*. Elsevier. doi:10.1016/B978-0-12-824010-6.00058-7

Wang, Q., & Chen, K. (2020). Multi-label zero-shot human action recognition via joint latent ranking embedding. *Neural Networks*, *122*, 1–23. doi:10.1016/j.neunet.2019.09.029 PMID:31675624

Wang, S., Jiang, X., Singh, S., Marmor, R., Bonomi, L., Fox, D., Dow, M., & Ohno-Machado, L. (2016). Genome privacy: Challenges, technical approaches to mitigate risk, and ethical considerations in the United States. *Annals of the New York Academy of Sciences*. doi:10.1111/nyas.13259 PMID:27681358

Wang, S., Ren, J., & Bai, R. (2020). A regularized attribute weighting framework for naive Bayes. *IEEE Access : Practical Innovations, Open Solutions*, *8*, 225639–225649. doi:10.1109/ACCESS.2020.3044946

Wang, W., Duan, Y., Cao, L., & Jiang, Z. (2023). Application of improved Naive Bayes classification algorithm in 5G signaling analysis. *The Journal of Supercomputing*, *79*(6), 6941–6964. doi:10.1007/s11227-022-04946-x

Wang, X., & Liu, F. (2020). Data-driven relay selection for physical-layer security: A decision tree approach. *IEEE Access : Practical Innovations, Open Solutions*, *8*, 12105–12116. doi:10.1109/ACCESS.2020.2965963

Wang, Y., Pan, Z., & Dong, J. (2022). A new two-layer nearest neighbor selection method for kNN classifier. *Knowledge-Based Systems*, *235*, 107604. doi:10.1016/j.knosys.2021.107604

Wang, Y., Xu, L., & Zou, Q. (2023). Deep learning methods for bioinformatics and biomedicine. *Methods (San Diego, Calif.)*, *216*, 1–2. doi:10.1016/j.ymeth.2023.06.003 PMID:37295580

Wang, Y., Yao, Q., Kwok, J. T., & Ni, L. M. (2020). Generalizing from a few examples: A survey on few-shot learning. *ACM Computing Surveys*, *53*(3), 1–34. doi:10.1145/3386252

Wang, Y., Yao, Q., Kwok, J. T., & Ni, L. M. (2020). Generalizing from a Few Examples: A Survey on Few-shot Learning. *ACM Computing Surveys*, *53*(6), 1–34. doi:10.1145/3214306

Waring, J., Lindvall, C., & Umeton, R. (2020). Automated machine learning: Review of the state-of-the-art and opportunities for healthcare. *Artificial Intelligence in Medicine*, *104*, 101822. doi:10.1016/j.artmed.2020.101822 PMID:32499001

Wdowiak, A., Mazurek, P. A., Wdowiak, A., & Bojar, I. (2017). Effect of electromagnetic waves on human reproduction. *Annals of Agricultural and Environmental Medicine*, *24*(1), 13–18. doi:10.5604/12321966.1228394 PMID:28378967

Wermuth, P. J., Piera-Velazquez, S., Rosenbloom, J., & Jimenez, S. A. (2018). Existing and novel biomarkers for precision medicine in systemic sclerosis. In Nature Reviews Rheumatology (Vol. 14, Issue 7, pp. 421–432). Nature Publishing Group. doi:10.1038/s41584-018-0021-9

Wolfgang, E. (2017). *Introduction to artificial intelligence*. Springer.

Wu, T., Rappaport, T. S., & Collins, C. M. (2015). Safe for generations to come: Considerations of safety for millimeter waves in wireless communications. *IEEE Microwave Magazine*, *16*(2), 65–84. doi:10.1109/MMM.2014.2377587 PMID:26539066

Xia, F., Yang, L. T., Wang, L., & Vinel, A. (2012). Internet of things. *International Journal of Communication Systems*, *25*(9), 1101–1102. doi:10.1002/dac.2417

Xia, J., Benner, M. J., & Hancock, R. (2014). NetworkAnalyst - integrative approaches for protein–protein interaction network analysis and visual exploration. *Nucleic Acids Research*, *42*(W1), W167–W174. Advance online publication. doi:10.1093/nar/gku443 PMID:24861621

Xia, Y. (2019). A novel reject inference model using outlier detection and gradient boosting technique in peer-to-peer lending. *IEEE Access : Practical Innovations, Open Solutions*, *7*, 92893–92907. doi:10.1109/ACCESS.2019.2927602

Xia, Z., Li, Y., Zhang, B., Li, Z., Hu, Y., Chen, W., & Gao, X. (2019). DeeReCT-PolyA: A robust and generic deep learning method for PAS identification. *Bioinformatics (Oxford, England)*, *35*(14), 2371–2379. doi:10.1093/bioinformatics/bty991 PMID:30500881

Xie, X., & Sun, S. (2019). Multi-view support vector machines with the consensus and complementarity information. *IEEE Transactions on Knowledge and Data Engineering*, *32*(12), 2401–2413. doi:10.1109/TKDE.2019.2933511

Xu, B., Zhao, K., Ying, Z., Sjöberg, D., He, W., & He, S. (2019). Analysis of impacts of expected RF EMF exposure restrictions on peak EIRP of 5G user equipment at 28 GHz and 39 GHz bands. *IEEE Access : Practical Innovations, Open Solutions*, *7*, 20996–21005. doi:10.1109/ACCESS.2019.2897271

Xu, X., Niu, S., Wang, Z., Li, D., Yang, H., & Du, W. (2022). Client selection based weighted federated few-shot learning. *Applied Soft Computing*, *128*, 109488. doi:10.1016/j.asoc.2022.109488

Yang, P., Wang, D., Zhao, W. B., Fu, L. H., Du, J. L., & Su, H. (2021). Ensemble of kernel extreme learning machine based random forest classifiers for automatic heartbeat classification. *Biomedical Signal Processing and Control*, *63*, 102138. doi:10.1016/j.bspc.2020.102138

Yang, X., Yang, S., Lian, X., Wuchty, S., & Zhang, Z. (2021). Transfer Learning via multiscale convolutional neural layers for human–virus protein–protein interaction prediction. *Bioinformatics (Oxford, England)*, *37*(24), 4771–4778. doi:10.1093/bioinformatics/btab533 PMID:34273146

Yarragunta, A. (2022). The Future of Pharmaceuticals: AI Drug Discovery & COVID-19 Treatments. *Journal of Student Research*, *11*(3). doi:10.47611/jsrhs.v11i3.3046

Yasin, N. (2023). *Amyotrophic Lateral Sclerosis (ALS): A Visual Representation*. McMaster Brain Research Society.

Yoon, J., Kim, T., Dia, O., Kim, S., Bengio, Y., & Ahn, S. (2018). Bayesian model-agnostic meta-learning. In Advances in Neural Information Processing Systems, (pp. 7343–7353). IEEE.

Yousefi, S., Amrollahi, F., Amgad, M., Dong, C., Lewis, J. E., Song, C., Gutman, D. A., Halani, S. H., Velazquez Vega, J. E., Brat, D. J., & Cooper, L. A. D. (2017). Predicting clinical outcomes from large scale cancer genomic profiles with deep survival models. *Scientific Reports*, *7*(1), 1–11. doi:10.1038/s41598-017-11817-6 PMID:28916782

Yousef, M., Ketany, M., Manevitz, L., Showe, L. C., & Showe, M. K. (2009). Classification and biomarker identification using gene network modules and support vector machines. *BMC Bioinformatics*, *10*(1), 1–7. doi:10.1186/1471-2105-10-337 PMID:19832995

Yu, K. H., Beam, A. L., & Kohane, I. S. (2018). Artificial intelligence in healthcare. In Nature Biomedical Engineering (Vol. 2, Issue 10, pp. 719–731). Nature Publishing Group. doi:10.1038/s41551-018-0305-z

Yuan, F., Zhang, Z., & Fang, Z. (2023). An effective CNN and Transformer complementary network for medical image segmentation. In *Pattern Recognition* (Vol. 136, p. 109228). Elsevier BV. doi:10.1016/j.patcog.2022.109228

Yue, Z., Zhang, H., Sun, Q., & Hua, X.-S. (2020). Interventional few-shot learning. *Advances in Neural Information Processing Systems*, *33*, 2734–2746.

Yu, H., Sun, C., Yang, X., Zheng, S., & Zou, H. (2019). Fuzzy support vector machine with relative density information for classifying imbalanced data. *IEEE Transactions on Fuzzy Systems*, *27*(12), 2353–2367. doi:10.1109/TFUZZ.2019.2898371

Yu, X., Zhou, S., Zou, H., Wang, Q., Liu, C., Zang, M., & Liu, T. (2023). Survey of deep learning techniques for disease prediction based on omics data. *Human Gene*, *35*, 201140. doi:10.1016/j.humgen.2022.201140

Zafar, I., Anwar, S., kanwal, F., Yousaf, W., Un Nisa, F., Kausar, T., ul Ain, Q., Unar, A., Kamal, M. A., Rashid, S., Khan, K. A., & Sharma, R. (2023). Reviewing methods of deep learning for intelligent healthcare systems in genomics and biomedicine. *Biomedical Signal Processing and Control*, *86*, 105263. doi:10.1016/j.bspc.2023.105263

Zell, A., Sumbul, G., & Demir, B. (2022). Deep Metric Learning-Based Semi-Supervised Regression With Alternate Learning. In *IEEE International Conference on Image Processing (ICIP)*. IEEE. 10.1109/ICIP46576.2022.9897939

Zeng, T., Zhang, W., Yu, X., Liu, X., Li, M., & Chen, L. (2016). Big-data-based edge biomarkers: Study on dynamical drug sensitivity and resistance in individuals. *Briefings in Bioinformatics*, *17*(4), 576–592. doi:10.1093/bib/bbv078 PMID:26411472

Zhang, F., Wang, Q., & Li, X. (2020). Deep meta-relation network for visual few-shot learning. In ICASSP 2020-2020 *IEEE International Conference on Acoustics, Speech and Signal Processing (ICASSP),* (pp. 1509–1513). IEEE. 10.1109/ICASSP40776.2020.9053330

Zhang, L., Xiang, T., & Gong, S. (2017). Learning a deep embedding model for zero-shot learning. *Proceedings - 30th IEEE Conference on Computer Vision and Pattern Recognition, CVPR 2017, 2017-Janua*, (pp. 3010–3019). IEEE. 10.1109/CVPR.2017.321

Zhang, M., Chen, Y., & Susilo, W. (2022). Decision tree evaluation on sensitive datasets for secure E-healthcare systems. *IEEE Transactions on Dependable and Secure Computing.*

Zhang, R., Sun, J., Peng, J., Qiu, Y., & Lo, B. (2023). Big data analytics for wearable sensor-based health applications. *IEEE Transactions on Biomedical Engineering.*

Zhang, R., Zong, Q., Dou, L., Zhao, X., Tang, Y., & Li, Z. (2021). Hybrid deep neural network using transfer learning for EEG motor imagery decoding. *Biomedical Signal Processing and Control*, *63*, 102144. doi:10.1016/j.bspc.2020.102144

Zhang, W., Chen, X., Liu, Y., & Xi, Q. (2020). A distributed storage and computation k-nearest neighbor algorithm based cloud-edge computing for cyber-physical-social systems. *IEEE Access : Practical Innovations, Open Solutions*, *8*, 50118–50130. doi:10.1109/ACCESS.2020.2974764

Zhao, H., Wang, R., Lei, Y., Liao, W. H., Cao, H., & Cao, J. (2022). Severity level diagnosis of Parkinson's disease by ensemble K-nearest neighbor under imbalanced data. *Expert Systems with Applications*, *189*, 116113. doi:10.1016/j.eswa.2021.116113

Zheng, X., Wang, M., Huang, K., & Zhu, E. (2024). Global and cross-modal feature aggregation for multi-omics data classification and application on drug response prediction. In *Information Fusion* (Vol. 102, p. 102077). Elsevier BV. doi:10.1016/j.inffus.2023.102077

Zhou, J., Huang, S., Tao, M., Khandelwal, M., Dai, Y., & Zhao, M. (2023). Stability prediction of underground entry-type excavations based on particle swarm optimization and gradient boosting decision tree. *Underground Space*, *9*, 234–249. doi:10.1016/j.undsp.2022.08.002

Zhou, W., Xu, Y., Lv, Q., Sheng, Y. H., Chen, L., Li, M., Shen, L., Huai, C., Yi, Z., Cui, D., & Qin, S. (2019). Genetic Association of Olanzapine Treatment Response in Han Chinese Schizophrenia Patients. *Frontiers in Pharmacology*, *10*, 422832. doi:10.3389/fphar.2019.00177 PMID:30886581

Zhou, Y., Jia, E., Shi, H., Liu, Z., Sheng, Y., Pan, M., Tu, J., Ge, Q., & Lu, Z. (2022). Prediction of Time-Series Transcriptomic Gene Expression Based on Long Short-Term Memory with Empirical Mode Decomposition. *International Journal of Molecular Sciences*, *23*(14), 7532. doi:10.3390/ijms23147532 PMID:35886880

Zikopoulos, P., & Eaton, C. (2011). *Understanding big data: Analytics for enterprise class hadoop and streaming data.* McGraw-Hill Osborne Media.

Ziska Fields, I. G. I. (Ed.). (2023). Human Creativity vs. Machine Creativity: Innovations and Challenges. Multidisciplinary Approaches in AI, Creativity, Innovation, and Green Collaboration (pp. 19–28). Global. doi:10.4018/978-1-6684-6366-6.ch002

About the Contributors

Umesh Kumar Lilhore is currently a Professor at the School of Computing Science & Engineering (CSE) at Galgotia University, Greater Noida. With over 19 years of teaching and 8 years of research experience, he has previously held positions at various renowned universities and colleges in India and abroad. Dr. Lilhore holds a Ph.D. and M.Tech in CSE and has completed his postdoctoral research at the Institute of Advanced Computing, University of Louisiana at Lafayette. He has a strong publication record with articles in reputed, peer-reviewed national and international Scopus journals and conferences.

Abhishek Kumar is currently working as an Assistant Director in the Computer Science & Engineering Department at Chandigarh University, Punjab, India. is doing Post-Doctoral Fellow in Ingenium Research Group Ingenium Research Group Lab, Universidad De Castilla- La Mancha, Ciudad Real, and Ciudad Real Spain. He has total Academic teaching experience of more than 12 years along with 2 years teaching assistantship. He has more than 160 publications in reputed, peer-reviewed National and International Journals, books & Conferences He has authored/Co-Authored 7 books published internationally and edited 39 books (Published & ongoing with IET, Elsevier, Wiley, IGI GLOBAL Springer, Apple Academic Press, De-Gruyter and CRC, etc. He is Patent holder and got Sir CV Raman National award for 2018 in young researcher and faculty Category from IJRP Group. He is acting as Series Editor for three books series, Quantum Computing with Degruyter Germany, Intelligent Energy System with Elsevier, & Mathematical Methods in the Digital Age: Computational Intelligence & Advancements.

Sarita Simaiya is a distinguished Professor at the School of Computing Science & Engineering (CSE) at Galgotias University, Greater Noida. Boasting over 17 years of teaching and 8 years of research experience, Dr. Sarita has held esteemed positions at various prestigious universities and colleges in India and abroad. With a Ph.D. and M.Tech, BE in CSE, Dr. Sarita completed postdoctoral research at the Institute of Advanced Computing, University of Louisiana at Lafayette. His extensive publication record includes articles in reputed, peer-reviewed national and international Scopus journals and conferences.

Narayan Vyas, a distinguished academic and expert in advanced technologies, is currently working at Chandigarh University, Punjab, India. He cleared the NTA UGC NET & JRF in Computer Science & Applications on his first attempt, underscoring his academic excellence. With profound knowledge of the Internet of Things (IoT) and Mobile Application Development, he has trained students worldwide and authored numerous articles in reputable national and international Scopus-indexed conferences and journals. His research spans IoT, Remote Sensing, Machine Learning, Deep Learning, and Computer Vision. A sought-after keynote speaker, Mr. Vyas collaborates with leading publishers like Wiley, IGI Global, and DeGruyter on various book projects, marking his significant contribution to the field.

Vishal Dutt is a renowned industry expert in Machine Learning with over eight years of academic teaching experience. He is currently working at Chandigarh University, Punjab, India. He has authored over 50 publications in prestigious national and international journals, SCI and Scopus journals, conferences, and book chapters. Vishal has contributed editorially to multiple books with Wiley, IGI Global, DeGruyter, and Eureka publications and is working on three additional publications with Wiley. A sought-after keynote speaker, he has significantly contributed to workshops and webinars across India. He provides peer review services for elite publishers like Elsevier, Springer, and IEEE Access and was a program committee member and reviewer for ICCIPS 2021. His research focuses on Data Science, Data Mining, Machine Learning, Deep Learning, and Remote Sensing, with extensive experience in data analytics using Rapid Miner, Tableau, and WEKA, as well as over six years in Java and Android development.

***

Abdullah Alzahrani is a lead rechearher and special lecturer in the Electrical and Computer Engineering Department at Oakland University in Rochester, MI, USA. Alzahrani recived his Ph.D in the Electrical and Computer Engineering Department at Oakland University. He received his B.S. in electrical engineering in 2014 from Northern Border University located in Arar, KSA, and in 2018 his M.S. in electrical and computer engineering from Rochester Institute of Technology, Rochester, NY, USA. His research interests are in the application of optimization techniques and policy analysis to smart grid technology, electrical power, and control power systems. He is a member of the Saudi Council of Engineering.

Baiju B V is currently an Assistant Professor Senior Grade at the School of Computer Science and Engineering, Vellore Institute of Technology, Vellore, Tamil Nadu. He holds a B.E. in Computer Science and Engineering, M.E. in Computer Science and Engineering, and Ph.D. in Computer Science and Engineering. Dr. Baiju B V has over 17 years of experience in teaching and research. His domains of interest include Artificial Intelligence, Networks, Software Engineering, and IoT. He has published over 38 papers in national and international journals and conferences.

Sumit Bansal is an active researcher who is always keen to learn new technologies. He is pursuing Ph.D. in computer science & engineering at Punjabi University, Patiala. He has research interests in Fog computing, Cloud computing, AWS. He is working as assistant professor in Gurukula Kangri (DU), Haridwar.

Pooja Dixit has done M.C.A in Computer Science from MDS University Ajmer, Rajasthan, India. She has been working as the Guest Faculty of Department of Computer science and Engineering MDS University Ajmer. She has total Academic teaching experience of more than 8 years. She has published more than 25 papers in reputable, peer-reviewed national and international journals, books, and conferences, including Wiley, IGI GLOBAL, Taylor & Francis Springer, and IEEE. Her research includes Artificial intelligence, Data Mining, Machine Learning, Internet of Things.

Dwijendra Nath Dwivedi is a professional with 20+ years of subject matter expertise creating right value propositions for analytics and AI. He currently heads the EMEA+AP AI and IoT team at SAS, a worldwide frontrunner in AI technology. He is a post-Graduate in Economics from Indira Gandhi Institute of Development and Research and is PHD from crackow university of economics Poland. He has presented his research in more than 20 international conference and published several Scopus indexed paper on AI adoption in many areas. As an author he has contributed to more than 8 books and has more than 25 publications in high impact journals. He conducts AI Value seminars and workshops for the executive audience and for power users.

Subathra G. is working as an Assistant Professor in the school of Computer Science and Engineering at Sathyabama Institute of Science and Technology, Chennai, TamilNadu. She did B.Tech in Information Technology, M.E in Computer Science and Engineering and Ph.D. in Computer Science and Engineering. She has 4years of experience in research and Teaching. Her domain of interest is Network security, Blockchain Technology and Machine Learning .She has published 11 papers in International Journals and Conferences.

Dankan Gowda V is currently working as an Assistant Professor in the Department of Electronics and Communication Engineering at BMS Institute of Technology and Management in Bangalore. Previously, he worked as a Research Fellow at ADA DRDO and as a Software Engineer at Robert Bosch. With a total experience of 14 years, including teaching and industry, Dr. Gowda has made significant contributions to both academia and research. He has published over 80 research papers in renowned international journals and conferences. In recognition of his innovative work, Dr. Gowda has been granted six patents, including four from Indian authorities and two international patents. His research interests primarily lie in the fields of IoT and Signal Processing, where he has conducted workshops and handled industry projects. Dr. Gowda is passionate about teaching and strives to create an engaging and effective learning environment for his students. He consistently seeks opportunities to enhance his knowledge and stay updated with the latest advancements in his field.

K. Logeswaran presently pursuing his Ph.D. in Information and Communication Engineering under Anna University, Chennai. He received his M.E. Degree in Computer Science and Engineering in 2011 and B.Tech. Degree in Information Technology from Anna University, Chennai, India in 2009. He is currently working as an Assistant Professor (Sr.G.) in the Department of Artificial Intelligence at Kongu Engineering College, Perundurai, India. He is having 12 years of experience in Teaching and 6 years of experience in Research. His research interests are in the areas of Data mining, Cloud Computing, Knowledge discovery and Networking. He has completed around 10 consultancy projects. He has authored over 30 research papers in refered journals and international conferences. He has received grants from various funding agencies like DRDO, SERB, and CSIR towards the conduct of seminars and workshops.

P. Keerthika is currently working as Associate Professor in the School of Computer Science and Engineering, Vellore Institute of Technology, Vellore. She has a work experience of 17 years in teaching. She has completed her Ph.D in Anna University, Chennai. She is contributing towards research in the area of Grid and Cloud Computing for the past 14 years. She has published around 50 research papers in reputed journals indexed by Scopus and SCI and presented 30 papers in National and International Conferences. Her areas of interest are Machine Learning, Data Science and Cloud Computing. She was one of the recognized Supervisors in the Faculty of Information and Communication Engineering under Anna University, Chennai. She is currently one of the recognized supervisors in VIT University. She has guided 5 Ph.D scholars in the areas of Machine Learning and Cloud under Anna University. She has received grants from various funding agencies like DRDO and DST. She is currently acting as Reviewer and Editorial Board member in reputed journals.

Suresh P is currently working as Associate Professor in the School of Computer Science and Engineering, Vellore Institute of Technology, Vellore, Tamil Nadu, India. He has 17 years of teaching and academic experience. He has completed Ph.D in Anna University, Chennai. He has published 50 research papers in reputed international journals indexed by Scopus/SCI/WoS and presented 35 papers in National and International Conferences. He has authored 6 books and 15 book chapters. His areas of interest include Networks, Cloud Computing, Data Science, Machine Learning and Deep Learning. He has received grants from various funding agencies like DRDO, ICMR, and CSIR towards the conduct of seminars and workshops. He has received Best Faculty Award and Best Performer in Academics, Research and consultancy from Kongu Engineering College, Erode, Tamilnadu. He is acting as Reviewer and Editorial Board member in reputed journals.

Pramod Singh Rathore has done M. Tech in Computer Sci. & Engineering from Government engineering college Ajmer, Rajasthan Technical University, Kota India. He has been working as the Assistant professor of Department of Computer and Communication Engineering Manipal University Jaipur. With over 11 years of academic teaching experience, he has published more than 55 papers in reputable, peer-reviewed national and international journals, books, and conferences, including Wiley, IGI GLOBAL, Taylor & Francis Springer, Elsevier Science Direct, Annals of Computer Science, Poland, and IEEE. He has co-authored and edited numerous books with well-known publishers such as Wiley, CRC Press, USA. His research interests include NS2, computer networks, mining, and DBMS. He serves on the editorial and advisory committees of the Global Journal Group and is also a member of various national and international professional societies in the field of engineering and research, including the International Association of Engineers (IAENG).

Kishor Kumar Sadasivuni is an accomplished Research Assistant Professor and the esteemed leader of the SmartNano Solutions group at the Center for Advanced Materials, Qatar University. With a strong foundation in analytical chemistry, he embarked on his academic journey, earning a Master's degree from Andhra University, India. His pursuit of knowledge led him to the University of South Brittany in Lorient, France, where he earned his Ph.D. in Materials Science and Engineering in 2012 under the guidance of Professor Yves Grohens. This marked the beginning of his impressive career in academia and research. Dr. Kishor Kumar Sadasivuni's contributions to the scientific community have been exceptional. His groundbreaking work earned him recognition as one of the world's top 2% scientists, a distinction conferred by Stanford University in 2019, 2020, 2021 and 2023. Qatar University has also acknowledged his remarkable achievements with honors. Boasting over 15 years of active research experience, Dr. Kishor has made significant strides in his field. His prolific publication record includes more than 450 research articles in esteemed international peer-reviewed journals, accumulating an impressive total citation count of 14000 and an h-index of 64. He has also authored 20 book chapters and served as the editor of 11 books, with several of his works featured in Springer's prestigious Top 25 e-book downloads of 2020. Dr. Kishor's innovative mindset is evident through his patent portfolio. He holds three US patents and two Indian patents, with five patent disclosures currently under consideration. As a visionary leader, he spearheads 13 research projects, encompassing grants from the Qatar National Research Fund, including NPRP, UREP, IRCC grants, as well as projects supported by Qatar University, collectively amounting to an impressive 5 million USD. A true collaborator, Dr. Kishor has actively engaged with researchers from diverse disciplines worldwide, boasting over 450 co-authors in fields such as computer science, biomedical sciences, industrial engineering, and electrical engineering. His global collaborations span countries like the USA, France, South Korea, Oman, Spain, Italy, Australia, and Malaysia.

Sajja Suneel, was awarded Ph.D. from Veer Bahadur Singh Purvanchal University, formerly Purvanchal University, Jaunpur, UP in 2023. He has 13 years of teaching experience in reputed institutions. He is currently working as Assistant Professor & PAT coordinator in the Department of CSE (Data Science) at Institute of Aeronautical Engineering. He has published 8 Research papers in National/International Journals and Conferences & has presented 2 conference papers around the globe, apart from these, he has been Participated many workshops, conferences and faculty development programmes and pedagogic training programmes. He has been a lifetime member in various professional bodies like CSI, IETE, & ISTE and his area of interest include Machine Learning, Artificial Intelligence, Cloud Computing & IoT.

Index

F

G

H

I

K

L

M

N

P

R

S

T

Z

www.ingramcontent.com/pod-product-compliance
Lightning Source LLC
LaVergne TN
LVHW060300011026
849161LV00011B/2745